Physical-Layer Security and Quantum Key Distribution

Ivan B. Djordjevic

Physical-Layer Security and Quantum Key Distribution

Springer

Ivan B. Djordjevic
Department of Electrical and Computer Engineering
University of Arizona
Tucson, AZ, USA

ISBN 978-3-030-27567-9 ISBN 978-3-030-27565-5 (eBook)
https://doi.org/10.1007/978-3-030-27565-5

This Springer imprint is published by the registered company Springer Nature Switzerland AG
The registered company address is: Gewerbestrasse 11, 6330 Cham, Switzerland

In memory of my father Blagoje

Preface

The growth of the internet and data traffic does not appear to be leveling off any time soon and it is projected to continue to grow exponentially in years to come. It is, however, necessary to make a dramatic improvement in the optical signal transmission rates in order to cope with the incoming bandwidth crunch. Although there are many proposals on how to improve the spectral efficiency, the security of optical networks seems to be almost completely neglected. By taping out the portion of dense wavelength-division multiplexing (DWDM) signal, this huge amount of data can be compromised. Therefore, the security of both optical networks and wireless networks is becoming one of the major issues to be addressed sooner rather than later. Public-key cryptography has several serious drawbacks such as it is difficult to implement it in devices with low memory and low process constraints. Internet and wireless technologies are becoming increasingly mobile, security schemes are based on unproven assumptions of intractability of certain functions, and the assumption of limiting computing resources of the Eve is often incorrect, to mention few. To solve all of these problems in simultaneous manner, new security concepts must be introduced, such as those described in this book. The purpose of this book is to introduce the reader to most advanced topics of physical-layer security (PLS), cryptography, covert/stealth communications, and quantum-key distribution (QKD), also known as the quantum cryptography. So far, these topics have been considered as separate disciplines, even though they are targeting the same security problems we are facing today.

This book integrates modern cryptography, physical-layer security, QKD, covert communication, and cyber-security technologies. Unique features of the book include the following:

- This book unifies the conventional cryptography, physical-layer security, and QKD.
- This book does not require any prior knowledge.
- This book does not require any prerequisite material; all background material is provided in the Appendix chapter.

- This book offers in-depth exposition on cryptography, information-theoretic approach to cryptography, physical-layer security, covert/stealth/low-probability of detection communications, quantum information theory, and QKD, to mention few.
- The successful reader will be prepared for further study in the corresponding area of interest and will be qualified to perform independent research in any of the areas listed above.
- Several either senior undergraduate or graduate courses can be offered by using this book.

The book is intended for very diverse group of readers in communications engineering, optical engineering, wireless communications, free-space optical communications, optical wireless communications, mathematics, physics, communication theory, information theory, photonics, as well as computer science.

The book is organized into ten chapters. In the introductory chapter (Chap. 1), the basic concepts of both physical-layer security and quantum-key distribution (QKD) are introduced. In Chap. 2, the concepts of classical information theory are provided together with corresponding application to fading channels and channels with memory. This chapter provides information and coding theory fundamentals to the level needed to easier follow the book. In Chap. 3, the conventional cryptography fundamentals are introduced. Chapter 4 provides a detailed description of the physical-layer security concepts. In Chap. 5, the basic concepts of quantum information processing, quantum information theory, and quantum error correction are provided to better understand the QKD systems. Chapter 6 is devoted to the QKD fundamentals, ranging from basic concepts, through various QKD protocols, to QKD security issues. Chapter 7 represents the continuation of Chap. 6, and it is devoted to a detailed description of the discrete variable (DV) QKD protocols. Chapter 8 is devoted to the detailed description of continuous variable (CV)-QKD schemes, in particular, those with Gaussian modulation and discrete modulation. Chapter 9 is devoted to the recently proposed both DV- and CV-QKD schemes, including measurement-device-independent (MDI), twin-field (TF), and floodlight (FL) QKD protocols, to mention few. Chapter 10 is devoted to covert communications, also known as low-probability of detection/intercept, as well as stealth communications, and how they can improve secret-key rate for QKD applications.

Author would like to thank his colleagues and former students, in particular, Xiaole Sun, John Gariano, and Tyan-Lin Wang. Further, the author would like to thank ONR, NSF, and Harris Co. for supporting in part the corresponding research.

Finally, special thanks are extended to Mary E. James and Zoe Kennedy of Springer US for their tremendous effort in organizing the logistics of the book, in particular, promotion and edition, which is indispensable to make this book happen.

Tucson, AZ, USA Ivan B. Djordjevic

Contents

Chapter 1
Introduction

Abstract In this chapter, the basic concepts of both physical-layer security (PLS) and quantum-key distribution (QKD) are introduced. The chapter starts with the role of PLS, following by a brief overview of conventional key-based cryptographic systems. The concept of information-theoretic security is introduced next, and the perfect secrecy condition is described. The computational security is described as a special case of information-theoretic security in which several relaxations are introduced. The concepts of strong and weak secrecy are then introduced. Further, the degraded wiretap channel model, introduced by Wyner, is described, and corresponding wiretap channel codes are defined. After that, the broadcast channel with confidential messages, introduced by Csiszár and Körner, is described then, together with corresponding stochastic code. The last topic in PLS section is devoted to the secret-key agreement protocol. The QKD section describes first how to break the RSA protocol with the help of Shor's factorization algorithm, followed by the brief description of foundations for both discrete variable (DV) and continuous variable (CV) QKD schemes. The key limitations of DV-QKD schemes are identified. Various QKD protocols are placed into three generic categories: device-dependent QKD, source-device-independent QKD, and measurement-device-independent (MDI) QKD. Further, the definition of the secrecy fraction for QKD protocols is provided, following by the brief description of individual (incoherent) and collective attacks, and explanation of how to calculate the corresponding secrecy fractions. In section on the organization of the book, the detailed description of the content of the chapters is provided.

1.1 Physical-Layer Security Basics

Public-key cryptography has several serious drawbacks such as it is difficult to implement it in devices with low memory and low process constraints, Internet is becoming more and more mobile, security schemes are based on unproven assumptions of intractability of certain functions, and the assumption of limiting computing resources of Eve is very often incorrect, to mention few. The open system interconnection (OSI) reference model defines seven layers. However, only five layers,

I. B. Djordjevic, *Physical-Layer Security and Quantum Key Distribution*,
https://doi.org/10.1007/978-3-030-27565-5_1

relevant to security issues, are provided in Fig. 1.1. The original OSI model does not even specify the security issues at all. The security issues are addressed in X.800 standard (security architecture for OSI) [1]. However, neither the physical-layer security (PLS) [2–6] nor quantum-key distribution (QKD) [7–11] have been discussed in this standard. Nevertheless, the services specified in these five layers can be significantly enhanced by employing the PLS and QKD. The PLS and QKD schemes can also operate independently.

The basic key-based cryptographic system [12–22] is provided in Fig. 1.2. The source emits the message (*plaintext*) $\boldsymbol{M}$ toward the encryption block, which with the help of key $\boldsymbol{K}$, obtained from key source, generates the *cryptogram (ciphertext)* $\boldsymbol{C}$. On receiver side, the cryptogram transmitted over insecure channel get processed by the decryption algorithm together with the key $\boldsymbol{K}$ obtained through the secure channel, which reconstructs the original plaintext to be delivered to the authenticated user.

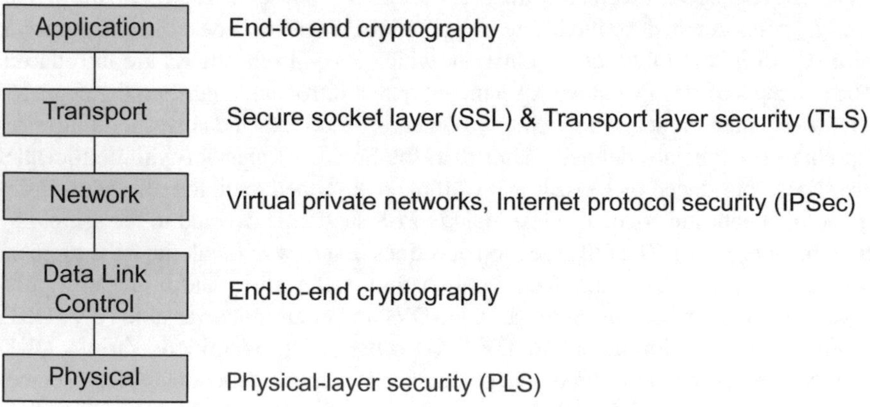

Fig. 1.1 Security mechanisms at different layers in OSI model (only security-relevant layers are shown)

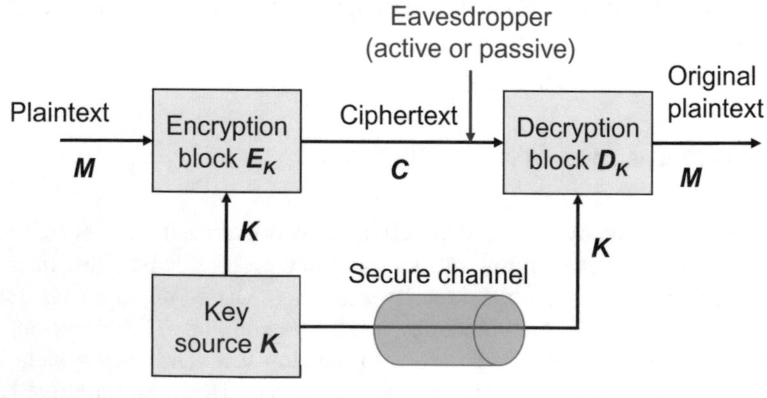

Fig. 1.2 The basic key-based cryptographic scheme

The encryption process can be mathematically described as $\boldsymbol{E_K}(\boldsymbol{M}) = \boldsymbol{C}$, while the decryption process by $\boldsymbol{D_K}(\boldsymbol{C}) = \boldsymbol{M}$. The composition of decryption and encryption functions yields to identity mapping $\boldsymbol{D_K}(\boldsymbol{E_K}(\boldsymbol{M})) = \boldsymbol{M}$. The key source typically generates the key randomly from the *keyspace* (the range of possible key values).

The key-based algorithms can be categorized into two broad categories:

- *Symmetric algorithms*, in which decryption key can be derived from encryption key and vice versa. Alternatively, the same key can be used for both encryption and decryption stages. Symmetric algorithms are also known as one-key (single-key) or secret-key algorithms. The well-known system employing this type of algorithms is digital encryption standard (DES) [13–18].
- *Asymmetric algorithms*, in which encryption and decryption keys are different. Moreover, the decryption key cannot be determined from encryption key, at least in any reasonable amount of time. Because of this fact, the encryption keys can be even made public, wherein the eavesdropper will not be able to determine the decryption key. The *public-key systems* [17] are based on this concept. In public-key systems, the encrypted keys have been made public, while the decryption key is known only to the intended user. The encryption key is then called the *public* key, while decryption the *secret (private)* key. The keys can be applied in arbitrary order to create the cryptogram from plaintext and to reconstruct the plaintext from the cryptogram.

The simplest private-key cryptosystem is the *Vernam cipher* also known as the *one-time pad*. In *one-time pad* [23], a completely random sequence of characters, with the sequence length being equal to the message sequence length, is used as a key. When for each new message another random sequence is used as a key, the one-time pad scheme provides so-called prefect security. Namely, the brute-force search approach would be required to verify m^n possible keys, where m is the employed alphabet size and n is the length of intercepted cryptogram. In practice, in digital and computer communications, we typically operate on binary alphabet {0, 1}. To obtain the key, we need a special random generator and to encrypt using one-time pad scheme we simply perform addition mod 2, i.e., XOR operation, as illustrated in Fig. 1.3. Even though that the one-time pad scheme offers so-called perfect security, it

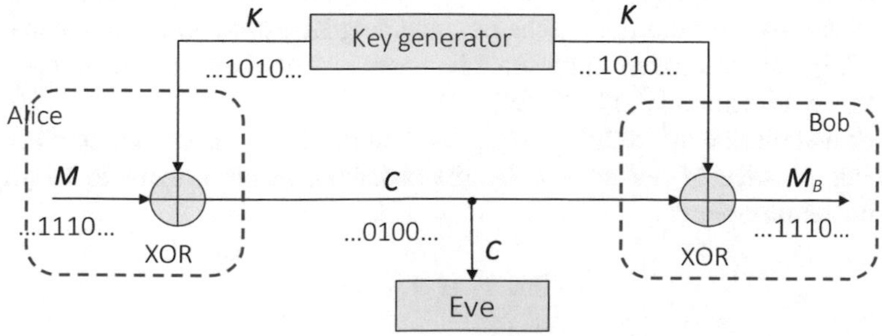

Fig. 1.3 The one-time pad encryption scheme

has several drawbacks [9–11]: it requires the secure distribution of the key, the length of the key must be at least as long as the message, the key bits cannot be reused, the keys must be delivered in advance, securely stored until used, and destroyed after the use.

According to Shannon [12], the *perfect security*, also known as *unconditional security*, has been achieved when the messages and cryptograms are statistically independent so that the corresponding mutual information between the message $\boldsymbol{M}$ and cryptogram $\boldsymbol{C}$ is equal to zero:

$$I(\boldsymbol{M}, \boldsymbol{C}) = H(\boldsymbol{M}) - H(\boldsymbol{M}|\boldsymbol{C}) = 0 \quad \Leftrightarrow \quad H(\boldsymbol{M}|\boldsymbol{C}) = H(\boldsymbol{M}), \tag{1.1}$$

where $H(\boldsymbol{M})$ is the entropy (uncertainty) about the message, while $H(\boldsymbol{M}|\boldsymbol{C})$ is conditional entropy of the message $\boldsymbol{M}$ given the cryptogram $\boldsymbol{C}$. The *perfect secrecy condition* can, therefore, be summarized as

$$H(\boldsymbol{M}) \leq H(\boldsymbol{K}). \tag{1.2}$$

In other words, the entropy (uncertainty) of the key cannot be lower than the entropy of the message, for an encryption scheme to be perfectly secure. Given that in Vernam cipher the length of the key is at least equal to the message length, it appears that one-time pad scheme is perfectly secure.

However, given that this condition is difficult to satisfy, in conventional cryptography, instead of information-theoretic security, the *computational security* is used [10, 13, 16, 22–25]. The computational security introduces two relaxations with respect to information-theoretic security [16]:

- Security is guaranteed against an efficient eavesdropper running the cryptanalytic attacks for certain limited amount of time. Of course, when eavesdropper has sufficient computational resources and/or sufficient time, he/she will be able to break the security of the encryption scheme.
- Eavesdroppers can be successful in breaking the security protocols, but with small success probability.

A reader interested to learn more about computational security is referred to as an excellent book due to Katz and Lindell [16]. However, by using quantum computing, any conventional cryptographic scheme, including Rivest–Shamir–Adleman (RSA) system [26], can be broken in reasonable amount of time by employing the Shor's factorization algorithm [9–11, 27–29].

Given that mutual information $I(\mathbf{M}, \mathbf{C})$ measures the average amount of information about message $\boldsymbol{M}$ leaked in $\boldsymbol{C}$, as the codeword length n tends to infinity, the following requirement

$$\lim_{n \to \infty} I(\boldsymbol{M}, \boldsymbol{C}) = 0 \tag{1.3}$$

is commonly referred to as the *strong secrecy condition*. From practical point of view, given that the strong secrecy condition is difficult to satisfy, instead of requesting the mutual information to vanish, we can soften the requirement and request that the *rate* of information leaked to Eve tends to zero:

$$\lim_{m\to\infty} \frac{1}{n} I(\boldsymbol{M}, \boldsymbol{C}) = 0 \tag{1.4}$$

This average information rate about the massage $\boldsymbol{M}$ leaked to $\boldsymbol{C}$ is well known as the *weak secrecy condition.*

Shannon's model is pessimistic as it assumes that no noise has been introduced during transmission. Wyner introduced so-called the *wiretap channel* [30], now also known as a *degraded wiretap channel model*, in which Eve's channel is degraded version of Alice–Bob channel (main channel), as indicated in Fig. 1.4. Alice encodes the message $\boldsymbol{M}$ into a codeword $\boldsymbol{X}^n$ of length n and sends it over the noisy channel, represented by conditional probability density function (PDF) $f(y|x)$ toward Bob. On the other hand, Eve observes the noisy version of the signal available to Bob. Therefore, the wiretap channel is degraded channel represented by the conditional PDF $f(z|y)$. Wyner suggested to use the *equivocation rate*, defined as $(1/n)H(\boldsymbol{M}|\boldsymbol{Z}^n)$, instead of the entropy of the message $H(\boldsymbol{M})$. So the *secrecy condition* in Wyner's sense will be

$$\frac{1}{n} H(\boldsymbol{M}) - \frac{1}{n} H(\boldsymbol{M}|\boldsymbol{Z}^n) = \frac{1}{n} I(\boldsymbol{M}, \boldsymbol{Z}^n) \underset{n\to\infty}{\to} 0, \tag{1.5}$$

which is clearly the weak secrecy condition. In addition to secrecy condition, the *reliability condition* must be satisfied as well:

$$\Pr(\boldsymbol{M}_B \neq \boldsymbol{M}|\boldsymbol{Y}^n) \underset{n\to\infty}{\to} 0. \tag{1.6}$$

In other words, the probability that Bob's message is different from the message sent by Alice tends to zero as $n \to \infty$. The channel codes to be used in this scenario must satisfy both reliability and secrecy conditions and the codes simultaneously satisfying both conditions are known as the *wiretap codes* [31]. For instance, LDPC, polar, and lattice codes can be used to design the wiretap codes. The (n, k) wiretap

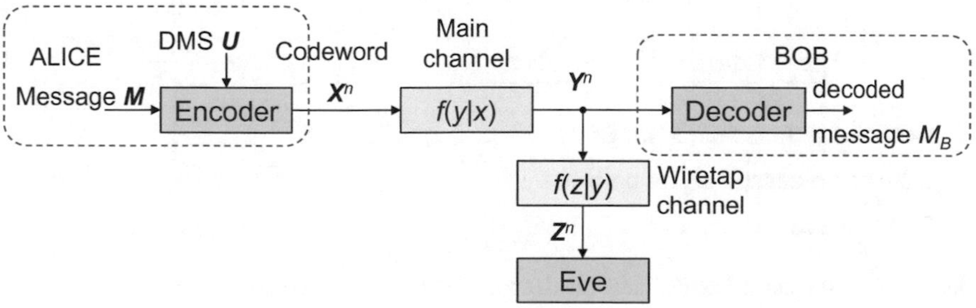

Fig. 1.4 Wyner's wiretap channel model. DMS: discrete memoryless source

code $\mathcal{C}_n$ of rate $R = k/n$ is specified by [3, 31]: (i) the set of messages $\mathcal{M}$ of size 2^{nR}, (ii) the local random source $\mathcal{U}$ with distribution $f_{\mathcal{U}}$, (iii) the encoder performing the mapping of the message and a random realization of the local source into a codeword, and (iv) the decoder performing the de-mapping of the received word into a message estimate. The largest transmission rate at which both reliability and secrecy conditions are simultaneously satisfied is commonly referred to as the *secrecy capacity*. For any distribution f_x of $\mathcal{X}$ from set of distributions $\mathcal{P}(R \geq 0)$ for which $I(X, Y) \geq R$ Wyner has defined the function, which can be called a secrecy rate:

$$SR(R) = \sup_{f_x \in \mathcal{P}(R)} [I(X, Y) - I(X, Z)]. \tag{1.7}$$

He also showed that $SR(R)$ is upper bounded by the capacity of the main channel C_m and lower bounded by $C_m - C_e$, where C_e is the capacity of the main-wiretap channel cascade, that is

$$C_m - C_e \leq SR(R) \leq C_m. \tag{1.8}$$

Wyner's wiretap channel gets generalized and refined by Csiszár and Körner [32], and the corresponding model, now known as the *broadcast channel with confidential messages* (BCC) , is provided in Fig. 1.5. The broadcast channel is assumed to be discrete and memoryless and characterized by input alphabet $\mathcal{X}$, and output alphabets $\mathcal{Y}$ and $\mathcal{Z}$ (corresponding to Bob and Eve, respectively), and transition PDF $f(yz|x)$. So, the channel itself is modeled by a joint PDF for Bob's and Eve's observations, $f(\mathbf{yz}|\mathbf{x})$, conditioned on the channel input. In this scenario, Alice wishes to broadcast a common message $\boldsymbol{M}_c$ to both Bob and Eve and a confidential message $\boldsymbol{M}$ to Bob. The corresponding stochastic code $\mathcal{C}_n$ of codeword length n is composed of the following:

- Two message sets: the common message set and the confidential message set.
- The encoding (stochastic) function that maps the confidential–common message pair into a codeword.

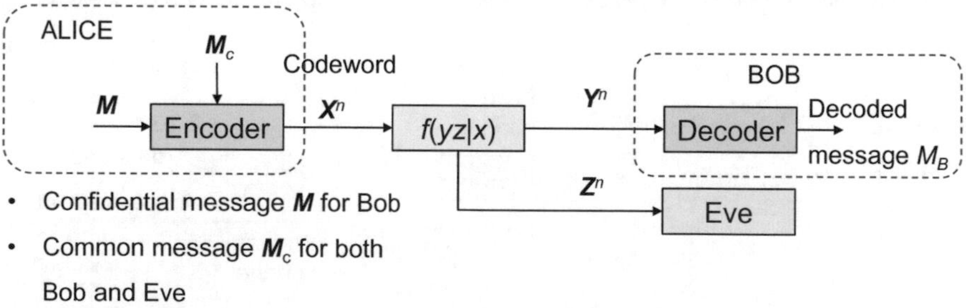

Fig. 1.5 The broadcast channel model with confidential messages (BCC)

- Two decoding functions: the first one mapping the observation vector $\boldsymbol{y}^n$ to the estimated message pair, while the second one mapping the observation z^n to common message estimate.

Csiszár and Körner proved the *corollary* [32] claiming that the *secrecy capacity* is determined as the difference of mutual information for Alice–Bob and Alice–Eve links, when the rate of the common message is set to zero, that is,

$$C_s = \max_{\substack{f_{VX} \\ V \to X \to YZ}} [I(V, Y) - I(V, Z)], \tag{1.9}$$

where the maximization is performed over all possible joint distributions $f_{VX}(v, x)$ and V, X, and YZ form a Markov chain $V \to X \to YZ$. Clearly, the secrecy capacity is strictly positive when Bob's channel is less noisy than Eve's channel, i.e., $I(X; Y) > I(X; Z)$. Namely, by setting $V = X$, the secrecy capacity expression becomes $C_s = \max_{f_X}[I(X, Y) - I(X, Z)]$, which is clearly strictly positive when $I(X; Y) > I(X; Z)$.

Compared to conventional cryptographic approaches where strong error control coding (ECC) schemes are used to provide reliable communication, the transmission in PLS scenario needs to be simultaneously reliable and secure. This indicates that different classes of channel codes must be developed. Alternatively, similar to QKD, the randomness of the channel can be exploited to generate the key, and this approach is commonly referred to as the *secret-key agreement* [2–5], and this concept is described in Fig. 1.6, inspired by [2, 3, 6]. Alice and Bob monitor Alice–Bob channel capacity (also known as the capacity of the main channel) C_M and the secrecy capacity C_S, defined as a difference between main channel capacity and eavesdropping channel capacity C_E. When the secrecy capacity is well above the threshold value $C_{S,\text{tsh}}$ and the main channel capacity is well above threshold value $C_{M,\text{tsh}}$, Alice transmits Gaussian-shaped symbols $\boldsymbol{X}$ to Bob. When the secrecy capacity and main channel capacity are both below corresponding thresholds due to deep fading in wireless channels or atmospheric turbulence effects in free-space optical channels, Alice and Bob perform *information reconciliation* of previously transmitted symbols, which is based strong ECC scheme to ensure that errors introduced by either channel or Eve can be corrected for. Similar to QKD schemes [7–11], a systematic low-density parity-check (LDPC) code can be used (that does not affect information bits but generates the parity-check bits algebraically related to the information bits) to generate the parity bits and transmit them over an authenticated public channel. There exist direct and reverse information reconciliation schemes. In *direct reconciliation*, shown in Fig. 1.6, Alice performs LDPC encoding and sends the parity bits to Bob. Bob performs the LDPC decoding to get the correct key $\boldsymbol{X}$. In *reverse reconciliation*, Bob performs LDPC encoding instead. *Privacy amplification* is then performed between Alice and Bob to distil from $\boldsymbol{X}$ a smaller set of bits $\boldsymbol{K}$ (final key), whose correlation with Eve's string is below the desired threshold [9–11, 33]. One way to accomplish privacy amplification is through the use of *universal hash*

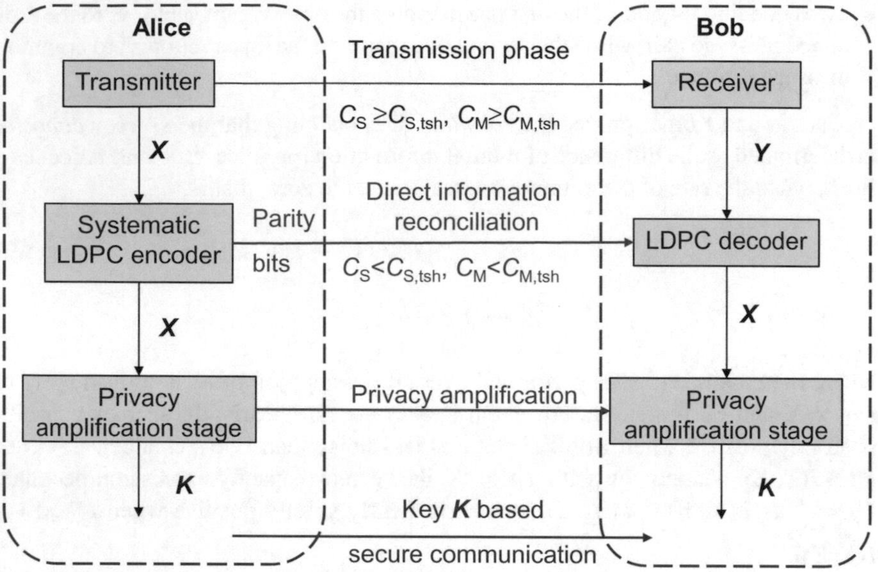

Fig. 1.6 Secret-key generation (agreement) protocol suitable for wireless as well as optical communications

functions $\mathcal{G}$ [9–11, 33], which map the set of n-bit strings $\boldsymbol{X}$ to the set of m-bit strings $\boldsymbol{K}$ such that for any distinct $\boldsymbol{X}_1$ and $\boldsymbol{X}_2$ from the set of corrected keys, when the mapping g is chosen uniformly at random from $\mathcal{G}$, the probability of having $g(\boldsymbol{X}_1) = g(\boldsymbol{X}_2)$ is very low. Two types of models are typically considered for secret-key agreement [34]:

- *Source-type model*, in which terminals observe the correlated output of the source of randomness without having control of it.
- *Channel-type model*, in which one terminal transmits random symbols to other terminals using a broadcast channel. This scenario is similar to the wiretap channel model with feedback channel, which is an authenticated public noiseless channel.

Both of these models are very similar to QKD [7–11, 35–38], except that raw key in PLS is transmitted over the classical channel, while in QKD over the quantum channel. The secret-key agreement protocols in addition to the *reliability condition* and *secrecy condition* must also satisfy the *uniformity condition*, which ensures that the secret key is uniformly distributed within the corresponding set. The rate at which secret key is generated can be called the same way as in QKD, the *secret-key rate* (SKR). If the protocols exploit the public messages sent in one direction only (from either Alice to Bob or Bob to Alice), the corresponding SKR is said to be achievable with *one-way communication*; otherwise, the SKR is said to be achievable with *two-way communication*. We say that the secret-key rate R is *achievable* if there exists a sequence of secret-key generation protocols satisfying all three conditions (constraints) as $n \to \infty$. The supremum of achievable SKRs is commonly referred to as the *secret-key capacity*, denoted here as C_{SK}. Given two-way communication over the authenticated public channel, it is difficult to derive an exact expression for C_{SK}; however, based on [34, 39], it can be bounded from both sides as follows:

$$\max_{f_X} \max[I(X,Y) - I(X,Z), I(Y,X) - I(Y,Z)] \leq C_{SK} \leq \max_{f_X}[I(X,Y|Z)]. \tag{1.10}$$

The upper bound term indicates the secret-key capacity when Bob has access to Eve's observations. The lower bound term $\max[I(X, Y) - I(X, Z)]$ indicates that direct reconciliation is employed, while the lower bound term $\max[I(Y, X) - I(Y, Z)]$ indicates that reverse reconciliation is employed instead.

To summarize, the PLS is related to different methods and algorithms to enable security by exploiting the properties of the physical medium. Additional details of various PLS schemes can be found in incoming chapters.

1.2 Quantum-Key Distribution (QKD) Basics

Significant achievements have been recently made in quantum computing [9–11]. There are many companies currently working on development of the medium-scale quantum computers. Given that the most of cryptosystems depend on the computational hardness assumption, the quantum computing represents a serious challenge to the modern cybersecurity systems. As an illustration, to break the RSA protocol [26], one needs to determine the period r of the function $f(x) = m^x \bmod n = f(x + r)$ ($r = 0, 1, \ldots, 2^l - 1$; m is an integer smaller than $n - 1$). This period is determined in one of the steps of the Shor's factorization algorithm [9–11, 27–29].

The QKD with symmetric encryption can be interpreted as one of the physical-layer security schemes that can provide the provable security against quantum computer-initiated attack [35]. The first QKD scheme was introduced by Bennett and Brassard, who proposed it in 1984 [7, 8], and it is now known as the BB84 protocol. The security of QKD is guaranteed by the quantum mechanics laws. Different photon degrees of freedom, such as polarization, time, frequency, phase, and orbital angular momentum, can be employed to implement various QKD protocols. Generally speaking, there are two generic QKD schemes, discrete variable (DV)-QKD, and continuous variable (CV)-QKD, depending on strategy applied on Bob's side. In DV-QKD schemes, a single-photon detector (SPD) is applied on Bob's side, while in CV-QKD the field quadratures are measured with the help of homodyne/heterodyne detection. The DV-QKD scheme achieves the unconditional security by employing no-cloning theorem and theorem on indistinguishability of arbitrary quantum states. The no-cloning theorem claims that arbitrary quantum states cannot be cloned, indicating that Eve cannot duplicate non-orthogonal quantum states even with the help of quantum computer. On the other hand, the second theorem claims that non-orthogonal states cannot be unambiguously distinguished. Namely, when Eve interacts with the transmitted quantum states, trying to get information on transmitted bits, she will inadvertently disturb the fidelity of the quantum states that will be detected by Bob. On the other hand, the CV-QKD employs the uncertainty principle claiming that both in-phase and quadrature components of a coherent state cannot be simultaneously measured with the complete precision. We can also classify different QKD schemes as either entanglement-assisted or prepare-and-measure types.

The research in QKD is getting momentum, in particular after the first satellite-to-ground QKD demonstration [36]. Recently, the QKD over 404 km of ultralow-loss optical fiber is demonstrated, however, with ultralow secure-key rate (3.2×10^{-4} b/s). Given that quantum states cannot be amplified, the fiber attenuation limits the distance. On the other hand, the deadtime (the time over which an SPD remains unresponsive to incoming photons due to long recovery time) of the SPDs, typically in 10–100 ns range, limits the baud rate and therefore the secure-key rate. The CV-QKD schemes, since they employ the homodyne/heterodyne detection, do not have deadtime limitation; however, the typical distances are shorter.

By transmitting non-orthogonal qubit states between Alice and Bob, and by checking for disturbance in transmitted state, caused by the channel or Eve's activity, they can establish an upper bound on noise/eavesdropping in their quantum communication channel [9]. The *threshold for maximum tolerable error rate* is dictated by the efficiency of the best postprocessing steps [9]. The QKD protocols can be categorized into several general categories:

- **Device-dependent QKD**, in which, typically, the quantum source is placed on Alice side and quantum detector at Bob's side. Popular classes include DV-QKD, CV-QKD, entanglement-assisted (EA) QKD, distributed phase reference, etc. For EA QKD, the entangled source can be placed in the middle of the channel to extend the transmission distance.
- **Source-device-independent QKD**, in which the quantum source is placed at Charlie's (Eve's) side, while the quantum detectors at both Alice and Bob's sides.
- **Measurement-device-independent QKD (MDI-QKD)**, in which the quantum detectors are placed at Charlie's (Eve's) side, while the quantum sources are placed at both Alice and Bob's sides. The quantum states get prepared at both Alice and Bob's sides and get transmitted toward Charlie's detectors. Charlie performs the partial Bell state measurements and announces when the desired partial Bell states are detected, with details to be provided in later chapters.

The QKD can achieve the unconditional security, which means that its security can be verified without imposing any restrictions on either Eve's computational power or eavesdropping strategy. The bounds on the fraction rate are dependent on the classical postprocessing steps. The most common is the one-way postprocessing, in which either Alice or Bob holds the reference key and sends the classical information to the other party through the public channel, while the other party performs certain procedure on data without providing the feedback. The most common one-way processing consists of two steps, the information reconciliation and privacy amplification. The expression for *secret fraction*, obtained by *one-way postprocessing* is very similar to that for the classical PLS schemes and it is given by

$$r = I(A;B) - \min_{\text{Eve's strategies}} (I_{EA}, I_{EB}), \tag{1.11}$$

where $I(A; B)$ is the mutual information between Alice and Bob, while the second term corresponds to Eve's information I_E about Alice or Bob's raw key, where minimization is performed over all possible eavesdropping strategies. Alice and Bob will decide to employ either direct or reverse reconciliation so that they can minimize Eve's information.

We now describe different eavesdropping strategies that Eve may employ, which determine Eve's information I_E. *Independent (individual) or incoherent attacks* represent the most constrained family of attacks, in which Eve attacks each qubit independently, and interacts with each qubit by applying the same strategy. Moreover, she measures the quantum states before the classical postprocessing takes place. The security bound for incoherent attacks is the same as that for classical PLS, wherein the mutual information between Alice and Eve is given by

$$I_{EA} = \max_{\text{Eve's strategies}} I(A;E), \tag{1.12}$$

where the maximization is performed over all possible incoherent eavesdropping strategies. The similar definition holds for I_{BE}.

The *collective attacks* represent generalization of the incoherent attacks given that Eve's interaction with each quantum bit, also known as qubit, is also independent and identically distributed (i.i.d). However, in these attacks, Eve's can store her ancilla qubits in a quantum memory until the end of classical postprocessing steps. The security bound for collective attacks, assuming one-way postprocessing, is given by Eq. (1.11), wherein Eve's information about Alice sequence is determined from *Holevo information* as follows [37, 38]:

$$I_{EA} = \max_{\text{Eve's strategies}} \chi(A;E), \tag{1.13}$$

where maximization is performed over all possible collective eavesdropping strategies. The similar definition holds for I_{BE}. This bound is also known as Devetak–Winter bound. The *Holevo information*, introduced in [40], is defined here as

$$\chi(A;E) = S(\rho_E) - \sum\nolimits_a S(\rho_{E|a})p(a), \tag{1.14}$$

where $S(\rho)$ is the von Neumann entropy defined as $S(\rho) = -\text{Tr}(\log(\rho)) = -\sum_i \lambda_i \log \lambda_i$, with λ_i being the eigenvalues of the density operator (state) ρ. The density operator is used to represent the ensemble of quantum states, each occurring with a given probability. (For additional details on density operators please refer to Chap. 5.) In (1.14), $p(a)$ represents the probability of occurrence of symbol a from Alice's classical alphabet, while $\rho_{\text{E|a}}$ is the corresponding density operator of Eve's ancilla. Finally, ρ_{E} is Eve's partial density state defined by $\rho_{\text{E}} = \sum_a p(a)\rho_{\text{E|a}}$. In other words, the Holevo information corresponds to the average reduction in von Neumann entropy given that we know how ρ_E get prepared.

1.3 Organization of the Book

This book is organized as follows. After the introduction, in Chap. 2, the concepts of classical information theory are provided together with corresponding application to fading channels and channels with memory. This chapter provides information and coding theory fundamentals to the level needed to easier follow the book. The chapter starts with definitions of entropy, joint entropy, conditional entropy, relative entropy, mutual information, and channel capacity, followed by the information capacity theorem. We discuss the channel capacity of discrete memoryless channels, continuous channels, and channels with memory. Regarding the wireless channels, we describe how to calculate the channel capacity of flat fading and frequency-selective channels. We also discuss different optimum and suboptimum strategies to achieve channel capacity including the water-filling method, multiplexed coding and decoding, channel inversion, and truncated channel inversion. We also study different strategies for channel capacity calculation depending on what is known about the channel state information. Further, we explain how to model the channel with memory and describe McMillan–Khinchin model for channel capacity evaluation. After that, the fundaments of linear blocks codes are introduced, followed by the binary LDPC coding fundamentals.

In Chap. 3, the conventional cryptography fundamentals are introduced. The basic terminology and cryptographic schemes, including symmetric and asymmetric cryptography, basic ciphers such as substitution and transposition ciphers, and one-time pads are introduced first. The concepts of secrecy, authentication, and non-repudiation are discussed then, followed by various cryptanalytic attacks such as ciphertext-only, known-plaintext, chosen-plaintext, chosen-ciphertext, and adaptive-chosen-plaintext attacks. The concept of perfect security is introduced next and compared against the computational security. In the same section, unicity distance is discussed as well as the role of compression in cryptography. After that, one-way functions and one-way hash functions are discussed. The chapter concludes with several relevant practical cryptographic systems including DES and RSA systems as well as Diffie–Hellman public-key distribution.

Chapter 4 is devoted to the physical-layer security. The chapter starts with the discussion on security issues, followed by the introduction of information-theoretic security, and comparison against the computational security. In the same section, various information-theoretic security measures are introduced, including strong secrecy and weak secrecy conditions. After that, the Wyner's wiretap channel model, also known as the degraded wiretap channel model, is introduced. In the same section, the concept of secrecy capacity is introduced as well as the nested wiretap coding. Further, the broadcast channel with confidential messages is introduced, and the secrecy capacity definition is generalized. The focus is then moved to the secret-key generation (agreement), the source and channel-type models are introduced, and corresponding secret-key generation protocols are described. The next section is devoted to the coding for the physical-layer security systems, including both coding for weak and strong secrecy systems. Regarding the coding for weak secrecy systems, the special attention is devoted to two-edge type LDPC coding, punctured LDPC coding, and polar codes. Regarding the coding for strong secrecy systems, the focus is on coset coding with dual of LDPC codes and hash functions/extractor-based coding. The attention is then moved to information reconciliation and privacy amplification. In wireless channels PLS section, the following topics are covered: MIMO fundamentals, wireless MIMO PLS, and secret-key generation in wireless networks. In section on optical channels PLS, both PLS for spatial division multiplexing (SDM)-fiber-based systems and free-space optical (FSO) systems is discussed.

In Chap. 5, the basic concepts of quantum information processing, quantum information theory, and quantum error correction are provided to better understand the QKD systems. The following topics from quantum information processing are covered: state vectors, operators, density operators, measurements, dynamics of a quantum system, superposition principle, quantum parallelism, no-cloning theorem, and entanglement. The following concepts from quantum information theory are provided: Holevo information, accessible information, Holevo bound, Shannon Entropy & von Neumann Entropy, Schumacher's noiseless quantum coding theorem, and Holevo–Schumacher–Westmoreland theorem. The basic concepts of quantum error correction are introduced as well.

Chapter 6 is devoted to the QKD fundamentals. The chapter starts with description of key differences between conventional cryptography, classical PLS, and QKD. In section on QKD basics, after historical overview, we review different QKD types and describe briefly common postprocessing steps, namely, information reconciliation and privacy amplification steps. In the same section, we provide two fundamental theorems on which QKD relies on, no-cloning theorem and the theorem of inability to unambiguously distinguish non-orthogonal quantum states. In section on discrete variable (DV)-QKD systems, we describe in detail BB84 and B92 protocols as well as Ekert (E91) and EPR protocols. In the same section, the time-phase encoding protocol is also described. Regarding, the BB84 protocols, different versions, suitable for different technologies, are described. In the section on QKD security, the secret-key rate is represented as the product of raw key rate and fractional rate, followed by the description of different limitations to the raw key rate. After that, the generic expression for the fractional rate is provided, followed by the description of different eavesdropping strategies including individual (independent or incoherent) attacks, collective attacks, and coherent attacks as well as the quantum hacking/side-channel attacks. For individual and coherent attacks, the corresponding secrete fraction expressions are described. The next section is devoted to various definitions of security, including the concept of ε-security. After that, the generic expressions for 2-D DV-QKD schemes are derived for both prepare-and-measure and decoy-state-based protocols. To facilitate the description of continuous variable (CV)-QKD protocols, the fundamentals of quantum optics are introduced first. In section on CV-QKD protocols, both squeezed state-based and coherent state-based protocols are described. Given that the coherent states are much easier to generate and manipulate, the coherent state-based protocols with both homodyne and heterodyne detections are described in detail. The secret fraction is derived for both direct and reverse reconciliation-based CV-QKD protocols. Furthermore, the details on practical aspects of GG02 protocol are provided. In the same section, the secret fraction calculation for collective attacks is discussed. After that, the basic concepts for measurement-device-independent (MDI)-QKD protocols are introduced. Then, final section in the chapter provides some relevant concluding remarks.

Chapter 7 represents the continuation of Chap. 6, and it is devoted to the discrete variable (DV) QKD protocols. The chapter starts with the description of BB84 and decoy-state-based protocols, and evaluation of their secrecy fraction performance in terms of achievable distance. The next topic is related to the security of DV-QKD protocols when the resources are finite. We introduce the concept of composable ε-security and describe how it can be evaluated for both collective and coherent attacks. We also discuss how the concept of correctness and secrecy can be combined to come up with tight security bounds. After that, we evaluate the BB84 and decoy-state protocols for finite key assumption over atmospheric turbulence effects. We also describe how to deal with time-varying free-space optical channel conditions. The focused is then moved to high-dimensional (HD) QKD protocols, starting with the description of mutually unbiased bases (MUBs) selection, followed by the introduction of the generalized Bell states. We then describe how to evaluate the security of HD QKD protocols for finite resources. We describe various HD QKD protocols, including time-phase encoding, time-energy encoding, OAM-based HD QKD, fiber Bragg grating (FBGs)-based HD QKD, and waveguide Bragg gratings (WBGs)-based HD QKD protocols.

Chapter 8 is devoted to the detailed description of CV-QKD schemes, in particular, with Gaussian modulation and discrete modulation. The chapter starts with the fundamentals of Gaussian quantum information theory, where the *P*-representation is introduced and applied to represent the thermal noise as well as the thermal noise plus the coherent state signal. Then quadrature operators are introduced, followed by the phase-space representation. Further, Gaussian and squeezed states are introduced, followed by the Wigner function definition as well as the definition of correlation matrices. The next subsection is devoted to the Gaussian transformation and Gaussian channels, with beam splitter operation and phase rotation operation being the representative examples. The thermal decomposition of Gaussian states is discussed next, and the von Neumann entropy for thermal states is derived. The focus is then moved to the nonlinear quantum optics fundamentals, in particular, the three-wave mixing and the four-wave mixing are described in detail. Further, the generation of the Gaussian states is discussed, in particular, the EPR state. The correlation matrices for two-mode Gaussian states are discussed next, and how to calculate the symplectic eigenvalues, relevant in von Neumann entropy calculation. The Gaussian states measurements and detection is discussed then, with emphasis on homodyne detection, heterodyne detection, and partial measurements. In section on CV-QKD protocols with Gaussian modulation, after the brief description of squeezed state-based protocols, the coherent state-based protocols are described in detail. We start the section with the description of both lossless and lossy transmission channels, followed by the description of how to calculate the covariance matrix under various transformations, including beam splitter, homodyne detection, and heterodyne detection. The equivalence between the prepare-and-measure (PM) and entanglement-assisted protocols with Gaussian modulation is discussed next. The focused is then move to the secret-key rate calculation under collective attacks. The calculation of mutual information between Alice and Bob is discussed first, followed by the calculation of Holevo information between Eve and Bob, in both cases assuming the PM protocol and reverse reconciliation. Further, entangling cloner attack is described, followed by the derivation of Eve–Bob Holevo information. The entanglement-assisted protocol is described next as well as the corresponding Holevo information derivation. In all these derivations, both homodyne detection and heterodyne detection are considered. Some illustrative SKR results, corresponding to the Gaussian modulation, are provided as well. In section on CV-QKD with discrete modulation, after the brief introduction, we describe both four-state and eight-state CV-QKD protocols. Both the PM and entanglement-assisted protocols are discussed. The SKR calculation for discrete modulation is discussed next, with illustrative numerical results. We also identify conditions under which the discrete modulation can outperform the Gaussian modulation. In section on RF-assisted CV-QKD scheme, we describe a generic RF-assisted scheme applicable to arbitrary two-dimensional modulation schemes, including M-ary PSK and M-ary QAM. This scheme exhibits better tolerance to laser phase noise and frequency offset fluctuations compared to conventional CV-QKD schemes with discrete modulation. We then discuss how to increase the SKR through the parallelization approach. The final section in the chapter provides some relevant concluding remarks.

Chapter 9 is devoted to the recently proposed discrete variable (DV) and continuous variable (CV)-QKD schemes. The chapter starts with the description of Hong–Ou–Mandel effect and photonic Bell state measurements (BSMs). Both polarization state-based and time-bin-state-based BSMs are introduced. After that, the BB84 and decoy-state protocols are briefly revisited. The next topic in the chapter is devoted to the measurement-device-independent (MDI)-QKD protocols. Both polarization state-based and time-phase state-based MDI-QKD protocols are described. Further, the twin-field (TF)-QKD protocols are described, capable of beating the Pirandola–Laurenza–Ottaviani–Banchi (PLOB) bound on a linear key rate. Floodlight (FL) CV-QKD protocol is then described, capable of achieving record secret-key rates. Finally, Kramers–Kronig (KK)-receiver-based CV-QKD scheme is introduced, representing high-SKR scheme of low-complexity.

Chapter 10 is devoted to covert communications, also known as low probability of detection/intercept, as well as stealth communications, and how they can improve secret-key rate for QKD applications. The chapter starts with brief introduction to covert communications, followed by the description of their differences with respect to steganography. One of the key technologies to enable covert communication over wireless channels, the spread spectrum concept, is introduced next. After that, the rigorous treatment of covert communication over an additive white Gaussian noise channel is provided, and the square root law is derived. The importance of hypothesis testing in covert communications is discussed as well. The covert communication over the discrete memoryless channels is discussed after that. The next topic is related to different approaches to overcome the square root law, including the use of friendly jammer that varies the noise power so that the square root law can be overcome, and positive covert rate be achieved. The concept of effective secrecy is introduced next, a recent secrecy measure, whose definition includes both strong secrecy and stealth communication conditions. After that, the covert/stealth communication concept is applied to optical communication systems. We further describe how the covert concept can be applied to information reconciliation step in QKD to simultaneously improve secret-key rate and extend the transmission distance.

In Appendix, some background material is provided, such as abstract algebra fundamentals, which helps unfamiliar reader to better understand both physical-layer security and QKD concepts.

References

1. X.800: Security architecture for open systems interconnection for CCITT applications, Recommendation X.800 (03/91). https://www.itu.int/rec/T-REC-X.800-199103-I
2. Bloch M (2008) Physical-layer security. PhD dissertation, School of Electrical and Computer Engineering, Georgia Institute of Technology
3. Bloch M, Barros J (2011) Physical-layer security: from information theory to security engineering. Cambridge University Press, Cambridge
4. Bloch M (2014) Fundamentals of physical layer security. In: Zhou X, Song L, Zhang Y (eds) Physical layer security in wireless communications. CRC Press, Boca Raton, London, New York, pp 1–16
5. Chorti A et al (2016) Physical layer security: a paradigm shift in data confidentiality. In Physical and data-link security techniques for future communications systems. Lecture notes in electrical engineering, vol 358. Springer, pp 1–15
6. Bloch M, Barros J, Rodrigues MRD, McLaughlin SW (2008) Wireless information-theoretic security. IEEE Trans Inform Theory 54(6):2515–2534
7. Bennet CH, Brassard G (1984) Quantum cryptography: public key distribution and coin tossing. In: Proceedings of the IEEE international conference on computers, systems, and signal processing, Bangalore, India, pp 175–179
8. Bennett CH (1992) Quantum cryptography: uncertainty in the service of privacy. Science 257:752–753
9. Neilsen MA, Chuang IL (2000) Quantum computation and quantum information. Cambridge University Press, Cambridge
10. Van Assche G (2006) Quantum cryptography and secrete-key distillation. Cambridge University Press, Cambridge, New York
11. Djordjevic IB (2012) Quantum information processing and quantum error correction: an engineering approach. Elsevier/Academic Press, Amsterdam, Boston
12. Shannon CE (1949) Communication theory of secrecy systems. Bell Syst Tech J 28:656–715
13. Schneier B (2015) Applied cryptography, second edition: protocols, algorithms, and source code in C. Wiley, Indianapolis, IN
14. Drajic D, Ivanis P (2009) Introduction to information theory and coding, 3rd edn. Akademska Misao, Belgrade, Serbia. (in Serbian)
15. Haykin S (2001) Communication systems, 4th edn. Wiley, Hamilton Printing Company, Canada
16. Katz J, Lindell Y (2015) Introduction to modern cryptography, 2nd edn. CRC Press, Boca Raton, FL

17. Diffie W, Hellman ME (1976) New direction in cryptography. IEEE Trans Inform Theory 22:644–654
18. Hellman ME (1977) An extension of the Shannon theory approach to cryptography. IEEE Trans Inform Theory 23:289–294
19. Rivest RL, Shamir A, Adleman L (1983) Cryptographic communications system and method. US Patent 4,405,829
20. Merkle M (1978) Secure communication over an insecure channel. Comm ACM 21:294–299
21. McEliece RJ (1978) A public key cryptosystem based on algebraic coding theory. JPL DSN Prog Rep 42–44:114–116
22. Aumasson J-P (2018) Serious cryptography: a practical introduction to modern encryption. No Starch Press, San Francisco, CA
23. Kahn D (1967) The codebreakers: the story of secret writing. Macmillan Publishing Co., Ney York
24. Sebbery J, Pieprzyk J (1989) Cryptography: an introduction to computer security. Prentice Hall, New York
25. Delfs H, Knebl H (2015) Introduction to cryptography: principles and applications (information security and cryptography), 3rd edn. Springer, Heidelberg, New York
26. Rivest RL, Shamir A, Adleman L (1978) A method for obtaining digital signatures and public-key cryptosystems. Comm ACM 21(2):120–126
27. Le Bellac M (2006) An introduction to quantum information and quantum computation. Cambridge University Press
28. Shor PW (1997) Polynomial-time algorithms for prime number factorization and discrete logarithms on a quantum computer. SIAM J Comput 26(5):1484–1509
29. Ekert A, Josza R (1996) Quantum computation and Shor's factoring algorithm. Rev Modern Phys 68(3):733–753
30. Wyner AD (1975) The wire-tap channel. Bell Syst Tech J 54(8):1355–1387
31. Lin F, Oggier F (2014) Coding for wiretap channels. In: Zhou X, Song L, Zhang Y (eds) Physical layer security in wireless communications. CRC Press, Boca Raton, London, New York, pp 17–32
32. Csiszár I, Körner J (1978) Broadcast channels with confidential messages. IEEE Trans Inf Theory 24(3):339–348
33. Bennett CH, Brassard G, Crepeau C, Maurer U (1995) Generalized privacy amplification. IEEE Inform Theory 41(6):1915–1923
34. Ahlswede R, Csiszár I (1993) Common randomness in information theory and cryptography-Part I: Secret sharing. IEEE Trans Inf Theory 39(4):1121–1132
35. Barnett SM (2009) Quantum information. Oxford University Press, Oxford
36. Liao S-K et al (2017) Satellite-to-ground quantum key distribution. Nature 549:43–47
37. Scarani V, Bechmann-Pasquinucci H, Cerf NJ, Dušek M, Lütkenhaus N, Peev M (2009) The security of practical quantum key distribution. Rev Mod Phys 81:1301
38. Devetak I, Winter A (2005) Distillation of secret key and entanglement from quantum states. Proc R Soc Lond Ser A 461(2053):207–235
39. Maurer UM (1993) Secret key agreement by public discussion from common information. IEEE Trans Inf Theory 39(3):733–742
40. Holevo AS (1973) Bounds for the quantity of information transmitted by a quantum communication channel. Probl Inf Transm 9(3):177–183

Chapter 2
Information Theory and Coding Fundamentals

Abstract This chapter is devoted to classical information theory fundamentals and its application to fading channels and channels with memory. The chapter starts with definitions of entropy, joint entropy, conditional entropy, relative entropy, mutual information, and channel capacity, followed by the information capacity theorem. We discuss the channel capacity of discrete memoryless channels, continuous channels, and channels with memory. Regarding the wireless channels, we describe how to calculate the channel capacity of flat fading and frequency-selective channels. We also discuss different optimum and suboptimum strategies to achieve channel capacity including the water-filling method, multiplexed coding and decoding, channel inversion, and truncated channel inversion. We also study different strategies for channel capacity calculation depending on what is known about the channel state information. Further, we explain how to model the channel with memory and describe McMillan–Khinchin model for channel capacity evaluation. After that, the fundamentals of linear blocks codes are introduced, followed by the binary LDPC coding fundamentals.

2.1 Entropy, Conditional Entropy, Relative Entropy, Mutual Information

Let us observe a discrete memoryless source (DMS), characterized by a finite alphabet $S = \{s_0, s_1, \ldots, s_{K-1}\}$, wherein each symbol get generated with probability $P(S = s_k) = p_k, k = 0, 1, \ldots, K-1$. At a given time instance, the DMS generates one symbol from the alphabet, so that we can write $\sum_{k=0}^{K-1} p_k = 1$. The generation of a symbol at a given time instance is independent of previously generated symbols. The amount of information that a given symbol carries is related to the surprise when it occurs, and it is, therefore, reversely proportional to the probability of its occurrence. Since there is uncertainty about which symbol will be generated by the source, it appears that terms uncertainty, surprise, and amount of information are interrelated. Given that certain symbols can occur with very low probability, the amount of information value will be huge if the reverse of probability is used to determine the amount of information. In practice, we use the logarithm of reverse probability of occurrence

I. B. Djordjevic, *Physical-Layer Security and Quantum Key Distribution*,
https://doi.org/10.1007/978-3-030-27565-5_2

as the amount of information to solve this problem:

$$I(s_k) = \log\left(\frac{1}{p_k}\right); \quad k = 0, 1, \ldots, K-1. \tag{2.1}$$

The most common base of logarithm is the base 2, and the unit for the amount of information is *bi*nary uni*t* (bit). When $p_k = 1/2$, the amount of information is $I(s_k) =$ 1 bit, indicating that 1 bit is the amount of information gained when one out of two equally likely events occurs. It is straightforward to show that amount of information is nonnegative, that is, $I(s_k) \geq 0$. Further, when $p_k > p_i$ then $I(s_i) > I(s_k)$. Finally, when two symbols s_i and s_k are independent that the joint amount of information is additive, that is, $I(s_k s_i) = I(s_k) + I(s_i)$.

The average information content per symbol is commonly referred to as the *entropy*:

$$H(S) = E_P I(s_k) = E\left[\log\left(\frac{1}{p_k}\right)\right] = \sum_{k=0}^{K-1} p_k \log\left(\frac{1}{p_k}\right). \tag{2.2}$$

It can be easily be shown that the entropy is upper and lower bounded as follows:

$$0 \leq H(S) \leq |S|. \tag{2.3}$$

The entropy of binary source $S = \{0, 1\}$ is given by

$$H(S) = p_0 \log\left(\frac{1}{p_0}\right) + p_1 \log\left(\frac{1}{p_1}\right) = -p_0 \log p_0 - (1 - p_0)\log(1 - p_0) = H(p_0), \tag{2.4}$$

where $H(p_0)$ is known as the binary entropy function.

The entropy definition is also applicable to any random variable X, namely, we can write $H(X) = -\sum_k p_k \log p_k$. The *joint entropy* of a pair of random variables (X, Y), denoted as $H(X, Y)$, is defined as

$$H(X, Y) = -E_{p(x,y)} \log p(X, Y) = -\sum_{x\in\mathcal{X}} \sum_{x\in\mathcal{Y}} p(x, y) \log p(x, y). \tag{2.5}$$

When a pair of random variables (X, Y) has the joint distribution $p(x, y)$, the *conditional entropy* $H(Y|X)$ is defined as

$$\begin{aligned} H(Y|X) &= -E_{p(x,y)} \log p(Y|X) = -\sum_{x\in\mathcal{X}} \sum_{y\in\mathcal{Y}} \underbrace{p(x, y)}_{p(y|x)p(x)} \log p(y|x) \\ &= -\sum_{x\in\mathcal{X}} p(x) \sum_{y\in\mathcal{Y}} p(y|x) \log p(y|x). \end{aligned} \tag{2.6}$$

Since $p(x, y) = p(x)p(y|x)$, by taking logarithm we obtain

$$\log p(X, Y) = \log p(X) + \log p(Y|X), \tag{2.7}$$

and now by applying the expectation operator, we obtain

$$\underbrace{E \log p(X, Y)}_{H(X,Y)} = \underbrace{E \log p(X)}_{H(X)} + \underbrace{E \log p(Y|X)}_{H(Y|X)}$$
$$\Leftrightarrow H(X, Y) = H(X) + H(Y|X). \tag{2.8}$$

The equation above is commonly referred to as the *chain rule*.

The *relative entropy* is a measure of distance between two distributions $p(x)$ and $q(x)$, and it is defined as follows:

$$D(p||q) = E_p \log\left[\frac{p(X)}{q(X)}\right] = \sum_{x\in\mathcal{X}} p(x) \log\left[\frac{p(X)}{q(X)}\right]. \tag{2.9}$$

The relative entropy is also known as the Kullback–Leibler (KL) distance, and can be interpreted as the measure of inefficiency of assuming that distribution is $q(x)$, when true distribution is $p(x)$. Now by replacing $p(X)$ with $p(X, Y)$ and $q(X)$ with $p(X)q(Y)$, the corresponding relative entropy is between the joint distribution and product of distributions, which is well known as *mutual information*:

$$D(p(X, Y)||p(X)q(Y)) = E_{p(X,Y)} \log\left[\frac{p(X, Y)}{p(X)q(Y)}\right]$$
$$= \sum_{x\in\mathcal{X}} \sum_{y\in\mathcal{Y}} p(x, y) \log\left[\frac{p(X, Y)}{p(X)q(Y)}\right] \doteq I(X, Y). \tag{2.10}$$

2.2 Mutual Information, Channel Capacity, Information Capacity Theorem

2.2.1 *Mutual Information and Information Capacity*

Figure 2.1 shows an example of a discrete memoryless channel (DMC) , which is characterized by channel (transition) probabilities. If $X = \{x_0, x_1, \ldots, x_{I-1}\}$ and $Y = \{y_0, y_1, \ldots, y_{J-1}\}$ denote the channel input alphabet and the channel output alphabet, respectively, the channel is completely characterized by the following set of transition probabilities:

$$p(y_j|x_i) = P(Y = y_j|X = x_i), \quad 0 \le p(y_j|x_i) \le 1, \tag{2.11}$$

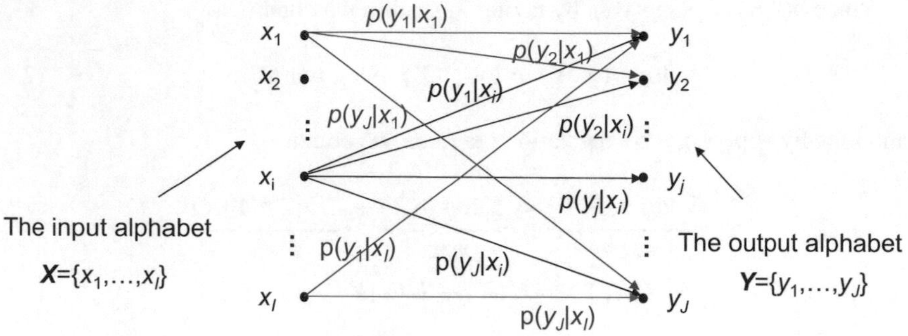

Fig. 2.1 Discrete memoryless channel (DMC)

where $i \in \{0, 1, \ldots, I-1\}, j \in \{0, 1, \ldots, J-1\}$, while I and J denote the sizes of input and output alphabets, respectively. The transition probability $p(y_j|x_i)$ represents the conditional probability that $Y = y_j$ for given input $X = x_i$.

One of the most important characteristics of the transmission channel is the *information capacity*, which is obtained by maximization of mutual information $I(\boldsymbol{X}, \boldsymbol{Y})$ over all possible input distributions:

$$C = \max_{\{p(x_i)\}} I(\boldsymbol{X}, \boldsymbol{Y}),\ I(\boldsymbol{X}, \boldsymbol{Y}) = H(\boldsymbol{X}) - H(\boldsymbol{X}|\boldsymbol{Y}), \tag{2.12}$$

where $H(\mathbf{U}) = -\langle \log_2 P(\boldsymbol{U})\rangle$ denotes the entropy of a random variable $\boldsymbol{U}$, and <.> denotes the expectation operator. The mutual information can be determined as

$$\begin{aligned} I(X;Y) &= H(X) - H(X|Y) \\ &= \sum_{i=1}^{M} p(x_i)\log_2\left[\frac{1}{p(x_i)}\right] - \sum_{j=1}^{N} p(y_j)\sum_{i=1}^{M} p(x_i|y_j)\log_2\left[\frac{1}{p(x_i|y_j)}\right]. \end{aligned} \tag{2.13}$$

In the above equation, $H(\boldsymbol{X})$ represents the uncertainty about the channel input before observing the channel output, also known as *entropy*, while $H(\boldsymbol{X}|\boldsymbol{Y})$ denotes the conditional entropy or the amount of uncertainty remaining about the channel input after the channel output has been received. (The log function from above equation relates to the base 2, and it will be like that throughout this chapter). Therefore, the mutual information represents the amount of information (per symbol) that is conveyed by the channel, which represents the uncertainty about the channel input that is resolved by observing the channel output. The mutual information can be interpreted by means of Venn diagram [1–9] shown in Fig. 2.2a. The left and right circles represent the entropy of the channel input, and channel output, respectively, while the mutual information is obtained as intersection area of these two circles. Another interpretation is illustrated in Fig. 2.2b [8]. The mutual information, i.e., the

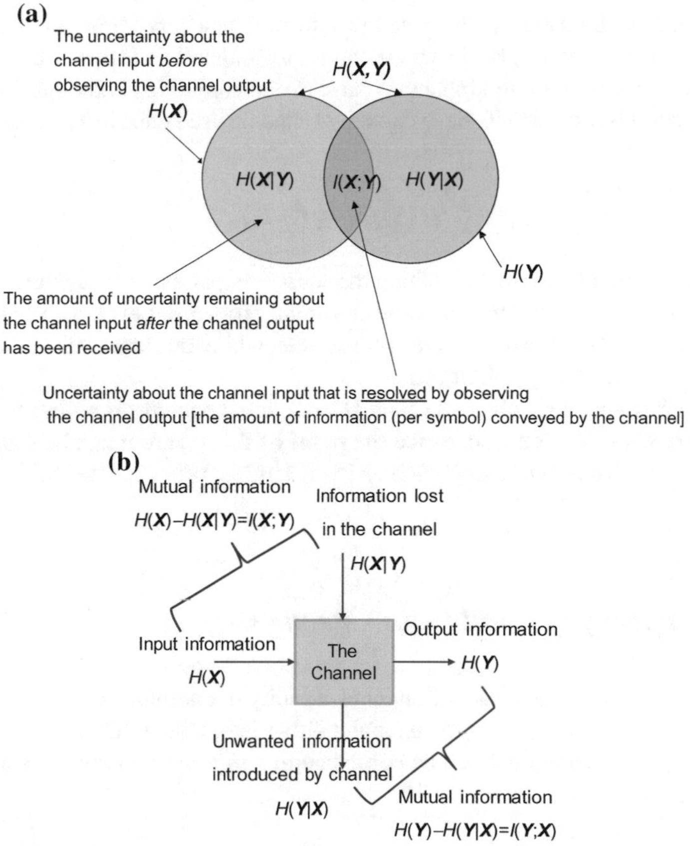

Fig. 2.2 Interpretation of the mutual information by using: **a** Venn diagrams, and **b** the approach due to Ingels

information conveyed by the channel, is obtained as the output information minus information introduced by the channel.

Since for M-ary input and M-ary output symmetric channel (MSC), we have that $p(y_j|x_i) = P_s/(M-1)$ and $p(y_j|x_j) = 1 - P_s$, where P_s is symbol error probability, the channel capacity, in bits/symbol, can be found as

$$C = \log_2 M + (1 - P_s)\log_2(1 - P_s) + P_s \log_2\left(\frac{P_s}{M-1}\right). \tag{2.14}$$

The channel capacity represents an important bound on data rates achievable by any modulation and coding schemes. It can also be used in comparison of different coded modulation schemes in terms of their distance to the maximum channel capacity curve.

Now we have built enough knowledge to formulate a very important theorem, the *channel coding theorem* [1–9], which can be formulated as follows. Let a discrete memoryless source with an alphabet $\boldsymbol{S}$ have the entropy $H(\boldsymbol{S})$ and emit the symbols every T_s seconds. Let a DMC have capacity C and be used once in T_c seconds. Then, if

$$H(\boldsymbol{S})/T_s \leq C/T_c,$$

there exists a coding scheme for which the source output can be transmitted over the channel and reconstructed with an arbitrary small probability of error. The parameter $H(\boldsymbol{S})/T_s$ is related to the average information rate, while the parameter C/T_c is related to the channel capacity per unit time.

For binary symmetric channel ($N = M = 2$), the inequality is reduced down to $R \leq C$, where R is the code rate. Since the proof of this theorem can be found in any textbook on information theory, such as [5–8, 10], the proof of this theorem will be omitted.

2.2.2 Capacity of Continuous Channels

In this section, we will discuss the channel capacity of continuous channels. Let $\boldsymbol{X} = [X_1, X_2, \ldots, X_n]$ denote an n-dimensional multivariate, with a PDF $p_1(x_1, x_2, \ldots, x_n)$, representing the channel input. The corresponding *differential entropy* is defined by [2, 3, 5, 6]

$$\begin{aligned} h(X_1, X_2, \ldots, X_n) &= -\underbrace{\int_{-\infty}^{\infty} \cdots \int_{-\infty}^{\infty}}_{n} p_1(x_1, x_2, \ldots, x_n) \log p_1(x_1, x_2, \ldots, x_n) dx_1 dx_2 \ldots dx_n \\ &= \langle -\log p_1(x_1, x_2, \ldots, x_n) \rangle, \end{aligned} \tag{2.15}$$

where we use $\langle\cdot\rangle$ to denote the expectation operator. In order to simplify explanations, we will use the compact form of Eq. (2.15), namely, $h(\boldsymbol{X}) = \langle -\log p_1(\boldsymbol{X})\rangle$, which was introduced in [6]. In similar fashion, the channel output can be represented as m-dimensional random variable $Y = [Y_1, Y_2, \ldots, Y_m]$ with a PDF $p_2(y_1, y_2, \ldots, y_m)$, while corresponding differential entropy is defined by

$$\begin{aligned} h(Y_1, Y_2, \ldots, Y_m) &= -\underbrace{\int_{-\infty}^{\infty} \cdots \int_{-\infty}^{\infty}}_{m} p_2(y_1, y_2, \ldots, y_m) \log p_1(y_1, y_2, \ldots, y_m) dy_1 dy_2 \ldots dy_m \\ &= \langle -\log p_2(y_1, y_2, \ldots, y_m) \rangle. \end{aligned} \tag{2.16}$$

In compact form, the differential entropy of output can be written as $h(\boldsymbol{Y}) = \langle -\log p_1(\boldsymbol{Y})\rangle$.

Example Let an n-dimensional multivariate $\boldsymbol{X} = [X_1, X_2, \ldots, X_n]$ with a PDF $p_1(x_1, x_2, \ldots, x_n)$ be applied to the nonlinear channel with the following nonlinear characteristic: $\boldsymbol{Y} = g(\boldsymbol{X})$, where $\boldsymbol{Y} = [Y_1, Y_2, \ldots, Y_n]$ represents the channel output with PDF $p_2(y_1, y_2, \ldots, y_n)$. Since the corresponding PDFs are related by the Jacobian symbol as follows: $p_2(y_1, \ldots, y_n) = p_1(x_1, \ldots, x_n)\left|J\left(\frac{X_1,\ldots,X_n}{Y_1,\ldots,Y_m}\right)\right|$, the output entropy can be determined as

$$h(Y_1, \ldots, Y_m) \cong h(X_1, \ldots, X_n) - \left\langle \log\left|J\left(\frac{X_1, \ldots, X_n}{Y_1, \ldots, Y_m}\right)\right|\right\rangle.$$

To account for the channel distortions and additive noise influence, we can observe the corresponding conditional and joint PDFs:

$$\begin{aligned} &P(y_1 < Y_1 < y_1 + dy_1, \ldots, y_m < Y_m < y_m + dy_m | X_1 = x_1, \ldots, X_n = x_n) = p(\tilde{y}|\tilde{x})d\tilde{y} \\ &P(y_1 < Y_1 < y_1 + dy_1, \ldots, x_n < X_n < x_n + dx_n) = p(\tilde{x}, \tilde{y})d\tilde{x}d\tilde{y}. \end{aligned} \tag{2.17}$$

The mutual information (also known as information rate) can be written in compact form as follows [6]:

$$I(\boldsymbol{X}; \boldsymbol{Y}) = \left\langle \log \frac{p(\boldsymbol{X}, \boldsymbol{Y})}{p(\boldsymbol{X})P(\boldsymbol{Y})}\right\rangle. \tag{2.18}$$

Notice that various differential entropies $h(\boldsymbol{X})$, $h(\boldsymbol{Y})$, $h(\boldsymbol{Y}|\boldsymbol{X})$ do not have direct interpretation as far as the information processed in the channel is concerned, as compared to their discrete counterparts, from the previous subsection. Some authors, such as Gallager in [7], prefer to define the mutual information directly by Eq. (2.18), without considering the differential entropies at all. The mutual information, however, has the theoretical meaning and represents the average information processed in the channel (or amount of information conveyed by the channel). The mutual information has the following important properties [5–8]: (i) it is symmetric: $I(X; Y) = I(Y; X)$; (ii) it is a nonnegative; (iii) it is finite; (iv) it is invariant under linear transformation; (v) it can be expressed in terms of the differential entropy of channel output by $I(X; Y) = h(Y) - h(Y|X)$, and (vi) it is related to the channel input differential entropy by $I(X; Y) = h(X) - h(X|Y)$.

The *information capacity* can be obtained by maximization of Eq. (2.18) under all possible input distributions, which is

$$C = \max I(\boldsymbol{X}; \boldsymbol{Y}). \tag{2.19}$$

Let us now determine the mutual information of two random vectors $\boldsymbol{X} = [X_1, X_2, \ldots, X_n]$ and $\boldsymbol{Y} = [Y_1, Y_2, \ldots, Y_m]$, which are normally distributed. Let $\boldsymbol{Z} = [\boldsymbol{X}; \boldsymbol{Y}]$ be the random vector describing the joint behavior. Without loss of generality, we further assume that $\overline{X}_k = 0\ \forall\ k$ and $\bar{Y}_k = 0\ \forall\ k$ where we used the overbar to denote the mean value operation. The corresponding PDFs for $\boldsymbol{X}$, $\boldsymbol{Y}$, and Z are, respectively, given as [6]

$$p_1(\boldsymbol{x}) = \frac{1}{(2\pi)^{n/2}(\det \boldsymbol{A})^{1/2}} \exp\left(-0.5(\boldsymbol{A}^{-1}\boldsymbol{x}, \boldsymbol{x})\right),$$
$$\boldsymbol{A} = [a_{ij}],\ a_{ij} = \int x_i x_j p_1(\boldsymbol{x}) d\boldsymbol{x}, \tag{2.20}$$

$$p_2(\boldsymbol{y}) = \frac{1}{(2\pi)^{n/2}(\det \boldsymbol{B})^{1/2}} \exp\left(-0.5(\boldsymbol{B}^{-1}\boldsymbol{y}, \boldsymbol{y})\right),$$
$$\boldsymbol{B} = [b_{ij}],\ b_{ij} = \int y_i y_j p_2(\boldsymbol{y}) d\boldsymbol{y}, \tag{2.21}$$

$$p_3(\boldsymbol{z}) = \frac{1}{(2\pi)^{(n+m)/2}(\det \boldsymbol{C})^{1/2}} \exp\left(-0.5(\boldsymbol{C}^{-1}\boldsymbol{z}, \boldsymbol{z})\right),$$
$$\boldsymbol{C} = [c_{ij}],\ c_{ij} = \int z_i z_j p_3(\boldsymbol{z}) d\boldsymbol{z}, \tag{2.22}$$

where $(\cdot,\cdot)$ denotes the dot product of two vectors. By substitution of Eqs. (2.20)–(2.22) into Eq. (2.28), we obtain [6]

$$I(\boldsymbol{X}; \boldsymbol{Y}) = \frac{1}{2} \log \frac{\det \boldsymbol{A} \det \boldsymbol{B}}{\det \boldsymbol{C}}. \tag{2.23}$$

The mutual information between two Gaussian random vectors can also be expressed in terms of their correlation coefficients [6]:

$$I(\boldsymbol{X}; \boldsymbol{Y}) = -\frac{1}{2} \log\left[(1-\rho_1^2)\ldots(1-\rho_l^2)\right], \quad l = \min(m, n), \tag{2.24}$$

where ρ_j is the correlation coefficient between X_j and Y_j.

In order to obtain the information capacity for additive Gaussian noise, we make the following assumptions: (i) the input X, output Y, and noise Z are n-dimensional random variables; (ii) $\bar{X}_k = 0$, $\overline{X_k^2} = \sigma_{x_k}^2\ \forall k$ and $\bar{Z}_k = 0$, $\overline{Z_k^2} = \sigma_{z_k}^2\ \forall k$; and (iii) the noise is additive: $Y = X + Z$. Since we have that

$$p_{\boldsymbol{x}}(\boldsymbol{y}|\boldsymbol{x}) = p_{\boldsymbol{x}}(\boldsymbol{x}+\boldsymbol{z}|\boldsymbol{x}) = \prod_{k=1}^{n}\left[\frac{1}{(2\pi)^{1/2}\sigma_{z_k}} e^{-z_k^2/2\sigma_{z_k}^2}\right] = p(\boldsymbol{z}), \tag{2.25}$$

the conditional differential entropy can be obtained as

$$H(\boldsymbol{Y}|\boldsymbol{X}) = H(\boldsymbol{Z}) = -\int_{-\infty}^{\infty} p(\boldsymbol{z}) \log p(\boldsymbol{z}) d\boldsymbol{z}. \tag{2.26}$$

The mutual information is then

$$I(\boldsymbol{X};\boldsymbol{Y}) = h(\boldsymbol{Y}) - h(\boldsymbol{Y}|\boldsymbol{X}) = h(\boldsymbol{Y}) - h(\boldsymbol{Z}) = h\boldsymbol{Z} - \frac{1}{2}\sum_{k=1}^{n}\log 2\pi e\sigma_{z_k}^2. \quad (2.27)$$

The information capacity, expressed in bits per channel use, is therefore obtained by maximizing $h(\boldsymbol{Y})$. Because the distribution maximizing the differential entropy is Gaussian, the information capacity is obtained as

$$\begin{aligned} C(\boldsymbol{X};\boldsymbol{Y}) &= \frac{1}{2}\sum_{k=1}^{n}\log 2\pi e\sigma_{y_k}^2 - \frac{1}{2}\sum_{k=1}^{n}\log 2\pi e\sigma_{z_k}^2 = \frac{1}{2}\sum_{k=1}^{n}\log\left(\frac{\sigma_{y_k}^2}{\sigma_{z_k}^2}\right) \\ &= \frac{1}{2}\sum_{k=1}^{n}\log\left(\frac{\sigma_{x_k}^2+\sigma_{z_k}^2}{\sigma_{z_k}^2}\right) = \frac{1}{2}\sum_{k=1}^{n}\log\left(1+\frac{\sigma_x^2}{\sigma_z^2}\right). \end{aligned} \quad (2.28)$$

For $\sigma_{x_k}^2 = \sigma_x^2, \sigma_{z_k}^2 = \sigma_z^2$, we obtain the following expression for information capacity:

$$C(\boldsymbol{X};\boldsymbol{Y}) = \frac{n}{2}\log\left(1+\frac{\sigma_x^2}{\sigma_z^2}\right), \quad (2.29)$$

where σ_x^2/σ_z^2 presents the signal-to-noise ratio (SNR). The expression above represents the maximum amount of information that can be transmitted per symbol.

From a practical point of view, it is important to determine the amount of information conveyed by the channel per second, which is the information capacity per unit time, also known as the *channel capacity*. For bandwidth-limited channels and Nyquist signaling employed, there will be $2W$ samples per second (W is the channel bandwidth) and the corresponding channel capacity becomes

$$C = W\log\left(1+\frac{P}{N_0W}\right)\ \text{[bits/s]}, \quad (2.30)$$

where P is the average transmitted power and $N_0/2$ is the noise power spectral density (PSD). Equation (2.30) represents the well-known *information capacity theorem*, commonly referred to as Shannon's third theorem [11].

Since the Gaussian source has the maximum entropy clearly, it will maximize the mutual information. Therefore, the equation above can be derived as follows. Let the n-dimensional multivariate $\boldsymbol{X} = [X_1, \ldots, X_n]$ represent the Gaussian channel input with samples generated from zero-mean Gaussian distribution with variance σ_x^2. Let the n-dimensional multivariate $\boldsymbol{Y} = [Y_1, \ldots, Y_n]$ represent the Gaussian channel output, with samples spaced $1/2W$ apart. The channel is additive with noise samples generated from zero-mean Gaussian distribution with variance σ_z^2. Let the PDFs of input and output be denoted by $p_1(\boldsymbol{x})$ and $p_2(\boldsymbol{y})$, respectively. Finally, let the joint PDF of input and output of channel be denoted by $p(\boldsymbol{x}, \boldsymbol{y})$. The maximum mutual information can be calculated from

$$I(X;Y) = \iint p(\boldsymbol{x},\boldsymbol{y}) \log \frac{p(\boldsymbol{x},\boldsymbol{y})}{p(\boldsymbol{x})P(\boldsymbol{x})} d\boldsymbol{x} d\boldsymbol{y}. \tag{2.31}$$

By following the similar procedure to that used in Eqs. (2.15)–(2.19), we obtain the following expression for Gaussian channel capacity:

$$C = W \log\left(1 + \frac{P}{N}\right), \ N = N_0 W, \tag{2.32}$$

where P is the average signal power and N is the average noise power.

2.3 Capacity of Flat Fading and Frequency-Selective Wireless Fading Channels

In this section, we study both flat fading and frequency-selective fading channels [10, 12–17] and describe how to calculate the corresponding channel capacities [11] as well as approaches to achieve these channel capacities.

2.3.1 Flat Fading Channel Capacity

The typical wireless communication system together with an equivalent channel model is shown in Fig. 2.3. The channel power gain $g[i]$, where i is the discrete-time instance, is related to the channel coefficient $h[i]$ by $g[i] = |h[i]|^2$ and follows a given probability density function (PDF) $f(g)$. For instance, as shown in this chapter, for Rayleigh fading the PDF $f(g)$ is exponential. The channel gain is commonly

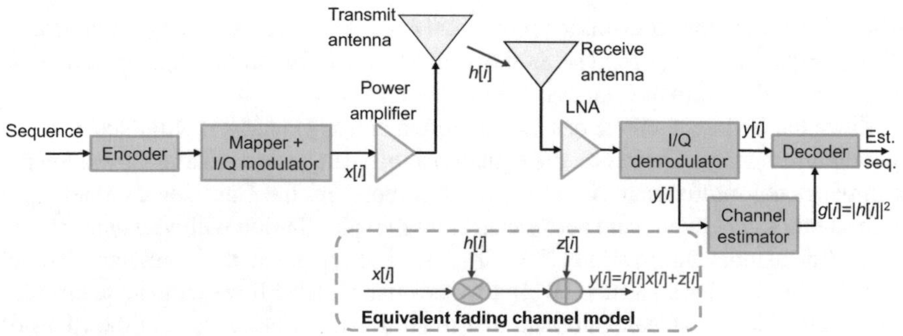

Fig. 2.3 Typical wireless communication system together with an equivalent fading channel model

referred to as *channel-side information* or *channel state information* (CSI). The samples of channel gain $g[i]$ change at each time instance i, and could be generated from an independent identically distributed (i.i.d.) process or could be generated from correlated sources. In a block fading channel, considered in this section, the channel gain $g[i]$ is constant over some block length T, and once this block length is over a new realization of channel gain $g[i]$ is generated based on PDF $f(g)$.

The channel capacity is strongly dependent on what is known about CSI at the transmitter and/or receiver sides. In this section, three scenarios of interest will be studied: (i) *channel distribution information* (CDI), (ii) CSI is available at receiver side (CSIR), and (iii) the CSI is available both at transmitter and receiver sides (full CSI) . In CDI scenario, the PDF of g is known to both transmitter and receiver sides. Determination of the capacity-achieving input distribution and corresponding capacity of fading channels under CDI scenario for any distribution $f(g)$, in closed form, is still an open problem, except specific fading channel models such as i.i.d. Rayleigh fading channels and FSMCs. Other two scenarios, CSIR and full CSI are discussed with more details below.

2.3.1.1 Channel-Side Information at Receiver

Two channel capacity definitions in CSIR scenario are of interest in the system design: (i) *Shannon capacity*, also known as the *ergodic capacity*, which represents the maximum data rate that can be sent over the channel with symbol error probability P_s tending to zero; and (ii) *capacity with outage*, which represents the maximum data rate that can be transmitted over a channel with some outage probability P_{outage}, defined as the probability that SNR falls below a given SNR threshold value ρ_{tsh} corresponding to the maximum tolerable symbol error probability P_s. Clearly, in the outage, the transmission cannot be completed with negligible symbol error probability. The capacity with outage corresponds to the highest data rate can be transmitted over the fading channel reliably except when the fading channel is in deep fade. To determine either the average symbol error probability or outage probability in the presence of fading, the distribution of SNR, $f(\rho)$, is needed. The *average symbol error probability* is then defined as

$$\overline{P}_s = \langle P_s(\gamma) \rangle = \int_0^{\infty} P_s(\rho) f(\rho) d\rho, \tag{2.33}$$

where $P_s(\rho)$ is SNR-dependent expression for symbol error probability, for a given modulation format. The *outage probability* is defined as

$$P_{\text{outage}} = P(\rho \le \rho_{\text{tsh}}) = \int_0^{\rho_{\text{tsh}}} f(\rho) d\rho, \tag{2.34}$$

where ρ_{tsh} is the threshold SNR corresponding to the maximum tolerable symbol error probability.

Shannon (Ergodic) Capacity
Under assumptions that the channel gain $g[i]$ originates from the flat fading, average power gain is equal to 1, and receiver knows in CSI (CSIR) we study the channel capacity calculation. When fading is fast, with a certain decoding delay requirement, we transmit the signal over the time duration that contains N coherence time periods T_c, namely, NT_c, wherein $N \gg 1$. In this scenario, since the transmission is sufficiently long and the fading is fast all possible channel gain realizations $g[i]$ come to the "picture" and the channel capacity can be calculated by averaging out channel capacities for different realizations. For nth coherence time period, we assume that channel gain is constant equal to $g[n]$, so the received SNR can be estimated as $g[n]\bar{\rho}$. The corresponding channel capacity, related to nth gain realization, will be $C[n] = W \log_2(1 + g[n]\bar{\rho})$, where W is the channel bandwidth. The Shannon (ergodic) channel capacity can be obtained by averaging over N coherence time periods as follows:

$$C_N = \frac{1}{N} \sum_{n=1}^{N} W \log_2(1 + g[n]\bar{\rho}). \tag{2.35}$$

Now by letting $N \to \infty$, we obtain

$$C = \langle W \log_2(1 + \rho) \rangle = \int_0^{\infty} W \log_2(1 + \rho) f(\rho) d\rho. \tag{2.36}$$

By applying the Jensen's inequality [5], the following upper limit is obtained for C:

$$C = \langle W \log_2(1 + \rho) \rangle \leq W \log_2(1 + \langle \rho \rangle) = W \log_2(1 + \bar{\rho}). \tag{2.37}$$

Channel Capacity with Outage
Capacity with outage is related to the *slowly varying fading channels*, where the instantaneous SNR ρ is constant for duration of transmission of large number of symbols, which is also known as *a transmission burst*, and changes to a new SNR value at the end of the burst period according to the given fading distribution. When the received SNR ρ is constant for the duration of the transmission burst, then the maximum possible transmission rate will be $W \log_2(1 + \rho)$, with symbol error probability tending to zero. Because the transmitter does not know the exact value of SNR ρ, it will be forced to transmit at the fixed date rate corresponding to the minimum tolerable SNR $\rho_{\min}$. The maximum data rate corresponding to SNR $\rho_{\min}$ will be $C = W \log_2(1 + \rho_{\min})$. As indicated earlier, the probability of outage is defined as $P_{\text{outage}} = P(\rho < \rho_{\min})$. The average data rate, reliably transmitted over many transmission bursts, can be defined as

$$C_{\text{outage}} = (1 - P_{\text{outage}})W\log_2(1 + \rho_{\min}), \tag{2.38}$$

Given that data are reliably transmitted in $1 - P_{\text{outage}}$ transmission bursts. In other words, this is the transmission rate that can be supported in $100(1 - P_{\text{outage}})\%$ of the channel realizations.

2.3.1.2 Full CSI

In full CSI scenario, both the transmitter and receiver have CSI available, and the transmitter can adapt its transmission strategy based on this CSI. Depending on the type of fading channel, fast- or slow-varying, we can define either Shannon capacity or capacity with outage.

Shannon Capacity

Let S denote the set of discrete memoryless wireless channels and let $p(s)$ denote the probability that the channel is in state $s \in S$. In other words, $p(s)$ is the fraction of the time for which the channel was in state s. The channel capacity, as defined by Wolfowitz [10], of the time-varying channel, in full CSI scenario (when both transmitter and receiver have the CSI available), is obtained as by averaging the channel capacity when the channel is in state s, denoted as C_s, as follows:

$$C = \sum\nolimits_{s \in S} C_s p(s). \tag{2.39}$$

By letting the cardinality of set S to tend to infinity, $|S| \to \infty$, the summation becomes integration and $p(s)$ becomes the PDF of ρ, so that we can write

$$C = \int_0^\infty C(\rho) f(\rho) d\gamma = \int_0^\infty W\log_2(1 + \rho) f(\rho) d\rho, \tag{2.40}$$

which is the same as in CSIR case. Therefore, the channel capacity does not increase in full CSI case unless some form of the *power adaptation* is employed, which is illustrated in Fig. 2.4.

The fading channel capacity, with power adaptation, is an optimization problem:

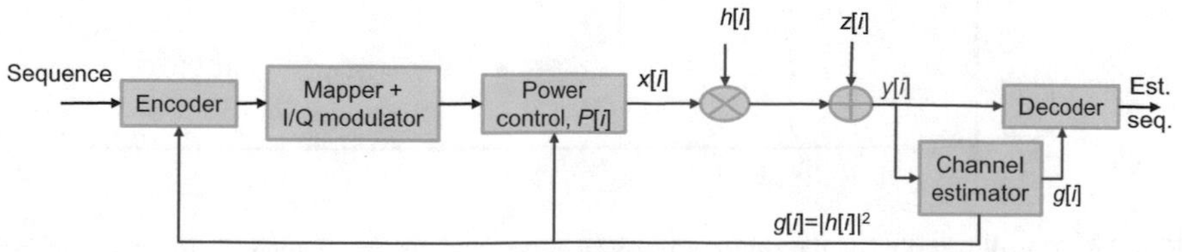

Fig. 2.4 The wireless communication system employing the power adaptation

$$C = \max_{P(\rho):\int_0^\infty P(\rho)f(\rho)d\rho=\overline{P}} \int_0^\infty W\log_2\left(1+\frac{P(\rho)\rho}{\overline{P}}\right)f(\rho)d\rho,\quad \rho=\frac{\overline{P}g}{N_0W}. \tag{2.41}$$

The optimum solution can be obtained by employing the Lagrangian method, wherein the Lagrangian $\mathfrak{L}$ is defined as

$$\mathfrak{L}(P(\rho)) = \int_0^\infty W\log_2\left(1+\frac{P(\rho)\rho}{\overline{P}}\right)f(\rho)d\rho - \mu\int_0^\infty P(\rho)f(\rho)d\rho, \tag{2.42}$$

which is subject to the following constraint:

$$\int_0^\infty P(\rho)f(\rho)d\rho \le \overline{P}. \tag{2.43}$$

By finding the first derivative of Lagrangian with respect to $P(\rho)$ and setting it to zero, we obtain

$$\frac{\partial\mathfrak{L}(P(\rho))}{\partial P(\rho)} = \left[\frac{W/\ln 2}{1+\rho P(\rho)/\bar{P}}\frac{\rho}{\bar{P}} - \mu\right]f(\rho) = 0. \tag{2.44}$$

By solving for $P(\rho)$ from (2.44), we obtain the optimum power allocation policy, also known as the *water-filling method*, which is illustrated in Fig. 2.5, as follows:

$$\frac{P(\rho)}{\bar{P}} = \begin{cases} 1/\rho_{\text{tsh}} - 1/\rho, & \rho \ge \rho_{\text{tsh}} \\ 0, & \rho < \rho_{\text{tsh}} \end{cases} \tag{2.45}$$

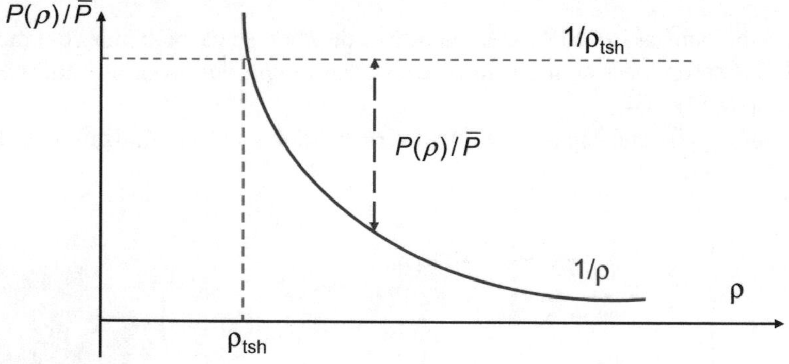

Fig. 2.5 The illustration of the optimum power allocation strategy, known as the water-filling method

The amount of power allocated on a transmitter side for a given SNR ρ is equal to $1/\rho_{\text{tsh}} - 1/\rho$. In other words, the amount of allocated power is proportional to the "water" filled between the bottom of the bowl ($1/\rho$) and the constant water line ($1/\rho_{\text{tsh}}$). The intuition behind the water-filling is to take the advantage of realizations when the channel conditions are good. Namely, when the channel conditions are good and SNR ρ is large, we allocate higher power and transmit at higher date rate over the channel. On the other hand, when the channel conditions deteriorate and ρ becomes small, we allocate less power and reduce the date rate. Once the instantaneous channel SNR ρ falls below the threshold (cutoff) value, we do not transmit at all. When the channel conditions are so bad, the reliable transmission is not possible and the average symbol error rate would be dominated by such transmissions.

After substituting Eq. (2.45) into (2.43), we obtain the following expression for the channel capacity corresponding to the water-filling method:

$$C = \int_{\rho_{\text{tsh}}}^{\infty} W \log_2\left(\frac{\rho}{\rho_{\text{tsh}}}\right) f(\rho) d\rho. \tag{2.46}$$

After substitution of Eq. (2.45) into (2.43), we obtain the following equation that needs to be solved numerically to determine the threshold SNR, ρ_{tsh}:

$$\int_{\rho_{\text{tsh}}}^{\infty} \left(\frac{1}{\rho_{\text{tsh}}} - \frac{1}{\rho}\right) f(\rho) d\rho = 1. \tag{2.47}$$

Let us now consider the optimum allocation policy in discrete-time domain. The channel model is similar to that of Fig. 2.3:

$$y[n] = h[n]x[n] + z[n], \quad |h[n]|^2 = g[n], \tag{2.48}$$

where n is the discrete-time index. The optimization problem can be formulated as

$$\max_{P_1, P_2, \ldots, P_N} \frac{1}{N} \sum_{n=1}^{N} \log_2\left(1 + \frac{P[n]|h[n]|^2}{N_0}\right) = \max_{P_1, P_2, \ldots, P_N} \frac{1}{N} \sum_{n=1}^{N} \log_2\left(1 + \frac{P[n]g[n]}{N_0}\right), \tag{2.49}$$

which is the subject to the following constraint:

$$\sum_{n=1}^{N} P[n] \le P. \tag{2.50}$$

The optimum solution can again be determined by the Lagrangian method:

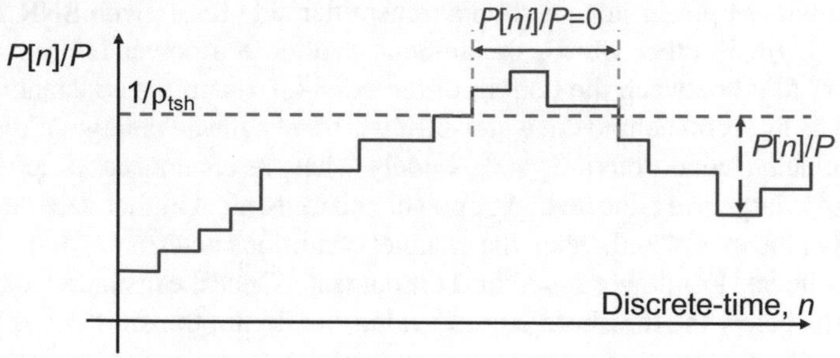

Fig. 2.6 The illustration of the optimum power allocation strategy in time domain

$$\frac{P_{\mathrm{opt}}[n]}{P}=\left(\underbrace{\frac{1}{\mu P}}_{\frac{1}{\rho_{\mathrm{tsh}}}}-\underbrace{\frac{N_0}{g[n]P}}_{\frac{1}{\rho}}\right)^+=\left(\frac{1}{\rho_{\mathrm{tsh}}}-\frac{1}{\rho}\right)^+,\quad (x)^+=\max(0,x), \tag{2.51}$$

and it is clearly the same allocation policy (the water-filling method) as given by Eq. (2.45), which is illustrated in Fig. 2.6. The threshold SNR equation can be obtained after substitution Eq. (2.51) into (2.50):

$$\frac{1}{N}\sum_{n=1}^{N}\left(\frac{1}{\rho_{\mathrm{tsh}}}-\frac{1}{\rho}\right)^+=1. \tag{2.52}$$

As $N\to\infty$, summation becomes integral, and we obtain from Eq. (2.52) the following power allocation policy:

$$\left\langle\left(\frac{1}{\rho_{\mathrm{tsh}}}-\frac{1}{\rho}\right)^+\right\rangle=1, \tag{2.53}$$

which is identical to Eq. (2.47). The corresponding channel capacity for the optimum power adaptation strategy in time domain will be

$$C=\left\langle W\log_2\left(1+\frac{P_{\mathrm{opt}}(g)}{P}\underbrace{\frac{Pg}{N_0}}_{\rho}\right)\right\rangle_h=\left\langle W\log_2\left(1+\frac{P_{\mathrm{opt}}(g)}{P}\rho\right)\right\rangle,\quad \frac{P_{\mathrm{opt}}(g)}{P}=\left(\frac{1}{\rho_{\mathrm{tsh}}}-\frac{1}{\rho}\right)^+. \tag{2.54}$$

The channel capacity-achieving coding scheme, inspired by this method, is *multiplexed coding and decoding scheme*, as shown in Fig. 2.7.

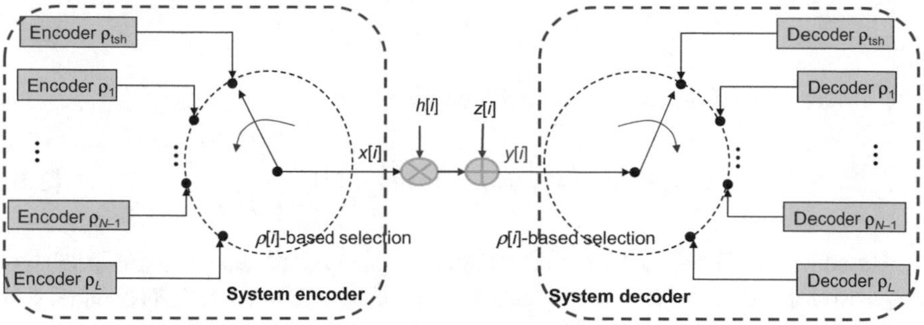

Fig. 2.7 The illustration of the multiplexed coding and decoding scheme

The starting point is the quantization of the range of SNRs for different fading effects to a finite set $\{\rho_n \mid 1 \leq n \leq L\}$. For each ρ_n, we design an encoder/decoder pair for an AWGN channel with SNR ρ_n achieving the channel capacity. The output x_n of encoder ρ_n is transmitted with an average power $P(\rho_n)$ and corresponding data rate is $R_n = C_n$, where C_n is the capacity of a time-invariant AWGN channel with received SNR $P(\rho_n)\ \rho_n/P$. These encoder/decoder pairs are selected according to the CSI, that is, SNR ρ_n. In other words, when $\rho[i] \cong \rho_n$, the corresponding pair of ports get connected through the wireless channel. OF course, we assume that CSI is perfect. Clearly, this multiplexed encoding and decoding scheme effectively replaces the time-varying channel with a set of time-invariant channels operating in parallel, wherein the nth channel is only used when $\rho[i] \cong \rho_n$. The average rate transmitted over the channel is determined by averaging, namely, $\sum_n C_n p_n$, where p_n is the probability of CSI being ρ_n; in other words, the percentage of time that the channel SNR was equal to ρ_n. Of course, the complexity and cost of this scheme are high. To reduce the system complexity and cost, the channel inversion can be used.

Channel Inversion and Zero-Outage Capacity

The *channel inversion* represents a suboptimal transmitter power adaptation strategy, where the transmitter employs the available CSI to maintain a constant received power; in other words, this scheme "inverts" the channel fading effects. The encoder and decoder in this scenario will see a time-invariant AWGN channel. The power loading is reversely proportional to the SNR:

$$P(\rho)/\overline{P} = \rho_0/\rho, \tag{2.55}$$

so that when the channel conditions are bad we load higher power such that the receiver sees the constant power. By substituting Eq. (2.55) into (2.43), we obtain

$$\int (\rho_0/\rho) f(\rho) d\rho = 1, \tag{2.56}$$

so that the constant of proportionality (SNR) ρ_0 can be determined as

$$\rho_0 = 1/\langle 1/\rho \rangle, \tag{2.57}$$

and the corresponding channel capacity becomes

$$C = W \log_2(1 + \rho_0) = W \log_2\left(1 + \frac{1}{\langle 1/\rho \rangle}\right). \tag{2.58}$$

Based on Eq. (2.58), we conclude that the capacity-achieving transmission strategy is to employ fixed-rate encoder and decoder designed for an AWGN channel with SNR equal to ρ_0, regardless of the wireless channel conditions. Given that the data rate is fixed under all channel conditions, even very bad ones, there is no channel outage when channel conditions are bad, and corresponding channel capacity is called *zero-outage capacity*. The main drawback of this strategy is that the zero-outage capacity can exhibit a large data rate reduction, compared to Shannon capacity in extremely bad fading conditions, when $\rho \to 0$ and subsequently $C \to 0$. To solve this problem, the truncated channel inversion is used.

Truncated Channel Inversion and Outage Capacity

As explained above, the requirement to maintain a constant data rate regardless of the fading state results in inefficacy of this method, so that the corresponding zero-outage capacity can be significantly smaller than Shannon capacity in a deep fade. When the fading state is very bad by not transmitting at all, we can ensure that data rates are higher in the other fading states and on such a way we significantly improve the channel capacity. The state when we do not transmit is an outage state, so that the corresponding capacity can be called the *outage capacity*, which is defined as the maximum data rate that can be kept constant, when there is no outage, times the probability of not having outage $(1 - P_{\text{outage}})$. Outage capacity is achieved with a so-called *truncated channel inversion policy* for power adaptation, in which we only transmit reversely proportional to SNR, when the SNR is larger than the threshold SNR ρ_{tsh}:

$$P(\rho)/\bar{P} = \begin{cases} \rho_0/\rho, & \rho \geq \rho_{\text{tsh}} \\ 0, & \rho < \rho_{\text{tsh}} \end{cases}. \tag{2.59}$$

The parameter ρ_0 can be determined as

$$\rho_0 = 1/\langle 1/\rho \rangle_{\rho_{\text{tsh}}}, \ \langle 1/\rho \rangle_{\rho_{\text{tsh}}} = \int_{\rho_{\text{tsh}}}^{\infty} \frac{1}{\rho} f(\rho) d\rho. \tag{2.60}$$

For the outage probability, defined as $P_{\text{outage}} = P(\rho < \rho_{\text{tsh}})$, the corresponding channel capacity will be

$$C(P_{\text{out}}) = W \log_2\left(1 + \frac{1}{\langle 1/\rho \rangle_{\rho_{\text{tsh}}}}\right) P(\rho \geq \rho_{\text{tsh}}). \tag{2.61}$$

The channel capacity can be obtained by the following maximization with respect to ρ_{tsh}:

$$C = \max_{\rho_{\text{tsh}}} W \log_2\left(1 + \frac{1}{\langle 1/\rho \rangle_{\rho_{\text{tsh}}}}\right)(1 - P_{\text{outage}}). \tag{2.62}$$

2.3.2 *Frequency-Selective Fading Channel Capacity*

Two cases of interest are considered here: (i) time-invariant and (ii) time-variant channels.

2.3.2.1 Time-Invariant Channel Capacity

We consider here the time-invariant channel, as shown in Fig. 2.8, and we assume that a total transmit power cannot be larger than P. In full CSI scenario, the transfer function $H(f)$ is known to both the transmitter and receiver. We first study the *block fading* assumption for $H(f)$ so that whole frequency band can be divided into subchannels each of the bandwidths equal to W, wherein the transfer function is constant for each block i, namely, $H(f) = H_j$, which is illustrated in Fig. 2.9.

Therefore, we decomposed the frequency-selective fading channel into a set of the parallel AWGN channels with corresponding SNRs being $|H_i|^2 P_i/(N_0 W)$ on the

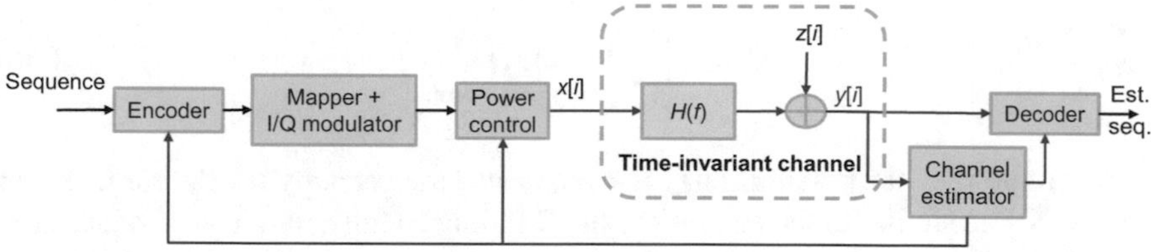

Fig. 2.8 The illustration of the wireless communication system operating over the time-invariant channel characterized by the transfer function $H(f)$

Fig. 2.9 The illustration of the block frequency-selective fading in which the channel is divided into subchannels each of them having the bandwidth W

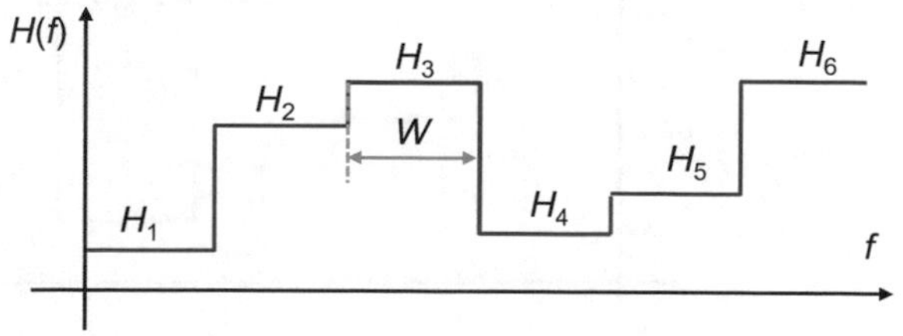

ith channel, where P_i is the power allocated to the ith channel in this parallel set of subchannels, subject to the power constraint $\sum_i P_i \le P$.

The *capacity* of this *parallel set of channels* is the sum of corresponding rates associated with each channel, wherein the power is optimally allocated over all subchannels:

$$C = \max_{P_i:\sum_i P_i \le P} \sum W \log_2 \left(1 + \underbrace{\frac{|H_i|^2 P_i}{N_0 W}}_{\rho_i} \right) = \max_{P_i:\sum_i P_i \le P} \sum W \log_2 (1 + \rho_i), \tag{2.63}$$

The optimum power allocation policy of the ith subchannel is the water filling:

$$\frac{P_i}{P} = \begin{cases} 1/\rho_{\text{tsh}} - 1/\rho_i, & \rho_i \ge \rho_{\text{tsh}} \\ 0, & \rho_i < \rho_{\text{tsh}} \end{cases}, \tag{2.64}$$

and the threshold SNR is obtained as the solution of the following equation:

$$\sum_i \left(\frac{1}{\rho_{\text{tsh}}} - \frac{1}{\rho_i} \right)^+ = 1. \tag{2.65}$$

The illustration of the water filling in the frequency-selective block fading channels is provided in Fig. 2.10. The corresponding channel capacity of the optimum distribution, obtained by substituting Eq. (2.64) into (2.63), is obtained as follows:

$$C = \sum_{i:\rho_i \ge \rho_{\text{tsh}}} W \log_2 \left(\frac{\rho_i}{\rho_{\text{tsh}}} \right). \tag{2.66}$$

When the transfer function $H(f)$ is *continuous*, the capacity for the same power constraint P is similar to the case of the block fading channel, but instead of summation we need to perform the integration so that the corresponding channel capacity

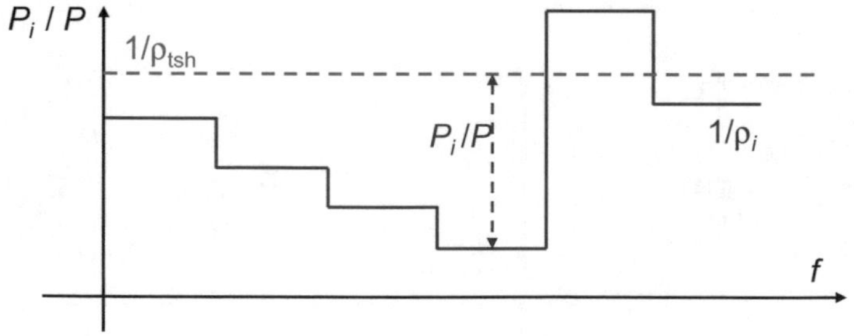

Fig. 2.10 The illustration of the water filling in the frequency-selective block fading channels

expression is given by

$$C = \max_{P(f):\int P(f)df \leq P} \int W \log_2\left(1 + \frac{|H(f)|^2 P(f)}{N_0}\right) df. \tag{2.67}$$

The optimum power allocation strategy is, not surprising, again the water filling:

$$\frac{P(f)}{P} = \begin{cases} 1/\rho_{\text{tsh}} - 1/\rho(f), & \rho(f) \geq \rho_{\text{tsh}} \\ 0, & \rho(f) < \rho_{\text{tsh}} \end{cases}, \quad \rho(f) = \frac{|H(f)|^2 P}{N_0 W}. \tag{2.68}$$

After substituting Eq. (2.68) into (2.67), we obtain the following channel capacity for the optimum power allocation case:

$$C = \int_{f:\rho(f) \geq \rho_{\text{tsh}}} W \log_2\left[\frac{\rho(f)}{\rho_{\text{tsh}}}\right] df. \tag{2.69}$$

Example Let us now consider the time-invariant frequency-selective block fading channel with three subchannels of bandwidth 2 MHz, with frequency response amplitudes being 1, 2, and 3, respectively. For the transmit power 10 mW and noise power spectral density of $N_0 = 10^{-9}$ W/Hz, we are interested to determine the corresponding Shannon capacity.

The signal power-to-noise power is given by $\rho_0 = P/(N_0 W) = 5$. The subchannels' SNRs can be expressed as $\rho_i = P|H_i|^2/(N_0 W) = \rho_0|H_i|^2$. For $H_1 = 1, H_2 = 2, H_3 = 3$, the corresponding subchannel SNRs are $\rho_1 = \rho_0|H_1|^2 = 5$, $\rho_2 = \rho_0|H_2|^2 = 20$, and $\rho_3 = \rho_0|H_3|^2 = 45$, respectively. Based on Eq. (2.65), the threshold SNR can be determined from the following equation:

$$\frac{3}{\rho_{\text{tsh}}} = 1 + \frac{1}{\rho_1} + \frac{1}{\rho_2} + \frac{1}{\rho_3} = 1 + \frac{1}{5} + \frac{1}{20} + \frac{1}{45},$$

as follows: $\rho_{\text{tsh}} = 2.358 < \rho_i$ for very i. The corresponding channel capacity is given by Eq. (2.66) as

$$C = \sum_{i=1}^{3} W \log_2(1 + \rho_i) = W \sum_{i=1}^{3} \log_2(1 + \rho_i) = 16.846 \text{ kb/s}.$$

2.3.2.2 Time-Variant Channel Capacity

The time-varying frequency-selective fading channel is similar to the time-invariant model, except that $H(f) = H(f, i)$, i.e., the channel varies over both frequency and time, and corresponding channel model is given in Fig. 2.11.

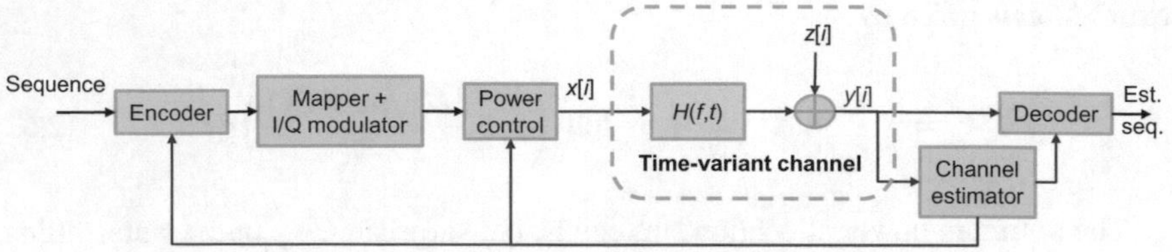

Fig. 2.11 The illustration of the wireless communication system operating over the time-varying channel characterized by the time-variant transfer function *H*(*f*, *t*)

We can approximate channel capacity in time-varying frequency-selective fading by taking the channel bandwidth *W* of interest and divide it up into subchannels of bandwidth equal to the channel coherence bandwidth B_c, as shown in Fig. 2.12. Under assumption that each of the resulting subchannels is independent, time-varying with flat fading within the subchannel $H(f, k) = H_i\ [k]$ on the *i*th subchannel at time instance *k*, we can obtain the capacity for each of these flat fading subchannels based on the average power P_i that we allocate to each subchannel, subject to a total power constraint *P*.

Since the channels are mutually independent, the *total channel capacity* is just equal to the *sum of capacities on the individual narrowband flat fading channels*, subject to the total average power constraint, averaged over both time and frequency domains:

$$C = \max_{\{\bar{P}_i\}:\sum_i \bar{P}_i \le \bar{P}} \sum_i C_i(\overline{P}_i),\ C_i = \int_{\rho_{\mathrm{tsh}}}^{\infty} B_c \log_2\left(\frac{\rho_i}{\rho_{\mathrm{tsh}}}\right) f(\rho_i) d\rho_i, \tag{2.70}$$

which requires the knowledge of the distribution functions of SNR for each subchannel, namely, $f(\rho_i)$.

When we fixed the average power per subchannel, the optimum power allocation policy is the water filling in the time domain. On the other hand, the optimum power allocation strategy among subchannels is again the water filling but now in the frequency domain. Therefore, the optimum power allocation strategy for time-varying channels is the two-dimensional water filling, over both time and frequency

Fig. 2.12 The illustration of the frequency-selective time-varying fading channels

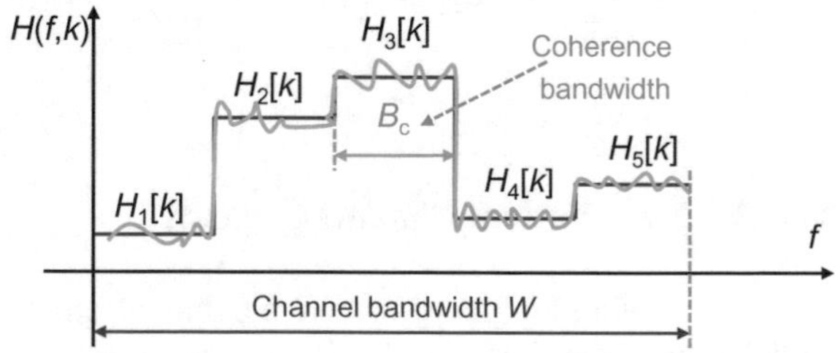

domains. The instantaneous SNR in the ith subchannel at the kth time interval is given by $\rho_i[k] = |H_i[k]|^2 \bar{P}/(N_0 B_c)$.

The Shannon capacity assuming the perfect full CSI will be achieved by optimizing the power allocation in both time domain, by fixing $\rho_i[k] = \rho_i$, and frequency domain over subchannel indices i as follows:

$$C = \max_{P_i(\rho_i):\sum_i \int P_i(\rho_i) p(\rho_i) d\rho_i \leq \bar{P}} \sum_i \int_0^\infty B_c \log_2\left(1 + \frac{P_i(\rho_i)}{\bar{P}}\right) f(\rho_i) d\rho_i, \tag{2.71}$$

which is subject to the following constraint:

$$\sum_i \int_0^\infty P_i(\rho_i) f(\rho_i) d\rho_i = \overline{P}. \tag{2.72}$$

The optimum power allocation strategy can be determined by the Lagrangian method, and corresponding solution is given by

$$\frac{P_i(\rho_i)}{\overline{P}} = \begin{cases} 1/\rho_{\text{tsh}} - 1/\rho_i, & \rho_i \geq \rho_{\text{tsh}} \\ 0, & \rho_i < \rho_{\text{tsh}} \end{cases} \tag{2.73}$$

The threshold SNR can be obtained as the solution of the following equation:

$$\sum_i \int_{\rho_{\text{tsh}}}^\infty (1/\rho_{\text{tsh}} - 1/\rho_i)^+ f(\rho_i) d\rho_i = 1. \tag{2.74}$$

After substituting Eq. (2.73) into (2.71), we obtain the following channel capacity expression for the optimum two-dimensional adaptation strategy:

$$C = \sum_i \int_{\rho_{\text{tsh}}}^\infty B_c \log_2\left[\frac{\rho_i}{\rho_{\text{tsh}}}\right] f(\rho_i) d\rho_i. \tag{2.75}$$

2.4 Capacity of Channels with Memory

In this section, we will describe Markov and McMillan's sources with memory [18], McMillan–Khinchin channel model with memory [19], and describe how to determine the entropies of sources with memory and mutual information of channels with memory [20–22]. All this serves as a baseline in analysis of optical channel capacity, in particular, for fiber optics communications [23–30].

2.4.1 Markov Sources and Their Entropy

The finite Markovian chain is commonly used model to describe both the sources and channels with memory. The Markovian stochastic process with finite number of states $\{S\} = \{S_1, \ldots, S_n\}$ is characterized by transition probabilities π_{ij} of moving from state S_i to state S_j $(i, j = 1, \ldots, n)$. The Markov chain is the sequence of states with transitions governed by the following transition matrix:

$$\Pi = [\pi_{ij}] = \begin{bmatrix} \pi_{11} & \pi_{12} & \cdots & \pi_{1n} \\ \pi_{21} & \pi_{22} & \cdots & \pi_{2n} \\ \cdots & \cdots & \ddots & \vdots \\ \pi_{n1} & \pi_{n2} & \cdots & \pi_{nn} \end{bmatrix}, \tag{2.76}$$

where $\sum_j \pi_{ij} = 1$.

Example Let us observe three-state Markov source shown in Fig. 2.13a. The corresponding transition matrix is given by

$$\Pi = [\pi_{ij}] = \begin{bmatrix} 0.6 & 0.4 & 0 \\ 0 & 1 & 0 \\ 0.3 & 0.7 & 0 \end{bmatrix}.$$

As we see, the sum in any row is equal to 1. From state S_1, we can move to either state S_2 with probability 0.4 or stay in state S_1 with probability 0.6. Once we enter the state S_2, we stay there forever. Such a state is called absorbing, and corresponding Markov chain is called *absorbing Markov chain*. The probability of moving from state S_3 to state S_2 in two steps can be calculated as $\pi_{32}^{(2)} = 0.3 \cdot 0.4 + 0.7 \cdot 1 = 0.82$. Another way to calculate this probability is to find the second power of transition matrix, and then read out the probability of desired transition:

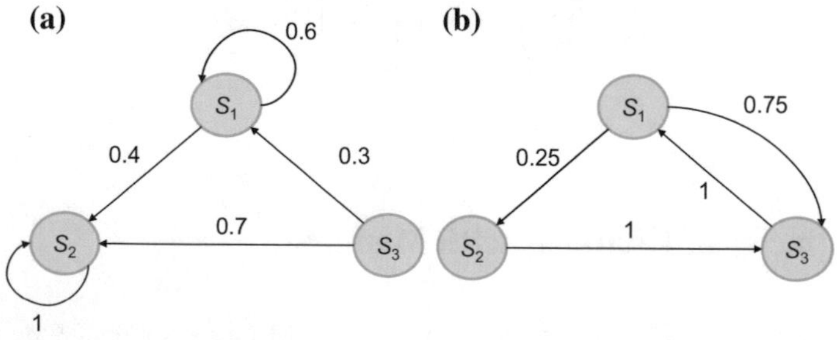

Fig. 2.13 Two three-state Markov chains: **a** irregular Markov chain, and **b** regular Markov chain

$$\mathbf{\Pi}^2 = \left[\pi_{ij}^{(2)}\right] = \mathbf{\Pi}\mathbf{\Pi} = \begin{bmatrix} 0.36 & 0.64 & 0 \\ 0 & 1 & 0 \\ 0.18 & 0.82 & 0 \end{bmatrix}.$$

The probability of reaching all states from initial states after k-steps can be determined by

$$\mathbf{\Pi}^{(k)} = \boldsymbol{P}^{(0)}\mathbf{\Pi}^k, \tag{2.77}$$

where $\boldsymbol{P}^{(0)}$ is row vector containing the probabilities of initial states.

If the transition matrix $\mathbf{\Pi}^k$ has only nonzero elements, we say that Markov chain is *regular*. Accordingly, if k_0th power of $\mathbf{\Pi}$ does not have any zero entry, any kth power of $\mathbf{\Pi}$ for $k > k_0$ will not have any zero entry either. The so-called *ergodic Markov chains* are the most important ones from communication system point of view. We can say that Markov chain is ergodic if it is possible to move from any specific state to any other state in finite number of steps with nonzero probability. The Markov chain from the example above is non-ergodic. It is also interesting to notice that transition matrix for this example has the following limit:

$$\boldsymbol{T} = \lim_{k\to\infty} \boldsymbol{P}^k = \begin{bmatrix} 0 & 1 & 0 \\ 0 & 1 & 0 \\ 0 & 1 & 0 \end{bmatrix}.$$

Example Let us now observe an example of regular Markov chain, which is shown in Fig. 2.13b. The transition matrix $\mathbf{\Pi}$, its fourth and fifth powers, and matrix $\mathbf{\Pi}$ limit as $k \to \infty$, are given, respectively, as

$$\mathbf{\Pi} = \begin{bmatrix} 0 & 0.25 & 0.75 \\ 0 & 0 & 1 \\ 1 & 0 & 0 \end{bmatrix}, \quad \mathbf{\Pi}^4 = \begin{bmatrix} 0.5625 & 0.0625 & 0.3750 \\ 0.7500 & 0 & 0.2500 \\ 0.2500 & 0.1875 & 0.5625 \end{bmatrix},$$

$$\mathbf{\Pi}^5 = \begin{bmatrix} 0.3750 & 0.1406 & 0.4844 \\ 0.2500 & 0.1875 & 0.5625 \\ 0.5625 & 0.0625 & 0.3750 \end{bmatrix}, \quad \boldsymbol{T} = \lim_{k\to\infty} \mathbf{\Pi}^k = \begin{bmatrix} 0.4444 & 0.1112 & 0.4444 \\ 0.4444 & 0.1112 & 0.4444 \\ 0.4444 & 0.1112 & 0.4444 \end{bmatrix}.$$

We can see that fourth power has one zero entry, while the fifth power and all higher powers do not have zero entries. Therefore, this Markov chain is both regular and ergodic one. We can also notice that stationary transition matrix $\boldsymbol{T}$ has identical rows.

It is evident from the example above that for regular Markov chain the transition matrix converges to stationary transition matrix $\boldsymbol{T}$ with all rows identical to each other:

$$T = \lim_{k\to\infty} \mathbf{\Pi}^k = \begin{bmatrix} t_1 & t_2 & \cdots & t_n \\ t_1 & t_2 & \cdots & t_n \\ \cdots & \cdots & \ddots & \vdots \\ t_1 & t_2 & \cdots & t_n \end{bmatrix}. \tag{2.78}$$

Additionally, the following is valid:

$$\lim_{k\to\infty} \mathbf{\Pi}^{(k)} = \lim_{k\to\infty} \boldsymbol{P}^{(0)} \mathbf{\Pi}^k = \boldsymbol{P}^{(0)} \boldsymbol{T} = \begin{bmatrix} t_1\ t_2\ \cdots\ t_n \end{bmatrix}, \tag{2.79}$$

so we can find stationary probabilities of states (or equivalently solve elements of $\boldsymbol{T}$) from equations

$$\begin{aligned} t_1 &= \pi_{11}t_1 + \pi_{21}t_2 + \cdots + \pi_{n1}t_n \\ t_2 &= \pi_{12}t_1 + \pi_{22}t_2 + \cdots + \pi_{n2}t_n \\ &\vdots \\ t_n &= \pi_{1n}t_1 + \pi_{2n}t_2 + \cdots + \pi_{nn}t_n \\ \sum_{i=1}^{n} t_i &= 1. \end{aligned} \tag{2.80}$$

For instance, for Markov chain from Fig. 2.13b, we can write that

$$t_1 = t_3, \quad t_2 = 0.25t_1, \quad t_3 = 0.75t_1 + t_2, \quad t_1 + t_2 + t_3 = 1,$$

while corresponding solution is given by $t_1 = t_3 = 0.4444$, $t_2 = 0.1112$.

The uncertainty of source associated with Markov source $\{S\} = \{S_1, \ldots, S_n\}$ when moving one step ahead from an initial state A_i, here denoted as $H_i^{(1)}$, can be expressed as

$$H_i^{(1)} = -\sum_{j=1}^{n} \pi_{ij} \log \pi_{ij}. \tag{2.81}$$

If the probability associated with state S_i is equal to p_i, we can obtain the entropy of Markov source by averaging over entropies associated with all states. The uncertainty of moving one step ahead becomes

$$H(X) = H^{(1)} = E\left\{H_i^{(1)}\right\} = \sum_{i=1}^{n} p_i H_i^{(1)} = -\sum_{i=1}^{n} p_i \sum_{j=1}^{n} \pi_{ij} \log \pi_{ij}. \tag{2.82}$$

In similar fashion, the entropy of Markov source for moving k-steps ahead from initial states is given by

$$H^{(k)} = E\left\{H_i^{(k)}\right\} = \sum_{i=1}^{n} p_i \underbrace{H_i^{(k)}}_{-\sum_{j=1}^{n}\pi_{ij}^{(k)}\log \pi_{ij}^{(k)}} = -\sum_{i=1}^{n} p_i \sum_{j=1}^{n} \pi_{ij}^{(k)} \log \pi_{ij}^{(k)}. \quad (2.83)$$

It can be shown that for ergodic Markov sources there is a limit defined as $H^{(\infty)} = \lim_{k\to\infty} H^{(k)}/k$. In order to prove this, we can use the following property, which should be proved as a homework problem:

$$H^{(k+1)} = H^{(k)} + H^{(1)}. \quad (2.84)$$

By applying this property in an iterative fashion, we obtain that

$$H^{(k)} = H^{(k-1)} + H^{(1)} = H^{(k-2)} + 2H^{(1)} = \cdots = kH^{(1)} = kH(X). \quad (2.85)$$

From Eq. (2.85), it is evident that

$$\lim_{k\to\infty} \frac{H^{(k)}}{k} = H^{(1)} = H(X). \quad (2.86)$$

Equation (2.86) can be now used as an alternative definition of entropy of Markov source, which is applicable to arbitrary *stationary* source as well.

Example Let us determine the entropy of Markov source shown in Fig. 2.13b. By using the definition from Eq. (2.82), we obtain that

$$\begin{aligned} H(X) &= -\sum_{i=1}^{n} p_i \sum_{j=1}^{n} p_{ij} \log p_{ij} \\ &= -0.4444(0.25 \log 0.25 + 0.75 \log 0.75) - 0.1111 \cdot 1 \log 1 - 0.4444 \cdot 1 \log 1 \\ &= 0.6605 \text{ bits}. \end{aligned}$$

2.4.2 McMillan Sources and Their Entropy

A McMillan's description [18] of the discrete source with memory is more general than that of Markov chain. In our case, we will pay attention to stationary sources. Let S represent a finite source alphabet with corresponding letters $\{s_1, s_2, \ldots, s_M\} = S$. The source emits one symbol at the time instance t_k. The transmitted sequence can be represented as $X = \{\ldots, x_{-1}, x_0, x_1, \ldots\}$, where $x_i \in S$. Among all members of ensemble $\{X\}$, we are interested just in those having specified source symbols at certain prespecified instances of time. All sequences with these properties create so-called *cylinder set*.

Example Let specified location be defined by indices -1, 0, 3 and k, with corresponding symbols at these locations equal to $x_{-1} = s_2$, $x_0 = s_5$, $x_3 = s_0$, $x_k = s_{n-1}$. The corresponding cylinder set will be given as $C_1 = \{\ldots, x_{-2}, s_2, s_5, x_1, x_2, s_0, \ldots, x_k = s_{n-1}, \ldots\}$. Since we observe the stationary processes, the statistical properties of cylinder will not change if we shift the cylinder for one-time unit in either direction (either by T or by T^{-1}). For instance, the time-shifted C_1 cylinder is given by $TC_1 = \{\ldots, x_{-1}, s_2, s_5, x_2, x_3, s_0, \ldots, x_{k+1} = s_{n-1}, \ldots\}$.

The *stationary property* for arbitrary cylinder C can be expressed as

$$P\{TC\} = P\{T^{-1}C\} = P\{C\}, \tag{2.87}$$

where $P\{\cdot\}$ denotes the probability measure.

Let us now specify n letters from alphabet S to be sent at positions $k+1, \ldots, k+n$. This sequence can be denoted as $x_{k+1}, \ldots, x_{k+n}$, and there is the total of M^n possible sequences. The entropy of all possible sequences is defined as

$$H_n = -\sum_C p_m(C) \log p_m(C), \tag{2.88}$$

where $p_m(\cdot)$ is the probability measure. The McMillan's definition of entropy of stationary discrete source is given by [18]

$$H(X) = \lim_{n\to\infty} \frac{H_n}{n}. \tag{2.89}$$

As we can see, the McMillan's definition of entropy is consistent with Eq. (2.86), which is applicable to stationary Markovian sources.

2.4.3 McMillan–Khinchin Model for Channel Capacity Evaluation

Let the input and output alphabets of the channel be finite and denoted by A and B, respectively, while the channel input and output sequences are denoted by X and Y. The noise behavior for memoryless channels is generally captured by a conditional probability matrix $P\{b_j|a_k\}$ for all $b_j \in B$ and $a_j \in A$. On the other side, in channels with finite memory (such as the optical channel), the transition probability is dependent on the transmitted sequences up to the certain prior finite instance of time. For instance, the transition matrix for channel described by Markov process has the form $P\{Y_k = b|\ldots, X_{-1}, X_0, X_1, \ldots, X_k\} = P\{Y_k = b|X_k\}$.

Let us consider a member x of input ensemble $\{X\} = \{\ldots, x_{-2}, x_{-1}, x_0, x_1, \ldots\}$ and its corresponding channel output y from ensemble $\{Y\} = \{\ldots, y_{-2}, y_{-1}, y_0, y_1, \ldots\}$. Let $\boldsymbol{X}$ denote all possible input sequences and $\boldsymbol{Y}$ denote all possible output sequences. By fixing a particular symbol at specific location, we can obtain the *cylinder*. For instance, cylinder $x^{4,1}$ is obtained by fixing symbol a_1 to position x_4, so it is $x^{4,1} = \ldots, x_{-1}, x_0, x_1, x_2, x_3, a_1, x_5, \ldots$ The output cylinder $y^{1,2}$ is obtained by fixing the output symbol b_2 to position 1, namely, $y^{1,2} = \ldots, y_{-1}, y_0, b_2, y_2, y_3, \ldots$ In order to characterize the channel, we have to determine transition probability $P(y^{1,2}|x^{4,1})$, which is probability that cylinder $y^{1,2}$ was received if cylinder $x^{4,1}$ was transmitted. Therefore, for all possible input cylinders $S_A \subset X$, we have to determine the probability that cylinder $S_B \subset Y$ was received if S_A was transmitted.

The channel is completely specified by the following: (i) input alphabet A, (ii) output alphabet B, and (iii) transition probabilities $P\{S_B|S_A\} = v_x$ for all $S_A \in X$ and $S_B \in Y$. Accordingly, the channel is specified by the triplet $[A, v_x, B]$. If transition probabilities are invariant with respect to time shift T, which means that $v_{Tx}(TS) = v_x(S)$, the channel is said to be *stationary*. If the distribution of Y_k depends only on the statistical properties of sequence $\ldots, x_{k-1}, x_k$, we say that the channel is without *anticipation*. If, furthermore, the distribution of Y_k depends only on $x_{k-m}, \ldots, x_k$, we say that channel has finite *memory* of m units.

The source and channel may be described as a new source $[C, \zeta]$, where C is the Cartesian product of input A and output B alphabets ($C = A \times B$), while ζ is a corresponding probability measure. The joint probability of symbol $(x, y) \in C$, where $x \in A$ and $y \in B$, is obtained as the product of marginal and conditional probabilities: $P(x \cap y) = P\{x\}P\{y|x\}$.

Let us further assume that both source and channel are stationary. Following description presented in [6, 19], it is useful to describe the concatenation of a stationary source and a stationary channel as follows:

1. If the source $[A, \mu]$ (μ is the probability measure of the source alphabet) and the channel $[A, v_x, B]$ are stationary, the product source $[C, \zeta]$ will also be stationary.
2. Each stationary source has an entropy, and therefore $[A, \mu]$, $[B, \eta]$ (η is the probability measure of the output alphabet), and $[C, \zeta]$ each have the finite entropies.
3. These entropies can be determined for all n-term sequences $x_0, x_1, \ldots, x_{n-1}$ emitted by the source and transmitted over the channel as follows [19]:

$$\begin{aligned} &H_n(X) \leftarrow \{x_0, x_1, \ldots, x_{n-1}\} \\ &H_n(Y) \leftarrow \{y_0, y_1, \ldots, y_{n-1}\} \\ &H_n(X, Y) \leftarrow \{(x_0, y_0), (x_1, y_1), \ldots, (x_{n-1}, y_{n-1})\}. \\ &H_n(Y|X) \leftarrow \{(Y|x_0), (Y|x_1), \ldots, (Y|x_{n-1})\} \\ &H_n(X|Y) \leftarrow \{(X|y_0), (X|y_1), \ldots, (X|y_{n-1})\} \end{aligned} \tag{2.90}$$

It can be shown that the following is valid:

$$\begin{aligned} H_n(X, Y) &= H_n(X) + H_n(Y|X) \\ H_n(X, Y) &= H_n(Y) + H_n(X|Y). \end{aligned} \tag{2.91}$$

The equations above can be rewritten in terms of entropies per symbol:

$$
\begin{aligned}
\frac{1}{n} H_n(X, Y) &= \frac{1}{n} H_n(X) + \frac{1}{n} H_n(Y|X) \\
\frac{1}{n} H_n(X, Y) &= \frac{1}{n} H_n(Y) + \frac{1}{n} H_n(X|Y).
\end{aligned} \tag{2.92}
$$

For sufficiently long sequences, the following channel entropies exist

$$
\begin{aligned}
&\lim_{n\to\infty} \frac{1}{n} H_n(X, Y) = H(X, Y) \quad \lim_{n\to\infty} \frac{1}{n} H_n(X) = H(X) \\
&\lim_{n\to\infty} \frac{1}{n} H_n(Y) = H(Y) \quad \lim_{n\to\infty} \frac{1}{n} H_n(X|Y) = H(X|Y) \\
&\lim_{n\to\infty} \frac{1}{n} H_n(Y|X) = H(Y|X).
\end{aligned} \tag{2.93}
$$

The mutual information also exist and it is defined as

$$
I(X, Y) = H(X) + H(Y) - H(X, Y). \tag{2.94}
$$

The *stationary information capacity* of the channel is obtained by maximization of mutual information over all possible information sources:

$$
C(X, Y) = \max I(X, Y). \tag{2.95}
$$

The results of analysis in this section will be applied later in evaluation the information capacity of optical channel with memory. But, before that, we will briefly describe the adopted model for signal propagation in single-mode optical fibers, which will also be used in evaluation of channel capacity.

2.5 Linear Block Codes Fundamentals

The *linear block code* (*n*, *k*) satisfies a linearity property, which means that a linear combination of arbitrary two codewords results in another codeword [1–4, 31]. If we use the terminology of vector spaces, it can be defined as a subspace of a vector space over finite (Galois) field, denoted as GF(q), with q being the prime power. Every space is described by its *basis*—a set of linearly independent vectors. The number of vectors in the basis determines the dimension of the space. Therefore, for an (*n*, *k*) linear block code the dimension of the space is *n*, and the dimension of the code subspace is *k*.

2.5.1 Generator and Parity-Check Matrices

Let $\boldsymbol{m} = (m_0, m_1, \ldots, m_{k-1})$ denote the k-bit message vector. Any codeword $\boldsymbol{x} = (x_0, x_1, \ldots, x_{n-1})$ from the (n, k) linear block code can be represented as a linear combination of k basis vectors $\boldsymbol{g}_i$ $(i = 0, 1, \ldots, k - 1)$ as follows:

$$\boldsymbol{x} = m_0\boldsymbol{g}_0 + m_1\boldsymbol{g}_1 + \cdots + m_{k-1}\boldsymbol{g}_{k-1} = \boldsymbol{m}\begin{bmatrix} \boldsymbol{g}_0 \\ \boldsymbol{g}_1 \\ \cdots \\ \boldsymbol{g}_{k-1} \end{bmatrix} = \boldsymbol{mG}; \quad \boldsymbol{G} = \begin{bmatrix} \boldsymbol{g}_0 \\ \boldsymbol{g}_1 \\ \cdots \\ \boldsymbol{g}_{k-1} \end{bmatrix}, \tag{2.96}$$

where $\boldsymbol{G}$ is the generator matrix (of dimensions $k \times n$), in which every row represents basis vector from the coding subspace. Therefore, in order to be encoded, the message vector $\boldsymbol{m}$ has to be multiplied with a generator matrix $\boldsymbol{G}$ to get the codeword, namely, $\boldsymbol{x} = \boldsymbol{mG}$.

The code may be transformed into a systematic form by elementary operations on rows in the generator matrix, i.e.,

$$\boldsymbol{G}_s = [\boldsymbol{I}_k | \boldsymbol{P}], \tag{2.97}$$

where $\boldsymbol{I}_k$ is the unity matrix of dimensions $k \times k$, and $\boldsymbol{P}$ is the matrix of dimensions $k \times (n - k)$ with columns denoting the positions of parity checks:

$$\boldsymbol{P} = \begin{bmatrix} p_{00} & p_{01} & \cdots & p_{0,n-k-1} \\ p_{10} & p_{11} & \cdots & p_{1,n-k-1} \\ \cdots & \cdots & \cdots & \cdots \\ p_{k-1,0} & p_{k-1,1} & \cdots & p_{k-1,n-k-1} \end{bmatrix}.$$

The codeword of a *systematic code* is obtained as

$$\boldsymbol{x} = [\boldsymbol{m}|\boldsymbol{b}] = \boldsymbol{m}[\boldsymbol{I}_k|\boldsymbol{P}] = \boldsymbol{mG}, \quad \boldsymbol{G} = [\boldsymbol{I}_k|\boldsymbol{P}], \tag{2.98}$$

and has the structure as shown in Fig. 2.14. The message vector stays unaffected during systematic encoding, while the vector of parity checks $\boldsymbol{b}$ is appended having the bits that are algebraically related to the message bits as follows:

$$b_i = p_{0i}m_0 + p_{1i}m_1 + \cdots + p_{k-1,i}m_{k-1}, \tag{2.99}$$

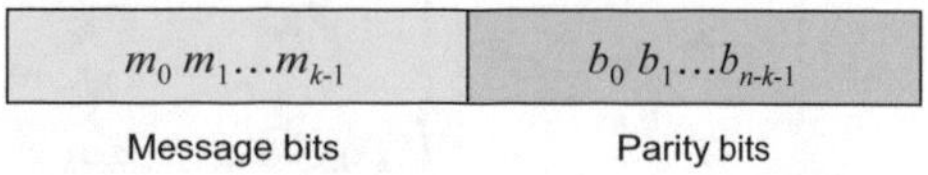

Fig. 2.14 Systematic codeword structure

where

$$p_{ij} = \begin{cases} 1, \text{ if } b_i \text{ depends on } m_j \\ 0, \text{ otherwise} \end{cases}.$$

The optical channel introduces the errors during transmission, and the received vector $\boldsymbol{r}$ can be represented as $\boldsymbol{r} = \boldsymbol{x} + \boldsymbol{e}$, where $\boldsymbol{e}$ is the error vector (error pattern) whose components are determined by

$$e_i = \begin{cases} 1, \text{ if an error occured in the } i\text{th location} \\ 0, \text{ otherwise} \end{cases}.$$

In order to verify if the received vector $\boldsymbol{r}$ is a codeword one, in incoming subsection we will introduce the concept of a *parity-check matrix* as another useful matrix associated with the linear block codes.

Let us expand the matrix equation $\boldsymbol{x} = \boldsymbol{mG}$ in a scalar form as follows:

$$\begin{aligned} x_0 &= m_0 \\ x_1 &= m_1 \\ &\ldots \\ x_{k-1} &= m_{k-1} \\ x_k &= m_0 p_{00} + m_1 p_{10} + \cdots + m_{k-1} p_{k-1,0} \\ x_{k+1} &= m_0 p_{01} + m_1 p_{11} + \cdots + m_{k-1} p_{k-1,1} \\ &\ldots \\ x_{n-1} &= m_0 p_{0,n-k-1} + m_1 p_{1,n-k-1} + \cdots + m_{k-1} p_{k-1,n-k-1}. \end{aligned} \tag{2.100}$$

By using the first k equalities, the last $n-k$ equations can be rewritten in terms of the first k codeword elements as follows:

$$\begin{aligned} & x_0 p_{00} + x_1 p_{10} + \cdots + x_{k-1} p_{k-1,0} + x_k = 0 \\ & x_0 p_{01} + x_1 p_{11} + \cdots + x_{k-1} p_{k-1,0} + x_{k+1} = 0 \\ & \ldots \\ & x_0 p_{0,n-k+1} + x_1 p_{1,n-k-1} + \cdots + x_{k-1} p_{k-1,n-k+1} + x_{n-1} = 0. \end{aligned} \tag{2.101}$$

The equations presented above can be rewritten through matrix representation as

$$\underbrace{\begin{bmatrix} x_0 \; x_1 \ldots x_{n-1} \end{bmatrix}}_{\boldsymbol{x}} \underbrace{\begin{bmatrix} p_{00} & p_{10} & \cdots & p_{k-1,0} & 1\;0 \ldots 0 \\ p_{01} & p_{11} & \cdots & p_{k-1,1} & 0\;1 \ldots 0 \\ \cdots & & \cdots & & \cdots \quad \cdots \\ p_{0,n-k-1} & p_{1,n-k-1} & \cdots & p_{k-1,n-k-1} & 0\;0 \ldots 1 \end{bmatrix}^T}_{\boldsymbol{H}^T} = 0$$

$$\Leftrightarrow \boldsymbol{x}\boldsymbol{H}^T = \boldsymbol{0}, \quad \boldsymbol{H} = \begin{bmatrix} \boldsymbol{P}^T & \boldsymbol{I}_{n-k} \end{bmatrix}_{(n-k)xn}. \tag{2.102}$$

The $\boldsymbol{H}$-matrix in Eq. (2.102) is known as *the parity-check* one. We can easily verify that $\boldsymbol{G}$ and $\boldsymbol{H}$ matrices satisfy equation

$$\boldsymbol{G}\boldsymbol{H}^T = \begin{bmatrix} \boldsymbol{I}_k & \boldsymbol{P} \end{bmatrix} \begin{bmatrix} \boldsymbol{P} \\ \boldsymbol{I}_{n-k} \end{bmatrix} = \boldsymbol{P} + \boldsymbol{P} = \boldsymbol{0}, \tag{2.103}$$

meaning that the parity-check matrix H of an (n, k) linear block code has rank $n - k$ and dimensions $(n - k) \times n$ whose null space is k-dimensional vector with basis forming the generator matrix $\boldsymbol{G}$.

Every (n, k) linear block code with generator matrix $\boldsymbol{G}$ and parity-check matrix $\boldsymbol{H}$ has a dual code, this time having generator matrix $\boldsymbol{H}$ and parity-check matrix $\boldsymbol{G}$. As an example, $(n, 1)$ repetition code and $(n, n - 1)$ single-parity-check code are dual ones.

2.5.2 *Minimum Distance and Error Correction Capability of Linear Block Code*

In order to determine the error correction capability of the linear block code, we have to introduce the concepts of Hamming distance and Hamming weight [1–4, 31]. The Hamming distance $d(\boldsymbol{x}_1, \boldsymbol{x}_2)$ between two codewords $\boldsymbol{x}_1$ and $\boldsymbol{x}_2$ is defined as the number of locations in which these two vectors differ. The Hamming weight $wt(\boldsymbol{x})$ of a codeword vector $\boldsymbol{x}$ is defined as the number of nonzero elements in the vector. The minimum distance $d_{\min}$ of a linear block code is defined as the smallest Hamming distance between any pair of code vectors in the code space. Since the zero vector is also a codeword, the minimum distance of a linear block code can be defined as the smallest Hamming weight of the nonzero code vectors in the code.

We can write the parity-check matrix in a form $\boldsymbol{H} = [\boldsymbol{h}_1, \boldsymbol{h}_2, \ldots, \boldsymbol{h}_n]$, where $\boldsymbol{h}_i$ presents the ith column in the matrix structure. Since every codeword $\boldsymbol{x}$ must satisfy the syndrome equation $\boldsymbol{x}\boldsymbol{H}^{\mathrm{T}} = \boldsymbol{0}$, the minimum distance of a linear block code is determined by the minimum number of columns in the $\boldsymbol{H}$-matrix whose sum is equal to zero vector. As an example, (7, 4) Hamming code discussed above has a minimum distance $d_{\min} = 3$ since the sum of first, fifth, and sixth columns leads to zero vector.

The codewords can be represented as points in n-dimensional space, as shown in Fig. 2.15. Decoding process can be visualized by creating the spheres of radius t around codeword points. The received word vector $\boldsymbol{r}$ in Fig. 9.2a will be decoded as a codeword $\boldsymbol{x}_i$ because its Hamming distance $d(\boldsymbol{x}_i, \boldsymbol{r}) \leq t$ is closest to the codeword $\boldsymbol{x}_i$. On the other hand, in example shown in Fig. 9.2b, the Hamming distance satisfies relation $d(\boldsymbol{x}_i, \boldsymbol{x}_j) \leq 2t$, and the received vector $\boldsymbol{r}$ that falls in intersection area of two spheres cannot be uniquely decoded. Therefore, (n, k) linear block code of minimum distance $d_{\min}$ can correct up to t errors if, and only if, $t \leq \lfloor 1/2(d_{\min} - 1) \rfloor$ or $d_{\min}$

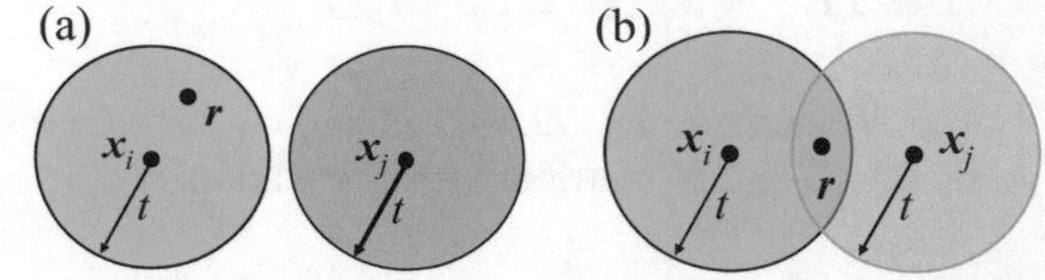

Fig. 2.15 The illustration of Hamming distance: **a** $d(\boldsymbol{x}_i, \boldsymbol{x}_j) \geq 2t + 1$, and **b** $d(\boldsymbol{x}_i, \boldsymbol{x}_j) < 2t + 1$

$\geq 2t + 1$ (where $\lfloor \cdot \rfloor$ denotes the largest integer smaller or equal to the enclosed quantity). If we are only interested in detecting e_d errors, then the minimum distance should be $d_{\min} \geq e_\mathrm{d} + 1$. However, if we are interested in detecting e_d errors and correcting e_c errors, then the minimum distance should be $d_{\min} \geq e_\mathrm{d} + e_\mathrm{c} + 1$. Accordingly, the Hamming (7, 4) code is a single-error-correcting and double-error-detecting code. More generally, Hamming codes are (n, k) linear block codes with the following parameters:

- Block length: $n = 2^m - 1$,
- Number of message bits: $k = 2^m - m - 1$,
- Number of parity bits: $n - k = m$, and
- $d_{\min} = 3$,

where $m >= 3$. Hamming codes belong to the class of perfect codes, the codes that satisfy the Hamming inequality given as [1–4, 31]

$$2^{n-k} \geq \sum_{i=0}^{t} \binom{n}{i}. \tag{2.104}$$

This bound gives how many errors t can be corrected with a specific (n, k) linear block code.

2.5.3 Coding Gain

The coding gain a linear (n, k) block is defined as the relative saving in the energy per information bit for a given bit error probability when coding is applied. Since the total information word energy kE_b must be the same as the total codeword energy nE_c (where E_c is the transmitted bit energy, while E_b is the information bit energy), we can establish the following relationship:

$$E_c = (k/n)E_b = RE_b, \tag{2.105}$$

where R defines the coding rate. As an example, the probability of error for BPSK in an AWGN channel, when coherent hard decision (bit-by-bit) demodulator is used, can be obtained as

$$p = \frac{1}{2}\text{erfc}\left(\sqrt{\frac{E_c}{N_0}}\right) = \frac{1}{2}\text{erfc}\left(\sqrt{\frac{RE_b}{N_0}}\right), \tag{2.106}$$

where erfc(x) function is the complementary error function. By using the Chernoff bound, we obtain the following expression for the coding gain when hard decision is applied in decoding process:

$$\frac{(E_b/N_0)_{\text{uncoded}}}{(E_b/N_0)_{\text{coded}}} \approx R(t+1), \tag{2.107}$$

where t is the error correction capability of the code. The corresponding soft decision coding gain can be estimated by [1–4]

$$\frac{(E_b/N_0)_{\text{uncoded}}}{(E_b/N_0)_{\text{coded}}} \approx Rd_{\text{min}}. \tag{2.108}$$

By comparing Eqs. (2.107) and (2.108), we can see that the soft decision coding gain is about 3 dB better than hard decision coding gain since the minimum distance $d_{\text{min}} \geq 2t + 1$.

2.5.4 *Coding Bounds*

In the rest of this section, we describe several important coding bounds including Hamming, Plotkin, Gilbert–Varshamov, and Singleton ones [2, 3, 31, 32]. The *Hamming* bound has already been introduced for binary linear block codes (LBC), and was expressed by Eq. (2.104). The Hamming bound for q-ary (n, k) LBC is given as

$$\left[1 + (q-1)\binom{n}{1} + (q-1)^2\binom{n}{2} + \cdots + (q-1)^i\binom{n}{i} + \cdots + (q-1)^t\binom{n}{t}\right]q^k \leq q^n, \tag{2.109}$$

where t is the error correction capability and $(q-1)^i\binom{n}{i}$ is the number of received words that differ from a given codeword in i symbols. The codes satisfying the Hamming bound with equality sign are known as *perfect codes*. Hamming codes are perfect codes since $n = 2^{n-k} - 1$, which is equivalent to $(1+n)2^k = 2^n$, so that the relationship between the right and left sides of Eq. (2.109) is expressed by equality sign. The (n, 1) repetition code is also example of a perfect code.

The *Plotkin* bound is defined by the following relation for minimum distance:

$$d_{\text{min}} \leq \frac{n2^{k-1}}{2^k - 1}. \tag{2.110}$$

Namely, if all codewords are written as the rows of a $2^k \times n$ matrix, each column will contain 2^{k-1} "zeros" and 2^{k-1} "ones", with the total weight of all codewords being equal to $n2^{k-1}$.

Gilbert–Varshamov bound is based on the property that the minimum distance $d_{\min}$ of a linear (n, k) block code can be determined as the minimum number of columns in $\boldsymbol{H}$-matrix that adds to zero, i.e.,

$$\binom{n-1}{1} + \binom{n-1}{2} + \cdots + \binom{n-1}{d_{\min}-2} < 2^{n-k} - 1. \tag{2.111}$$

The *Singleton bound* is defined by inequality:

$$d_{\min} \le n - k + 1. \tag{2.112}$$

This bound is straightforward to prove. Let only one bit of value 1 be present in information vector. If it is involved in $n - k$ parity checks, the total number of ones in codeword cannot be larger than $n - k + 1$. The codes satisfying the Singleton bound with equality sign are known as the *maximum-distance separable* (MDS) codes (e.g., RS codes are MDS codes).

2.6 Binary LDPC Coding Fundamentals

Because LDPC codes belong to the class of linear block codes, they can be described as a k-dimensional subspace C of the n-dimensional vector space of n-tuples, F_2^n, over the binary field $F_2 = \mathrm{GF}(2)$ [1–4, 31]. For this k-dimensional subspace, we can find the basis $B = \{g_0, g_1, \ldots, g_{k-1}\}$ that spans C so that every codeword v can be written as a linear combination of basis vectors $v = m_0 g_0 + m_1 g_1 + \cdots + m_{k-1} g_{k-1}$ for message vector $m = (m_0, m_1, \ldots, m_{k-1})$; or in compact form we can write $v = mG$, where G is so-called generator matrix with the ith row being g_i. The $(n - k)$-dimensional null space $C^{\perp}$ of G comprises all vectors x from F_2^n such that $xG^{\mathrm{T}} = 0$ and it is spanned by the basis $B^{\perp} = \{h_0, h_1, \ldots, h_{n-k-1}\}$. Therefore, for each codeword v from C, $vh_i^{\mathrm{T}} = 0$ for every i, or in compact form we can write $vH^{\mathrm{T}} = 0$, where H is so-called parity-check matrix whose ith row is h_i.

2.6.1 *Bipartite (Tanner) Graph*

An LDPC code can now be defined as an (n, k) linear block code whose parity-check matrix $\boldsymbol{H}$ has a low density of 1's. A *regular LDPC code* is a linear block code whose $\boldsymbol{H}$-matrix contains exactly w_c 1's in each column and exactly $w_r = w_c n/(n - k)$ 1's in each row, where $w_c \ll n - k$. The code rate of the regular LDPC code is determined by $R = k/n = 1 - w_c/w_r$. The graphical representation of LDPC codes, known as bipartite (Tanner) graph representation, is helpful in efficient description of LDPC decoding algorithms. A *bipartite (Tanner) graph* is a graph whose nodes may be separated into two classes (*variable* and *check* nodes), and where *undirected edges* may only connect two nodes not residing in the same class. The Tanner graph of a code is drawn according to the following rule: check (function) node c is connected to variable (bit) node v whenever element h_{cv} in a parity-check matrix $\boldsymbol{H}$ is a 1. In an $m \times n$ parity-check matrix, there are $m = n - k$ check nodes and n variable nodes. LDPC codes were invented by Robert Gallager (from MIT) in 1960, in his PhD dissertation [33], but received no attention from the coding community until 1990s [34]. As an illustrative example, consider the $\boldsymbol{H}$-matrix of the Hamming (7, 4) code:

$$\boldsymbol{H} = \begin{bmatrix} 1\,0\,0\,1\,0\,1\,1 \\ 0\,1\,0\,1\,1\,1\,0 \\ 0\,0\,1\,0\,1\,1\,1 \end{bmatrix}.$$

For any valid codeword $\boldsymbol{v} = [v_0\, v_1 \ldots v_{n-1}]$, the checks used to decode the codeword are written as

- Equation (c_0): $v_0 + v_3 + v_5 + v_6 = 0$ (mod 2),
- Equation (c_1): $v_1 + v_3 + v_4 + v_5 = 0$ (mod 2), and
- Equation (c_2): $v_2 + v_4 + v_5 + v_6 = 0$ (mod 2).

The bipartite graph (Tanner graph) representation of this code is given in Fig. 2.16a.

The circles represent the bit (variable) nodes while squares represent the check (function) nodes. For example, the variable nodes v_0, v_3, v_5, and v_6 are involved in (c_0), and therefore connected to the check node c_0. A closed path in a bipartite graph comprising l edges that closes back on itself is called a *cycle* of length l. The shortest cycle in the bipartite graph is called the *girth*. The girth influences the minimum distance of LDPC codes, correlates the extrinsic LLRs, and therefore affects the decoding performance. The use of large (high) girth LDPC codes is preferable because the large girth increases the minimum distance and prevents early correlations in the extrinsic information during decoding process. To improve the iterative decoding performance, we have to avoid cycles of length 4 and 6, which are illustrated in Fig. 2.16b and c, respectively.

The code description can also be done by the *degree distribution polynomials* $\mu(x)$ and $\rho(x)$, for the variable node (v-node) and the check node (c-node), respectively [31],

$$\mu(x) = \sum_{d=1}^{d_v} \mu_d x^{d-1}, \quad \rho(x) = \sum_{d=1}^{d_c} \rho_d x^{d-1}, \tag{2.113}$$

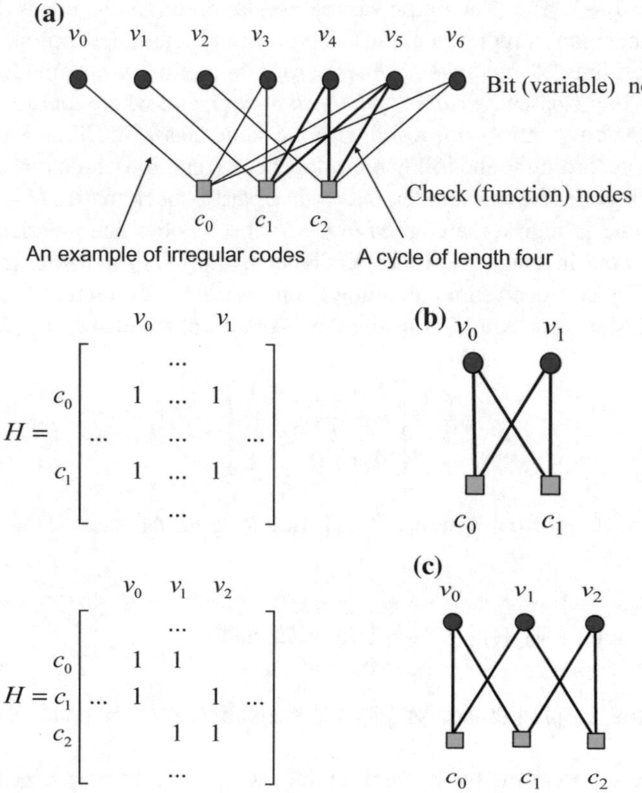

Fig. 2.16 **a** Bipartite graph of Hamming (7, 4) code described by ***H***-matrix above. Cycles in a Tanner graph: **b** cycle of length 4, and **c** cycle of length 6

where μ_d and ρ_d denote the fraction of the edges that are connected to degree-d v-nodes and c-nodes, respectively, and d_v and d_c denote the maximum v-node and c-node degrees, respectively.

2.6.2 *LDPC Codes Design*

The most obvious way to design LDPC codes is to construct a low-density parity-check matrix with prescribed properties. Some important LDPC designs, among others, include Gallager codes (semi-random construction) [33], MacKay codes (semi-random construction) [34], combinatorial design-based LDPC codes [35], finite-geometry-based LDPC codes [36, 37], and array [also known as quasi-cyclic (QC)] LDPC codes [38], to mention a few. The generator matrix of a QC-LDPC code can be represented as an array of circulant sub-matrices of the same size B indicating that QC-LDPC codes can be encoded in linear time using simple shift-register-based architectures [39]. A QC-LDPC code can be defined as an LDPC code for which every sectional cyclic shift to the right (or left) for $l \in [0, B-1]$ places of a codeword $\boldsymbol{v} = [\boldsymbol{v}_0\ \boldsymbol{v}_1\ \ldots\ \boldsymbol{v}_{B-1}]$ (each section $\boldsymbol{v}_i$ contains B elements) results in another codeword. Additional details on LPDC codes designs are postponed for Sect. 9.10.5.

2.6.3 Decoding of Binary LDPC Codes

The sum–product algorithm (SPA) is an iterative LDPC decoding algorithm in which extrinsic probabilities are iterated forward and back between variable and check nodes of bipartite (Tanner) graph representation of a parity-check matrix [1–4, 31, 32, 34, 40]. To facilitate the explanation of the various versions of the SPA, we use $N(v)$ [$N(c)$] to denote the neighborhood of v-node v (c-node c), and introduce the following notations:

- $N(c) = \{v\text{-nodes connected to } c\text{-node } c\}$;
- $N(c)\backslash\{v\} = \{v\text{-nodes connected to } c\text{-node } c \text{ except } v\text{-node } v\}$;
- $N(v) = \{c\text{-nodes connected to } v\text{-node } v\}$;
- $N(v)\backslash\{c\} = \{c\text{-nodes connected to } v\text{-node } v \text{ except } c\text{-node } c\}$;
- $P_v = \Pr(v = 1|y)$;
- $E(v)$: the event that the check equations involving variable node v are satisfied;
- $M(c')\backslash\{c\} = \{\text{messages from all } c'\text{-nodes except node } c\}$;
- $q_{vc}(b) = \Pr(v = b \mid E(v), y, M(c')\backslash\{c\})$;
- $M(v')\backslash\{v\} = \{\text{messages from all } v'\text{-nodes except node } v\}$;
- $r_{cv}(b) = \Pr(\text{check equation } c \text{ is satisfied} \mid v = b, M(v')\backslash\{v\})$;
- L_{vc}: the extrinsic likelihood to be sent from v-node v to c-node c; and
- L_{cv}: the extrinsic likelihood to be sent from c-node c to v-node v.

We are interested in computing the APP that a given bit in a transmitted codeword $\boldsymbol{v} = [v_0\ v_1\ \ldots\ v_{n-1}]$ equals 1, given the received word $\boldsymbol{y} = [y_0\ y_1\ \ldots\ y_{n-1}]$. Let us focus on decoding of the variable node v and we are concerned in computing LLR:

$$L(v) = \log\left[\frac{\Pr(v = 0|\boldsymbol{y})}{\Pr(v = 1|\boldsymbol{y})}\right],\ v \in \{v_0, \ldots, v_{n-1}\}. \tag{2.114}$$

The SPA, as indicated above, is an iterative decoding algorithm based on Tanner graph description of an LDPC code. We interpret the v-nodes as one type of processors and c-nodes as another type of processors, while the edges as the message paths for LLRs.

The subgraph illustrating the passing of messages (extrinsic information) from c-node to v-nodes in the check-node update half-iteration is shown in Fig. 2.17a. The information passed is the probability that parity-check equation c_0 is satisfied. The information passed from node c_0 to node v_1 represents the extrinsic information it had received from nodes v_0, v_4, and v_5 on the previous half-iteration. We are concerned in calculating the APP that a given bit v equals 1, given the received word $\boldsymbol{y}$ and

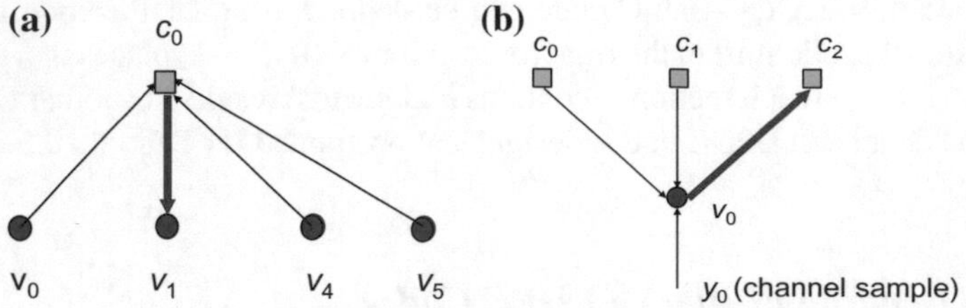

Fig. 2.17 Illustration of half-iterations of the SPA: **a** subgraph of bipartite graph corresponding to the H-matrix with the zeroth row [1 1 0 0 1 1 0 … 0], **b** subgraph of bipartite graph corresponding to the H-matrix with the zeroth column [1 1 1 0 … 0]

the fact that all parity-check equations involving bit v are satisfied, with this APP denoted as $\Pr(v = 1|\boldsymbol{y}, E(v))$. Instead of APP we can use log-APP or LLR defined as $\log\left[\frac{\Pr(v=0|\boldsymbol{y},E(v))}{\Pr(v=1|\boldsymbol{y},E(v))}\right]$.

In the v-nodes update half-iteration, the messages are passed in opposite direction from v-nodes to c-nodes, as depicted in subgraph of Fig. 2.17b. The information passed concerns $\log[\Pr(v_0 = 0|\boldsymbol{y})/\Pr(v_0 = 1|\boldsymbol{y})]$. The information being passed from node v_0 to node c_2 is information from the channel (via y_0) and extrinsic information node v_0 had received from nodes c_0 and c_1 on a previous half-iteration. This procedure is performed for all v-node/c-node pairs.

The calculation of the probability that cth parity-check equation is satisfied given $v = b$, $b \in \{0, 1\}$, denoted as $r_{cv}(b)$, is illustrated in Fig. 2.18a. On the other hand, the calculation of the probability that $v = b$ given extrinsic information from all check nodes, except node c, and given channel sample y, denoted as $q_{vc}(b)$, is illustrated in Fig. 2.18b.

In the first half-iteration, we calculate the extrinsic LLR to be passed from node c to variable node v, denoted as $L_{vc} = L(r_{cv})$, as follows:

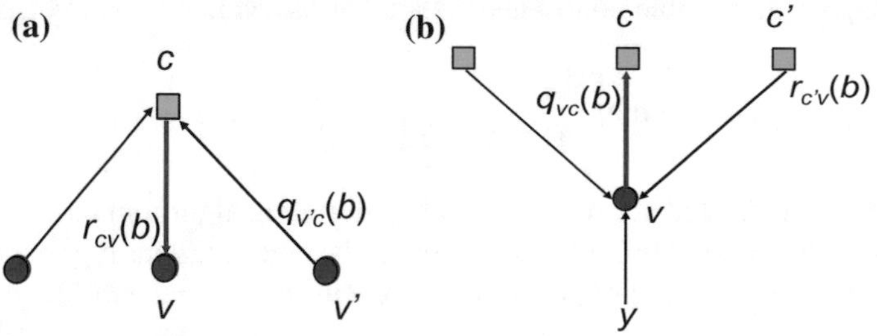

Fig. 2.18 Illustration of calculation of: **a** extrinsic information to be passed from node c to variable node v, **b** extrinsic information (message) to be passed from variable node v to check node c regarding the probability that $v = b$, $b \in \{0, 1\}$

$$L_{cv} = \left(\prod_{v'} \alpha_{v'c}\right)\phi\left[\sum_{v' \in N(c)\backslash v} \phi(\beta_{v'c})\right]; \alpha_{vc} = \text{sign}[L_{vc}], \beta_{vc} = |L_{vc}|, \quad (2.115)$$

where with $\phi(x)$ we denoted the following function:

$$\phi(x) = -\log\tanh\left(\frac{x}{2}\right) = \log\left[\frac{e^x + 1}{e^x - 1}\right], \quad (2.116)$$

which is plotted in Fig. 9.26. In the second half-iteration, we calculate the extrinsic LLR to be passed from variable node v to function node c regarding the probability that $v = b$, denoted as $L_{vc} = L(q_{vc})$, as

$$L_{vc} = L(q_{vc}) = L_{\text{ch}}(v) + \sum_{c' \in N(v)\backslash c} L(r_{c'v}). \quad (2.117)$$

For the derivation of Eqs. (2.115) and (2.117), an interested reader is referred to Appendix chapter.

Now we can summarize the *log-domain SPA* as follows.

(0) *Initialization*: For $v = 0, 1, \ldots, n-1$, initialize the messages L_{vc} to be sent from v-node v to c-node c to channel LLRs $L_{\text{ch}}(v)$, namely, $L_{vc} = L_{\text{ch}}(v)$.

(1) *c-node update rule*: For $c = 0, 1, \ldots, n-k-1$; compute $L_{cv} = \underset{N(c)\backslash\{v\}}{\boxplus} L_{vc}$. The box-plus operator is defined by

$$L_1 \boxplus L_2 = \prod_{k=1}^{2} \text{sign}(L_k) \cdot \phi\left(\sum_{k=1}^{2} \phi(|L_k|)\right),$$

where $\phi(x) = -\log\tanh(x/2)$. The box operator for $|N(c)\backslash\{v\}|$ components is obtained by recursively applying two-component version defined above.

(2) *v-node update rule*: For $v = 0, 1, \ldots, n-1$; set $L_{vc} = L_{\text{ch}}(v) + \sum_{N(v)\backslash\{c\}} L_{cv}$ for all c-nodes for which $h_{cv} = 1$.

(3) *Bit decisions*: Update $L(v)$ ($v = 0, \ldots, n-1$) by $L(v) = L_{\text{ch}}(v) + \sum_{N(v)} L_{cv}$ and set $\hat{v} = 1$ when $L(v) < 0$ (otherwise, $\hat{v} = 0$). If $\hat{\boldsymbol{v}}\boldsymbol{H}^T = \boldsymbol{0}$ or predetermined number of iterations has been reached then stop, otherwise go to step 1).

2.6.4 Min-Sum-Plus-Correction-Term Algorithm

Because the c-node update rule involves log and tanh functions, it is computationally intensive, and there exist many approximations. The very popular is the *min-sum-plus-correction-term approximation* [41]. Namely, it can be shown that "box-plus" operator $\boxplus$ can also be calculated by

$$L_1 \boxplus L_2 = \prod_{k=1}^{2} \text{sign}(L_k) \cdot \min(|L_1|, |L_2|) + c(x, y), \tag{2.118}$$

where $c(x, y)$ denotes the correction factor defined by

$$c(x, y) = \log\left(1 + e^{-|x+y|}\right) - \log\left(1 + e^{-|x-y|}\right), \tag{2.119}$$

commonly implemented as a look-up table (LUT). Given the fact that $|c(x, y)| < 0.693$, very often this term can be ignored. Alternatively, the following approximation can be used:

$$c(x, y) \simeq \begin{cases} -d, & |x-y| < 2 \cap |x+y| > 2|x-y| \\ d, & |x+y| < 2 \cap |x-y| > 2|x+y| \\ 0, & \text{otherwise} \end{cases}, \tag{2.120}$$

with typical d being 0.5.

Another popular decoding algorithm is *min-sum algorithm (MSA)* in which we simply ignore the correction term in (2.118). Namely, the shape of $\phi(x)$, shown in Fig. 2.19, suggests that the smallest β_{vc} in the summation (2.115) dominates and we can write

$$\phi\left[\sum_{v' \in N(c) \backslash v} \phi(\beta_{v'c})\right] \cong \phi\left(\phi\left(\min_{v'} \beta_{v'c}\right)\right) = \min_{v'} \beta_{v'c}. \tag{2.121}$$

Therefore, the min-sum algorithm is thus the log-domain algorithm with step (1) replaced by

$$L_{cv} = \left(\prod_{v' \in N(c) \backslash v} \alpha_{c'v}\right) \cdot \min_{v'} \beta_{v'c}. \tag{2.122}$$

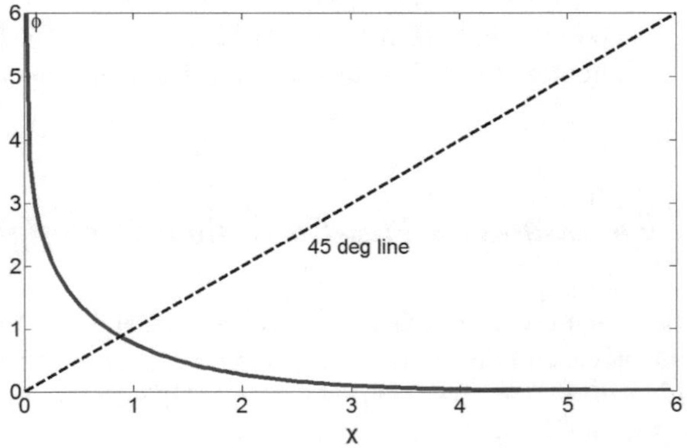

Fig. 2.19 The plot of function $\phi(x)$

2.7 Concluding Remarks

This chapter has been devoted to classical information theory fundamentals, including wireless channel capacity. The chapter starts with definitions entropy, joint entropy, conditional entropy, relative entropy, and mutual information in Sect. 2.1. In Sect. 2.2, the mutual information has been defined from the channel point of view followed by the channel capacity definition. In the same section, the channel capacity for discrete memoryless channels has been described in Sect. 2.2.1, while the capacity of continuous channels and information capacity theorem have been studied in Sect. 2.2.2. Regarding the wireless channels capacity, described in Sect. 2.3, we describe how to calculate the channel capacity of flat fading channels in Sect. 2.3.1 and frequency-selective channels in Sect. 2.3.2. We also discuss different optimum and suboptimum strategies to achieve channel capacity including the water-filling method, multiplexed coding and decoding, channel inversion, and truncated channel inversion. Depending on what is known about the channel state information (CSI) , different strategies are described for channel capacity evaluation. When CSI is known on receiver side only (CSI) , we describe Shannon (ergodic) channel capacity and capacity with outage in Sect. 2.3.1.1. On the other hand, when CSI is available on both transmitter and receiver sides (full CSI) , we describe Shannon capacity, zero-outage capacity, and outage capacity in Sect. 2.3.1.2. Regarding the frequency-selective fading channel capacity, described in Sect. 2.3.2, we describe how to calculate the channel capacity for both time-invariant channels (Sect. 2.3.2.1) and time-variant channels (Sect. 2.3.2.2). Additionally, we describe the optimum power adaption strategies in the same sections. For instance, the optimum power adaptation strategy for time-variant fading channels is the two-dimensional water-filling method in both time and frequency domains. Further, we explain how to model the channel with memory and describe how to calculate its capacity in Sect. 2.4. The key topics of this section include Markov sources (Sect. 2.4.1), McMillan sources (2.4.2), together with corresponding entropies, followed by the McMillan–Khinchin model for channel capacity evaluation (Sect. 2.4.3). The fundamentals of linear blocks are introduced in Sect. 2.5, while the fundamentals of binary LDPC codes in Sect. 2.5.

References

1. Djordjevic IB (2013) Advanced coding for optical communication systems. In: Kaminow I, Li T, Willner A (eds) Optical fiber telecommunications VI. Elsevier/Academic Press, pp 221–296
2. Djordjevic IB, Ryan W, Vasic B (2010) Coding for optical channels. Springer, New York-Dordrecht-Heidelberg-London
3. Cvijetic M, Djordjevic IB (2013) Advanced optical communication systems and networks. Artech House, Norwood, MA
4. Djordjevic IB (2016) On advanced FEC and coded modulation for ultra-high-speed optical transmission. IEEE Commun Surv Tutor 18(3):1920–1951. https://doi.org/10.1109/comst.2016.2536726
5. Cover TM, Thomas JA (1991) Elements of information theory. Wiley, New York

6. Reza FM (1961) An introduction to information theory. McGraw-Hill, New York
7. Gallager RG (1968) Information theory and reliable communication. Wiley, New York
8. Ingels FM (1971) Information and coding theory. Intext Educational Publishers, Scranton
9. Haykin S (2004) Communication systems. Wiley
10. Wolfowitz J (1964) Coding theorems of information theory, 2nd edn. Springer, Berling, New York
11. Shannon CE (1948) A mathematical theory of communication. Bell Syst Tech J 27:379–423, 623–656
12. Goldsmith A (2005) Wireless communications. Cambridge University Press, Cambridge
13. Tse D, Viswanath P (2005) Fundamentals of wireless communication. Cambridge University Press, Cambridge
14. Benedetto S, Biglieri E (1999) Principles of digital transmission with wireless applications. Kluwer Academic/Plenum Publishers, New York
15. Biglieri E (2005) Coding for wireless channels. Springer, New York
16. Duman TM, Ghrayeb A (2007) Coding for MIMO communication systems. Wiley, Chichester, UK
17. Proakis JG (2001) Digital communications. McGraw-Hill, Boston, MA
18. McMillan B (1952) The basic theorems of information theory. Ann Math Stat 24:196–219
19. Khinchin AI (1957) Mathematical foundations of information theory. Dover Publications, New York
20. Arnold D, Kavcic A, Loeliger H-A, Vontobel PO, Zeng W (2003) Simulation-based computation of information rates: upper and lower bounds. In: Proceedings of the IEEE international symposium on information theory (ISIT 2003), p 119
21. Arnold D, Loeliger H-A (2001) On the information rate of binary-input channels with memory. In: Proceedings of the 2001 international conference on communications, Helsinki, Finland, 11–14 June 2001, pp 2692–2695
22. Pfitser HD, Soriaga JB, Siegel PH (2001) On the achievable information rates of finite state ISI channels. In: Proceedings of the Globecom 2001, San Antonio, TX, 25–29 Nov 2001, pp 2992–2996
23. Djordjevic IB, Vasic B, Ivkovic M, Gabitov I (2005) Achievable information rates for high-speed long-haul optical transmission. IEEE/OSA J Lightw Technol 23:3755–3763
24. Ivkovic M, Djordjevic IB, Vasic B (2007) Calculation of achievable information rates of long-haul optical transmission systems using instanton approach. IEEE/OSA J Lightw Technol 25:1163–1168
25. Ivkovic M, Djordjevic I, Rajkovic P, Vasic B (2007) Pulse energy probability density functions for long-haul optical fiber transmission systems by using instantons and edgeworth expansion. IEEE Photon Technol Lett 19(20):1604–1606
26. Djordjevic IB, Alic N, Papen G, Radic S (2007) Determination of achievable information rates (AIRs) of IM/DD systems and AIR loss due to chromatic dispersion and quantization. IEEE Photon Technol Lett 19(1):12–14
27. Djordjevic IB, Minkov LL, Batshon HG (2008) Mitigation of linear and nonlinear impairments in high-speed optical networks by using LDPC-coded turbo equalization. IEEE J Sel Areas Comm Opt Commun Netw 26(6):73–83
28. Djordjevic IB, Minkov LL, Xu L, Wang T (2009) Suppression of fiber nonlinearities and PMD in coded-modulation schemes with coherent detection by using turbo equalization. IEEE/OSA J Opt Commun Netw 1(6):555–564
29. Essiambre R-J, Foschini GJ, Kramer G, Winzer PJ (2008) Capacity limits of information transport in fiber-optic networks. Phys Rev Lett 101:163901-1–163901-4
30. Essiambre R-J, Kramer G, Winzer PJ, Foschini GJ, Goebel B (2010) Capacity limits of optical fiber networks. J Lightw Technol 28(4):662–701
31. Djordjevic IB (2017) Advanced optical and wireless communications systems. Springer International Publishing, Switzerland
32. Ryan WE, Lin S (2009) Channel codes: classical and modern. Cambridge University Press
33. Gallager RG (1963) Low density parity check codes. MIT Press, Cambridge, MA

34. MacKay DJC (1999) Good error correcting codes based on very sparse matrices. IEEE Trans Inform Theory 45:399–431
35. Vasic B, Djordjevic IB, Kostuk R (2003) Low-density parity check codes and iterative decoding for long haul optical communication systems. J Lightw Technol 21(2):438–446
36. Djordjevic IB, Sankaranarayanan S, Vasic B (2004) Projective plane iteratively decodable block codes for WDM high-speed long-haul transmission systems. J Lightw Technol 22:695–702
37. Djordjevic IB, Vasic B (2004) Performance of affine geometry low-density parity-check codes in long-haul optical communications. Eur Trans Telecommun 15:477–483
38. Fan JL (2000) Array-codes as low-density parity-check codes. In: Proceedings of the second international symposium on turbo codes, Brest, France, pp 543–546
39. Li Z-W, Chen L, Zeng L-Q, Lin S, Fong W (2006) Efficient encoding of quasi-cyclic low-density parity-check codes. IEEE Trans Commun 54(1):71–81
40. Ryan WE (2004) An introduction to LDPC codes. In: Vasic B (ed) CRC handbook for coding and signal processing for recording systems. CRC Press
41. Chen J, Dholakia A, Eleftheriou E, Fossorier M, Hu X-Y (2005) Reduced-complexity decoding of LDPC codes. IEEE Trans Commun 53:1288–1299

Chapter 3
Conventional Cryptography Fundamentals

Abstract In this chapter, the conventional cryptography fundamentals are introduced. The chapter starts with basic terminology and cryptographic schemes, including symmetric and asymmetric cryptography, basic ciphers such as substitution and transposition ciphers, and one-time pads. The concepts of secrecy, authentication, and non-repudiation are introduced then, followed by various cryptanalytic attacks such as ciphertext-only, known-plaintext, chosen-plaintext, chosen-ciphertext, and adaptive-chosen-plaintext attacks. In section on information-theoretic approach to cryptography, the concept of perfect security is introduced and compared against the computational security. In the same section, unicity distance is discussed as well as the role of compression in cryptography. After that, one-way functions and one-way hash functions are discussed. The chapter concludes with several relevant practical cryptographic systems including DES and RSA systems as well as Diffie–Hellman public-key distribution.

3.1 Basic Terminology and Cryptographic Schemes

This section describes both the basic terminology used in cryptography and the basic cryptographic schemes [1–11].

3.1.1 Basics Cryptographic Schemes

A generic cryptographic system is illustrated in Fig. 3.1.

The source of information emits the message $\boldsymbol{M}$ to be encrypted, commonly referred to as the *plaintext*. The encryption block with the help of an algorithm $\boldsymbol{E}$ generates the *cryptogram* $\boldsymbol{C}$, also known as the *ciphertext*. The ciphertext get transmitted over insecure channel toward the *decryption* block, which with the help of a decryption algorithm $\boldsymbol{D}$ deciphers the original message $\boldsymbol{M}$ and delivers it to an authorized user. The encryption process can be described as the following mapping

I. B. Djordjevic, *Physical-Layer Security and Quantum Key Distribution*,
https://doi.org/10.1007/978-3-030-27565-5_3

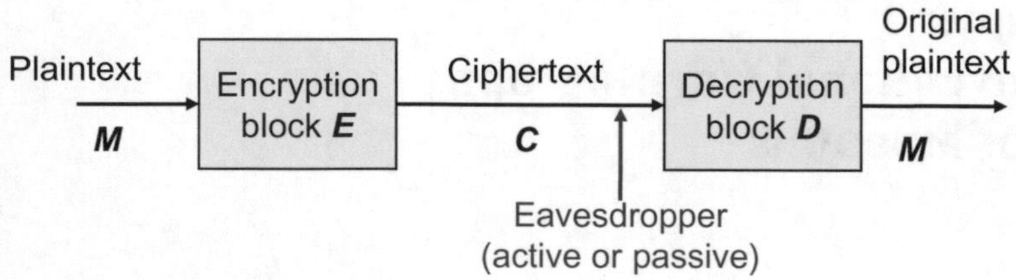

Fig. 3.1 A generic cryptographic system

$\boldsymbol{E}(\boldsymbol{M}) = \boldsymbol{C}$; in other words, the encryption algorithm $\boldsymbol{E}$ operates on plaintext message $\boldsymbol{M}$ to generate the cryptogram $\boldsymbol{C}$. On the other hand, the operation of decryption block can be described as the following mapping $\boldsymbol{D}(\boldsymbol{C}) = \boldsymbol{M}$, that is, the decryption algorithm $\boldsymbol{D}$ operates on cryptogram $\boldsymbol{C}$ to reproduce the message $\boldsymbol{M}$. Given that the main purpose of cryptographic system is to encrypt the message so that eavesdropper is enable to get it, and to decrypt the cryptogram so that end user can receive the intended message, the following composition property must be true $\boldsymbol{D}(\boldsymbol{E}(\boldsymbol{M})) = \boldsymbol{M}$. The cryptographic system prevents an eavesdropper from extracting information being transmitted over insecure channel, thus offering solution to the *secrecy* problem. At the same time, the cryptographic system prevents the eavesdropper from impersonating the message sender, thus providing the solution to the *authentication* problem.

A *cryptographic algorithm*, used in encryption and decryption process, commonly referred to as a *cipher*, is a mathematical function used to perform encryption and decryption. The corresponding algorithm is called *restricted*, if the security relies on keeping the operation of algorithm secret. However, the restricted algorithms are not suitable for standardization. To solve this problem, the modern cryptography relies on concept of the *key*, denoted as $\boldsymbol{K}$. The key-based cryptographic scheme is provided in Fig. 3.2.

The source emits the message (plaintext) $\boldsymbol{M}$ toward the encryption block, which with the help of key $\boldsymbol{K}$, obtained from key source, generates the cryptogram. On

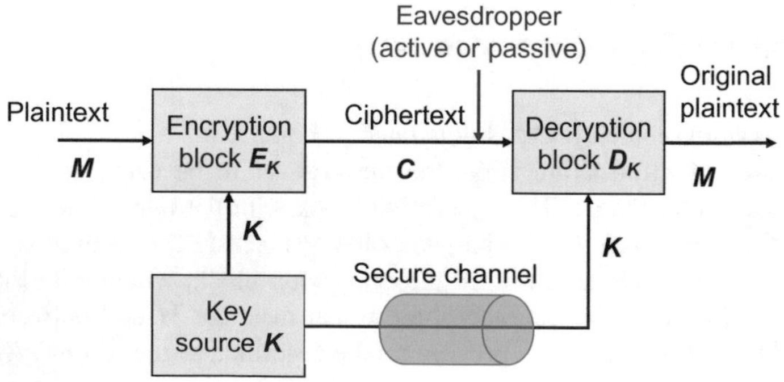

Fig. 3.2 A key-based cryptographic system

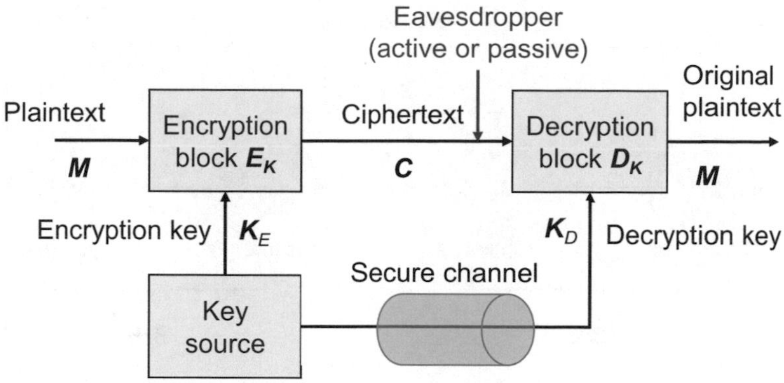

Fig. 3.3 A key-based cryptographic system in which encryption and decryption keys are different

the receiver side, the cryptogram transmitted over insecure channel get processed by the decryption algorithm together with the key $\boldsymbol{K}$ obtained through the secure channel, which reconstructs the original plaintext to be delivered to the authenticated user. The encryption process can be mathematically described as $\boldsymbol{E_K}(\boldsymbol{M}) = \boldsymbol{C}$, while the decryption process by $\boldsymbol{D_K}(\boldsymbol{C}) = \boldsymbol{M}$. Similarly as before, the composition of decryption and encryption functions yields to identity mapping $\boldsymbol{D_K}(\boldsymbol{E_K}(\boldsymbol{M})) = \boldsymbol{M}$. The key source typically generates the key randomly from the *keyspace* (the range of possible key values). The encryption and decryption key can be different as illustrated in Fig. 3.3. The corresponding encryption and decryption functions can be mathematically described as follows:

$$\boldsymbol{E}_{\boldsymbol{K}_E}(\boldsymbol{M}) = \boldsymbol{C}, \quad \boldsymbol{D}_{\boldsymbol{K}_D}(\boldsymbol{C}) = \boldsymbol{M}$$

Similarly, the composition of encryption and decryption functions is an identity mapping:

$$\boldsymbol{D}_{\boldsymbol{K}_D}\left(\boldsymbol{E}_{\boldsymbol{K}_E}(\boldsymbol{M})\right) = \boldsymbol{M}.$$

The key-based algorithms can be categorized into two broad categories:

- *Symmetric algorithms*, in which decryption key can be derived from the encryption key and vice versa. Alternatively, the same key can be used for both encryption and decryption stages. Symmetric algorithms are also known as one-key (single-key) or secret-key algorithms. The well-known system employing this type of algorithms is digital encryption standard (DES) [2, 5].
- *Asymmetric algorithms*, in which encryption and decryption keys are different. Moreover, the decryption key cannot be determined from encryption key, at least in any reasonable amount of time. Because of this fact, the encryption keys can be even made public, wherein the eavesdropper will not be able to determine the

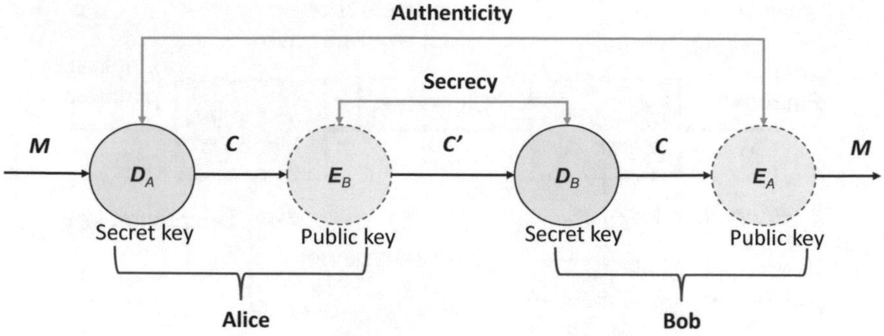

Fig. 3.4 Achieving simultaneously the secrecy and authenticity in public-key systems

decryption key. The *public-key systems* [6] are based on this concept. In public-key systems, the encrypted keys have been made public, while the decryption key is known only to the intended user. The encryption key is then called the *public* key, while decryption the *secret (private)* key. The keys can be applied in arbitrary order to create the cryptogram from plaintext, and to reconstruct the plaintext from the cryptogram. As illustrated in Fig. 3.4, the secrecy is achieved as follows. When Alice wants to send a message to Bob, she employs Bob's public key to encrypt, while Bob is only one capable of reconstructing the message by employing his secret key. The authenticity is achieved by employing the secret key by Alice to encrypt, and Bob performing decryption by using Alice's public key (see Fig. 3.4). The public–private key pair must satisfy the following two conditions [4]: (i) whichever message gets encrypted with one key from the pair, it can be decrypted with another key from the pair; and (ii) given the public-key knowledge, it is computationally infeasible to determine the private key.

Symmetric algorithms can be further categorized into (i) *block ciphers* (algorithms) and (ii) *stream ciphers* (algorithms). In block ciphers, the message gets split into blocks of fixed size (such as 64 bits), and each block gets then encrypted separately. The stream ciphers operate on a plaintext by processing a single bit at the time. Therefore, the block ciphers have a certain similarity to the block codes, while the stream ciphers are, on the other hand, similar to the convolutional codes in channel coding theory.

3.1.2 Basic Ciphers

In his 1949 paper, Shannon proposed two practical methods as general approaches to design a good cipher [1], namely, diffusion and confusion. Their purpose is to frustrate the statistical analysis of the cryptogram, and therefore make the breaking of the cipher extremely difficult. In the method of *diffusion*, the statistical structure of the message is hidden within cryptogram by spreading the single symbol of plaintext into many symbols of ciphertext. In the method of *confusion*, the substitutions are introduced in such a way to make the relationship between plaintext and ciphertext as complex as possible. Moreover, Shannon proposed to combine several cryptosystems by creating their *product*. In the rest of this section, we describe basic ciphers, including the *transposition* ciphers and *substitution* ciphers.

In the *transposition ciphers*, belonging to the class of block ciphers, the plaintext stays the same but order of characters is shuffled around. We can interpret the transposition cipher as being written according to a certain rule on two-dimensional geometric figure such as matrix and being read out according to another rule. The transposition can be combined with *permutations*. As an illustration, let us parse the word CRYPTOGRAPHY into group of four characters and then apply 3-1-4-2 permutation to obtain the following:

Plaintext: CRYP TOGR APHY
Ciphertext: YCPR GTRO HAYP

The transmission of ciphertext is without the space. Let us now write the groups of four characters in CRYPTOGRAPHY 2ND E as a 4 $\times$ 4 matrix and apply the 3-1-4-2 permutation columnwise; the corresponding ciphertext will be then

CRYP
TOGR
APHY
2NDE
Ciphertext: AC2T PRNO HYDG YPER

By applying the English language rules, the period of permutation and the order of elements in permutation can be determined, and the encryption protocol can be broken.

In *substitution cipher*, each character in plaintext is substituted with another character in ciphertext. Four types of substitution ciphers are commonly used [2, 3]:

- *Monoalphabetic cipher*, in which each character in plaintext is substituted with corresponding character of ciphertext. By numerating the characters in English alphabet (i = 0, 1, … 25), the plaintext can be encrypted by adding to each character the alphabet index shifts for j characters observed per modulo 26, that is, $(i + j)$ mod 26. For instance, in famous *Caesar cipher*, the shift was $j = 3$. ROT13 encryption program used in UNIX operating system to hide potentially offensive text is a monoalphabetic cipher in which every letter is rotated for 13 positions in the alphabet.
- *Polyalphabetic cipher*, in which multiple monoalphabetic ciphers are applied. This cipher was introduced by Battista in 1568 [2, 12] and was used by the Union Army during the American Civil War [2]. One popular version, the *Vigenère cipher*, applies successively several monoalphabetic ciphers.
- *Homophonic substitution cipher*, in which a single character from plain text is mapped to one of several possible characters in ciphertext.

- *Polygram substitution cipher*, in which a group of plain characters is mapped into a group of ciphertext characters.

Rotor machine ciphers were introduced in 1920s to automate the encryption process [2]. The basic ingredient was a rotor, a mechanical device based on the wheel wired to perform simple substitution. Rotor machine was composed of a keyboard and several rotors, with each rotor performing arbitrary permutation in the alphabet and thus introducing the substitution. Clearly, the rotor machine cipher is an instance of the Vigenère cipher.

As an illustration, let us perform the encryption by employing the Vigenère cipher employing the word *key*:

CRY PTO GRA PHY
+KEY KEY KEY KEY
Ciphertext: MVW ZXM QUY ZLW

3.1.2.1 One-Time Pads

In *one-time pad*, also known as *Vernam cipher* [12], a completely random sequence of characters, with the sequence length being equal to the message sequence length, is used as a key. When for each new message another random sequence is used as a key, the one-time pad scheme provides the perfect security (to be discussed in Sect. 3.2.1), and it is not possible to break. Namely, the brute-force search approach would require to verify m^n possible keys, where m is the employed alphabet size and n is the length of intercepted cryptogram. To illustrate the Vernam cipher, let us again observe the plaintext CRYPTOGRAPHY and apply the randomly selected key:

CRYPTOGRAPHY
+ZFAHMCYCIWDP
Ciphertext: BWYWFQENILKN

The use of such cipher requires huge memory resources. One of the polyalphabetic ciphers that can be used instead is so-called *running-key cipher* in which the text from a book is used to provide very long keystream. The book to be used is agreed on ahead of time, and the randomly selected passage to be used as key is secretly indicated in the message. To generate a very simple running key, we can use so-called *autokey* cipher. In this encryption scheme, the initial key to start with is important, and the message itself is used for the rest of the key. To illustrate, let us use the word *key* as initial key in the following example:

CRYPTOGRAPHY…
+KEYCRYPTOGRAPHY…
Ciphertext: MVWRKMVJOVXY…

So far, the English alphabet and the ciphers operating on the text have been observed. In practice, in digital and computer communications, we typically operate

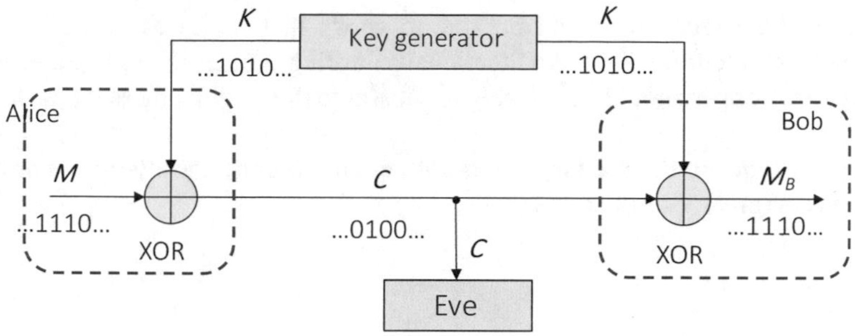

Fig. 3.5 One-time pad encryption scheme

on binary alphabet GF(2) = {0, 1}. To obtain the key, we need a special random generator and to encrypt using one-time pad scheme we simply perform addition mod 2, i.e., XOR operation, as illustrated in Fig. 3.5. Even though that the one-time pad scheme offers perfect security, it has several drawbacks [13–15]: it requires the secure distribution of the key, the length of the key must be at least as long as the message, the key bits cannot be reused, the keys must be delivered in advance, securely stored until used, and destroyed after the use. The one-time pad encryption scheme is very similar to the scrambling procedure used in digital communications, wherein transmitted bits are combined with pseudonoise (PN) sequence mod 2. However, the maximum-length sequence (or m-sequence) [16] cannot be used for encryption purposes. The nonlinear PN sequences as well sequences derived from m-sequences but processed properly should be used instead [2]. The *random sequences suitable for cryptography* must satisfy the following *requirements* [2]:

- They must pass all the existing tests of randomness,
- They are unpredictable (based on previously generated random bits it must be computationally unfeasible to predict next bit), and
- They cannot be reliably reproduced (when the random generator gets run twice for the same input, the output of random generator must not be related to previous outputs).

3.1.3 Secrecy, Authentication, Integrity, and Non-repudiation

The purpose of a cryptographic system, as discussed above, is to provide confidentiality, and therefore, solve the secrecy problem. In addition to confidentiality, the cryptographic system should solve the following problems:

- *Authentication*, the eavesdropper should not be able to insert fake cryptogram without being noticed by the receiver. In other words, the receiver should be able to determine the originator of the message.

- *Integrity*, the receiver should be able to verify that the message has not been altered (be intentionally or unintentionally) during transmission. In other words, the eavesdropper should not be able to substitute the original message with a fake one.
- *Non-repudiation*, the sender should not be able to deny falsely later that he/she was the originator of the message.

3.1.4 Cryptoanalysis

The *cryptanalysis* is the science/art of recovering the plaintext without having access to the key. The practitioners performing the cryptoanalysis are commonly referred to as *cryptanalysts*. An act of an attempted cryptanalysis is typically called the *attack*. If the cryptanalyst has the complete knowledge of the encryption algorithm, the attacks can be classified into five generic categories:

- *Ciphertext-only attack* assumes that eavesdropper cryptanalyst has access to either portion or whole ciphertext only. His/her task is to recover the plaintext of as many massages as possible or to determine the keys used to encrypt the messages. This cryptanalytic attack can be described as follows:
  Given the cryptograms: $\boldsymbol{C}_1 = \boldsymbol{E_K}(\boldsymbol{M}_1), \ldots, \boldsymbol{C}_L = \boldsymbol{E_K}(\boldsymbol{M}_L)$
  Determine either messages $\boldsymbol{M}_1, \ldots, \boldsymbol{M}_L$ and key $\boldsymbol{K}$ or the algorithm $\boldsymbol{E_K}$ so that the incoming messages $\boldsymbol{M}_{L+1}, \boldsymbol{M}_{L+2}, \ldots$ can be recovered from mappings $\boldsymbol{C}_{L+1} = \boldsymbol{E_K}(\boldsymbol{M}_{L+1}), \boldsymbol{C}_{L+2} = \boldsymbol{E_K}(\boldsymbol{M}_{L+2}), \ldots$
- *Known-plaintext attack* assumes that eavesdropper cryptanalyst has access to some of ciphertext–plaintext pairs. His/her task is to determine the keys used to encrypt the messages or deduct the decryption algorithm. This cryptanalytic attack can be described as follows:
  Given the plaintext–ciphertext pairs: $(\boldsymbol{M}_1, \boldsymbol{C}_1 = \boldsymbol{E_K}(\boldsymbol{M}_1)), \ldots, (\boldsymbol{M}_L, \boldsymbol{C}_L = \boldsymbol{E_K}(\boldsymbol{M}_L))$, determine either the key $\boldsymbol{K}$ or the algorithm $\boldsymbol{E_K}$ so that the incoming messages $\boldsymbol{M}_{L+1}, \boldsymbol{M}_{L+2}, \ldots$ can be recovered from mappings $\boldsymbol{C}_{L+1} = \boldsymbol{E_K}(\boldsymbol{M}_{L+1}), \boldsymbol{C}_{L+2} = \boldsymbol{E_K}(\boldsymbol{M}_{L+2}), \ldots$
- *Chosen-plaintext attack* assumes that eavesdropper cryptanalyst has been able to submit any plaintext and receive in return the corresponding ciphertext when the actual key is used. His/her task is to determine the keys used to encrypt the messages or deduct the decryption algorithm. This cryptanalytic attack can be described as follows:
  Given the plaintext–ciphertext pairs: $(\boldsymbol{M}_1, \boldsymbol{C}_1 = \boldsymbol{E_K}(\boldsymbol{M}_1)), \ldots, (\boldsymbol{M}_L, \boldsymbol{C}_L = \boldsymbol{E_K}(\boldsymbol{M}_L))$, wherein the cryptanalyst can chose plaintext messages $\boldsymbol{M}_1, \ldots, \boldsymbol{M}_L$,
  determine either the key $\boldsymbol{K}$ or the algorithm $\boldsymbol{E_K}$ so that the incoming messages $\boldsymbol{M}_{L+1}, \boldsymbol{M}_{L+2}, \ldots$ can be recovered from mappings $\boldsymbol{C}_{L+1} = \boldsymbol{E_K}(\boldsymbol{M}_{L+1}), \boldsymbol{C}_{L+2} = \boldsymbol{E_K}(\boldsymbol{M}_{L+2}), \ldots$

Clearly, this much more powerful attack than known-plaintext attack because cryptanalysts can submit carefully chosen messages to reveal more information about the key.

- *Chosen-ciphertext attack* assumes that eavesdropper cryptanalyst has been able to choose arbitrary ciphertext and has the access to the decrypted plaintext. His/her task is to determine the keys being used to encrypt the messages. This cryptanalytic attack can be described as follows:
  Given the plaintext–ciphertext pairs: $(\boldsymbol{M}_1, \boldsymbol{C}_1 = \boldsymbol{E}_{\boldsymbol{K}}(\boldsymbol{M}_1)), \ldots, (\boldsymbol{M}_L, \boldsymbol{C}_L = \boldsymbol{E}_{\boldsymbol{K}}(\boldsymbol{M}_L))$, determine the key $\boldsymbol{K}$ being used.
  This attack is suitable for public-key algorithms.
- *Adaptive-chosen-plaintext attack* assumes that eavesdropper cryptanalyst has been able not only to submit any plaintext and receive in return the corresponding ciphertext when the actual key is used but is also able to adapt the plaintext based on previous encryption result. Therefore, the cryptanalyst is able to deduct the key in an adaptive and iterative fashion.

For other cryptanalytic attacks, an interested reader is referred to [2, 5].

3.2 Information-Theoretic Approach to Cryptography

In this section, we discuss cryptography from information theory point of view. The information-theoretic model under consideration is provided in Fig. 3.6. Alice with the help of k-bit message $\boldsymbol{M}$ and key $\boldsymbol{K}$ performs the encryption and generates the cryptogram $\boldsymbol{C} = f(\boldsymbol{M}, \boldsymbol{K})$, which is available to both Bob and Eve. Bob based on the cryptogram and key decrypts the original message $\boldsymbol{M}$.

3.2.1 Perfect Security Versus Computational Security

According to Shannon [1], the *perfect security*, also known as *unconditional security*, has been achieved when the messages and cryptograms are statistically independent so that the corresponding mutual information is equal to zero:

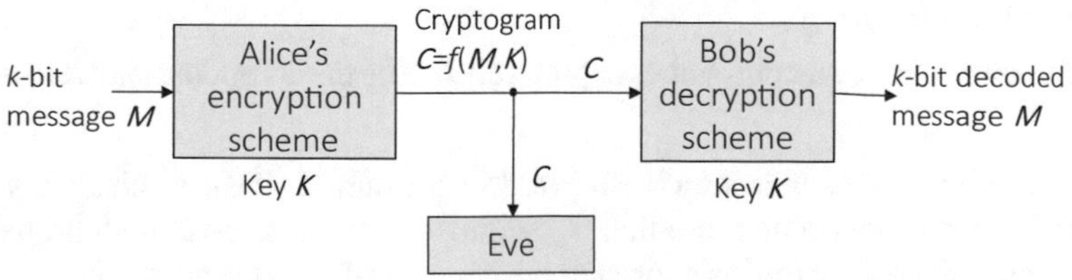

Fig. 3.6 Information-theoretic cryptographic model

$$I(\boldsymbol{M}, \boldsymbol{C}) = H(\boldsymbol{M}) - H(\boldsymbol{M}|\boldsymbol{C}) = 0. \tag{3.1}$$

In other words, the entropy of the message (uncertainty of the message) stays the same regardless of the availability of cryptogram:

$$H(\boldsymbol{M}|\boldsymbol{C}) = H(\boldsymbol{M}). \tag{3.2}$$

By employing the chain rule from previous chapter, we can write:

$$\begin{aligned} H(\boldsymbol{M}) &= H(\boldsymbol{M}|\boldsymbol{C}) \\ &\le H(\boldsymbol{M}, \boldsymbol{K}|\boldsymbol{C}) \\ &= H(\boldsymbol{K}|\boldsymbol{C}) + \underbrace{H(\boldsymbol{M}|\boldsymbol{C}, \boldsymbol{K})}_{=0} \\ &= H(\boldsymbol{K}|\boldsymbol{C}) = H(\boldsymbol{K}), \end{aligned} \tag{3.3}$$

where we used the fact that message is completely specified when both the cryptogram and the key are known, that is $H(\boldsymbol{M}|\boldsymbol{K}, \boldsymbol{C}) = 0$. Therefore, the *perfect secrecy condition* is given by:

$$H(\boldsymbol{M}) \le H(\boldsymbol{K}). \tag{3.4}$$

In conclusion, the entropy (uncertainty) of the key cannot be lower than the entropy of the message, for an encryption scheme to be perfectly secured. Given that in Vernam cipher the length of the key is at least equal to the message length, it appears that one-time pad scheme is perfectly secure.

However, given that this condition is difficult to satisfy, in conventional cryptography instead of information-theoretic security the *computational security* is used [2, 5, 11, 14, 17, 18]. The computational security introduces two relaxations with respect to information-theoretic security [5]:

- Security is guaranteed against an efficient eavesdropper running the cryptanalytic attacks for certain limited amount of time. Of course, when eavesdropper has sufficient computational resources and/or sufficient time he/she will be able to break the security of the encryption scheme.
- Eavesdroppers can be successful in breaking the security protocols, but with small success probability.

There are two general approaches to precisely define the computational relaxations [5]:

- The *concrete approach*, in which an eavesdropper running the eavesdropping algorithm for time interval no more than T seconds can be successful with probability at most ε. We could call this approach the concept of (T, ε)-security.
- The *asymptotic approach*, in which a probabilistic polynomial-time eavesdropper caring out an eavesdropping strategy (cryptanalytic attack) for every polynomial

$p(n)$ there exists an integer N such that when $n > N$ the success attack probability is smaller than $1/p(n)$.

A reader interested to learn more about computational security is referred to as an excellent book due to Katz and Lindell [5].

3.2.1.1 Unicity Distance

In this section, we are trying to determine the required cryptogram length for an eavesdropper to determine the key. This parameter is called the *unicity distance* and represents the average cryptogram length after which an enemy cryptanalyst is able uniquely to determine the used key. It can be determined from conditional entropy of the key when the cryptogram is known $H(\boldsymbol{K}|\boldsymbol{C})$ and can be defined as the smallest length N such that the conditional entropy is approximately zero. For the "random cipher," as defined by Shannon [1], the unicity distance can be approximately determined as [1]

$$N_u \simeq \frac{H(\boldsymbol{K})}{R \log L_c}, \; R = 1 - \frac{H(\boldsymbol{M})}{N \log L_c}, \tag{3.5}$$

where L_c is the ciphertext alphabet size and R is the redundancy of the message information contained in N-bit ciphertext. The entropy of the key is upper limited by

$$H(\boldsymbol{K}) \leq \log\left(L_k^{|\boldsymbol{K}|}\right), \tag{3.6}$$

where L_k is the size of key alphabet, and equality sign is applicable when the key is completely random. After substituting (3.6) into (3.5) and assuming that $L_k = L_c$, we obtain the following result for the unicity distance:

$$N_u \simeq |\boldsymbol{K}|/R. \tag{3.7}$$

Given that typical English text redundancy is 75%, the unicity distance is only 1.3334 $|\boldsymbol{K}|$.

Let us now assume that both message and key alphabets are binary [1, 3, 17]. Let the message symbols be statistically independent with probability of occurrence of 0 being p, that is $P(0) = p$, $P(1) = 1 - p$. Let the key symbols be statistically independent occurring equally likely, i.e., $P(K = 0) = P(K = 1) = 0.5$. Further, we assume the cryptogram is created by mod 2 addition of the message and the key. We are concerned determining the conditional entropy of the key as a function of cryptogram length and parameter p. Since $K \in \{0, 1\}$, the probability of the cryptogram having exactly n zeros and $N - n$ ones, regardless of the order, is given by

$$P\left(C_{n,N}\right) = \frac{1}{2}\binom{N}{n}\left[p^n(1-p)^{N-n} + (1-p)^n p^{N-n}\right]. \tag{3.8}$$

The conditional probability of $K = 0$ given $C_{n,N}$ can be determined by applying the Bayes' rule:

$$P\left(K = 0|C_{n,N}\right) = \frac{P\left(C_{n,N}|K = 0\right)P(K = 0)}{P\left(C_{n,N}\right)}. \tag{3.9}$$

When $K = 0$, the conditional probability $P(C_{n,N}|K = 0)$ is equal to the probability of having n zeros and N-n ones:

$$P\left(C_{n,N}|K = 0\right) = \binom{N}{n}p^n(1-p)^{N-n}. \tag{3.10}$$

Substituting (3.10) into (3.9), we obtain the following expression for conditional probability of $K = 0$ given $C_{n,N}$:

$$\begin{aligned} P\left(K = 0|C_{n,N}\right) &= \frac{\binom{N}{n}p^n(1-p)^{N-n}\frac{1}{2}}{\frac{1}{2}\binom{N}{n}\left[p^n(1-p)^{N-n} + (1-p)^n p^{N-n}\right]} \\ &= \frac{1}{1 + \frac{(1-p)^n p^{N-n}}{p^n(1-p)^{N-n}}} = \frac{1}{1 + \underbrace{\left(\frac{p}{1-p}\right)^{N-2n}}_{q}} = \frac{1}{1+q}, q = \left(\frac{p}{1-p}\right)^{N-2n}. \end{aligned} \tag{3.11}$$

On the other hand, the conditional probability of $K = 1$ given $C_{n,N}$ is simply

$$P\left(K = 1|C_{n,N}\right) = 1 - P\left(K = 0|C_{n,N}\right) = 1 - \frac{1}{1+q} = \frac{q}{1+q}. \tag{3.12}$$

The conditional entropy of the key given $C_{n,N}$ will be then:

$$\begin{aligned} H\left(K|C_{n,N}\right) &= -P\left(K = 1|C_{n,N}\right)\log P\left(K = 1|C_{n,N}\right) \\ &\quad - P\left(K = 0|C_{n,N}\right)\log P\left(K = 0|C_{n,N}\right) \\ &= -\frac{q}{1+q}\log q + \log(1+q). \end{aligned} \tag{3.13}$$

Now by averaging Eq. (3.13) out with respect to $P(C_{n,N})$, we obtain the following expression for conditional entropy (equivocation) of the key given cryptogram of N:

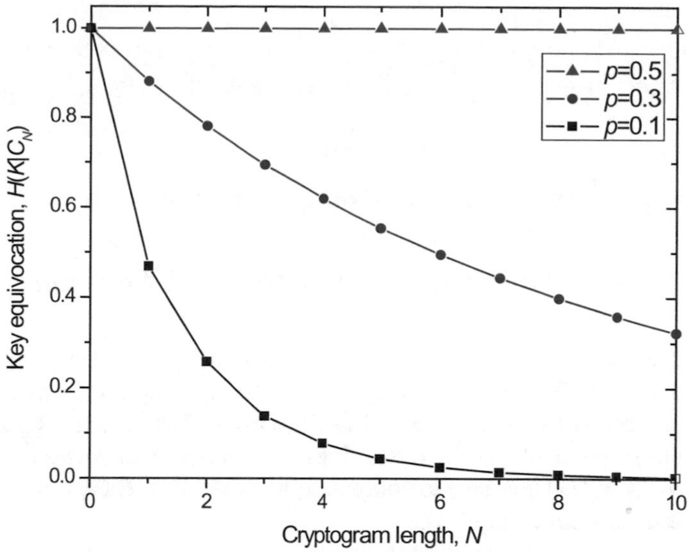

Fig. 3.7 Key equivocation as a function of cryptogram length

$$H(K|C_N) = \sum_{n=0}^{N} H(K|C_{n,N}) P(C_{n,N})$$
$$= \frac{1}{2} \sum_{n=0}^{N} \left[-\frac{q}{1+q} \log q + \log(1+q) \right] (1+q) \binom{N}{n} p^n (1-p)^{N-n}. \tag{3.14}$$

As an illustration, in Fig. 3.7, we show the key equivocation $H(K|C_N)$ dependence versus cryptogram length for probability p being as parameter. When $p = 0.5$, there is no redundancy in the cryptogram and the secrecy is perfect. However, when p = 0.1, only several bits in cryptogram are needed to determine if $K = 1$ or 0.

3.2.1.2 Compression and Cryptography

The data compaction (lossless data compression) can be employed as a useful tool to improve security. Namely, by employing the data compaction, we can reduce the redundancy R in Eq. (3.5) and increase the unicity distance N_u. If redundancy in the message tends to zero, the unicity distance would tend to infinity. However, it is not realistic to implement a device to perform this operation. In addition to increasing the unicity distance, the compression can speed up the encryption process which is time-consuming. However, the data compaction is sensitive to errors introduced by channel, so it is a good idea to combine it with channel coding as illustrated in Fig. 3.8.

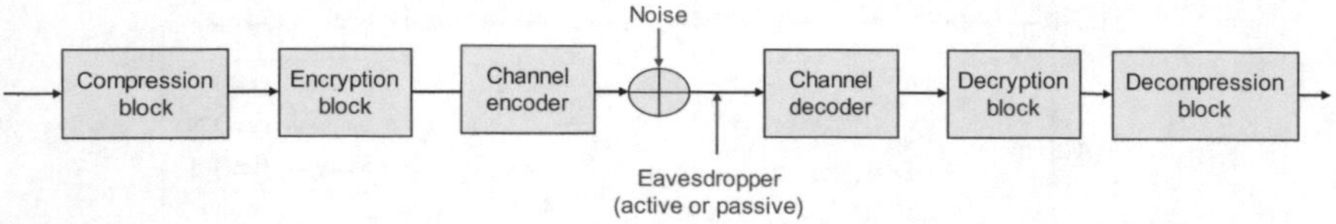

Fig. 3.8 Combining compression and channel coding with encryption

3.2.2 One-Way Functions and One-Way Hash Functions

One-way function represents an important concept employed in many cryptographic protocols. One-way functions $f(x) = y$ are easy to calculate for a given argument (message) x; however, it is extremely difficult to invert these functions (to calculate $f^{-1}(y)$). One instance of one-way functions is a *trapdoor one-way function*, for which given y and $f(x)$ it is easy to compute x; however, it is still extremely difficult to calculate inverse function $f^{-1}(y)$.

One-way hash function represents another relevant concept in cryptography. A hash function maps a variable-length input string, also known as *pre-image*, into a smaller fixed-length output string, known as a *hash value*. One-way hash function is easy to calculate in one direction, based on pre-image it is straightforward to calculate the hash value. However, it is impossible to determine the pre-image based on the hash value. The hash function must be *collision-free*, i.e., it is hard to determine two different pre-images based on the same hash value. The cyclic redundancy check (CRC) algorithm, used in error detection, has certain similarities with the hash functions. Namely, based on k information bits, we can generate the n-k parity bits such that $n\text{-}k \ll k$. The parity bits are obtained from reminder of information polynomial and generating polynomial. From n-k parity bits, it is impossible to determine k information bits. However, the CRC codes do not satisfy the collision-free condition, there exist multiple information words corresponding to the same parity bits, and therefore, cannot be used to create the hash value. To summarize, the hash function based on a pre-image generates the "digested" shortened version of fixed length, which can be interpreted as a "fingerprint," a unique hash value. Mathematically, we can define the one-way hash function as follows [2]:

$$\boldsymbol{h} = H(\boldsymbol{M}), \tag{3.15}$$

where the message $\boldsymbol{M}$ length is variable, while hash value $\boldsymbol{h}$ is of a fixed length. The one-way hash functions must satisfy the following properties [2, 19]:

- Given $\boldsymbol{M}$ it is easy to calculate $H(\boldsymbol{M})$.
- Given $\boldsymbol{h}$ it is hard to calculate the original message; in other words, it is hard to find the inverse function $H^{-1}(\boldsymbol{h})$.
- Probability of finding two messages $\boldsymbol{M}$ and $\boldsymbol{M}'$ having the same hash value $\boldsymbol{h}$ is extremely small.

Implementing one-way hash function is not a trivial task. To simplify the problem, the *compression function* is used, which outputs a hash value of length m for an input of larger length n [2, 20]. The input to this function is the previous hash value of previous message blocks $\boldsymbol{h}_{i-1}$ and the current message block $\boldsymbol{M}_i$, while the output is the hash value of all message blocks received so far:

$$\boldsymbol{h}_i = H(\boldsymbol{M}_i, \boldsymbol{h}_{i-1}). \tag{3.16}$$

One of the first serious hash algorithms, MD4 message-digest algorithm, was introduced by Rivest [21]. The improved version, MD5 message-digest algorithm, was proposed by the same author [22], and will be described here briefly. Even though it does not satisfy the collision condition, it is a good illustration of how the hash function should be implemented. The MD5 processes the input text in 512-bit blocks, divided into 16 32-bit sub-blocks, and generates 128-bit hash value, composed of four 32-bit blocks. Each message is padded by 100 … 0 so that the total length is $k \cdot 512$-64, where k is an integer. Then the 64-bit message is added, so that the total length is a multiple of 512. Four 32-bit registers are initialized to (in hexadecimal notation)

$$A = 0 \times 01234567,\ B = 0 \times 89\text{abcdef},\ C = 0 \times \text{fedcba89},\ \text{and } D = 0 \times 76543210.$$

Message gets split into 512-bit blocks and gets processed in four rounds each containing 16 operations. There are four nonlinear functions used in each round, respectively,

$$\begin{aligned} &F(X, Y, Z) = (X \wedge Y) \vee \left(\bar{X} \wedge Z\right), G(X, Y, Z) = (X \wedge Z) \vee \left(Y \wedge \bar{Z}\right) \\ &H(X, Y, Z) = X \oplus Y \oplus Z, I(X, Y, Z) = Y \oplus \left(X \vee \bar{Z}\right), \end{aligned} \tag{3.17}$$

where we use $\wedge$, $\vee$, $\oplus$, and overbar to denote AND, OR, XOR, and NOT operations. In these operations, 32-bit constant t_i is used determined as $2^{32} \cdot |\sin(i)|$, where i is in radians, together with the message sub-block $\boldsymbol{M}_i$. We use $\lll$ to denote the left-circular shift. As an illustration, in Fig. 3.9, we provide one MD5 operation in round 1. This operation is repeated 16 times with cyclically shifted inputs to the F-function. Then the G-function is employed in round 2, H-function in round 3, and I-function in round 4. After 64 MD5 operations, the content of output ABCD register denotes the hash value.

To solve the collision problem, secure hash algorithm (SHA) has been proposed for use in the digital signature standard [23]. SHA generates 160-bit-long hash. The message is padded by 100 … 0 sequence into a multiple of 512–64 bits, and then 64 information bit sequence is added, the same as in MD5. SHA employs five 32-bit variables, instead of four in MD5. There are four nonlinear functions and four constants in SHA. The details of SHA can be found in [2]. SHA solves the integrity problem of MD5.

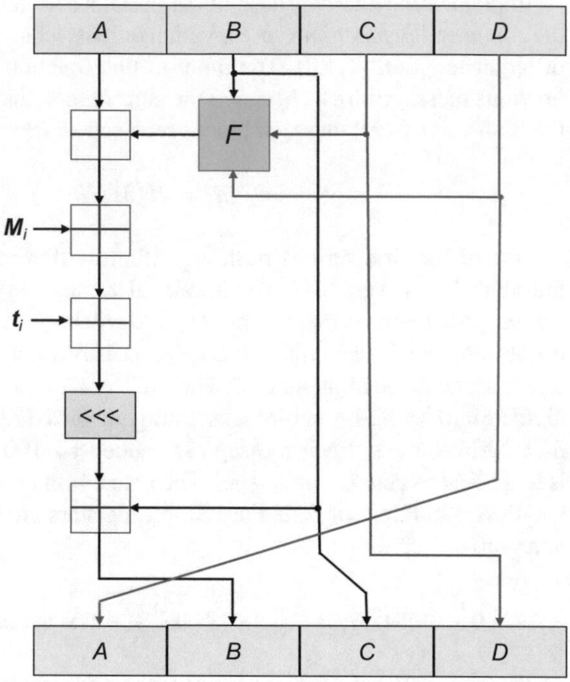

Fig. 3.9 One MD5 operation in round 1

3.3 Some Practical Cryptography Systems

In this section, two relevant cryptography systems, the digital encryption standard (DES) [24, 25] and Rivest–Shamir–Adleman (RSA) system, are discussed [26].

3.3.1 Digital Encryption Standard (DES)

In 1973, the National Bureau for Standards (NBS) issued a public request for a standard cryptographic algorithm, with specific design criteria [2]:

- The algorithm must be highly secure;
- The algorithm must be completely specified and easy to understand;
- The security of algorithm must reside in the key, and it should not be dependent on the secrecy of algorithm;
- The algorithm must be available for all users;
- The algorithm must be adaptable for use in different applications;
- The algorithm must be economical for hardware implementation;
- The algorithm must be efficient to use;
- The algorithm must be exportable; and
- The algorithm must be suitable for validation.

The IBM proposal, based on Feistel cipher, has been chosen, and with small modification requests by National Security Agency (NSA), it was adopted as a federal standard in 1976. The American National Standards Institute (ANSI) approved DES for private use in 1981 [25].

The *Feistel cipher* [27, 28], employed in DES, is composed of S-box and P-box building blocks. The S-box performs the substitutions controlled by the key. It is secure for enough internal states but difficult to realize for larger number of states. The P-box performs blockwise permutations of digits. The Feistel network is obtained by properly combining the S- and P-boxes, thus reviving the concept of product ciphers introduced by Shannon [1]. The main idea is to achieve the strong polyalphabetic substitution in multiple rounds. The operational principle of a *practical Feistel cipher* is illustrated in Fig. 3.10. In the figure, the $F(K_i, R_i)$-function is the round function and K_i ($i = 0, 1, \ldots, n$) denotes the ith round sub-key. The plaintext gets split into two equal (left and right) parts, L_0 and R_0, and in the ith round ($i = 0, 1, \ldots, n$) the following key-dependent computation is performed:

$$\boldsymbol{L}_{i+1} = \boldsymbol{R}_i, \quad \boldsymbol{R}_{i+1} = \boldsymbol{L}_i \oplus F(\boldsymbol{K}_i, \boldsymbol{R}_i). \tag{3.18}$$

The ciphertext is obtained in the nth round as $\boldsymbol{C} = (\boldsymbol{L}_{n+1}, \boldsymbol{R}_{n+1})$. The decryption process is performed by performing the same operation on the ciphertext, but with the sub-keys being applied in reverse order:

$$\boldsymbol{R}_i = \boldsymbol{L}_{i+1}, \boldsymbol{L}_i = \boldsymbol{R}_{i+1} \oplus F(\boldsymbol{K}_i, \boldsymbol{L}_{i+1}); i = n, n-1, \ldots, 1, 0. \tag{3.19}$$

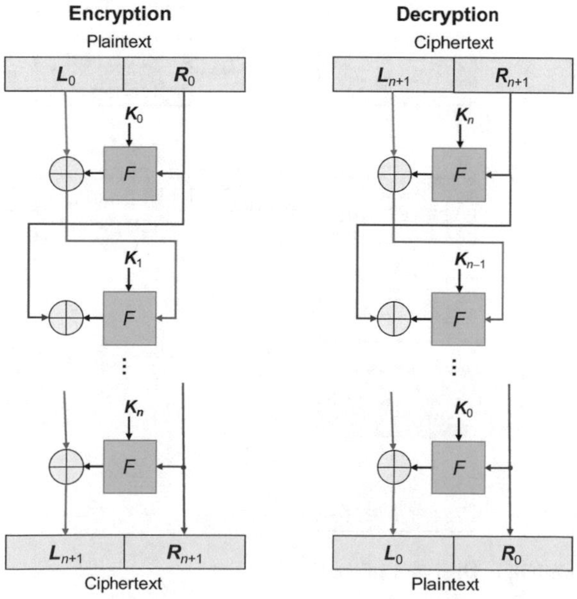

Fig. 3.10 The operational principle of a practical Feistel cipher

Evidently, compared to substitution–permutation networks, the round function F does not need to be invertible. Luby and Rackoff have shown that when F-function is cryptographically secure pseudorandom function, three rounds are enough for the cipher to perform pseudorandom permutation [29].

The DES applies a series of substitution and transposition operations as well as mod 2 additions and employs the Feistel cipher operation principle. The DES overview is provided in Fig. 3.11. The details of one DES round are provided in Fig. 3.12. The key used for encryption is composed of 56 bits; the remaining 8 bits, on locations 8, 16, …, 64, are used to create the parity checks for error detection. The permuted choice 1 disregards the parity-check positions in 64-bit key $\boldsymbol{K}_0$ and permutes the remaining 56 bits. Different keys for each round are derived from the key used in previous round by one or two left cyclic shifts, followed by the compression permutation (denoted as permuted choice 2). During the compression permutation, 48 bits are selected (out of 56). The key-dependent round computation in data encryption portion is the same as in Feistel cipher:

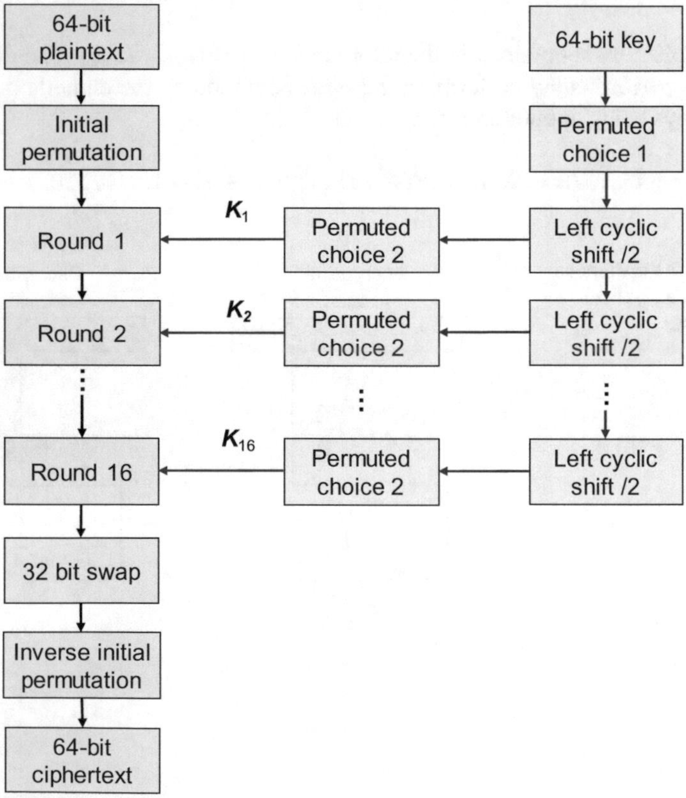

Fig. 3.11 Overview of the DES

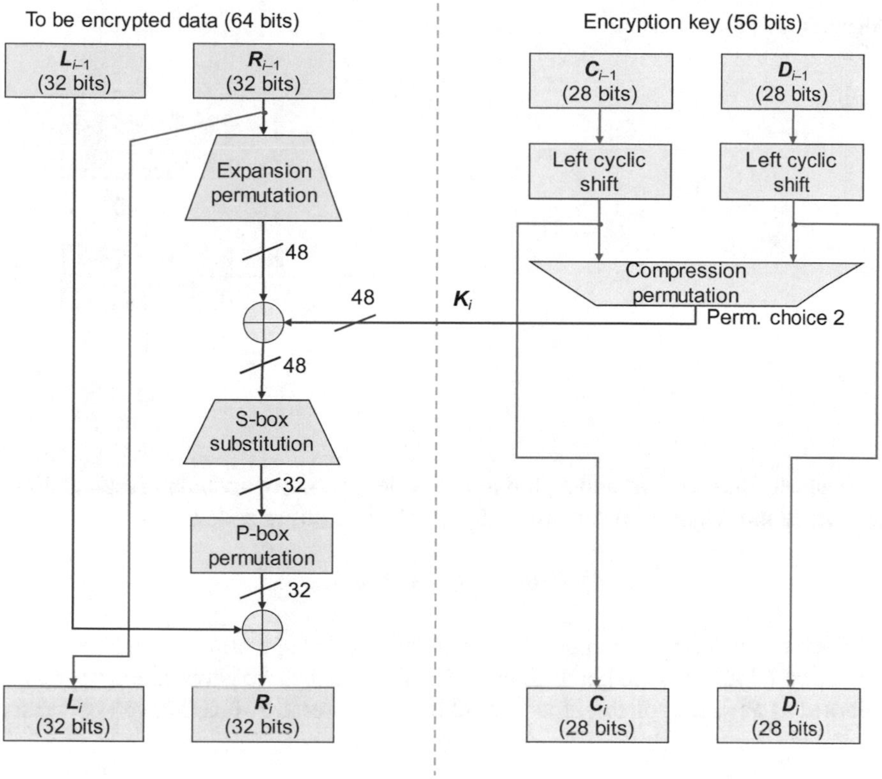

Fig. 3.12 Details of one DES round

$$\boldsymbol{L}_i = \boldsymbol{R}_{i-1}, \boldsymbol{R}_i = \boldsymbol{L}_{i-1} \oplus F(\boldsymbol{K}_i, \boldsymbol{R}_{i-1}); i = 1, 2, \ldots, 16. \tag{3.20}$$

The F-function is computed as provided by flowchart in Fig. 3.12. The first step is the expansion permutation, wherein 32-bit input is mapped into 48-bit permuted block as follows. The 32-bit word is split into eight 4-bit sub-words. Then each sub-word is expanded for two bits by appending one bit at both sides, which are in fact copy-pasted bits from neighboring sub-words. In other words, if original 32-bit word was

$$\boldsymbol{R} = \overbrace{r_1 r_2 r_3 r_4}^{\text{the first 4 - bit sub-word}} \overbrace{r_5 r_6 r_7 r_8}^{\text{the second 4-bit sub - word}} \cdots \overbrace{r_{29} r_{30} r_{31} r_{32}}^{\text{the eight 4-bit sub-word}}, \tag{3.21}$$

the 48-bit word $\boldsymbol{R}'$ will be

$$\boldsymbol{R}' = \overbrace{r_{32} r_1 r_2 r_3 r_4 r_5}^{\text{the first 6-bit sub-word}} \overbrace{r_4 r_5 r_6 r_7 r_8 r_9}^{\text{the second 6-bit sub-word}} \cdots \overbrace{r_{28} r_{29} r_{30} r_{31} r_{32} r_1}^{\text{the eight 6-bit sub-word}}. \tag{3.22}$$

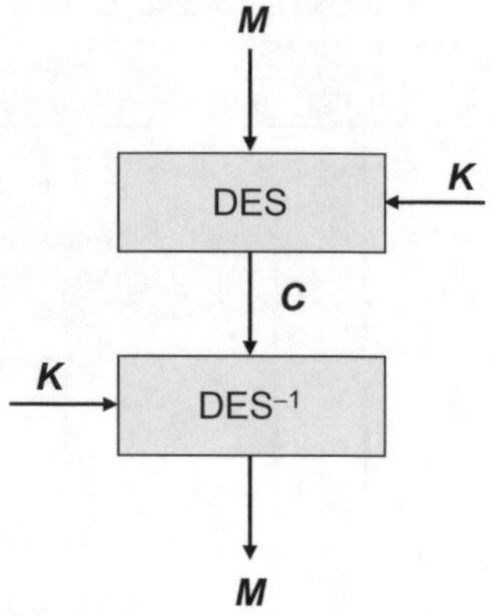

Fig. 3.13 The use of DES in ECB

Then 48-bit words $\boldsymbol{R}'$ and $\boldsymbol{K}_i$ are mod 2 added, and corresponding result $\boldsymbol{R}' \oplus \boldsymbol{K}_i$ is divided into eight 6-bit sub-words $\boldsymbol{B}_j$ ($j = 1, 2, \ldots, 8$) as follows:

$$\boldsymbol{B}_1\boldsymbol{B}_2\cdots\boldsymbol{B}_8 = \boldsymbol{R}' \oplus \boldsymbol{K}_i. \tag{3.23}$$

Each sub-word $\boldsymbol{B}_j$ undergoes the S-box substitution operation S_j implemented as a look-up table (LUT) to get 4-bit output $S_j(\boldsymbol{B}_j)$. Finally, the permutation operation, denoted as $P[\cdot]$, is applied outputting 32 bits, and overall F-function can be written as

$$F(\boldsymbol{K}_i, \boldsymbol{R}) = P[S_1(\boldsymbol{B}_1)S_2(\boldsymbol{B}_2)\cdots S_8(\boldsymbol{B}_8)]. \tag{3.24}$$

The DES can be implemented either in hardware or software. The DES can be used in different modes: electronic codebook (ECB), cipher block chaining (CBC), output feedback (OFB), and cipher feedback (CFB) as specified in FIPS PUB 81 [30]. The ECB is used to create the block code, as illustrated in Fig. 3.13. The cryptograms are independent, and the transmission error affects only the currently transmitted cryptogram. However, the identical messages for the same key generate the same cryptogram, indicating that ECB is suitable only for short messages.

The employment of DES in CBC has certain similarities with the Vigenère cipher; however, the key is used only the first time, and thereafter the autokey concept is employed as shown in Fig. 3.14, in which the cryptogram itself serves as the key.

Regarding the security of DES, Diffie and Hellman argued in 1977 that with technology of that time to break the key for one day, the corresponding compute machine would cost \$20 million [2, 31]. In 1998, the Electronic Frontier Foundation (EFF), a cyberspace civil rights group, invested \$250,000 into computer machine called *DES Cracker*, which was able to decrypt messages employing DES algorithm in 56 h [3, 32]. Given that in DES algorithm cryptogram length is fixed, it is possible to apply DES successively several times, as illustrated in Fig. 3.15. The effective length of the key in triple DES is 112, and the corresponding cryptogram is much

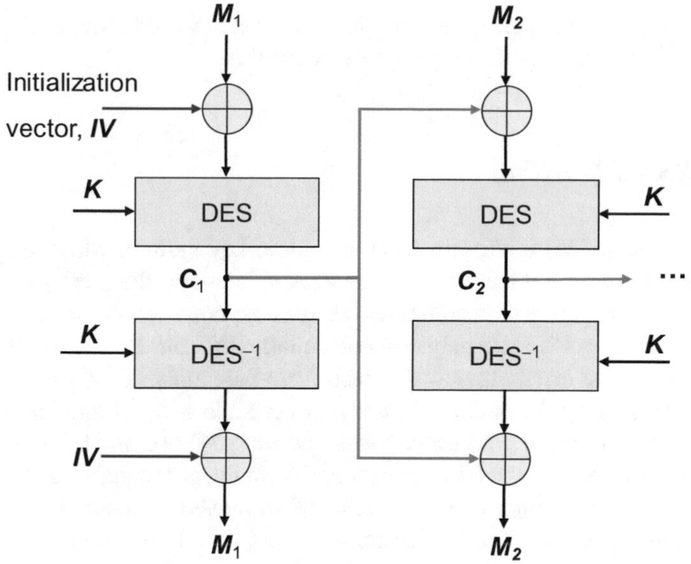

Fig. 3.14 The use of DES in CBC

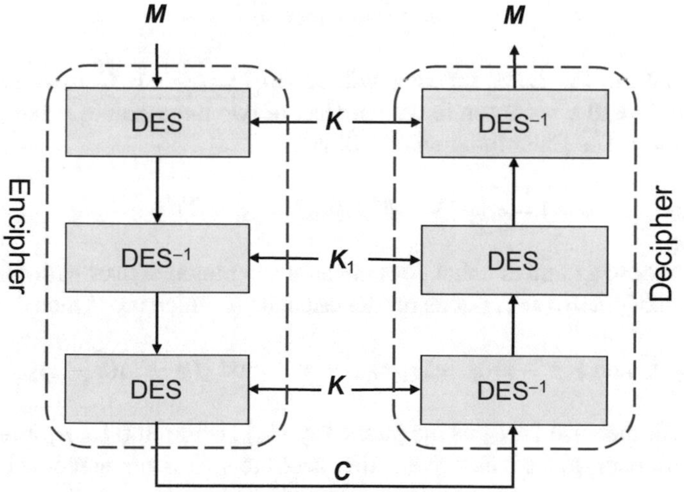

Fig. 3.15 The triple DES

more difficult to break since the brute-force approach would require 2^{112} attempts instead of 2^{56} attempts in the original DES algorithm.

3.3.2 RSA Algorithm

The RSA system [26] is the most known public-key system, which employs the concepts of classical number theory. The RSA is based on the fact that finding the prime number of large size is computationally easy; however, factoring the product of such two numbers is extremely computationally difficult. Before we describe the RSA algorithm, we briefly review few facts from the number theory [33].

An integer p is a *prime number* if it is only divisible with ± 1 and $\pm p$. A positive integer c is said to be the *greatest common divisor* (gcd) of a and b if c is divisor of a and b, any divisor of a and b is a divisor of c. Two integers a and b *relatively prime* if gcd(a, b) $\equiv$ 1. The *Euler totient function* of an integer n, denoted as $\phi(n)$, is the number of integers less than n being relatively coprime to n. Two integers a and b are *congruent modulo n*, denoted as $a \equiv b \bmod n$, if $(a \bmod n) = (b \bmod n)$. If a and n are relatively prime, the *Euler's theorem* claims

$$a^{\phi(n)} \equiv 1 \bmod n. \tag{3.25}$$

If n is prime, the Euler function will be $\phi(n) = n - 1$. In RSA system, the cryptogram length n is chosen to be a product of two large primes p and q ($p \neq q$), that is, $n = pq$. The Euler function of n is then

$$\phi(n) = \phi(pq) = \phi(p)\phi(q) = (p-1)(q-1). \tag{3.26}$$

Let us choose a number e that does not have a common divisor with $\phi(n)$, that is, for which gcd(e, $\phi(n)$) = 1. Let us further calculate d, which is $e^{-1} \bmod \phi(n)$, that is,

$$ed \equiv 1 \bmod [(p-1)(q-1)] \Rightarrow d = e^{-1} \bmod [(p-1)(q-1)]. \tag{3.27}$$

Alice will then use $\{e, n\}$ as the public key K_{pub}, while Bob $\{d, n\}$ as the private key K_{priv}. To encrypt the message m, Alice needs to split it into numerical blocks m_i smaller than n (for binary data, the length would be 2^l, where l is the largest power such that $2^l < n$). The encrypted message c will be composed of cryptograms c_i of comparable sizes. The ith cryptogram c_i is obtained by

$$c_i = (m_i)^e \bmod n. \tag{3.28}$$

On the receiver side, Bob decrypts the cryptogram by

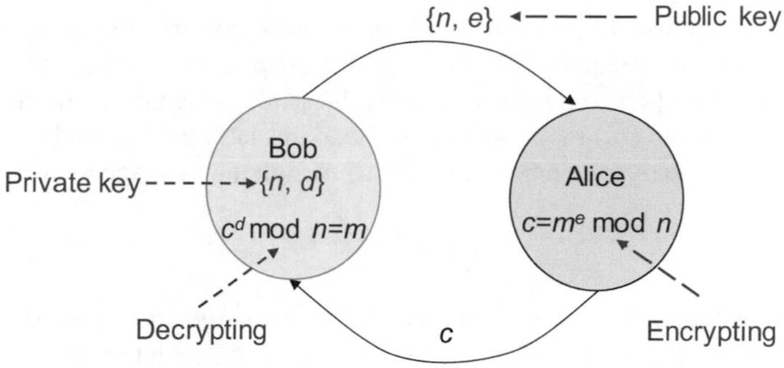

Fig. 3.16 The RSA encryption protocols

$$(c_i)^d \bmod n = (m_i)^{ed} \bmod n \underbrace{=}_{ed=1+k\phi(n),\exists k} (m_i)^{1+k\phi(n)} \bmod n = m_i \underbrace{(m_i)^{k\phi(n)} \bmod n}_{1} = m_i, \tag{3.29}$$

where we use the fact that $ed \bmod \phi(n)$ is equivalent to $ed = 1 + k\phi(n)$, where k is an arbitrary integer.

The RSA encryption protocol has been summarized in Fig. 3.16. Bob selects two primes p and q (and keeps them secret) to create $n = pq$, and selects e that is relatively prime to $\phi(n)$. He further calculates $d = e^{-1} \bmod \phi(n)$ and keeps $\{d, n\}$ as a private key. He then sends $\{n, e\}$ as the public key to Alice. Alice encrypts the message m by $m^e \bmod n = c$ and sends the cryptogram c to Bob. Based on cryptogram c, Bob decrypts Alice's message by $c^d \bmod n = m$.

To break the protocol, Eve needs to determine first the period r of the function $f(x) = m^x \bmod n = f(x + r)$ $(r = 0, 1, \ldots, 2^l - 1)$. From the public channel, Eve learns Bob's public-key $\{n, e\}$ and the cryptogram c. She then calculates d' as follows:

$$d' = e^{-1} \bmod r, \tag{3.30}$$

and decrypts Alice message as follows:

$$\begin{aligned} c^{d'} \bmod n &= \left(m^e\right)^{d'} \bmod n = m^{ed'} \bmod n \overbrace{=}^{ed'=1+kr,\exists k} m^{1+kr} \bmod n \\ &= m \underbrace{m^{kr}}_{1 \bmod n} \bmod n \\ &= m \bmod n, \end{aligned} \tag{3.31}$$

and, therefore, breaks the RSA protocol. However, the main problem for Eve is the determination of the period of function $f(x)$, which requires $O(n)$ steps, which is not feasible in reasonable amount of time. However, by applying the Shor's factorization algorithm, with the help of quantum computer [34, 15], Eve will be able to break RSA protocol, once the quantum computer of practical importance becomes available.

The RSA can also be used in *digital signatures* systems, to provide the proof about the origin of the message. The digital signature must satisfy the following three requirements [4]: (i) it is easy to verify the sender's signature, (ii) the signature is not forgeable, and (iii) the sender is unable to disclaim to be a sender.

The sender employs the private key $\{d, n\}$ to sign the message m as follows:

$$s = m^d \bmod n. \tag{3.32}$$

Clearly, only a given sender can use this key to sign the message. Based on discussion above the security of RSA, the signature is difficult to forge. The receiver employs the public key of the sender $\{e, n\}$ to validate the sender's signature:

$$s^e \bmod n = \left(m^d\right)^e \bmod n = m^{de} \bmod n = m^{(1+k\phi(n))} \bmod n = m \bmod n. \tag{3.33}$$

Evidently, all three requirements for RSA to be used to produce the digital signatures are satisfied.

Before concluding this chapter, we briefly describe Diffie–Hellman public-key distribution scheme [35], which is one of the oldest in the class, but still relevant.

3.3.3 Diffie–Hellman Public-Key Distribution

The Diffie–Hellman public-key distribution scheme [2, 6, 35] relies on the fact that it is computationally easy to calculate a discrete exponential; but difficult to calculate a discrete logarithm. Alice and Bob first need to agree on a large prime numbers n and g, wherein g is a primitive element mod n (different powers of g generate elements mod n); and these are not kept secret. The protocol can be summarized as follows:

- Alice selects randomly a large integer x and calculates

$$X = g^x \bmod n, \tag{3.34}$$

  and sends X to Bob.
- Bob selects randomly a large integer y and calculates

$$Y = g^y \bmod n, \tag{3.35}$$

  and sends Y to Bob.
- Alice calculates the key K_A by

$$K_A = Y^x \bmod n. \tag{3.36}$$

- Bob calculates the key K_B by

$$K_B = X^y \bmod n. \tag{3.37}$$

Clearly, both keys are identical since

$$\begin{aligned} K_A &= Y^x \bmod n = \left(g^y\right)^x \bmod n = g^{xy} \bmod n, \\ K_B &= X^y \bmod n = \left(g^x\right)^y \bmod n = g^{xy} \bmod n, \end{aligned} \tag{3.38}$$

and only Alice and Bob know the key. Eve can learn the key only if she computes discrete logarithm such as

$$\log_g X \bmod n = x, \tag{3.39}$$

which is computationally difficult. This protocol is easy to generalize to arbitrary number of participants. Let assume that ith and jth users want to determine the secret key to communicate with each other. Both will randomly generate large integers x_i and x_j, and then calculate corresponding discrete exponentials X_i and X_j by

$$X_i = g^{x_i} \bmod n, \quad X_j = g^{x_j} \bmod n, \tag{3.40}$$

and deposit them in a *public directory*. Finally, they will create corresponding keys as follows:

$$\begin{aligned} K_{ij} &= X_j^{x_i} \bmod n = \left(g^{x_j}\right)^{x_i} \bmod n = g^{x_i x_j} \bmod n, \\ K_{ji} &= X_i^{x_j} \bmod n = \left(g^{x_i}\right)^{x_j} \bmod n = g^{x_i x_j} \bmod n, \end{aligned} \tag{3.41}$$

which are clearly identical. Only way for Eve to determine the key is to perform the following computation:

$$K_{ji} = X_i^{\log_g X_j} \bmod n, \tag{3.42}$$

which involves the discrete logarithm computation.

3.4 Concluding Remarks

In this chapter, the conventional cryptography fundamentals have been introduced. In Sect. 3.1, the basic terminology and cryptographic schemes have been introduced, including symmetric and asymmetric cryptography, basic ciphers such as substitution and transposition ciphers, and one-time pads. The concepts of secrecy, authentication, and non-repudiation have been introduced as well, followed by various cryptanalytic attacks such as ciphertext-only, known-plaintext, chosen-plaintext, chosen-ciphertext, and adaptive-chosen-plaintext attacks. In Sect. 3.2, the information-theoretic approach to cryptography has been introduced, including the concepts of perfect security and computational security. In the same section, unicity distance has been discussed as well as the role of compression in cryptography. After that, one-way functions and one-way hash functions have been discussed. In Sect. 3.3, several relevant practical cryptographic systems have been described such as DES system, RSA system, and Diffie–Hellman public-key distribution system.

References

1. Shannon CE (1949) Communication theory of secrecy systems. Bell Syst Tech J 28:656–715
2. Schneier B (2015) Applied cryptography, second edition: protocols, algorithms, and source code in C. Wiley, Indianapolis, IN
3. Drajic D, Ivanis P (2009) Introduction to information theory and coding, 3rd edn. Akademska Misao, Belgrade, Serbia (in Serbian)
4. Haykin S (2001) Communication systems, 4th edn. Wiley, Hamilton Printing Company, Canada
5. Katz J, Lindell Y (2015) Introduction to modern cryptography, 2nd edn. CRC Press, Boca Raton, FL
6. Diffie W, Hellman ME (1976) New direction in cryptography. IEEE Trans Inform Theory IT 22:644–654
7. Hellman ME (1977) An extension of the Shannon theory approach to cryptography. IEEE Trans Inform Theory IT 23:289–294
8. Rivest RL, Shamir A, Adleman L (1983) Cryptographic communications system and method. US Patent 4,405,829
9. Merkle M (1978) Secure communication over an insecure channel. Comm ACM 21:294–299
10. McEliece RJ (1978) A public key cryptosystem based on algebraic coding theory. JPL DSN Prog Rep 42(44):114–116
11. Aumasson J-P (2018) Serious cryptography: a practical introduction to modern encryption. No Starch Press, San Francisco, CA
12. Kahn D (1967) The codebreakers: the story of secret writing. Macmillan Publishing Co., New York
13. Neilsen MA, Chuang IL (2010) Quantum computation and quantum information. Cambridge University Press, Cambridge
14. Van Assche G (2006) Quantum cryptography and secrete-key distillation. Cambridge University Press, Cambridge, New York
15. Djordjevic IB (2012) Quantum information processing and quantum error correction: an engineering approach. Elsevier/Academic Press, Amsterdam, Boston
16. Djordjevic IB (2017) Advanced optical and wireless communications systems. Springer International Publishing, Switzerland
17. Sebbery J, Pieprzyk J (1989) Cryptography: an introduction to computer security. Prentice Hall, New York
18. Delfs H, Knebl H (2015) Introduction to cryptography: principles and applications (Information Security and Cryptography), 3rd edn. Springer, Heidelberg, New York
19. Merckle RC (1979) Secrecy, authentication, and public key systems. PhD dissertation. Stanford University

20. Merckle RC (1990) One way hash functions and DES. In: Proceedings of Advances in Cryptology-CRYTPO '89. Springer, pp 428–446
21. Rivest RL (1991) The MD4 message digest algorithm. In: Proceedings of Advances in Cryptology-CRYTPO '90. Springer, pp 303–311
22. Rivest RL (1992) The MD5 message digest algorithm. RFC 1321. https://tools.ietf.org/html/rfc1321
23. National Institute of Standards and Technology, NIST FIPS PUB 186. Digital Signature Standard. US Department of Commerce (May 1994)
24. Diffie W, Hellman ME (1979) Privacy and authentication: an introduction to cryptography. Proc IEEE 67(3):397–427
25. ANSI X3.92, American National Standard for Data Encryption Algorithm (DEA). American National Standards Institute (1981)
26. Rivest RL, Shamir A, Adleman L (1978) A method for obtaining digital signatures and public-key cryptosystems. Comm ACM 21(2):120–126
27. Feistel H (1973) Cryptography and computer privacy. Sci Am 228(5):15–23
28. Feistel H (1974) Block cipher cryptographic system. US Patent 3,798,359
29. Luby M, Rackoff C (1988) How to construct pseudorandom permutations from pseudorandom functions. SIAM J Comput 17(2):373–386
30. National Bureau of Standards (1980) NBS FIPS PUB 81. DES modes of operation. US Department of Commerce
31. Menezes AJ, van Oorschot PC, Vanstone SA (1997) Handbook of applied cryptography. CRC Press, Boca Raton
32. Electronic Frontier Foundation (1998) Cracking DES—Secrets of Encryption Research. Wiretap Politics & Chip Design. Oreilly & Associates Inc. ISBN 1-56592-520-3
33. Andrews GE (1994) Number theory. Dover Publications, New York
34. Le Bellac M (2006) A short introduction to quantum information and quantum computation. Cambridge University Press, Cambridge, New York
35. Hellman ME (2002) An overview of public key cryptography. IEEE Commun Mag 40(5): 42–49

Chapter 4
Physical-Layer Security

Abstract This chapter is devoted to the physical-layer security. The chapter starts with discussion on security issues, followed by the introduction of information-theoretic security, and comparison against the computational security. In the same section, various information-theoretic security measures are introduced, including strong secrecy and weak secrecy conditions. After that, the Wyner's wiretap channel model, also known as the degraded wiretap channel model, is introduced. In the same section, the concept of secrecy capacity is introduced as well as the nested wiretap coding. Further, the broadcast channel with confidential messages is introduced, and the secrecy capacity definition is generalized. The focus is then moved to the secret-key generation (agreement), the source and channel-type models are introduced, and corresponding secret-key generation protocols are described. The next section is devoted to the coding for the physical-layer security systems, including both coding for weak and strong secrecy systems. Regarding the coding for weak secrecy systems, the special attention is devoted to two-edge-type LDPC coding, punctured LDPC coding, and polar codes. Regarding the coding for strong secrecy systems, the focus is on coset coding with dual of LDPC codes and hash functions/extractor-based coding. The attention is then moved to information reconciliation and privacy amplification. In wireless channels physical-layer security (PLS) section, the following topics are covered: MIMO fundamentals, wireless MIMO PLS, and secret-key generation in wireless networks. In the section on optical channels PLS, both PLS for spatial division multiplexing (SDM)-fiber-based systems and free-space optical (FSO) systems are discussed.

4.1 Security Issues

Public-key cryptography has several serious drawbacks such as it is difficult to implement it in devices with low memory and low process constraints, Internet is becoming more and more mobile, security schemes are based on unproven assumptions of intractability of certain functions, and the assumption of limiting computing resources of Eve is not always applicable, to mention few. The open system interconnection (OSI) reference model defines seven layers. However, only five layers,

I. B. Djordjevic, *Physical-Layer Security and Quantum Key Distribution*,
https://doi.org/10.1007/978-3-030-27565-5_4

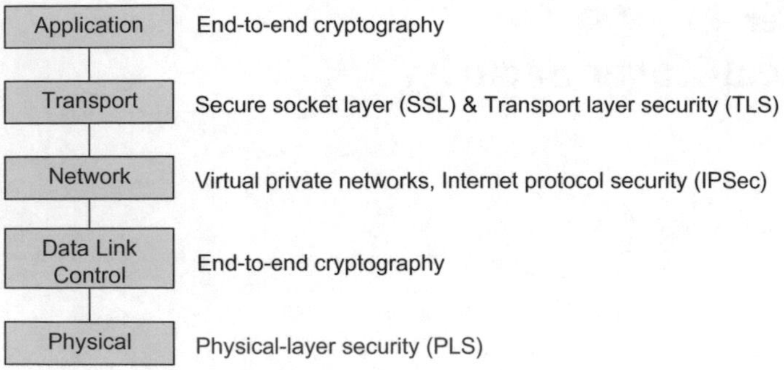

Fig. 4.1 Security mechanisms at different layers in OSI model (only security-relevant layers are shown)

relevant to security issues are provided in Fig. 4.1. The original OSI model does not specify the security issues at all. The security issues are addressed in X.800 standard (security architecture for OSI) [1]. Even though that the physical-layer security (PLS) is not discussed in this standard, the services specified in these five layers can be enhanced by employing the PLS as discussed in [2–4]. The PLS scheme can also be operated independently. The distortions and noise effects introduced by channel can be exploited to reduce the number of bits extracted by Eve. Compared to conventional cryptographic approaches where strong error control coding (ECC) schemes are used to provide reliable communication, the transmission in PLS scenario needs to be simultaneously reliable and secure. This indicates that different classes of ECC must be developed, and these codes will be discussed later in this chapter. Alternatively, similar to QKD, the randomness of the channel can be exploited to generate the key, and this approach is commonly referred to as the *secret-key agreement* [2–5], and this concept is described in Fig. 4.2, inspired by [2, 3, 6].

Alice and Bob monitor Alice–Bob channel capacity (also known as the capacity of the main channel) C_M and the secrecy capacity C_S, defined as a difference between main channel capacity and eavesdropping channel capacity C_E. When the secrecy capacity is sufficiently above the threshold value $C_{S,\mathrm{tsh}}$ and the main channel capacity is well above threshold value $C_{M,\mathrm{tsh}}$, Alice transmits Gaussian-shaped symbols $\boldsymbol{X}$ to Bob. When the secrecy capacity and main channel capacity are both below corresponding thresholds due to deep fading, Alice and Bob perform *information reconciliation* of previously transmitted symbols, which is based strong ECC scheme to ensure that errors introduced by either channel or Eve can be corrected for. Similar to QKD schemes [7–14], a systematic low-density parity-check (LDPC) code can be used (that does not affect information bits but generates the parity-check bit algebraically related to the information bits) to generate the parity bits and transmit them over an authenticated public channel. There exist direct and reverse information reconciliation schemes. In *direct reconciliation*, shown in Fig. 4.2, Alice performs LDPC encoding and sends the parity bits to Bob. Bob performs the LDPC decoding to get the correct key $\boldsymbol{X}$. In *reverse reconciliation*, Bob performs LDPC

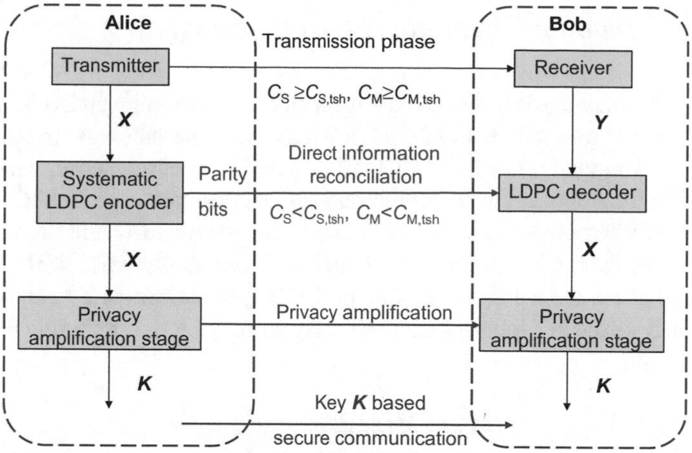

Fig. 4.2 Secret-key generation (agreement) protocol suitable for wireless communications

encoding instead. *Privacy amplification* is then performed between Alice and Bob to distil from $\boldsymbol{X}$ a smaller set of bits $\boldsymbol{K}$ (secure key), whose correlation with Eve's string is below a desired threshold. One way to accomplish privacy amplification is through the use of *universal hash functions* G [9, 12, 14, 15], which map the set of n-bit strings $\boldsymbol{X}$ to the set of m-bit strings $\boldsymbol{K}$ such that for any distinct $\boldsymbol{X}_1$ and $\boldsymbol{X}_2$ from the set of corrected keys, when the mapping g is chosen uniformly at random from G, the probability of having $g(\boldsymbol{X}_1) = g(\boldsymbol{X}_2)$ is very low.

4.2 Information-Theoretic Versus Computational Security

Shannon's model of secrecy system [16] is depicted in Fig. 4.3. The purpose of this system is to reliably to transmit the message $\boldsymbol{M}$, which is at the same time secret for Eve's point of view. To do so, Alice and Bob have access to the random key $\boldsymbol{K}$, which is not known to Eve, and it is used by Alice to encrypt the message into the cryptogram $\boldsymbol{C}$. On the receiver side, Bob decrypts the cryptogram $\boldsymbol{C}$ with the help of key $\boldsymbol{K}$ and determines the transmitted message $\boldsymbol{M}$.

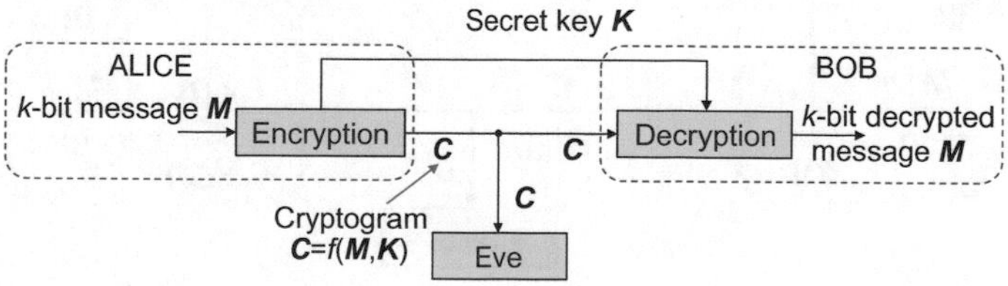

Fig. 4.3 Shannon's model of secrecy system

4.2.1 Information-Theoretic (Perfect) Security

We say that the message is transmitted with *perfect security*, if given Eve's knowledge about the cryptogram it does not help her deciphering the message. In other words, Pr($\boldsymbol{M}$ | Eve's knowledge) = Pr($\boldsymbol{M}$). Clearly, in perfect security sense, the codeword $\boldsymbol{C}$ is statistically independent of the message $\boldsymbol{M}$, and we can write $H(\boldsymbol{M}|\boldsymbol{C}) = H(\boldsymbol{M})$. Equivalently, the mutual information between the codeword $\boldsymbol{C}$ and the message $\boldsymbol{M}$ is zero, that is, $I(\boldsymbol{M};\boldsymbol{C}) = H(\boldsymbol{M}|\boldsymbol{C}) - H(\boldsymbol{M}) = 0$. In this scenario, the best strategy for Eve is to guess the message, and the probability of success is 2^{-k}, and decreases exponentially as the key length k increases. By employing the chain rule from Chap. 2, we can write

$$\begin{aligned} H(\boldsymbol{M}) &= H(\boldsymbol{M}|\boldsymbol{C}) \\ &\le H(\boldsymbol{M},\boldsymbol{K}|\boldsymbol{C}) \\ &= H(\boldsymbol{K}|\boldsymbol{C}) + \underbrace{H(\boldsymbol{M}|\boldsymbol{C},\boldsymbol{K})}_{=0} \\ &= H(\boldsymbol{K}|\boldsymbol{C}) = H(\boldsymbol{K}), \end{aligned} \tag{4.1}$$

where we used the fact that message is completely specified when both the cryptogram and the key are known, that is, $H(\boldsymbol{M}|\boldsymbol{K},\boldsymbol{C}) = 0$. Therefore, the *perfect secrecy condition* is given by

$$H(\boldsymbol{M}) \le H(\boldsymbol{K}). \tag{4.2}$$

In conclusion, the entropy (uncertainty) of the key cannot be lower than the entropy of the message, for an encryption scheme to be perfectly secure.

One-time pad encryption scheme (Vernam cipher) [17, 18], shown in Fig. 4.4, operates by adding mod 2 message $\boldsymbol{M}$ bits and uniform random key bits $\boldsymbol{K}$, and satisfies the perfect security condition [19]. Namely, the key bits randomize the encoding process and ensure that the statistical distribution of cryptogram is independent of the message. On the other hand, the key length is the same as the message length and

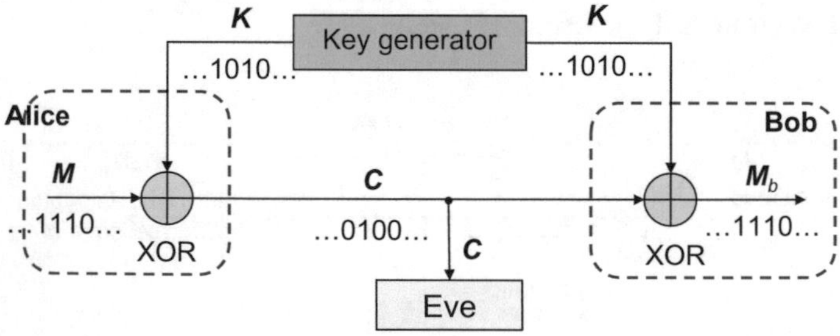

Fig. 4.4 One-time pad scheme

given complete randomness and statistical independence of the cryptogram and the message we have that $H(\boldsymbol{C}) = -\log 2^k = k = H(\boldsymbol{M})$, satisfying the inequality (4.2) with equality sign. Unfortunately, one-time pad scheme has several drawbacks [9, 13]: (i) it requires the secure distribution of the key, (ii) the length of the key must be at least as long as the message, (iii) the key bits cannot be reused, and (iv) the keys must be delivered in advance, securely stored until used, and destroyed after the use.

Given that the condition (4.2) is difficult to satisfy, in conventional cryptography, instead of perfect security, the *computational security* is used [14, 20–24].

4.2.2 Computational Security

The computational security introduces two relaxations with respect to information-theoretic security [21]:

- Security is guaranteed against an efficient eavesdropper running the cryptanalytic attacks for certain limited amount of time. Of course, when eavesdropper has enough computational resources and/or enough time he/she will be able to break the security of the encryption scheme.
- Eavesdroppers can be successful in breaking the security protocols, but with small success probability.

There are two general approaches to precisely define the computational relaxations [21]:

- The *concrete approach*, in which an eavesdropper running the eavesdropping algorithm for time interval no more than T seconds can be successful with probability at most ε. We could call this approach the concept of (T, ε)-security.
- The *asymptotic approach*, in which a probabilistic polynomial-time eavesdropper caring out an eavesdropping strategy (cryptanalytic attack) for every polynomial $p(n)$ there exists an integer N such that when $n > N$ the success attack probability is smaller than $1/p(n)$.

A reader interested to learn more about computational security is referred to as an excellent book due to Katz and Lindell [21].

4.2.3 Information-Theoretic Secrecy Metrics

Before concluding this section, we describe several relevant information-theoretic secrecy metrics [3, 4, 25]. The generic metric $D(\cdot,\cdot)$, measuring the distance between the joint distribution of $\boldsymbol{M}$ and $\boldsymbol{C} = [C_1, C_2, \ldots, C_k]$, denoted as $f_{\boldsymbol{M},\boldsymbol{C}}$, and the product of independent distributions for $\boldsymbol{M}$ and $\boldsymbol{C}$, denoted as $f_{\boldsymbol{M}}$ and $f_{\boldsymbol{C}}$, can be used to define the secrecy requirement as follows:

$$\lim_{k\to\infty} D\big(f_{M,C}, f_M f_C\big) = 0. \tag{4.3}$$

As an illustration, the Kullback–Leibler (KL) distance [26], introduced in Chap. 2, representing the measure of inefficiency of assuming that distribution is $f_M f_C$ when true distribution is $f_{M,C}$, leads to the well-known *mutual information*:

$$D\big(f_{M,C} \| f_M f_C\big) = E_{f_{M,C}} \log\left[\frac{f_{M,C}}{f_M f_C}\right] = I(\boldsymbol{M}, \boldsymbol{C}), \tag{4.4}$$

where E denotes an expectation operator. The L_1-distance between the joint distribution $f_{M,C}$ and the product of independent distribution $f_{M,C}$, known as the *variational distance* [4] can also be used as the secrecy metric:

$$V\big(f_{M,C}, f_M f_C\big) = E_{f_{M,C}}\big|f_{M,C} - f_M f_C\big|. \tag{4.5}$$

Given that mutual information $I(\mathbf{M}, \mathbf{C})$ measures the average amount of information about message $\boldsymbol{M}$ leaked in $\boldsymbol{C}$, the following requirement:

$$\lim_{k\to\infty} I(\boldsymbol{M}, \boldsymbol{C}) = 0 \tag{4.6}$$

is commonly referred to as the *strong secrecy condition*. From the practical point of view, given that the strong secrecy condition is difficult to satisfy, instead of requesting the mutual information to vanish, we can soften the requirement and request that the *rate* of information leaked to Eve tends to zero:

$$\lim_{k\to\infty} \frac{1}{k} I(\boldsymbol{M}, \boldsymbol{C}) = 0. \tag{4.7}$$

This average information rate about the massage $\boldsymbol{M}$ leaked to $\boldsymbol{C}$ is well known as the *weak secrecy condition*. To illustrate the difference, let us consider the following illustrative example, provided in Fig. 4.5. The message $\boldsymbol{M}$ of length k, denoted as $\boldsymbol{M}^k$, is to be encrypted. To encrypt the message, we apply the one-time pad on first $k - l$ bits, while the remaining l bits are unprotected, where l is fixed number. The strong secrecy metric can be expressed as

$$I\big(\boldsymbol{M}, C^k\big) = H(\boldsymbol{M}) - H\big(\boldsymbol{M}|C^k\big) = k - (k - l) = l,$$

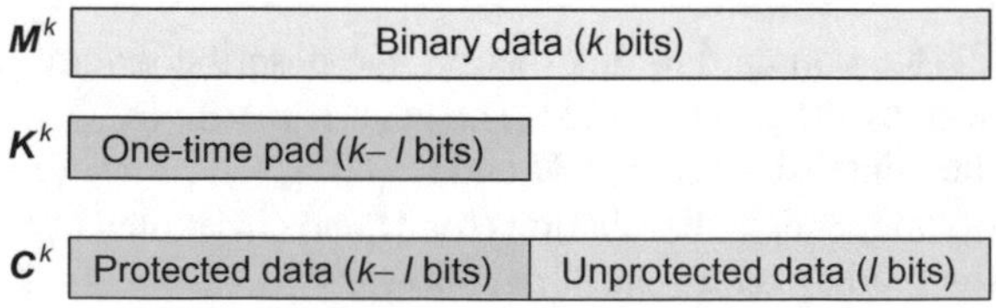

Fig. 4.5 Illustration of the weak secrecy concept

and since the mutual information does not tend to zero as $k \to \infty$, the strong secrecy condition is not satisfied. On the other hand, the weak secrecy metric will be

$$\frac{1}{k}I\left(\boldsymbol{M}, \boldsymbol{C}^k\right) = \frac{1}{k}\left[H(\boldsymbol{M}) - H\left(\boldsymbol{M}|\boldsymbol{C}^k\right)\right] = \frac{k-(k-l)}{k} = \frac{l}{k},$$

and clearly tends to zero as $k \to \infty$, indicating that weak secrecy condition is satisfied.

Form this example, we conclude that some secrecy metrics are mathematically stronger than others. For instance, the mutual information metric is stronger than variational distance secrecy metric, while variational metric is stronger than weak secrecy metric as shown in [27].

4.3 Wyner's Wiretap Channel

Shannon's model is too pessimistic as it assumes that no noise has been introduced during transmission. Wyner introduced the so-called *wiretap channel* [28], now also known as a *degraded wiretap channel model*, in which Eve's channel is degraded version of Alice-to-Bob channel (main channel), as indicated in Fig. 4.6. Alice encodes the message $\boldsymbol{M}$ into a codeword $\boldsymbol{X}^n$ of length n and sends it over the noisy channel, represented by conditional probability density function (PDF) $f(y|x)$ toward Bob. On the other hand, Eve observes the noisy version of the signal available to Bob. Therefore, the wiretap channel is degraded channel represented by the conditional PDF $f(z|y)$. More formally, the discrete memoryless degraded wiretap channel is specified by input alphabet $\mathcal{X}$, two output alphabets $\mathcal{Y}$, $\mathcal{Z}$ and transition probabilities $f(y|x)$ and $f(z|y)$ such that joint distributions for main and wiretap channels are independent, that is, we can write

$$f_{YZ}\left(\boldsymbol{y}^n \boldsymbol{z}^n|\boldsymbol{x}^n\right) = \prod_{m=1}^{n} f(y_m|x_m) \prod_{m=1}^{n} f(z_m|y_m). \tag{4.8}$$

Wyner suggested to use the *equivocation rate*, defined as $(1/n)H(M|\boldsymbol{Z}^n)$, instead of the entropy of the message $H(\boldsymbol{M})$. So the *secrecy condition* in Wyner's sense will be

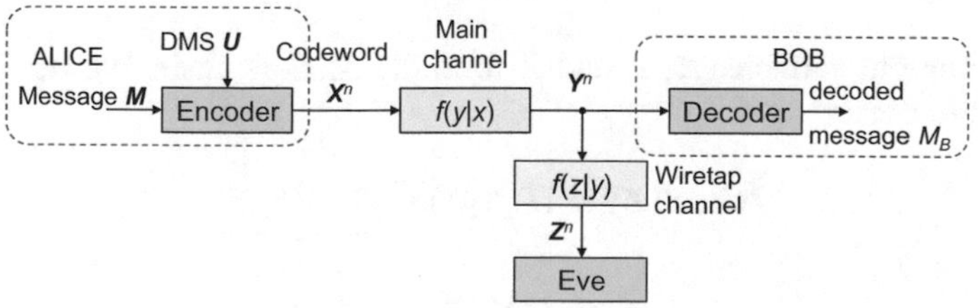

Fig. 4.6 Wyner's wiretap channel model. DMS: discrete memoryless source

$$\frac{1}{n}H(\boldsymbol{M}) - \frac{1}{n}H\big(\boldsymbol{M}|\boldsymbol{Z}^n\big) = \frac{1}{n}I\big(\boldsymbol{M}, \boldsymbol{Z}^n\big) \underset{n\to\infty}{\to} 0, \tag{4.9}$$

which is clearly the weak secrecy condition. In addition to secrecy condition, the *reliability condition* must be satisfied as well:

$$\Pr\big(\boldsymbol{M}_B \neq \boldsymbol{M}|\boldsymbol{Y}^n\big) \underset{n\to\infty}{\to} 0. \tag{4.10}$$

In other words, the probability that Bob's message is different from the message sent by Alice tends to zero as the codeword length $n \to \infty$. The channel codes to be used in this scenario must satisfy both reliability and secrecy conditions and sometimes referred to as the *wiretap codes* [29]. For instance, LDPC, polar, and lattice codes can be used to design the wiretap codes. The (n, k) wiretap code $\mathcal{C}_n$ of rate $R = k/n$ is specified by [3, 29]: (i) the set of messages $\mathcal{M}$ of size 2^{nR}, (ii) the local random source $\mathcal{U}$ with distribution $f_{\mathcal{U}}$, (iii) the encoder performing the mapping $\mathcal{M} \times \mathcal{U} \to \mathcal{X}^n$ (mapping the message and a random realization of the local source into a codeword), and (iv) the decoder performing the following mapping $\mathcal{Y}^n \to \mathcal{M} \cup \{Er\}$ (mapping the received word into a message or an error Er).

The transmission rate R-equivocation rate R_{eq} pair (R, R_{eq}) is *achievable* in Wyner's sense for the sequence of codes $\{\mathcal{C}_n\}$ of code rate R if both reliability and secrecy conditions are satisfied, defined, respectively, as

$$\begin{aligned} &\Pr(M_B \neq M|C_n) = P_e(\mathcal{C}_n) \underset{n\to\infty}{\to} 0, \\ &\underbrace{\lim_{n\to\infty} \frac{1}{n} H\big(M|Z^n\mathcal{C}_n\big)}_{ER(C_n)} = \lim_{n\to\infty} ER(\mathcal{C}_n) \geq R_{eq}, \end{aligned} \tag{4.11}$$

where the first line represents the probability of error when wiretap code $\mathcal{C}_n$ is used, while the second term represents the equivocation rate (ER). The largest transmission rate at which both reliability and secrecy conditions are simultaneously satisfied is commonly referred to as the *secrecy capacity*. For any distribution f_x of $\mathcal{X}$ from set of distributions $\mathcal{P}(R \geq 0)$ for which $I(X, Y) \geq R$ Wyner has defined the merit function, which can be called a *secrecy rate*, as follows:

$$SR(R) = \sup_{f_x \in \mathcal{P}(R)} I(X, Y|Z). \tag{4.12}$$

Since random variables X, Y, and Z form the Markov chain, we can apply the chain rule to obtain

$$\begin{aligned} I(X, Y|Z) &= H(X|Z) - \underbrace{H(X|Y, Z)}_{H(X|Y)} \\ &= I(X, Y) - I(X, Z), \end{aligned} \tag{4.13}$$

so that the alternative expression for $C(R)$ is obtained by

$$SR(R) = \sup_{f_x \in \mathcal{P}(R)} [I(X, Y) - I(X, Z)]. \tag{4.14}$$

Wyner further proved the following *lemma* [28]: The secrecy rate $SR(R)$, $0 \le R \le C_m$(C_m is the capacity of the main channel) satisfies the following properties:

- The supremum in definitions above is in fact maximum, i.e., for each R there exists distribution from $\mathcal{P}(R)$ such that $I(X, Y|Z) = SR(R)$,
- $SR(R)$ is concave and continuous function of R,
- $SR(R)$ is nonincreasing in R, and
- $SR(R)$ is upper bounded by C_m and lower bounded by $C_m - C_e$, where C_e is the capacity of the main wiretap channel cascade, that is,

$$C_m - C_e \le SR(R) \le C_m.$$

Wyner also proved the following *theorem* [28]: The achievable region $\mathcal{R}$ is determined by

$$\mathcal{R} = \left\{\left(R, R_{eq}\right) : 0 \le R \le C_m, 0 \le ER \le H(\boldsymbol{M}), RR_{eq} \le SR(R)H(\boldsymbol{M})\right\}. \tag{4.15}$$

The typical achievable region is provided in Fig. 4.7. As expected from the lemma, the achievable region is not convex and there are two critical corner points: (C_m, $H_m = SR(R)H(\boldsymbol{M})/C_m$) and ($C_M$, $H(\boldsymbol{M})$). The *secrecy capacity* can now be defined as

$$C_S = \max_{(R, H(\boldsymbol{M})) \in \mathcal{R}} R. \tag{4.16}$$

Leung-Yan-Cheong has shown in [30] that the secrecy capacity is equal to the lower bound in lemma, that is, $C_S = C_m - C_e$, when both main and wiretap channels are *weakly symmetric*. The channel ($\mathcal{X}$, $\mathcal{Y}$, $f_{\mathcal{Y}|\mathcal{X}}$) is weakly symmetric when the rows of the channel transition matrix are permutations of each other, while the summations per column are independent of y. The binary symmetric channel (BSC) is a weakly symmetric channel. As an illustration, let us assume that both main and wiretap

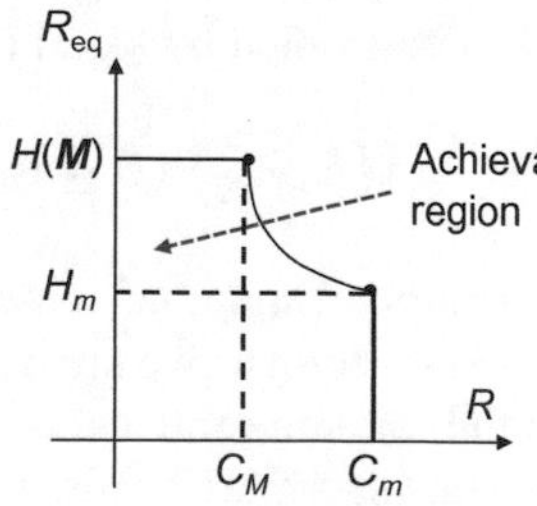

Fig. 4.7 Typical achievable region

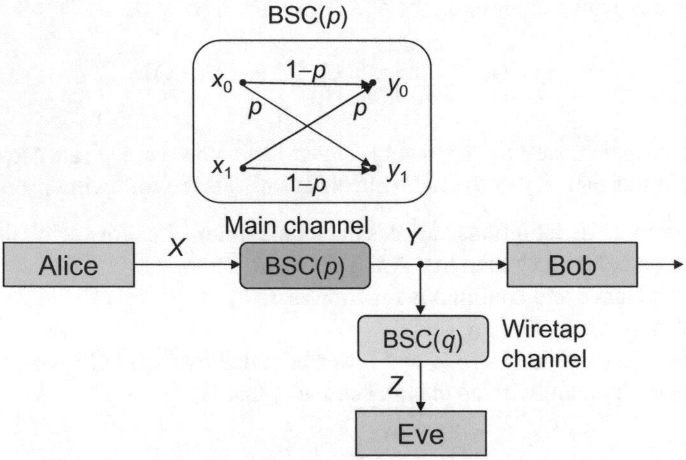

Fig. 4.8 The wiretap channel model when both main and wiretap channels are BSCs

channels can be both modeled as binary symmetric channels, as shown in Fig. 4.8, but with two different error probabilities (p and q). From Chap. 2, we know that the capacity of BSC is given by $C_m = 1 - h(p)$, where $h(p)$ is the binary entropy function $h(p) = -p \log p - (1 - p)\log p$. Given that cascade of two BSC is also a BSC with equivalent error probability of $p + q - 2pq$, the secrecy capacity for this case will be

$$\begin{aligned} C_S = C_m - C_e &= 1 - h(p) - [1 - h(p + q - 2pq)] \\ &= h(p + q - 2pq) - h(p). \end{aligned}$$

If we instead use the *strong secrecy condition*, as it was done in [3], defined as follows,

$$\lim_{n \to \infty} \left[H\left(M | \mathbf{Z}^n \mathcal{C}_n\right) - nR_{eq} \right] \geq 0, \tag{4.17}$$

together with the same reliability condition, the corresponding achievable rate-equivocation region $\mathcal{R}'$ becomes convex and can be defined as a sequence of (n, nR) codes $\{\mathcal{C}_n\}$ satisfying both strong secrecy constraint and reliability constraint and be determined by [3, 31]

$$\mathcal{R}' = \left\{ \left(R, R_{eq}\right) : 0 \leq R_{eq} \leq R \leq I(X, Y), 0 \leq R_{eq} \leq I(X, Y|Z) \right\}. \tag{4.18}$$

The typical shape of achievable region in strong secrecy scenario is provided in Fig. 4.9. As long as we are below the $I(X, Y|Z)$, it is possible to find codes satisfying both full transmission and equivocation rates. However, above $I(X, Y|Z)$ the equivocation rate saturates, while the transmission rate can be increased up to $I(X, Y)$. The

Fig. 4.9 The typical rate-equivocation achievable region in strong secrecy scenario

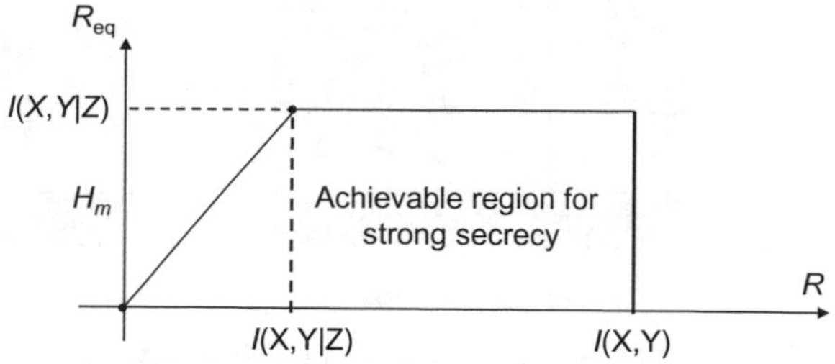

secrecy capacity in strong secrecy scenario can be determined by [3]

$$C_S = \max_{f_x}[I(X, Y) - I(X, Z)]. \tag{4.19}$$

For the corresponding proofs of (4.18) and (4.19), an interested reader can refer to [3].

Before completing this section, we discuss the *wiretap codes* with more details. Let us observe an example of the wiretap channel model, mentioned in Wyner's paper [28], in which the main channel is noiseless, while the wiretap channel is the binary erasure channel (BEC) with erasure probability α (Fig. 4.10). The BEC has two inputs x_0 and x_1, and three outputs y_0, y_1, and Er. The output symbol Er is called the erasure. A fraction α of the incoming bits get erased by the channel. The channel capacity of such channel is $1-\alpha$. The secrecy capacity for this wiretap channel model will be then $C_S = 1-(1-\alpha) = \alpha$. Therefore, out of n transmitted bits, $n\alpha$ bit will be erased. Clearly, since the secrecy capacity is lower than 1, Eve is capable to learn the content of relevant number of bits. To solve this problem, Alice can apply the *stochastic encoding* as follows. Let assume that Alice wants to transmit a binary message $M \in \{0, 1\}$, by encoding the message zero as codeword $\boldsymbol{x}_0^n = (x_{00}, x_{01}, \ldots, x_{0,n-1})$, while encoding the message one to codeword $\boldsymbol{x}_1^n = (x_{10}, x_{11}, \ldots, x_{1,n-1})$. Since these codewords get transmitted over the noiseless channel, Bob will receive the codeword correctly and will be able to conclude which message was transmitted by simply taking the parity check. However, since Eve's channel is BEC and in her received word there will be $n\alpha$ erasures, so she will not be able to determine what

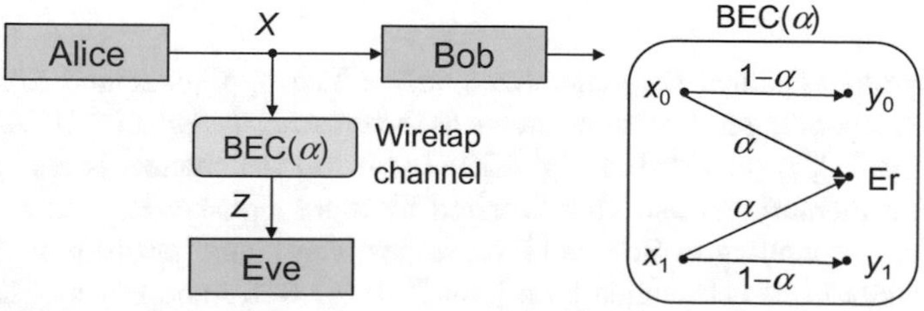

Fig. 4.10 An example of wiretap channel model in which the main channel is noiseless while the wiretap channel is a BEC

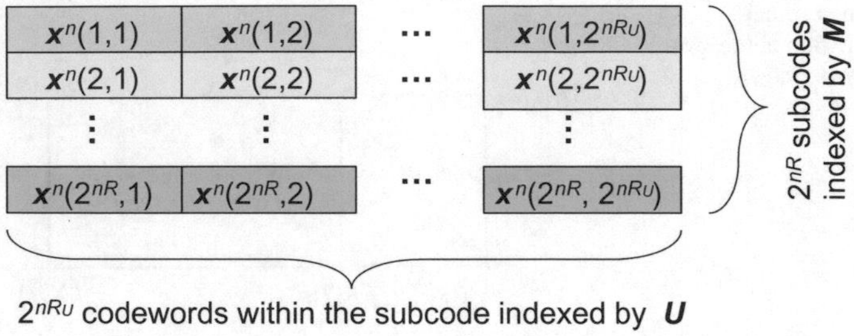

Fig. 4.11 Illustrating the nested structure of a wiretap code

was transmitted by taking the parity check. So the conditional entropy will be lower bounded by

$$H\left(M|Z^n\right) \geq 1 - (1-\alpha)^n,$$

and tends to the entropy of the message 1 as $n \to \infty$ (for sufficiently small erasure probability). On the other hand, the code rate of such stochastic encoding scheme is only $1/n$. However, this concept can be generalized as shown in Fig. 4.11. For each of 2^{nR} possible messages indexed by $\boldsymbol{M}$, there exist 2^{nRU} codewords indexed by a local random generator $\boldsymbol{U}$. The set of codewords corresponding to a message forms a subcode of the wiretap code. Given that the main channel is noiseless, we need to ensure that the total code rate $R + R_U \leq 1$.

4.4 Broadcast Channel with Confidential Messages and Wireless Channel Secrecy Capacity

We first define the broadcast channel with confidential messages.

4.4.1 Broadcast Channel with Confidential Messages

Wyner's wiretap channel get generalized and refined by Csiszár and Körner [31], and the corresponding model, now known as the *broadcast channel with confidential messages* (BCC), is provided in Fig. 4.12. The broadcast channel is assumed to be discrete and memoryless and characterized by input alphabet $\mathcal{X}$, output alphabets $\mathcal{Y}$ and $\mathcal{Z}$ (corresponding to Bob and Eve, respectively), and transition PDF $f(yz|x)$. So the channel itself is modeled by a joint PDF for Bob's and Eve's observations, $f(\mathbf{yz}|\mathbf{x})$, given the channel input. Since the BCC is memoryless channel, the PDF can be decomposed as

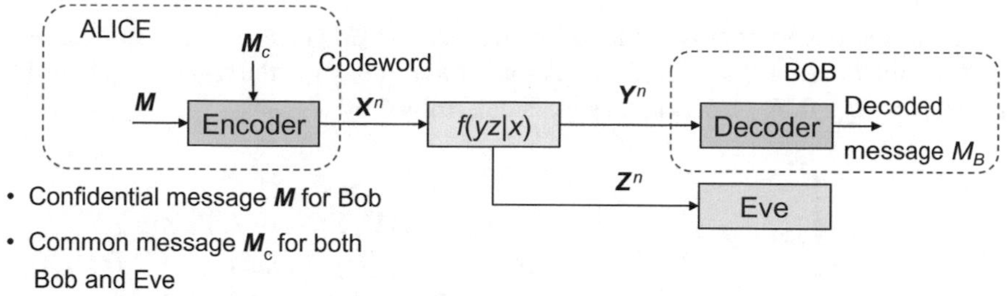

Fig. 4.12 The broadcast channel model with confidential messages (BCC)

$$f\left(\boldsymbol{y}^n \boldsymbol{z}^n|\boldsymbol{x}^n\right) = \prod_{m=1}^{n} f_{YZ|X}(y_m z_m|x_m). \tag{4.20}$$

In this scenario, Alice wishes to broadcast a common message $\boldsymbol{M}_c$ to both Bob and Eve and a confidential message $\boldsymbol{M}$ to Bob. The corresponding stochastic code $\mathcal{C}_n$ of codeword length n is composed of

- Two message sets: the common messages set $\mathcal{M}_c$ of size 2^{nRc} and the confidential message set $\mathcal{M}$ of size 2^{nR}.
- The encoding (stochastic) function e_n: $\mathcal{M}_c \times \mathcal{M} \to \mathcal{X}^n$, which maps the message pair $(\boldsymbol{m}_c, \boldsymbol{m}) \in \mathcal{M}_c \times \mathcal{M}$ into a codeword $\boldsymbol{x}^n \in \mathcal{X}^n$.
- The decoding functions d_n: $\mathcal{Y}^n \to \mathcal{M}_c \times \mathcal{M}$, $d_{E,n} : \mathcal{Z}^n \to \mathcal{M}_c$, with the first one mapping the observation vector $\boldsymbol{y}^n$ to a message pair $\left(\hat{\boldsymbol{m}}_c, \hat{\boldsymbol{m}}\right)$, while the second one the observation z^n to common message estimate $\hat{\boldsymbol{m}}_c$.

Similar to Wyner's approach, the confidential message $\boldsymbol{M}$ secrecy with respect to Eve is measured in terms of the *equivocation rate* as follows:

$$ER(\mathcal{C}_n) = \frac{1}{n} H\left(\boldsymbol{M}|\boldsymbol{Z}^n \mathcal{C}_n\right). \tag{4.21}$$

The probability of error is defined in similar fashion as for Wyner's approach:

$$P_e(\mathcal{C}_n) = \Pr\left(d_n\left(\boldsymbol{Y}^n\right) = \left(\hat{\boldsymbol{M}}_c, \hat{\boldsymbol{M}}\right) \neq (\boldsymbol{M}_c, \boldsymbol{M}) \,\text{or}\, d_{E,n}\left(\boldsymbol{Y}^n\right) = \hat{\boldsymbol{M}}_c \neq \boldsymbol{M}_c \,|\mathcal{C}_n\right). \tag{4.22}$$

We further request that the (stochastic) code $\mathcal{C}_n$ must satisfy both *reliability* and *secrecy constraints*, specified, respectively, as

$$\lim_{n\to\infty} P_e(\mathcal{C}_n) = 0, \lim_{n\to\infty} ER(\mathcal{C}_n) \geq R_{eq}. \tag{4.23}$$

The rate tuple (R_c, R, R_{eq}) for the BCC is *achievable* if and if there exists a (stochastic) wiretap code $\mathcal{C}_n$ of length n satisfying simultaneously both reliability and security constraints, defined above.

Csiszár and Körner proved the following *theorem* [31] (see also [3]): For the joint distribution f_{UVX} on $U \times V \times X$ that factorizes as $f_U f_{U|V} f_{X|V}$, the region of achievable rate tuples (R_c, R, R_{eq}), denoted as $\mathfrak{R}$, is determined by

$$\mathfrak{R} = \bigcup_{f_U f_{V|U} f_{X|V}} \left\{ \left(R_c, R, R_{eq}\right) : \begin{array}{c} 0 \le R_{eq} \le R \\ R_{eq} \le I(V, Y|U) - I(V, Z|U) \\ R + R_c \le I(V, Y|U) + \min[I(U, Y), I(U, Z)] \\ 0 \le R_c \le \min[I(U, Y), I(U, Z)] \end{array} \right. . \tag{4.24}$$

U and V are auxiliary random variables so that random variables U, V, X, and YZ are for the Markov chain $U \rightarrow V \rightarrow X \rightarrow YZ$. Finally, the cardinalities of U and V maybe limited to $|\mathcal{U}| \le |\mathcal{X}| + 3$ and $|\mathcal{V}| \le |\mathcal{X}|^2 + 4|\mathcal{X}| + 3$.

Csiszár and Körner provided the following *corollary* [31] (see also [3]): *Secrecy capacity* is determined as the difference of mutual information for Alice–Bob and Alice–Eve links, when the rate of the common message is set to zero, that is,

$$C_s = \max_{\substack{f_{VX} \\ V \rightarrow X \rightarrow YZ}} [I(V, Y) - I(V, Z)], \tag{4.25}$$

where the maximization is performed over all possible joint distributions $f_{VX}(v, x)$ and V, X, and YZ form a Markov chain $V \rightarrow X \rightarrow YZ$.

Clearly, the secrecy capacity is strictly positive when Bob's channel is less noisy than Eve's channel, i.e., $I(X; Y) > I(X; Z)$. Namely, by setting $V = X$, the secrecy capacity expression becomes $C_s = \max_{f_X}[I(X, Y) - I(X, Z)]$, which is clearly strictly positive when $I(X; Y) > I(X; Z)$.

Similarly, as in Wyner's wiretap channel case, the secrecy metric can be replaced with strong secrecy metric.

4.4.2 Wireless Channel Secrecy Capacity

Wireless communication scenario is illustrated in Fig. 4.13. Due to fading effects [32–34], the signal-to-noise ratios (SNRs) in both main and Eve's channels are time-varying. For certain time periods, the SNR in the main channel is higher than that in Eve's channel and vice versa. Moreover, if Alice and Bob have access to Eve's channel state information (CSI), they can exploit this knowledge to get an advantage over Eve.

As indicated in [35, 36], the general goal is to maximize the transmission rate between Alice and Bob $R = H(\boldsymbol{M}^k)/n$, while minimizing Eve's equivocation rate $ER = H(\boldsymbol{M}^k|\boldsymbol{Z}^n)/H(\boldsymbol{M}^k)$, wherein the error probability tends to zero. Secrecy capacity is defined as the maximum transmission rate R for which Eve's equivocation rate is equal to 1. The generic fading wireless BCC model is provided in Fig. 4.14.

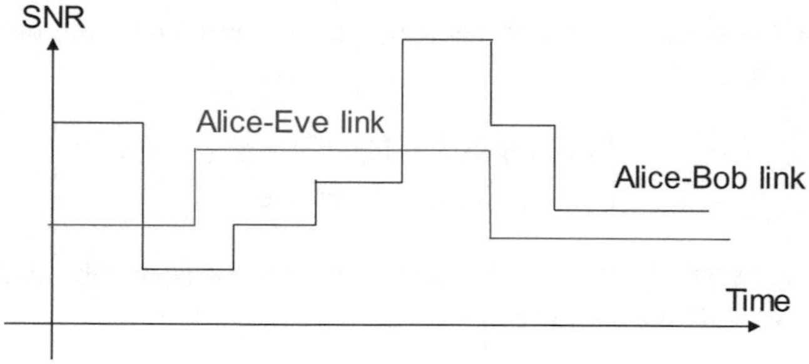

Fig. 4.13 Illustration of the wireless communication scenario

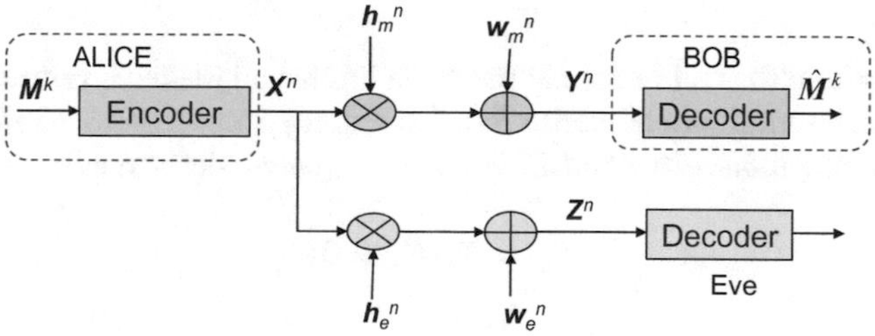

Fig. 4.14 The generic wireless BCC

Clearly, both main and Eve's channels are simultaneously additive and multiplicative. The additive noise is a zero-mean Gaussian, with variance in main (Alice–Bob) channel being σ_m^2 and variance in Eve's channel being σ_e^2.

The multiplicative components are described by (i) main channel coefficients $\boldsymbol{h}_m^n = [h_m[1], \ldots, h_m[n]]$ and (ii) Eve's channel coefficients $\boldsymbol{h}_e^n = [h_e[1], \ldots, h_e[n]]$. For *ergodic fading* channel model, channel coefficients change every channel use, that is, $h_m[i]$ and $h_e[i]$ ($i = 1, \ldots, n$) are mutually independent. For *quasi-static* channel model, employed in the rest of this section, the channel coefficients do not change for duration of whole codeword, that is, $h_m = h_m[i]$, $h_e = h_e[i]$. The *block fading* channel model describes the scenario between the two cases just described, in which the channel coefficients do not change for certain number of symbols, but still change many times during the codeword transmission. For additional details on wireless channel models, an interested reader is referred to [32–34]. The SNRs for the main channel, denoted as ρ_m, and Eve's channel, denoted as ρ_e, assuming quasi-static fading channel are simply

$$\rho_m = \underbrace{|h_m|^2}_{g_m} P/\sigma_m^2 = g_m P/\sigma_m^2, \quad \rho_e = \underbrace{|h_e|^2}_{g_e} P/\sigma_e^2 = g_e P/\sigma_e^2, \tag{4.26}$$

wherein P is the transmit power, g_m is the gain coefficient in the main channel ($g_m = |h_m|^2$), and g_e is the gain coefficient in Eve's channel ($g_e = |h_e|^2$). Based on discussion related to (4.25), the instantaneous secrecy capacity for quasi-static fading channel is determined by

$$C_S = \begin{cases} \log(1+\rho_m) - \log(1+\rho_e), \rho_m > \rho_e \\ 0, \quad \text{otherwise} \end{cases} \tag{4.27}$$

The *average secrecy capacity* for quasi-static channel can be obtained by averaging out the gain coefficients as follows:

$$\bar{C}_S = E_{g_m,g_e}\{C_s(g_m,g_e)\},\ C_s(g_m,g_e) = \left[\log\left(1+\frac{g_m P}{\sigma_m^2}\right) - \log\left(1+\frac{g_e P}{\sigma_e^2}\right)\right]^+, \tag{4.28}$$

where $[x]^+ = \max(0, x)$. For formal proof, an interested reader is referred to [35].

Another relevant figure of merit related to secrecy, important in wireless communications, is the *probability of positive secrecy capacity* defined as

$$P_S^+ = \Pr(C_S > 0). \tag{4.29}$$

4.5 Secret-Key Generation (Agreement) Protocols

Alice and Bob can employ the noisy channel to generate correlated random sequences and subsequently use a public authenticated error-free feedback channel to agree on a secret key [37–40], and this approach is commonly referred to as the *secret-key agreement* [2–5, 41]. Two types of models are typically considered for secret-key agreement [38]:

- *Source-type model*, in which terminals observe the correlated output of the source of randomness without having control of it.
- *Channel-type model*, in which one terminal transmits random symbols to other terminals using a broadcast channel. This scenario is similar to the wiretap channel model with feedback channel, which is an authenticated public noiseless channel.

Both of these models are very similar to QKD [7–14], except that raw key is transmitted over classical channel, while in QKD over the quantum channel. The *source-type scenario* is illustrated in Fig. 4.15. Even if the feedback message $\boldsymbol{F}$ is available to Eve, it is still possible to generate secure key such that $H(\boldsymbol{K}|\boldsymbol{Z}^n\boldsymbol{F})$ is arbitrary close to $H(\boldsymbol{K})$. In this specific example, Alice, Bob, and Eve get correlated noisy observations, $\boldsymbol{X}^n$, $\boldsymbol{Y}^n$, and $\boldsymbol{Z}^n$, respectively. Alice and Bob are able to generate the common key $\boldsymbol{K}$ based on their observations and a set of feedback messages $\boldsymbol{F}$. Even if Eve's channel is better than Bob's channel, it is possible to achieve positive secrecy capacity, as explained by Maurer [37]. The corresponding procedure that

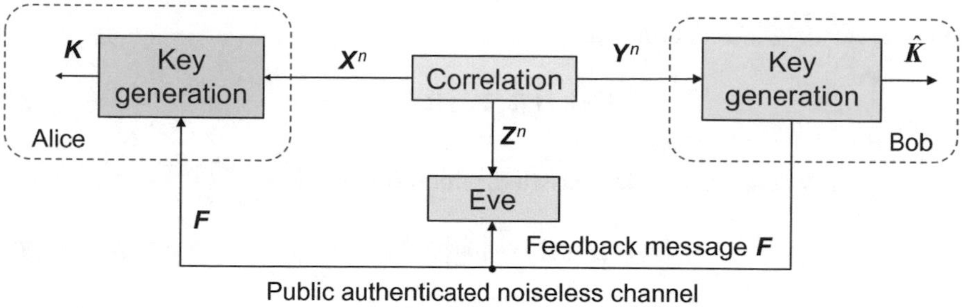

Fig. 4.15 The generic source model secret-key generation (agreement) scenario

Alice and Bob need to follow to determine the secret key can be called secret-key agreement (generation) protocol. The rate at which secret key is generated can be called the same way as in QKD, the *secret-key rate* (SKR). If the protocols exploit the public messages sent in one direction only (from either Alice-to-Bob or Bob-to-Alice), the corresponding secret-key rate (SKR) is said to be achievable with *one-way communication*; otherwise, the SKR is said to be achievable with *two-way communication*.

The generic *channel-type scenario* is described in Fig. 4.16, which is clearly the generalization (extension) of the Csiszár–Körner wiretap model. The channel is characterized by conditional PDF $f(yz|x)$, and its input is controlled by Alice by generating n symbols $\boldsymbol{X}^n$. Bob and Eve observe the channel outputs and get $\boldsymbol{Y}^n$ and $\boldsymbol{Z}^n$ realizations. Additionally, Alice and Bob communicate over two-way authenticated noiseless channel of unlimited capacity. By exchanging the random symbols over the channel and feedback symbols $\boldsymbol{F}$ over the public channel, Alice and Bob are able to generate the secret key not known to Eve. To get the secret key, Alice and Bob apply a set of steps that can also be called the secret-key generation protocol.

This protocol should be properly designed such that in addition to *reliability condition*, defined as

$$P_e = \Pr\left(\boldsymbol{K} \neq \hat{\boldsymbol{K}}\right) \underset{n\to\infty}{\rightarrow} 0, \tag{4.30}$$

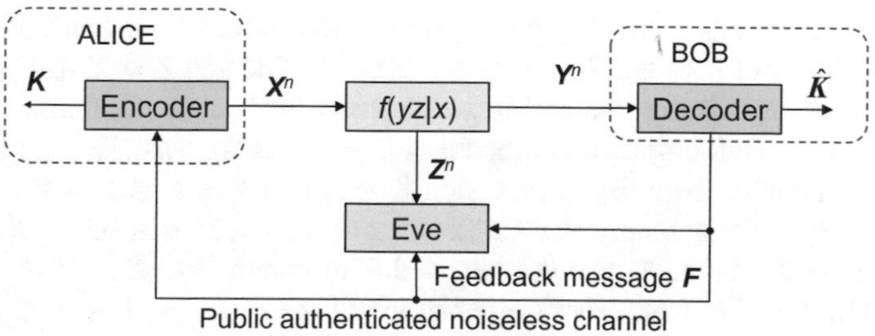

Fig. 4.16 The generic channel model secret-key generation (agreement) scenario

and *secrecy condition*, defined as

$$L = I(\boldsymbol{K}, \boldsymbol{Z}^n \boldsymbol{F}) \underset{n\to\infty}{\to} 0, \tag{4.31}$$

the *uniformity condition* is also satisfied, which is defined as

$$U = \left|H(\boldsymbol{K}) - \log\lceil 2^{nR}\rceil\right| \underset{n\to\infty}{\to} 0, \tag{4.32}$$

where R is the SKR. As before, the reliability condition tells us that Alice and Bob must agree on secret key with the high probability. The security condition is important to ensure that the key Alice and Bob finally agreed is indeed secret with respect to Eve. Finally, the uniformity condition ensures that the secret key is uniformly distributed within the corresponding keyspace. The corresponding parameters defined above are known as *probability of error* (P_e), the *leakage of information* (L), and *uniformity of the keys* (U), respectively. We say that the secret-key rate R is *achievable* if there exist a sequence of secret-key generation protocols satisfying all three constraints as $n \to \infty$. The supremum of achievable SKRs is commonly referred to as the *secret-key capacity*, denoted here as C_{SK}. Given two-way communication over the authenticated public channel, it is difficult to derive and exact expression for C_{SK}; however, based on [37, 38], it can be bounded from both sides as follows:

$$\max_{f_X} \max[I(X,Y) - I(X,Z), I(Y,X) - I(Y,Z)] \le C_{SK} \le \max_{f_X}[I(X,Y|Z)]. \tag{4.33}$$

The upper bound term indicates the secret-key capacity when Bob has access to Eve's observations. The lower bound term max[$I(X, Y) - I(X, Z)$] indicates that direct reconciliation is employed, while the lower bound term max[$I(Y, X) - I(Y, Z)$] indicates that reverse reconciliation is employed instead.

Let us now provide the following *example* due to Maurer [37] to illustrate how the noiseless feedback channel can help to get positive secrecy capacity even when Eve's channel is less noisy. We assume that both main and Eve's channels are BSCs with bit error probabilities p_m and p_e, respectively. Alice transmits bit X over BSC(p_m) to Bob, and this transmission was received by Eve over BSC(p_e). The errors introduced by channels can be represented by corresponding Bernoulli random variables E_m and E_e, added mod 2 to X to get $Y = X \oplus E_m$ at Bob's side and $Z = X \oplus E_e$ on Eve's side. Let us assume Bob's channel is noisier than Eve's channel, that is, $\Pr(E_m = 1) > \Pr(E_e = 1)$. Bob employs the feedback channel as follows. To send a bit V to Alice, Bob first adds the noisy observation Y, to obtain $F = V \oplus Y = V \oplus X \oplus E_m$ and sends it over the authenticated public channel as a feedback bit F. Since Alice already knows X, she adds it to the received F to obtain $F \oplus X = V \oplus Y \oplus X = V \oplus E_m$. On the other hand, Eve's estimation will be $F \oplus Z = V \oplus Y \oplus Z = V \oplus X \oplus E_m \oplus Z = V \oplus E_m \oplus E_e$, which is more noisy than Bob's channel given that equivalently V was sent over cascade of main BSC and Eve's BSC, as illustrated in Fig. 4.17. In the absence of feedback, the schemes become the wiretap channel

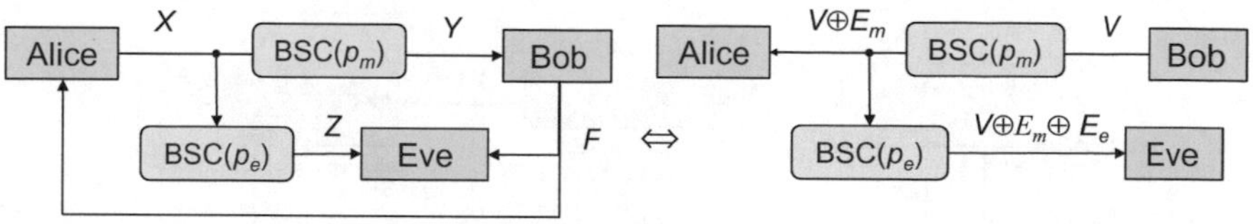

Fig. 4.17 The illustration of the channel-type secret-key generation scheme (on the left) and an equivalent wiretap channel model (on the right)

model, and we have already shown that secrecy capacity of such channel will be

$$C_S = [C_m - C_e]^+ = [1 - h(p_m) - (1 - h(p_e))]^+ = [h(p_e) - h(p_m)]^+. \quad (4.34)$$

So if $p_m > p_e$, clearly, the secrecy capacity will be zero. On the other hand, when feedback is used, based on Fig. 4.17(right), we conclude that corresponding secrecy capacity will be

$$\begin{aligned} C_S^{(F)} &= \left[C_m^{(F)} - C_e^{(F)}\right]^+ = [1 - h(p_m) - (1 - h(p_m + p_e - 2p_m p_e))]^+ \\ &= [h(p_m + p_e - 2p_m p_e) - h(p_m)]^+, \end{aligned} \quad (4.35)$$

which is strictly positive unless $p_m = 1/2, p_e = 0$.

To distil for a shorter shared secret key, Bennet [15] (see also [5, 9–14]) proposed the following three-phase approach:

- *Advantage distillation* phase, in which Alice and Bob out of a set of correlated observations, select the positions over which they have an advantage over Eve. In QKD, this step is also known as a sifting procedure.
- *Information reconciliation* phase, in which these "advantage" observations are further processed to reconcile the discrepancies by employing the error correction.
- *Privacy amplification* phase, in which Alice and Bob perform further processing of corrected key to remove redundancy and get a uniformly distributed key sequence with no leakage to Eve, and this is done with the help of hash functions, as discussed in introductory part of this chapter.

We can extend these three phases with the zeroth phase representing the *raw key transmission*, which in the source-type model can be called the *randomness sharing* [3]. Now, we provide more details on both source-type and channel-type secret-key agreement scenarios.

4.5.1 Source-Type Secret-Key Generation

The *source model for secret-key agreement scenario* is provided in Fig. 4.18. Here, we use the interpretation due to Bloch and Barros [3], which is slightly different than

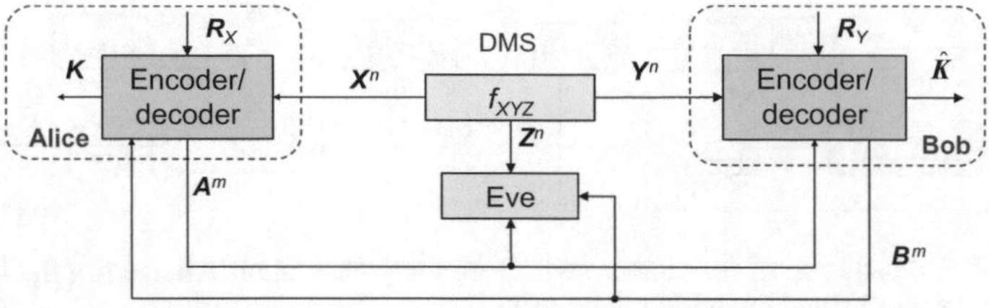

Fig. 4.18 The source model for secret-key generation (agreement) scenario

originally proposed in [38], but it has some similarities with [15]. This protocol is very similar to the *entanglement-assisted QKD* [12–14] to be described in Chaps. 6 and 7. The main point of this scenario is discrete memoryless source (DMS), not controlled by involved parties, described by joint PDF f_{XYZ}, with three output ports $\mathfrak{X}$, $\mathfrak{Y}$, $\mathfrak{Z}$ corresponding to Alice, Bob, and Eve, respectively. In other words, Alice, Bob, and Eve have access to correlated *n*-component realizations $\boldsymbol{X}^n$, $\boldsymbol{Y}^n$, and $\boldsymbol{Z}^n$, governed by the joint PDF. Additionally, Alice and Bob communicate over two-way public authenticated noiseless channel of unlimited capacity. This side channel does not contribute to the secret key but helps Alice and Bob to distil for the shorter secure key. As mentioned above, this feedback channel is authenticated to prevent Eve tampering with the key. Finally, Alice and Bob employ the local sources of randomness, denoted as $\mathfrak{R}_x$ and $\mathfrak{R}_y$, respectively; to randomize the message they transmit. These sources of randomness are described by corresponding PDFs f_{Rx} and f_{Ry}. As already indicated above, the set of steps Alice and Bob need to follow constitutes the *secret-key generation protocol*, which can also be called the key-distillation protocol (strategy) [3].

The generic *key-generation (key-distillation) protocol* P_n for *DMS*, denoted as $(\mathfrak{X}, \mathfrak{Y}, \mathfrak{Z}, f_{XYZ})$, of secret-key rate R and codeword length n, with m rounds over the authenticated public channel, employs.

The following *alphabets*:

- A key alphabet K of cardinality 2^{nR};
- Alice alphabet A used to communicate over the authenticated public channel; and
- Bob alphabet B used to communicate over the authenticated public channel.

The following sources of the local randomness:

- A source of local randomness for Alice $(\mathfrak{R}_X, f_{RX})$;
- A source of local randomness for Bob $(\mathfrak{R}_Y, f_{RY})$.

The following *encoding functions*:

- The m Alice encoding functions $e_i^{(A)}$:

$$e_i^{(A)} : \boldsymbol{X}^n \times \boldsymbol{B}^{i-1} \times \boldsymbol{R}_X \to \boldsymbol{A},\ i \in \{1, 2, \ldots, m\}; \tag{4.36}$$

- The m Bob encoding functions $e_i^{(B)}$:

$$e_i^{(B)} : \boldsymbol{Y}^n \times \boldsymbol{A}^{i-1} \times \boldsymbol{R}_Y \rightarrow \boldsymbol{B},\ i \in \{1, 2, \ldots, m\}. \tag{4.37}$$

The following *key-distillation functions*:

- Alice key-distillation function k_A, defined as the following mapping:

$$k_A : \boldsymbol{X}^n \times \boldsymbol{B}^m \times \boldsymbol{R}_X \rightarrow \boldsymbol{K}; \tag{4.38}$$

- Bob key-distillation function k_B, defined as the following mapping:

$$k_B : \boldsymbol{Y}^n \times \boldsymbol{A}^m \times \boldsymbol{R}_Y \rightarrow \boldsymbol{K}. \tag{4.39}$$

The key-distillation protocol P_n is composed of the following *steps*:

- Alice observes n realizations of the source $\boldsymbol{x}^n$, while Bob observes $\boldsymbol{y}^n$;
- Alice generates local random realization r_x, while Bob generates r_y;
- In round $i \in [1, m]$, Alice transmits a_i and Bob b_i by

$$a_i = e_i^{(A)}\left(\boldsymbol{x}^n, \boldsymbol{b}^{i-1}, r_x\right),\ b_i = e_i^{(B)}\left(\boldsymbol{y}^n, \boldsymbol{a}^{i-1}, r_y\right); \tag{4.40}$$

- After round m Alice computes key k_A and Bob key k_B by

$$\boldsymbol{k} = k_B\left(\boldsymbol{x}^n, \boldsymbol{b}^m, r_x\right),\ \hat{\boldsymbol{k}} = k_B\left(\boldsymbol{y}^n, \boldsymbol{a}^m, r_y\right). \tag{4.41}$$

- Alice and Bob perform information reconciliation, by employing the error correction, to correct for the discrepancies in symbols at different positions on the raw keys.
- Finally, Alice and Bob perform privacy amplification to remove any correlation with Eve.

At the end of protocol P_n Alice and Bob end with the common key $\boldsymbol{k}_c = \hat{\boldsymbol{k}}_c$, which is secret with respect to Eve.

This source-type key-generation protocol P_n is evaluated in terms of

- The *average probability of error*:

$$P_e(P_n) = \Pr\left(\boldsymbol{K} \neq \hat{\boldsymbol{K}} | P_n\right); \tag{4.42}$$

- The *information leakage to Eve*:

$$L(P_n) \triangleq I\left(\boldsymbol{K}; \boldsymbol{Z}^n \boldsymbol{A}^r \boldsymbol{B}^r | P_n\right); \tag{4.43}$$

- The *uniformity of the key*:

$$U(P_n) \triangleq \log\lceil 2^{nR}\rceil - H(\boldsymbol{K}|P_n). \tag{4.44}$$

Similarly as before, we can define both strong and weak SKRs. A *strong secret-key rate R* is achievable for a source/channel model if there exist a sequence of key-distillation protocols $\{P_n\}$ of length $n = 1, 2, \ldots$ such that the reliability, secrecy, and uniformity conditions for $\{P_n\}$ are simultaneously satisfied; in other words, the following constraints are valid:

$$P_e(P_n) \underset{n\to\infty}{\to} 0, \; L(P_n) \underset{n\to\infty}{\to} 0, \; U(P_n) \underset{n\to\infty}{\to} 0. \tag{4.45}$$

On the other hand, a *weak secret-key rate R* is achievable for a source/channel model if there exist a sequence of key-distillation protocols $\{P_n\}$ of length $n = 1, 2, \ldots$ such that the reliability constraint, *weak secrecy* constraint, and *weak uniformity constraint* are simultaneously satisfied. The reliability constraint is the same as in strong SKR case, while weak secrecy and uniformity constraints are defined, respectively, as follows:

$$\frac{L(P_n)}{n} \underset{n\to\infty}{\to} 0, \; \frac{U(P_n)}{n} \underset{n\to\infty}{\to} 0. \tag{4.46}$$

The corresponding keys are weak secret keys. Given that we introduced both weak and strong secrecy constraints, we can also define the weak and strong secret-key capacity. The *weak secret-key capacity* of a source-type model with joint PDF f_{XYZ} is defined as

$$C_{SK}^{(\text{weak})} \triangleq \sup\{R : \; R \text{ is an achievable weak SKR}\}. \tag{4.47}$$

On the other hand, the *strong secret-key capacity* of a source-type model with joint PDF f_{XYZ} is defined as

$$C_{SK}^{(\text{strong})} \triangleq \sup\{R : \; R \text{ is an achievable strong SKR}\}. \tag{4.48}$$

The following *theorem* has been proved *by Maurer, Ahlswede, and Csiszár* [37, 38] (see also [3]): The weak secret-key capacity of a source-type model (XYZ, f_{XYZ}) satisfies the following lower and upper bounds:

$$\max(I(X;Y) - I(X;Z), I(Y;X) - I(Y;Z)) \le C_{SK}^{(\text{weak})} \le \min(I(X;Y), I(X,Y|Z)). \tag{4.49}$$

The lower bound can also be rewritten as $I(X;Y) - \min(I(X;Z), I(Y;Z))$. Moreover, for the one-way communication, the following SKR is achievable [37, 38]:

$$C_{SK}^{\text{(one-way communication)}} = \max(I(X;Y) - I(X;Z), I(Y;X) - I(Y;Z)). \quad (4.50)$$

Proof As discussed above, the lower bound corresponds to the Csiszár–Körner wire-tap channel model with a feedback. We have also shown in example related to the Fig. 4.17 that this channel model can be equivalently represented as a Wyner's channel model, by following the similar procedure explained in text related to Fig. 4.17, which starts with Alice observing the channel realization x and add uniformly generated integer u mod-$|\mathcal{X}|$. We know that the secrecy capacity of Wyner's channel model is higher than $I(X, Y) - I(X, Z)$, that is, we can write

$$C_{SK}^{\text{(weak)}} \geq I(X;Y) - I(X;Z). \quad (4.51)$$

Since the communication over authenticated public channel is two-way, the roles of Alice and Bob can be reversed by applying the reverse reconciliation so that the secret-key capacity is now lower limited by

$$C_{SK}^{\text{(weak)}} \geq I(Y;X) - I(Y;Z). \quad (4.52)$$

By combining these two inequalities, we obtain the lower bound of secret-key capacity as follows:

$$C_{SK}^{\text{(weak)}} \geq \max(I(X;Y) - I(X;Z), I(Y;X) - I(Y;Z)). \quad (4.53)$$

To determine the upper bound, we observe the scenario in which Bob has access to Eve's observations and can create new random variable $Y' = YZ$. Since X, Y', and Y form the Markov chain, we can apply the chain rule to obtain

$$\begin{aligned} I(X,Y|Z) &= \underbrace{I(X,YZ)}_{I(X,Y)+I(X,Z|Y)} - I(X,Z) = I(X,Y) + \underbrace{I(X,Z|Y)}_{0} - I(X,Z) \\ &= I(X,Y) - I(X,Z), \end{aligned} \quad (4.54)$$

which is clearly the upper bound of secrecy capacity. Moreover, from this equation, we conclude that in this scenario $I(X,Y|Z) \leq I(X,Y)$. However, since the participants do not have control of DMS, there are realizations for which $I(X, Y) > I(X, Y|Z)$. Let us take the following example due to Maurer [37]. When X, Y are binary independent with $\Pr(X = 0) = \Pr(Y = 0) = 1/2$ and Z is formed by $Z = X \oplus Y$, we have that $I(X, Y) = 0$ and $I(X, Y|Z) = H(Z) = 1$. So to account for both scenarios, our upper bound becomes

$$C_{SK}^{\text{(weak)}} \leq \min(I(X,Y), I(X,Y|Z)). \quad (4.55)$$

4.5.2 Channel-Type Secret-Key Generation

The *channel model for secret-key agreement scenario* is provided in Fig. 4.19. Here, we again use the interpretation due to Bloch and Barros [3] for consistency. In this scenario, Alice provides the input X to the discrete memoryless channel (DMC) described by the conditional PDF $f_{YZ|X}$. Bob and Eve do not have any control of the channel outputs; they can just observe random realizations at corresponding output ports. We again assume that Alice and Bob can communicate over authenticated public noiseless channel to which Eve can have access. As before, we assume that Alice and Bob employ the local sources of randomness, denoted as $\mathcal{R}_x$ and $\mathcal{R}_y$, respectively, to randomize their communication.

In analogy with the key-generation protocols for DMS, we can define the corresponding generic protocol for DMC, which is briefly described in the text related to Fig. 4.2. This scenario is very similar to *weak coherent state-based QKD* [12–14] to be described in Chaps. 6 and 7. The generic *key-generation (key-distillation) protocol P_n* for *DMC*, described by the conditional PDF $f_{YZ|X}$, of secret-key rate R and codeword length n, with m rounds over the authenticated public noiseless channel and n transmissions over DMC, employs

The following *alphabets*:

- A key alphabet K of cardinality 2^{nR};
- Alice alphabet A used to communicate over the authenticated public channel; and
- Bob alphabet B used to communicate over the authenticated public channel.

The following sources of the local randomness:

- A source of local randomness for Alice $(\mathcal{R}_X, f_{RX})$;
- A source of local randomness for Bob $(\mathcal{R}_Y, f_{RY})$.

The following *encoding functions* to be used over the public channel:

- The m Alice encoding functions $e_i^{(A)}$:

$$e_i^{(A)} : \boldsymbol{B}^{i-1} \times \boldsymbol{R}_X \to \boldsymbol{A},\ i \in \{1, 2, \ldots, m\}; \tag{4.56}$$

- The m Bob encoding functions $e_i^{(B)}$:

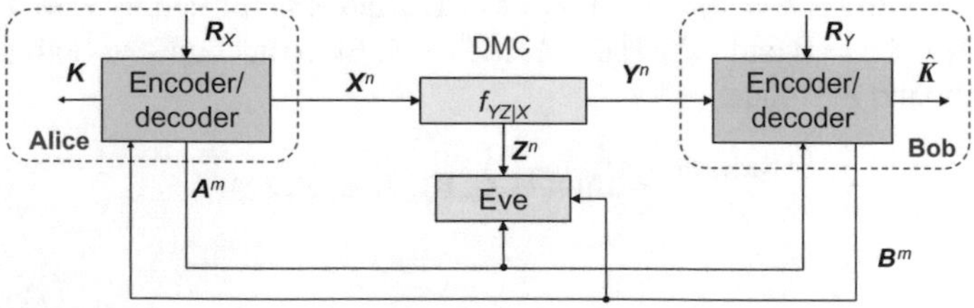

Fig. 4.19 The channel model for secret-key generation (agreement) scenario

$$e_i^{(B)} : \boldsymbol{Y}^j \times \boldsymbol{A}^{i-1} \times \boldsymbol{R}_Y \to \boldsymbol{B}, \ i \in \{1, 2, \ldots, m\}. \tag{4.57}$$

The index j is used to denote the number of available DMC channel outputs up to now.

The following *encoding functions* to be used over the DMC:

$$e_l^{(DMC)} : \boldsymbol{B}^k \times \boldsymbol{R}_X \to \boldsymbol{X}, \ l \in \{1, 2, \ldots, n\}. \tag{4.58}$$

The index k is used to denote the number of bits received from Bob over the public channel up to now.

The following *key-distillation functions*:

- Alice key-distillation function k_A, defined as the following mapping:

$$k_A : \boldsymbol{X}^n \times \boldsymbol{B}^m \times \boldsymbol{R}_X \to \boldsymbol{K}. \tag{4.59}$$

- Bob key-distillation function k_B, defined as the following mapping:

$$k_B : \boldsymbol{Y}^n \times \boldsymbol{A}^m \times \boldsymbol{R}_Y \to K. \tag{4.60}$$

The key-distillation protocol P_n is composed of the following *steps*:

- Alice generates local random realization r_x, while Bob generates r_y;
- In round $i \in [1, m]$ Alice transmits a_i and Bob b_i by

$$a_i = e_i^{(A)}\left(\boldsymbol{b}^{i-1}, r_x\right), \ b_i = e_i^{(B)}\left(\boldsymbol{a}^{i-1}, r_y\right); \tag{4.61}$$

- In the lth DMC transmission interval, Alice transmits symbol x_l over DMC:

$$x_l = e_l^{(DMC)}\left(\boldsymbol{b}^k, r_x\right). \tag{4.62}$$

Bob receives y_l and Eve receives z_l. (Once more, k denotes the number of bits received from Bob over the public channel up to now.) The transmission over DMC takes place when channel conditions are favorable.

- After all DMC and public channel transmissions are completed, Alice computes key k_A and Bob key k_B by

$$\boldsymbol{k} = k_A\left(\boldsymbol{x}^n, \boldsymbol{b}^m, r_x\right), \ \hat{\boldsymbol{k}} = k_B\left(\boldsymbol{y}^n, \boldsymbol{a}^m, r_y\right). \tag{4.63}$$

- Alice and Bob perform information reconciliation, by employing the error correction to correct for the errors introduced by the DMC and Eve.
- Finally, Alice and Bob perform information reconciliation to remove any correlation with Eve.

At the end of protocol P_n, Alice and Bob end up with the common key $\boldsymbol{k}_c = \hat{\boldsymbol{k}}_c$, which is not known to Eve. This channel-type key-generation protocol P_n is evaluated in terms of probability of error, information leakage to Eve, and uniformity of the key.

The *weak secret-key capacity for DMC-type scenario* can be defined as follows:

$$C_{SK}^{(\text{weak})} \triangleq \sup\{R : R \text{ is an achievable weak SKR}\}. \tag{4.64}$$

On the other hand, the *strong secret-key capacity* of a channel-type model is defined as

$$C_{SK}^{(\text{strong})} \triangleq \sup\{R : R \text{ is an achievable strong SKR}\}. \tag{4.65}$$

The following *theorem* has been proved *by Ahlswede, and Csiszár* in [38] (see also [3]): The weak secret-key capacity of a channel-type model satisfies the following lower and upper bounds:

$$\begin{aligned} &\max\left\{\max_{f_X}[I(X;Y) - I(X;Z)], \max_{f_X}[I(Y;X) - I(Y;Z)]\right\} \\ &\leq C_{SK}^{(\text{weak})} \leq \max_{f_X} \min(I(X;Y), I(X,Y|Z)). \end{aligned} \tag{4.66}$$

By employing the information reconciliation and privacy amplification steps, the strong secrecy condition can be obtained from the weak secrecy condition.

4.6 Coding for Physical-Layer Security Systems

We have already discussed the concept of *nested code* in Sect. 4.3. In this coding scenario, the message $\boldsymbol{M}$ is used to select the subcode (n, nR). After that, the source of local randomness $\boldsymbol{U}$ is used to select the codeword within the subcode. For each message, there are 2^{nR_U} possible codewords selected at random. In this section, we discuss (i) coding for weak secrecy, (ii) coding for strong secrecy, and (iii) information reconciliation.

4.6.1 Coding for Weak Secrecy Systems

Ozarow and Wyner considered in [42] the wiretap channel model in which the main channel is noiseless, while the wiretap (Eve's) channel is noisy. Clearly, assuming the binary transmission, the secrecy of this wiretap channel model $1 - C_w$, where C_w is the capacity of the wiretap channel. Let the binary linear block code of code rate R_0 be denoted by $\mathcal{C}_0$. For the parity-check matrix $\boldsymbol{H}$, each codeword $\boldsymbol{x}^n$ satisfies the parity-check equations: $\boldsymbol{H}\boldsymbol{x}^n = \boldsymbol{0}$. For each message $\boldsymbol{m}$, we define the *coset* $C_{\boldsymbol{m}}$ as follows:

$$C_{\boldsymbol{m}} = \left\{ \boldsymbol{x}^n : \ \boldsymbol{H}\boldsymbol{x}^n = \boldsymbol{m} \right\}. \tag{4.67}$$

To encode, Alice randomly chooses the n-tuple $\boldsymbol{x}^n$ from the coset $C_{\boldsymbol{m}}$. Since there are 2^n n-tuples and the cardinality of the cosset $C_{\boldsymbol{m}}$ is 2^{nR_0}, the number of possible cosets is $2^n/2^{nR_0} = 2^{n(1-R_0)}$, indicating that the code rate of this scheme is $1 - R_0$. Clearly, when $\mathcal{C}_0$ is the capacity-achieving code, the equivocation and wiretap rate can be achieved arbitrary close to the secrecy capacity. The nested codes get extended to multiterminal scenarios in [43].

The *nested wiretap code idea* gets extended to noisy main channel in [44] as follows. Let $\boldsymbol{H}$ be $(1 - R_{12})n \times n$ parity–check matrix of full rank representing the code $\mathcal{C}$, composed of two sub-matrices $\boldsymbol{H}_1$ and $\boldsymbol{H}_2$:

$$\boldsymbol{H} = \begin{bmatrix} \boldsymbol{H}_1 \\ \boldsymbol{H}_2 \end{bmatrix}, \tag{4.68}$$

where $\boldsymbol{H}_1$ is $(1 - R_1)n \times n$ submatrix representing the code $\mathcal{C}_1$, with $R_1 > R_{12}$, and $\boldsymbol{H}_2$ is $Rn \times n$ submatrix. Clearly, $(1 - R_1)n + Rn = (1 - R_{12})n$ so that we can write $R = R_1 - R_{12}$. The linear block code (LBC) $\mathcal{C}$ is a subcode of the LBC $\mathcal{C}_1$, while the distinct cossets of $\mathcal{C}$ in $\mathcal{C}_1$ represent the partition of $\mathcal{C}$. Alice wants to send the nR-bit message $\boldsymbol{M}$ to Bob. To do so she randomly selects n-tuple $\boldsymbol{X}^n$ from the following coset:

$$C_{\boldsymbol{M}} = \left\{ \boldsymbol{X}^n : \ \begin{bmatrix} \boldsymbol{H}_1 \\ \boldsymbol{H}_2 \end{bmatrix} \boldsymbol{X}^n = \begin{bmatrix} \boldsymbol{0} \\ \boldsymbol{M} \end{bmatrix} \right\}. \tag{4.69}$$

To decode the transmitted message, Bob performs the syndrome decoding as follows. By using the observation vector $\boldsymbol{Y}^n$, he obtains the estimate of transmitted n-tuple, denoted as $\hat{\boldsymbol{X}}^n$, by running the corresponding decoding algorithm to satisfy the parity-check equations $\boldsymbol{H}_1\hat{\boldsymbol{X}}^n = \boldsymbol{0}$. Finally, Bob computes the message estimate by $\boldsymbol{H}_2\hat{\boldsymbol{X}}^n = \hat{\boldsymbol{M}}$. The wiretap code just described can be denoted by $\mathcal{C}_n$.

Clearly, we can partition $\mathcal{C}_1$, using the cosets of $\mathcal{C}$ into 2^{nR} disjoint subsets and thus represents the generalization of Wyner's construction, since $\mathcal{C}_1$ is equivalent to $\mathcal{C}_0$. Assuming that $\mathcal{C}_1$ comes from the capacity-achieving channel code over the main channel and $\mathcal{C}$ forms the channel capacity-achieving code over the wiretap channel, authors in [44] (see also [45]) were able to show that this coset encoding scheme satisfies both the reliability condition ($P_e \to 0$ as $n \to \infty$) and the secrecy capacity condition $(1/n)I(\boldsymbol{M}, \boldsymbol{Z}^n) \to 0$ as $n \to \infty$. Because both codes $\mathcal{C}_1$ and $\mathcal{C}$ are capacity-achieving, the probability of error condition is satisfied. To prove the weak secrecy requirement, let us apply the chain rule for $I(\boldsymbol{M}, \boldsymbol{X}^n, \boldsymbol{Z}^n)$ in two different ways [26]:

$$I(M, Z^n) + I(X^n, Z^n|M) = I(X^n, Z^n) + \underbrace{I(M, Z^n|X^n)}_{=0} = I(X^n, Z^n), \quad (4.70)$$

wherein $I(M, Z^n|X^n) = 0$, since $M \to X^n \to Z^n$ is a Markov chain. We can rewrite the previous equation as

$$\begin{aligned} I(M, Z^n) &= I(X^n, Z^n) - \underbrace{I(X^n, Z^n|M)}_{H(X^n|M)-H(X^n|M,Z^n)} \\ &= I(X^n, Z^n) - H(X^n|M) + H(X^n|M, Z^n). \end{aligned} \quad (4.71)$$

Because $H(X^n|M) = nR_{12}$, $I(X^n, Z^n) \leq nC_w$, and from Fano's inequality [26, 45] $H(X^n|Z^n, M) \leq h(P_e) + P_e nR_{12}$, the previous equation becomes the following inequality:

$$I(M, Z^n) \leq nC_w - nR_{12} + h(P_e) + P_e nR_{12}, \quad (4.72)$$

end from the weak secrecy requirement definition, we obtain

$$\lim_{n\to\infty} \frac{1}{n} I(M, Z^n) = \lim_{n\to\infty} \left[C_w - R_{12} + \frac{1}{n} h(P_e) + P_e R_{12} \right] = 0, \quad (4.73)$$

where we employed the reliability condition ($P_e \to 0$ as $n \to \infty$).

In incoming subsections, we describe several codes' designs capable of achieving the weak secrecy capacity.

4.6.1.1 Two-Edge-Type LDPC Coding

The two-edge-type LDPC coding [46] is the most natural way to implement the nested coding. It can be considered as an instance of generalized LDPC coding [47–53]. As discussed in Chap. 2, we can describe an LDPC code in terms of the *degree distribution polynomials* $\mu(x)$ and $\rho(x)$, for the variable node (v-node) and the check node (c-node), respectively, as follows:

$$\mu(x) = \sum_{d=1}^{d_v} \mu_d x^{d-1}, \quad \rho(x) = \sum_{d=1}^{d_c} \rho_d x^{d-1}, \quad (4.74)$$

where μ_d and ρ_d denote the fraction of the edges that are connected to degree-d v-nodes and c-nodes, respectively, and d_v and d_c denote the maximum v-node and c-node degrees, respectively. For sufficiently large codeword length n, the most of parity-check matrices in a $\{\mu, \rho\}$ ensemble of LDPC codes of the same girth exhibit similar performance under the log-domain sum–product algorithm (SPA) described in Chap. 2. Now we can define the *two-edge LDPC code* ensemble as a $\{\mu_1, \mu_2,$

$\rho_1, \rho_2\}$ two-edge-type LDPC coding ensemble of codeword length n, wherein the variable nodes are connected to μ_1 (μ_2) check nodes of type 1 (2) and all type 1 (2) check nodes have degree ρ_1 (ρ_2) [46]. The parity-check matrix of two-edge type LDPC codes has the form introduced above, namely, $\boldsymbol{H} = \begin{bmatrix} \boldsymbol{H}_1 \\ \boldsymbol{H}_2 \end{bmatrix}$, where $\boldsymbol{H}_1$ ($\boldsymbol{H}_2$) define connections between the variable nodes and the check nodes of type 1 (2). Since the two-edge LPDC codes represent the instances of LBC codes described above, they achieve the weak secrecy capacity. Let $\boldsymbol{G}$ be the generator matrix that corresponds to the parity-check matrix $\boldsymbol{H}$. As discussed above, each coset C_m of $\mathcal{C}$ in $\mathcal{C}_1$ contains solutions of $\boldsymbol{x}\boldsymbol{H}^{\mathrm{T}} = [\boldsymbol{x}\boldsymbol{H}_1^{\mathrm{T}}\ \boldsymbol{x}\boldsymbol{H}_2^{\mathrm{T}}] = [\boldsymbol{0}\ \boldsymbol{m}]$ for message $\boldsymbol{m}$. Let $\boldsymbol{G}^* = [\boldsymbol{g}_1^{\mathrm{T}}\ \boldsymbol{g}_2^{\mathrm{T}} \ldots \boldsymbol{g}_{nR}^{\mathrm{T}}]^{\mathrm{T}}$ be the submatrix with rows being linearly independent, and when augmented with $\boldsymbol{G}$ to form the basis for code with parity-check matrix $\boldsymbol{H}_1$. To encode nR-bit message $\boldsymbol{m}$, Alice randomly chooses nR_{12}-bit auxiliary message $\boldsymbol{m}'$. By calculating,

$$\boldsymbol{x} = \begin{bmatrix} \boldsymbol{m} & \boldsymbol{m}' \end{bmatrix} \begin{bmatrix} \boldsymbol{G}^* \\ \boldsymbol{G} \end{bmatrix}. \tag{4.75}$$

Alice randomly chooses $\boldsymbol{x}$ among all possible solutions of $\boldsymbol{x}\boldsymbol{H}^{\mathrm{T}} = [\boldsymbol{0}\ \boldsymbol{m}]$. Bob applies the decoding procedure described in the paragraph below Eq. (4.69) to estimate the transmitted message.

4.6.1.2 Punctured LDPC Coding

The puncturing in communication systems is typically used to change the code rate of original code. The punctured bits are not transmitted. So the key idea of using puncturing in secrecy coding is to assign the punctured bits to message bits, denoted by $\boldsymbol{m}$ [54, 55], which is illustrated in Fig. 4.20.

Further auxiliary dummy bits get generated, denoted as $\boldsymbol{m}'$, and then parity bits get generated by LDPC encoding to get $\boldsymbol{p}$. Alice then sends the codeword $\boldsymbol{x}' = [\boldsymbol{m}'\ \boldsymbol{p}]$ to Bob over the noisy channel, and Bob performs log-domain SPA trying to reconstruct the punctured (message) bits. Clearly, the number of message bits k_m is smaller than the number of information bits k used in LDPC encoder of rate $R = k/n$. The number of dummy bits is $k - k_m$. Therefore, the secrecy rate $R_s = k_m/n$ is lower than the LDPC code rate R. The key problem here is to determine the optimized puncturing

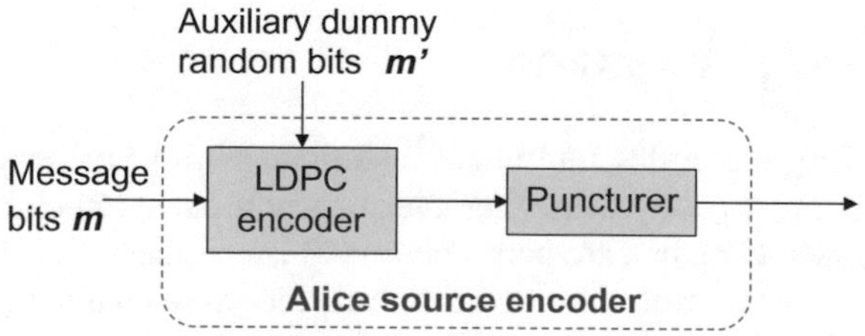

Fig. 4.20 The punctured LDPC encoding for wiretap channels

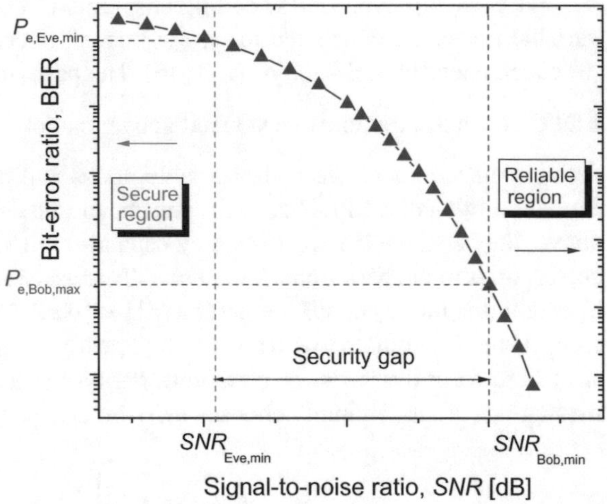

Fig. 4.21 Defining the secrecy gap

distribution, which is subject of investigation in [55]. For Gaussian wiretap channel, the *reliability region* is defined by the smallest signal-to-noise ratio ($SNR_{\text{Bob,min}}$) for Bob for which the probability of error ($P_{\text{e,Bob,max}}$) is maximally tolerable, as illustrated in Fig. 4.21. The *secure region* is defined as the highest SNR of Eve ($SNR_{\text{Eve,max}}$) for which the probability of error ($P_{\text{e,Eve,min}}$) is minimum that can be tolerated (it is close to 0.5). The security gap is defined as a difference (in dB scale), $SNR_{\text{Bob,min}} - SNR_{\text{Eve,min}}$. The goal is to optimize the puncturing function to minimize the security gap. For additional details, interested reader is referred to [55].

Another approach advocated in [54] was to choose which bits to puncture in an LDPC codeword such that the codewords with the same punctured bits form the subcodebook, thus creating the nested structure. Let us consider (n', k') LPDC code $\mathcal{C}'$ whose parity-check matrix can be represented in the form $\boldsymbol{H} = [\boldsymbol{H}_1\ \boldsymbol{H}_2]$, where $\boldsymbol{H}_2$ is a lower triangular matrix of size $(n' - k') \times (n' - k')$. The codewords for $\mathcal{C}'$ are represented in the form $\boldsymbol{x} = [\boldsymbol{m}\boldsymbol{m}'\ \boldsymbol{p}]$, where $\boldsymbol{m}$ is the message vector of length $k < k'$, $\boldsymbol{m}'$ is the axillary dummy message vector of length $k' - k$, and $\boldsymbol{p}$ is the vector of parity checks of length $n' - k'$. The (n, k) punctured code $\mathcal{C}$ is composed of punctured codewords $\boldsymbol{x}' = [\boldsymbol{m}'\ \boldsymbol{p}]$, which get transmitted toward Bob. The partition of code $\mathcal{C}$ can be done based on punctured bits' values in the message $\boldsymbol{m}$. Bob based on the received word performs LDPC decoding to determine the punctured bits.

4.6.1.3 Polar Codes

The polar codes, introduced by Arikan in [56], are low-complexity linear block codes being capacity-achieving over binary input symmetric channels (rows in transition channel matrix are permutations of each other). Let $W : \{0, 1\} \to \mathcal{Y}$, $(x \to y)$, be a generic binary input discrete memoryless channel (BI-DMC) characterized with transition probabilities $W(y|x)$. For the block length $n = 2^m$ (m is a positive integer), we define the $n \times n$ matrix $\boldsymbol{G}_n$ as follows:

$$\boldsymbol{G}_n = \boldsymbol{G}_2^{\otimes m}, \ \boldsymbol{G}_2 \doteq \begin{bmatrix} 1 & 0 \\ 1 & 1 \end{bmatrix}, \tag{4.76}$$

where $\otimes$ denotes the Kronecker product. Let us consider the sequence $\boldsymbol{u}$ of n-tuples from the binary field $\mathcal{F}_2$, that is, $\boldsymbol{u} \in \mathcal{F}_2^n$. Each bit x_i in a codeword $\boldsymbol{x} = \boldsymbol{u}\boldsymbol{G}_n$ is transmitted over BI-DMC with transition probability $W(y|x)$, and the resulting output vector is denoted by $\boldsymbol{y}$. For each bit u_i in $\boldsymbol{u}$, we define the bit channel $W^{(i)} : \mathcal{U} \to \mathcal{Y}^n \times \mathcal{U}^{i-1}$, indicating that the bit u_i can be decoded by using the received sequence $\boldsymbol{y}$ and previously decoded bits $u_1, u_2, \ldots, u_{i-1}$. The ith bit channel function is defined by

$$W^{(i)}(\boldsymbol{y}, \boldsymbol{u}^{i-1}|u_i) = \sum_{\boldsymbol{u}_{i+1}^N \in \{0,1\}^{n-i}} W^n(\boldsymbol{y}|\boldsymbol{u}\boldsymbol{G}_n), \tag{4.77}$$

where as before we use the notation $\boldsymbol{u}^{i-1}$ to denote the sequence $[u_1 \ u_2 \ \ldots \ u_{i-1}]$. Similarly, we use the notation $\boldsymbol{u}_{i+1}^n$ to denote the sequence $[u_{i+1} \ u_{i+2} \ \ldots \ u_n]$. Since the channel is memoryless, the $W^n(\boldsymbol{y}|\boldsymbol{x})$ term above can be written as product of individual transition probabilities, that is, $W^n(\boldsymbol{y}|\boldsymbol{x}) = \prod_{i=1}^n W(y_i|x_i)$. Arikan has shown that these bit channels "polarize" in the sense that they become either almost noiseless or almost completely noisy. This "polarization" effect can be described in terms of the *Bhattacharya parameter* $Z(W)$ of the binary input channel W, which is defined as

$$Z(W) = \sum\nolimits_y \sqrt{W(y|0)W(y|1)} \in [0, 1], \tag{4.78}$$

with $Z(W) \to 0$ meaning that the channel is noiseless, while for $Z(W) \to 1$ meaning that the channel is totally noisy.

Let $C(W)$ denote the channel capacity of channel W. Arikan has shown in [56] that for any fixed $\delta \in (0, 1)$ as $n \to \infty$, the fraction of indices $i \in \{1, 2, \ldots, n\}$ for which $Z(W^{(i)}) \in (1 - \delta, 1]$ tends to $C(W)$, while the fraction of indices for which $Z(W^{(i)}) \in [0, \delta)$ tends to $1 - C(W)$. The bit channels of the first type (noiseless) can be called good bit channels and denoted by $\mathcal{G}$, while the bit channels of the second type can be called bad bit channels denoted by $\mathcal{B} = \{1, 2, \ldots, n\} \backslash \mathcal{G}$. This polarization phenomenon can be employed to design the polar codes. Given that bad bit channels are too noisy, they should not be used to transmit any reliable information; instead, they should be frozen to a fixed value. On the other hand, the good bit channels are almost noise-free and reliable for transmission of information. So the main task in the polar code design is to identify the indices of good bit channels $\mathcal{G}$. By using the Bhattacharya parameter, we can select the indices of the bit channels with the k least Bhattacharya parameter to design an (n, k) polar code.

Regarding the decoding procedures of polar codes, we can use either sum–product algorithm or sequential decoding. The *sequential decoding* operates as follows. For indices belonging to the good bit channels, namely, $i \in \mathcal{G}$ we apply the following decoding rule:

$$\hat{u}_i = \begin{cases} 1, & \frac{W^{(i)}\left(y^n, \hat{u}^{i-1}|1\right)}{W^{(i)}\left(y^n, \hat{u}^{i-1}|0\right)} > 1 \\ 0, & \text{otherwise} \end{cases} \tag{4.79}$$

A typical polar code n-combiner architecture for $n \geq 4$, as proposed by Arikan [56] (see also [29]), is shown in Fig. 4.22. Clearly, the combining process proceeds in two steps. In step 1, we pair $\boldsymbol{u}^n$ into sequence of pairs $\left(\left(\boldsymbol{u}_1^2, \boldsymbol{u}_3^4, \ldots, \boldsymbol{u}_{n-1}^n\right)\right)$ and then apply $\boldsymbol{G}_2$ to each pair. This step can be described by the following transformation [29]:

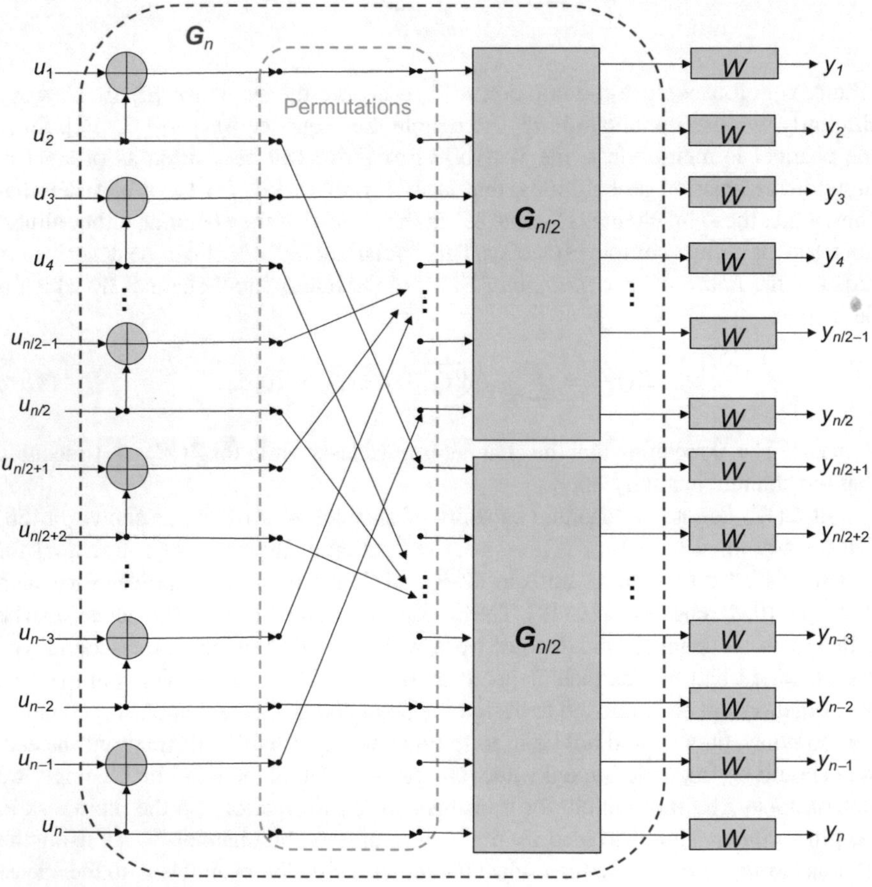

Fig. 4.22 Description of Arikan's combiner in the polar code

$$\boldsymbol{I}_{n/2} \otimes \boldsymbol{G}_2 = \begin{bmatrix} \boldsymbol{G}_2 & \boldsymbol{0} & \boldsymbol{0} \\ \boldsymbol{0} & \ddots & \boldsymbol{0} \\ \boldsymbol{0} & \boldsymbol{0} & \boldsymbol{G}_2 \end{bmatrix}. \tag{4.80}$$

In step two, we feed even bits ($u_2, u_4, \ldots, u_n$) to the odd positions of the bottom $n/2$-stage of the combiner. At the same time, we feed $n/2$ sum bits ($u_1 \oplus u_2, u_2 \oplus u_4, \ldots, u_{n-1} \oplus u_n$) to the even positions of the top $n/2$-stage of the combiner. This step can be described as $\boldsymbol{R}_n(\boldsymbol{I}_2 \otimes \boldsymbol{G}_{n/2})$, where $\boldsymbol{R}_n$ is the permutation matrix separating the even bits and the sum bits.

If Eve's channel is degraded, it can be shown that Eve's set of good bits, denoted as $\mathcal{G}_E$, is the subset of Bob's set of good bit $\mathcal{G}_B$ [57, 58]. Clearly, in this scenario, Eve will be able to decode properly only a subset of bits that Bob is able to decode. So we can group the bit channels into three groups: (i) bit channels decodable by Bob only, denoted as $\mathcal{G}_B\backslash\mathcal{G}_E$, (ii) bit channels decodable by both Bob and Eve, denoted as $\mathcal{G}_{BE}$, and (iii) Bob's set of bad bits denoted as $\mathcal{B}_B$. We can now reorder information word as follows: $\left[\boldsymbol{u}_{\mathcal{G}_\mathrm{B}\backslash\mathcal{G}_\mathrm{E}}\ \boldsymbol{u}_{\mathcal{G}_\mathrm{BE}}\ \boldsymbol{u}_{\mathcal{G}_\mathrm{B}}\right]$, wherein the Bob's bad bits are frozen (fixed). The codewords $\left[\boldsymbol{u}_{\mathcal{G}_\mathrm{B}\backslash\mathcal{G}_\mathrm{E}}\ \boldsymbol{u}_{\mathcal{G}_\mathrm{BE}}\ \boldsymbol{u}_{\mathcal{G}_\mathrm{B}}\right]\boldsymbol{G}_n$ can now be partitioned into different cosets characterized by different values of the $\boldsymbol{u}_{\mathcal{G}_\mathrm{B}\backslash\mathcal{G}_\mathrm{E}}$, thus representing the nested structure. To encode, Alice will place the message bits $\boldsymbol{m}$ into $\boldsymbol{u}_{\mathcal{G}_\mathrm{B}\backslash\mathcal{G}_\mathrm{E}}$ information bits block, auxiliary dummy bits $\boldsymbol{m}'$ into $\boldsymbol{u}_{\mathcal{G}_\mathrm{BE}}$ block, and transmit the codeword $\left[\boldsymbol{m}\ \boldsymbol{m}'\ \boldsymbol{u}_{\mathcal{G}_\mathrm{B}}\right]\boldsymbol{G}_n$. On the receiver side, Bob will perform the sequential decoding as described above.

Some other classes of codes suitable for use in physical-layer security are provided in [59–61].

4.6.2 *Coding for Strong Secrecy Systems*

This subsection is devoted to the coding for strong secrecy [62–65], and we describe two classes of codes capable of achieving the strong secrecy requirement.

4.6.2.1 Coset Coding with Dual of LDPC Codes

Let us consider an (n, k) LDPC code $\mathcal{C}$, with the parity-check matrix $\boldsymbol{H}$ belonging to the ensemble of codes with distribution (μ, λ) and with the BEC threshold $\alpha^*(\mu, \lambda)$ [62, 63], suitable for the binary erasure wiretap channel model shown in Fig. 4.10. (As a reminder, for BEC all errors with error probability α below the threshold α^* are correctable with the sum–product (belief-propagation) algorithm.) Let $\mathcal{C}^\perp$ be the dual of LDPC code C; in other words, the (n, $n - k$) low-density generator matrix (LDGM) code with generator matrix being $\boldsymbol{H}$. The 2^k cosets of $\mathcal{C}^\perp$ in n-tuple space form a partition, thus representing the nested structure. To encode a k-bit message $\boldsymbol{m}$, the encoder chooses the coset with syndrome $\boldsymbol{m} = \boldsymbol{x}\boldsymbol{G}^\mathrm{T}$ and then randomly selects a codeword from that coset. To simplify the codeword selection process, we can proceed as follows. The linearly independent rows of $\boldsymbol{H}$ (representing the basis for $\mathcal{C}^{\perp}$ code) can be augmented with k linearly independent vectors $\{\boldsymbol{g}_1, \boldsymbol{g}_2, \ldots, \boldsymbol{g}_k\}$ to form the basis for the n-tuple space (in $\{0, 1\}^n$), representing the matrix $\boldsymbol{G}^* = \left[\boldsymbol{g}_1^T \cdots \boldsymbol{g}_k^T\right]^T$, and the encoding process can be represented as follows:

$$\boldsymbol{x} = \left[\boldsymbol{m}\ \boldsymbol{m}'\right]\begin{bmatrix}\boldsymbol{G}\\ \boldsymbol{H}\end{bmatrix}, \tag{4.81}$$

where $\boldsymbol{m}'$ is the dummy auxiliary vector. The decoding process is of a reasonable complexity as discussed in [62].

Let $P_e^{(\mathcal{C},\, BEC(1-\alpha))}$ denote the block error probability of the code $\mathcal{C}$ over the BEC with erasure probability $1-\alpha$. It can be shown that the following inequality holds [64]:

$$\frac{1}{n}I\left(M, Z^n\right) \le \frac{k}{n}P_e^{(\mathcal{C},\, BEC(1-\alpha))}, \tag{4.82}$$

indicating that for $1-\alpha < \alpha^*(\mu, \lambda)$ this construction guarantees the weak secrecy. However, it has been shown in [63], when an ensemble of large-girth LDPC codes is used, whose girth grows exponentially with the codeword length n, the probability of error has the desired decay for $1-\alpha < \alpha^*(\mu, \lambda)$, and strong secrecy can be achieved.

When the main channel is noiseless, while the wiretap channel is binary additive, it has been shown in [66] that coset coding with parity-check matrix generated uniformly at random can also achieve the strong secrecy.

4.6.2.2 Hash Functions and Extractors Based Coding

The code constructions based on hash functions and extractors can achieve strong secrecy as discussed in [67–69]. The key idea behind these code constructions is to employ the universal$_2$ families of hash functions [15, 25, 67–70]. We say that the family of functions $\mathcal{F}$: $(0, 1)^r \to (0, 1)^k$ is a *universal$_2$ family of hash functions* if for any two distincts $\boldsymbol{x}_1$ and $\boldsymbol{x}_2$ from $(0, 1)^r$, the probability that $f(\boldsymbol{x}_1) = f(\boldsymbol{x}_2)$ when f is chosen uniformly at random from $\mathcal{F}$ is smaller than 2^{-k}. Let us now consider (n, r) reliability code $\mathcal{C}$ and choose f uniformly at random from $\mathcal{F}$. To encode a secret message $\boldsymbol{m}$ from $(0, 1)^k$, Alice selects the sequence $\boldsymbol{u}$ uniformly at random from $(0, 1)^r$ in the set $f^{-1}(\boldsymbol{m})$ and encodes it using the code $\mathcal{C}$. It has been shown in [67] that any choice of f guarantees the strong secrecy when the code rate k/n is lower than secrecy capacity. The problem with this approach is that function $f(\boldsymbol{x})$ is not inevitable or it is very difficult to invert. To solve this problem, the use of invertible extractors is advocated in [68, 69].

4.6.3 Information Reconciliation

We turn our attention now to the source-type secret-key generation (agreement) protocol, depicted in Fig. 4.18 and described in Sect. 4.5.1. In this model, Alice, Bob, and Eve have access to the respective output ports $\mathcal{X}$, $\mathcal{Y}$, $\mathcal{Z}$ of the source characterized by the joint PDF f_{XYZ}. Each of them sees the i.i.d. realizations of the source denoted, respectively, as $\boldsymbol{X}^n$, $\boldsymbol{Y}^n$, and $\boldsymbol{Z}^n$. Given that Alice's and Bob's realizations are not perfectly correlated, they need to correct the discrepancies' errors. This step in secret-key generation as well as in QKD is commonly referred to as the *information reconciliation* (or just reconciliation). The reconciliation process can be considered as a special case of the *source coding with the side information* [26, 71]. To encode the source $\mathcal{X}$, we know from Shannon's source coding theorem that the rate R_x must be as small as possible but not lower than the entropy of the source; in other words, $R_x > H(X)$. To jointly encode the source $(\mathcal{X}\mathcal{Y}, f_{XY})$ we need the rate $R > H(X, Y)$. If we separately encode the sources, the required rate would be $R > H(X) + H(Y)$. However, Slepian and Wolf have shown in [71] that the rate $R = H(X, Y)$ is still sufficient even for separate encoding of correlated sources. This claim is commonly referred to as the *Slepian–Wolf theorem*, and can be formulated as follows [26, 71]: The *achievable rate region* $\mathcal{R}$ (for which the error probability tends to zero) for the separate encoding of the correlated source $(\mathcal{X}\mathcal{Y}, f_{XY})$ is given by

$$\mathcal{R} \doteq \left\{ (R_x, R_y) : \begin{array}{r} R_x \geq H(X|Y) \\ R_y \geq H(Y|X) \\ R_x + R_y \geq H(X, Y) \end{array} \right\}. \tag{4.83}$$

The typical shape of the Slepian–Wolf region is provided in Fig. 4.23. Now if we assume that (X, f_X) should be compressed when (Y, f_Y) is available as the side information, then $H(X|Y)$ is sufficient to describe X.

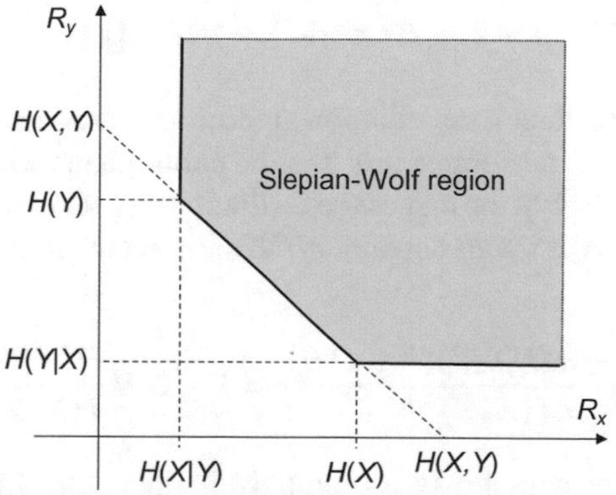

Fig. 4.23 The typical Slepian–Wolf achievable rate region for a correlated source $(\mathcal{X}\mathcal{Y}, f_{XY})$

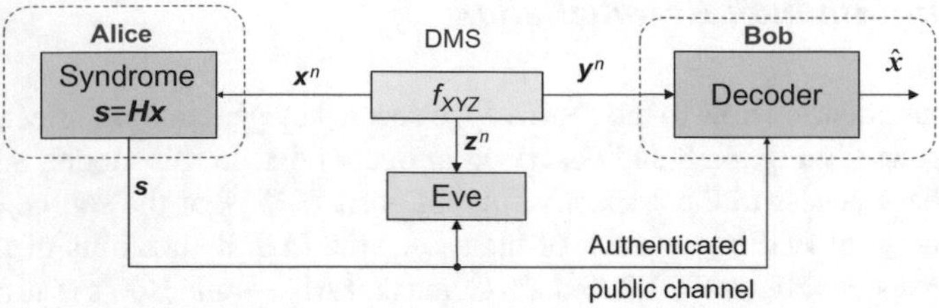

Fig. 4.24 The reconciliation problem represented as a source coding with syndrome-based side information

Going back to our reconciliation problem, Alice compresses her observations $\boldsymbol{X}^n$ and Bob decodes them with the help of correlated side information $\boldsymbol{Y}^n$, which is illustrated in Fig. 4.24. The source encoder can be derived from capacity-achieving channel code. For an (n, k) LDPC code, the parity-check matrix $\boldsymbol{H}$ of size $(n - k) \times n$ can be used to generate the syndrome by $\boldsymbol{s} = \boldsymbol{Hx}$, where $\boldsymbol{x}$ is Alice's DMS observation vector of length n. Since $\boldsymbol{x}$ is not a codeword, the syndrome vector is different from the all-zeros vector. The syndrome vector is of length $n - k$ and can be represented as $\boldsymbol{s} = [s_1\ s_2\ \ldots\ s_{n-k}]^{\mathrm{T}}$. Clearly, the syndrome vector is transmitted toward Bob over the (error-free) authenticated public channel.

Once Bob receives the syndrome vector $\boldsymbol{s}$, his decoder tries to determine the vector $\boldsymbol{x}$ based on $\boldsymbol{s}$ and Bob's observation vector $\boldsymbol{y}$, which maximizes the posterior probability:

$$\hat{\boldsymbol{x}} = \max\ P(\boldsymbol{y}|\boldsymbol{x}, \boldsymbol{s}), \tag{4.84}$$

and clearly, the decoding problem is similar to the maximum a posteriori probability (MAP) decoding. The compression rate is $(n - k)/n = 1 - k/n$, while the linear code rate is k/n. The number of syndrome bits required is

$$n - k \geq H\left(\boldsymbol{X}^n|\boldsymbol{Y}^n\right) = nH(X|Y). \tag{4.85}$$

In practice, the practical reconciliation algorithms introduce the overhead $OH > 0$, so that the number of transmitted bits over the public channel is $nH(X|Y)(1 + OH)$. Given that Alice and Bob at the end of the reciliation step share with high probability the common sequence $\boldsymbol{X}^n$ with entropy $nH(X)$, the *reconciliation efficiency* β will be then

$$\beta = \frac{nH(X) - nH(X|Y)(1 + OH)}{nI(X, Y)} = 1 - OH\frac{H(X|Y)}{I(X, Y)} \leq 1. \tag{4.86}$$

The MAP decoding complexity is prohibitively high when LDPC coding is used, instead the low-complexity sum–product algorithm (SPA) (also known as belief-propagation algorithm) should be used as described in [72]. Given that Alice observation $\boldsymbol{x}$ is not a codeword, the syndrome bits are nonzero. To account for this problem, we need to change the sign of log-likelihood to be sent from the cth check node to the vth variable node when the corresponding syndrome bit s_c is 1. Based on Chap. 2 and Ref. [73], we can summarize the *log-domain SPA* below. To improve the clarity of presentation we use the index $v = 1, \ldots, n$ to denote variable nodes and index $c = 1, \ldots, n - k$ to denote the check nodes. Additionally, we use the following notation (introduced in Chap. 2): (i) $N(c)$ to denote $\{v$-nodes connected to c-node $c\}$, (ii) $N(c)\backslash\{v\}$ denoting $\{v$-nodes connected to c-node c except v-node $v\}$, (iii) $N(v)$ to denote $\{c$-nodes connected to v-node $v\}$, and (iv) $N(v)\backslash\{c\}$ representing $\{c$-nodes connected to v-node v except c-node $c\}$.

The formulation of the *log-domain SPA for source coding with the side information* is provided now.

0) *Initialization*: For the variable nodes $v = 1, 2, \ldots, n$; initialize the extrinsic messages L_{vc} to be sent from variable node (v-node) v to the check node (c-node) c to the channel log-likelihood ratios (LLRs), denoted as $L_{\text{ch}}(v)$, namely, $L_{vc} = L_{\text{ch}}(v)$.
(1) *The check-node (c-node) update rule*: For the check nodes $c = 1, 2, \ldots, n - k$; compute $L_{cv} = (1-2s_c) \underset{N(c)\backslash\{v\}}{\boxplus} L_{vc}$. The box-plus operator is defined by

$$L_1 \boxplus L_2 = \prod_{k=1}^{2} \text{sign}(L_k) \cdot \phi\left(\sum_{k=1}^{2} \phi(|L_k|)\right),$$

where $\phi(x) = -\log \tanh(x/2)$. The box operator for $|N(c)\backslash\{v\}|$ components is obtained by recursively applying 2-component version defined above.
(2) *The variable node (v-node) update rule*: For $v = 1, 2, \ldots, n$; set $L_{vc} = L_{\text{ch}}(v) + \sum_{N(v)\backslash\{c\}} L_{cv}$ for all c-nodes for which the cth row/vth column element of the parity-check matrix $\boldsymbol{H}$ is one, that is, $h_{cv} = 1$.
(3) *Bit decisions*: Update $L(v)$ ($v = 1, \ldots, n$) by $L(v) = L_{\text{ch}}(v) + \sum_{N(v)} L_{cv}$ and make decisions as follows: $\hat{x}_v = 1$ when $L(v) < 0$ (otherwise set $\hat{x}_v$=0).

So the only difference with respect to conventional log-domain SPA is the introduction of the multiplication factor $(1 - 2s_c)$ in the c-node update rule. Moreover, the channel likelihoods $L_{\text{ch}}(v)$ should be calculated from the joint distribution f_{XY} by $L_{\text{ch}}(v) = \log[f_{XY}(x_v = 0|y_v)/f_{XY}(x_v = 1|y_v)]$. Because the c-node update rule involves log and tanh functions, it can be computationally intensive, and to lower the complexity the reduced complexity approximations should be used as described in Chap. 2.

4.7 Privacy Amplification

The purpose of the privacy amplification is to extract the secret key from the reconciled (corrected) sequence. The key idea is to apply a well-chosen compression function, Alice and Bob have agreed on, to the reconciled binary sequence of length n_r as follows g: $(0, 1)^n \rightarrow (0, 1)^k$, where $k < n$, so that Eve gets negligible information about the secret key [2, 3, 14, 15, 74]. Typically, g is selected at random from the set of compression functions $\mathcal{G}$ so that Eve does not know which g has been used. In practice, the set of functions $\mathcal{G}$ is based on the family of universal hash functions, introduced by Carter and Wegman [75, 76].

We have already introduced the concept of hash functions but repeat the definition here for completeness of the section. We say that the family of functions $\mathcal{G} : (0, 1)^n \rightarrow (0, 1)^k$ is a *universal*$_2$ *family of hash functions* if for any two distinct $\boldsymbol{x}_1$ and $\boldsymbol{x}_2$ from $(0, 1)^n$, the probability that $g(\boldsymbol{x}_1) = g(\boldsymbol{x}_2)$ when g is chosen uniformly at random from $\mathcal{G}$ is smaller than 2^{-k}.

As an illustrative example, let us consider the multiplication in GF(2^n) [15]. Let a and x be two elements from GF(2^n). Let the g-function be defined as the mapping $(0, 1)^n \rightarrow (0, 1)^k$ that assigns to argument x the first k bits of the product $ax \in$ GF(2^n). It can be shown that the set of such functions is a universal class of functions for $k \in [1, n]$.

The analysis of privacy amplification does not rely on Shannon entropy, but Rényi entropy [77–79]. For discrete random variable X, taking the values from the sample space $\{x_1, \ldots, x_n\}$ and distributed according to $P_X = \{p_1, \ldots, p_n\}$, the *Rényi entropy of X of order* α is defined by [78]:

$$R_\alpha(X) = \frac{1}{1-\alpha} \log_2 \left\{ \sum_x [P_X(x)]^\alpha \right\}. \tag{4.87}$$

The Réniy entropy of order two is also known as the *collision entropy*. From Jensen's inequality [26] we conclude that the collision entropy is upper limited by the Shannon entropy, that is $R_2(X) \leq H(X)$. In limit when $\alpha \rightarrow \infty$, the Réniy entropy becomes the *min-entropy*:

$$\lim_{\alpha \to \infty} R_\alpha(X) = -\log_2 \max_{x \in X} P_X(x) = H_\infty(X). \tag{4.88}$$

Finally, by applying the l'Hôpital's rule as $\alpha \rightarrow 1$, the Réniy entropy becomes the Shannon entropy:

$$\lim_{\alpha \to 1} R_\alpha(X) = -\sum\nolimits_x P_X(x) \log_2 P_X(x) = H(X). \tag{4.89}$$

The properties of Réniy entropy are listed in [78]. Regarding the conditional Réniy entropy there exist different definitions, three of them have been discussed in detail in [79]. We adopt the first one. Let the joint distribution for (X, Y) be denoted as $P_{X,Y}$, and the distribution for X denoted as P_X. Further, let the Rényi entropy of the conditional random variables $X|Y = y$, distributed according to $P_{X|Y=y}$ be denoted as $R_\alpha(X|Y = y)$. Then the conditional Rényi entropy can be defined by

$$R_\alpha(X|Y) = \sum\nolimits_y P_Y(y) R_\alpha(X|Y=y) = \frac{1}{1-\alpha} \sum\nolimits_y P_Y(y) \log_2 \sum\nolimits_x P_{X|Y}(x|y)^\alpha. \tag{4.90}$$

The mutual Rényi information can be defined then as

$$I_\alpha(X,Y) = R_\alpha(X) - R_\alpha(X|Y), \tag{4.91}$$

and it is not symmetric to $R_\alpha(Y) - R_\alpha(Y|X)$. Moreover, it can even be negative as discussed in [78, 79].

Bennett et al. have proved the following *privacy amplification theorem* in [15] (see Theorem 3): Let X be a random variable with distribution P_X and Rényi entropy of order two $R_2(X)$. Further, let G be the random variable representing uniform selection at random of a member of universal hash functions $(0, 1)^n \to (0, 1)^k$. Then the following inequalities are satisfied [15]:

$$\begin{aligned} H(G(X)|G) &\geq R_2(G(X)|G) \geq k - \log_2\left[1 + 2^{k-R_2(X)}\right] \\ &\geq k - \frac{2^{k-R_2(X)}}{\ln 2}. \end{aligned} \tag{4.92}$$

Bennett et al. have also proved the following *corollary of the privacy amplification theorem* [15] (see also [3]): Let $S \in \{0, 1\}^n$ be a random variable representing the sequence shared between Alice and Bob, and let E be a random variable representing the knowledge Eve was able to get about the S, with a particular realization of E denoted by e. If the conditional Rényi entropy of order two $R_2(S|E = e)$ is known to be at least r_2, and Alice and Bob choose the secret key by $K = G(S)$, where G is a hash function chosen uniformly at random from the universl$_2$ family of hash functions $(0, 1)^n \to (0, 1)^k$ then the following is valid:

$$H(K|G, E=e) \geq k - \frac{2^{k-r_2}}{\ln 2}. \tag{4.93}$$

Cachin has proved the following Rényi entropy *lemma* in [80] (see also [81]): Let X and Q be two random variables and $s > 0$. Then with probability of at least $1 - 2^{-s}$ the following is valid [80, 81]:

$$R_2(X) - R_2(X|Q=q) \leq \log|Q| + 2s + 2. \tag{4.94}$$

The total information available to Eve is composed of her observations $\mathbf{Z}^n$ from the source of common randomness and additional bits exchanged in information reconciliation phase over the authenticated public channel, represented by random variable Q. Based on Cachin's lemma, we can write

$$R_2\left(S|\mathbf{Z}^n = z^n\right) - R_2\left(S|\mathbf{Z}^n = z^n, Q=q\right) \leq \log|Q| + 2s + 2 \text{ with probability } 1 - 2^{-s}. \tag{4.95}$$

We know that $R_2(X) \leq H(X)$, so we conclude that $R_2(S|\mathbf{Z}^n = z^n) \leq H(S|\mathbf{Z}^n = z^n)$ and therefore we can upper bound $R_2(S|\mathbf{Z}^n = z^n)$ as follows:

$$R_2\big(S|\mathbf{Z}^n = z^n\big) \leq nH(X|Z). \tag{4.96}$$

Based on (4.95), we can lower bound $R_2(S|\mathbf{Z}^n = z^n, Q = q)$ by

$$R_2\big(S|\mathbf{Z}^n = z^n, Q = q\big) \geq nH(X|Z) - \log|Q| - 2s - 2 \text{ with probability } 1 - 2^{-s}. \tag{4.97}$$

Since the number of bits exchanged over the public channel $\log|Q|$ is approximately $nH(X|Y)(1 + OH)$, for sufficiently large n, the previous inequality becomes

$$R_2(S|\mathbf{Z}^n = z^n, Q = q) \geq \underbrace{nH(X|Z) - nH(X|Y)(1 + OH) - 2s - 2}_{r_2} \\ \text{with probability } 1 - 2^{-s}. \tag{4.98}$$

From Eq. (4.93), we conclude that $k - r_2 = -k_2 < 0$, which guarantees that for $k_2 > 0$ Eve's uncertainty of the key is lower bounded by

$$H(K|E) \geq k - 2^{-k_2}/\ln 2 \text{ with probability } 1 - 2^{-s}. \tag{4.99}$$

4.8 Wireless Channels' Physical-Layer Security

The wireless channel secrecy capacity has been already discussed in Sect. 4.4.2, while the corresponding coding schemes have been described in Sect. 4.6. Here, we concentrate on (i) PLS for MIMO channels and (ii) key-generation techniques for wireless channels. We first introduce the wireless MIMO systems fundamentals.

4.8.1 Wireless MIMO Channels Fundamentals

Multiple transmitters can be used to either to improve the performance through array and diversity gains or to increase data rate through multiplexing of independent data streams [32, 33]. In multiple-input–multiple-output (MIMO) systems, the transmitters and receivers can exploit the knowledge about the channel to determine the rank of so-called channel matrix $\boldsymbol{H}$, denoted as rank($\boldsymbol{H}$), which is related to the number of independent data streams that can be simultaneously transmitted. In MIMO systems, we can define the *multiplexing gain* as the asymptotic slope in SNR of the outage capacity:

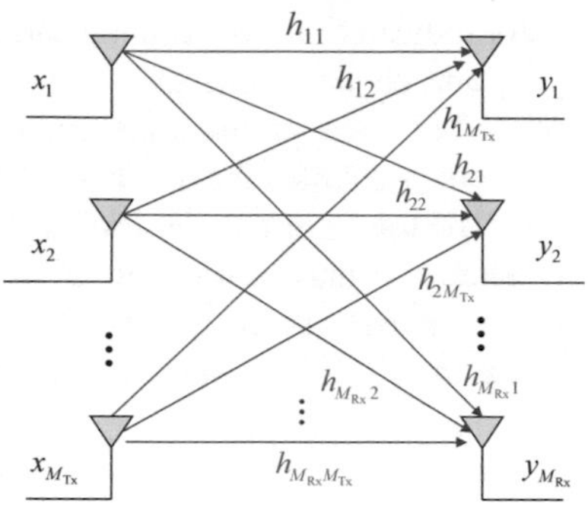

Fig. 4.25 The wireless MIMO concept

$$MG = -\lim_{\rho\to\infty}\frac{C_{\text{outage},p}(\rho)}{\log_2\rho}, \tag{4.100}$$

where $C_{\text{outage},p}\ (\rho)$ is the p percentage outage capacity at single-to-noise ratio ρ. The p percentage capacity is defined as the transmission rate that can be supported in $(100 - p)\%$ of the channel realizations. As an illustration, in Fig. 4.25, we provide the basic concept of wireless MIMO communication, with the number of transmit antennas denoted as M_{Tx} and the number of receive antennas denoted as M_{Rx}. The 2-D symbol transmitted on the mth transmit antennas is denoted by x_m ($m = 1, 2, \ldots, M_{\text{Tx}}$), while the received symbol on the nth received antenna is denoted as y_n. The channel coefficient relating the received symbol on nth receive antenna and transmitted symbol on the mth transmit antenna is denoted as h_{mn}. By using this model, the output of the nth receive antenna can be represented as

$$y_n = \sum_{m=1}^{M_{\text{Tx}}} h_{nm}x_m + z_n, \tag{4.101}$$

where z_n is the zero-mean Gaussian noise sample with variance $\sigma^2 = N_0B$, with $N_0/2$ being the double-sided power spectral density and B being the channel bandwidth.

A narrowband *point-to-point wireless communication system* composed of M_{Tx} transmit and M_{Rx} receive antennas can be represented by the following discrete-time model, which is matrix representation of Eq. (4.101):

$$\boldsymbol{y} = \boldsymbol{Hx} + \boldsymbol{z}, \tag{4.102}$$

where $\boldsymbol{x}$, $\boldsymbol{y}$, and $\boldsymbol{z}$ represent the transmitted, received, and noise vectors defined as

$$\boldsymbol{x} = \begin{bmatrix} x_1 \\ x_2 \\ \cdots \\ x_{M_{Tx}} \end{bmatrix}, \quad \boldsymbol{y} = \begin{bmatrix} y_1 \\ y_2 \\ \cdots \\ y_{M_{Rx}} \end{bmatrix}, \quad \boldsymbol{z} = \begin{bmatrix} z_1 \\ z_2 \\ \cdots \\ z_{M_{Rx}} \end{bmatrix}. \tag{4.103}$$

As indicated above, channel noise samples z_n are complex Gaussian with zero-mean and covariance matrix $\sigma^2 I_{Mr}$, where $\sigma^2 = N_0 B$. If the noise variance is normalized to one than the transmitted power P can be normalized as follows: $P/\sigma^2 = \rho$, where ρ denotes the average SNR per receiver antenna, assuming unity channel gain. The total normalized transmitted power would be then $\sum_{m=1}^{M_{Tx}} \langle x_i x_i^* \rangle = \rho$. The correlation matrix of transmitted vector $\boldsymbol{x}$ is given by $\boldsymbol{R}_x = \langle \boldsymbol{x}\boldsymbol{x}^\dagger \rangle$. The trace of correlation matrix is clearly $\mathrm{Tr}(\boldsymbol{R}_x) = \rho$. We use $\boldsymbol{H} = [h_{mn}]_{M_{Tx} \times M_{Rx}}$ in Eq. (4.103) to denote the so-called *channel matrix*:

$$\boldsymbol{H} = \begin{bmatrix} h_{11} & h_{12} & \ldots & h_{1M_{Tx}} \\ h_{21} & h_{22} & \ldots & h_{2M_{Tx}} \\ \ldots & \ldots & \ldots & \ldots \\ h_{M_{Rx}1} & h_{M_{Rx}2} & \ldots & h_{M_{Rx}M_{Tx}} \end{bmatrix}. \tag{4.104}$$

Let us perform the *singular value decomposition* (SVD) of channel matrix $\boldsymbol{H}$ given by Eq. (4.104). In singular value decomposition, we have the following: (i) the $\sum$-matrix corresponds to scaling operation, (ii) the columns of matrix $\boldsymbol{U}$, obtained as eigenvectors of the Wishart matrix $\boldsymbol{H}\boldsymbol{H}^\dagger$, correspond to the rotation operation, and (iii) the other rotation matrix $\boldsymbol{V}$ that has for columns the eigenvectors of $\boldsymbol{H}^\dagger\boldsymbol{H}$. The rank of matrix $\boldsymbol{H}$ corresponds to the multiplexing gain. The *parallel decomposition* of the wireless channel, illustrated in Fig. 4.26, is obtained by introducing the following two transformations to the channel input vector $\boldsymbol{x}$ and channel output vector $\boldsymbol{y}$: (i) *transmit precoding*, in which the input vector of symbols $\boldsymbol{x}$ to transmitter is linearly transformed by pre-multiplying with rotation matrix $\boldsymbol{V}$; (ii) *receiver shaping*, in which the receiver vector $\boldsymbol{y}$, upon optical coherent detection, is multiplied by rotation matrix $\boldsymbol{U}^\dagger$. With these two operations, we effectively performed the following manipulation:

$$\tilde{\boldsymbol{y}} = \boldsymbol{U}^\dagger(\underbrace{\boldsymbol{H}}_{\boldsymbol{U}\boldsymbol{\Sigma}\boldsymbol{V}^\dagger}\boldsymbol{x} + \boldsymbol{z}) = \boldsymbol{U}^\dagger\boldsymbol{U}\boldsymbol{\Sigma}\boldsymbol{V}^\dagger\boldsymbol{V}\tilde{\boldsymbol{x}} + \underbrace{\boldsymbol{U}^\dagger\boldsymbol{z}}_{\tilde{\boldsymbol{z}}} = \boldsymbol{\Sigma}\tilde{\boldsymbol{x}} + \tilde{\boldsymbol{z}}, \tag{4.105}$$

and successfully decomposed the optical MIMO optical channel into $R_H = \mathrm{rank}(\boldsymbol{H})$ parallel single-input–single-output (SISO) channels. The corresponding equivalent parallel wireless MIMO channel model is provided in Fig. 4.27. If there is no con-

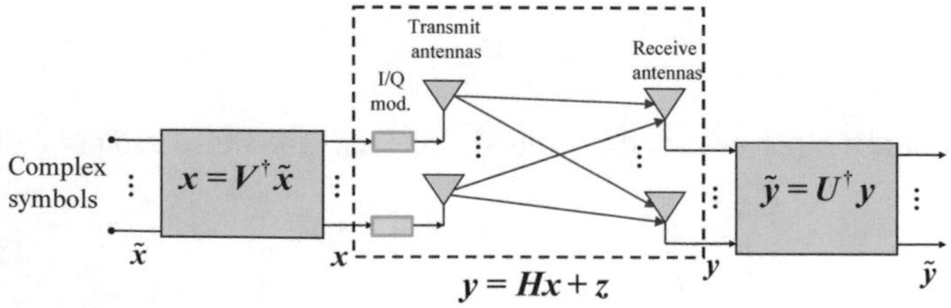

Fig. 4.26 The parallel decomposition of wireless MIMO channel. The received vector $\boldsymbol{y}$ can be written as $\boldsymbol{y} = \boldsymbol{H}\boldsymbol{x} + \boldsymbol{z}$ with $\boldsymbol{z}$ denoting the equivalent noise process

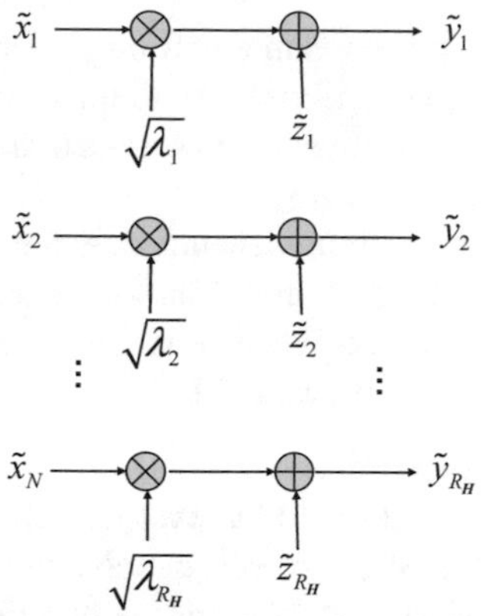

Fig. 4.27 Equivalent parallel wireless MIMO channel. The λ_i denotes the ith eigenvalue of the Wishart matrix $\boldsymbol{HH}^\dagger$

cern about multiplexing gain, a simpler scheme known in wireless communication literature as *beamforming* [32–34] can be used instead. In this scheme, illustrated in Fig. 4.28, the same symbol x, weighted by a complex scale factor, is sent over all spatial modes. Therefore, by applying the beamforming strategy, the precoding and receiver matrices become just column vectors, so it is $\boldsymbol{V} = \boldsymbol{v}$ and $\boldsymbol{U} = \boldsymbol{u}$. The resulting received signal is now given by

$$y = \boldsymbol{u}^\dagger \boldsymbol{H}\boldsymbol{v}x + \boldsymbol{u}^\dagger \boldsymbol{z}, \quad \|\boldsymbol{u}\| = \|\boldsymbol{v}\| = 1, \tag{4.106}$$

where with ||·|| we denoted the norm of a vector. When the channel matrix $\boldsymbol{H}$ is known at the receiver side, the received SNR can be optimized by choosing $\boldsymbol{u}$ and $\boldsymbol{v}$ as the principal left and right singular vectors of the channel matrix $\boldsymbol{H}$, while the corresponding SNR is given by $\sigma_{\max}^2\rho$, where $\sigma_{\max}$ is the largest eigenvalue of the Wishart matrix $\boldsymbol{HH}^\dagger$. The channel capacity beamforming scheme can be expressed as $C = B\log_2(1 + \sigma_{\max}^2\mathrm{SNR})$, which corresponds to the capacity of a single SISO

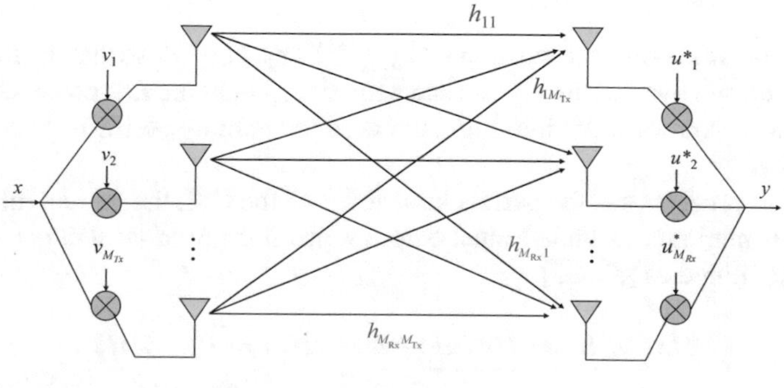

Fig. 4.28 The beamforming strategy in wireless MIMO

channel having the channel power gain equal σ^2_{max}. On the other hand, when the channel matrix is not known, we will need to do maximization only with respect to $\boldsymbol{u}$. The array gain of beamforming diversity is between $\max(M_{\text{Tx}}, M_{\text{Rx}})$ and $M_{\text{Tx}}M_{\text{Rx}}$, while the diversity gain equals $M_{\text{Tx}}M_{\text{Rx}}$.

MIMO channel capacity evaluation depends on what is known about the channel matrix or its distribution at the transmitter and/or receiver sides. We assume that channel state information at the receiver side is always available, denoted as CSIR. We consider the following relevant cases [32, 82]:

- The channel matrix $\boldsymbol{H}$ is deterministic (static).
- The channel matrix $\boldsymbol{H}$ is random, and at every channel use an independent realization of $\boldsymbol{H}$ is generated; in other words, the channel is *ergodic*.
- The channel matrix is random, but fixed for the duration of transmission; in other words, the channel is *non-ergodic*.

The MIMO channel capacity for *deterministic (static) channel* can be defined as the following maximization problem:

$$C = \max_{f_X(x_1,\ldots,x_{M_{Tx}})} \left[H(\boldsymbol{y}) - H(\boldsymbol{y}|\boldsymbol{x})\right], \tag{4.107}$$

where the maximization is performed over the multivariate PDF $f_X(\boldsymbol{x})$. Given that components of noise vector are i.i.d. and complex circular symmetric Gaussian, the corresponding distribution of z is a multivariate Gaussian [32, 82]:

$$f_Z(z) = \frac{1}{\det(\pi \boldsymbol{R}_z)} e^{-(z-\boldsymbol{\mu}_z)^\dagger \boldsymbol{R}_z^{-1}(z-\boldsymbol{\mu}_z)}, \tag{4.108}$$

where $\mu_z \doteq E(z)$ and R_z is the covariance matrix of z, defined as $\boldsymbol{R}_z \doteq E\big((z-\boldsymbol{\mu}_z)(z-\boldsymbol{\mu}_z)^\dagger\big) = E\big(zz^\dagger\big) - \boldsymbol{\mu}_z\boldsymbol{\mu}_z^\dagger$. Following derivation described in [32], we obtain the following expression for MIMO channel capacity:

$$C = \max_{\boldsymbol{R}_x:\text{Tr}(\boldsymbol{R}_x)=P} \log_2 \det\left(\boldsymbol{I}_{M_{Rx}} + \frac{1}{\sigma^2}\boldsymbol{H}\boldsymbol{R}_x\boldsymbol{H}^\dagger\right)[\text{bits/channel use}], \tag{4.109}$$

where the transmit power is constrained by $\sum_{i=1}^{M_{Tx}} \langle \boldsymbol{x}_i\boldsymbol{x}_i^* \rangle \le P$, so that the maximization is performed over all input covariance matrices satisfying this power constraint. If we want to convert units into bit/s, we need to multiply (4.109) by the channel bandwidth B.

When the receiver has the perfect knowledge of the CSI, the mutual information between channel input $\boldsymbol{x}$ and channel output $\boldsymbol{y}$, given the *random channel matrix* $\boldsymbol{H}$, can be determined as [32, 83]

$$\begin{aligned} I(\boldsymbol{x};\boldsymbol{y},\boldsymbol{H}) &= \underbrace{I(\boldsymbol{x};\boldsymbol{H})}_{=0} + I(\boldsymbol{x};\boldsymbol{y}|\boldsymbol{H}) = I(\boldsymbol{x};\boldsymbol{y}|\boldsymbol{H}) \\ &= E_{\boldsymbol{H}'} I\big(\boldsymbol{x};\boldsymbol{y}|\boldsymbol{H}=\boldsymbol{H}'\big), \end{aligned} \tag{4.110}$$

where we use the subscript $\boldsymbol{H}'$ in expectation operator to denote different realizations $\boldsymbol{H}'$ of the random channel matrix $\boldsymbol{H}$. The term $I(\boldsymbol{x}; \boldsymbol{H}) = 0$ as the input vector and the channel matrix are independent of each other. The expression for *ergodic MIMO capacity* is given by [32]

$$C = E_{\boldsymbol{H}} \max_{\boldsymbol{R}_x:\mathrm{Tr}(\boldsymbol{R}_x)=P} \log_2 \det\left(\boldsymbol{I}_{M_{Rx}} + \frac{1}{\sigma^2}\boldsymbol{H}\boldsymbol{R}_x\boldsymbol{H}^\dagger\right). \tag{4.111}$$

In *non-ergodic wireless channel*, the channel matrix is generated at random but kept fixed for the duration of the transmission. Since the average channel capacity does not have any meaning here, we can define the instantaneous channel capacity as a function of the channel matrix:

$$C(\boldsymbol{H}) = \log_2 \det\left(\boldsymbol{I}_{M_{Rx}} + \frac{P/\sigma^2}{M_{Tx}}\boldsymbol{H}\boldsymbol{H}^\dagger\right), \tag{4.112}$$

and determine the *outage probability* as the probability that the transmission rate exceeds instantaneous channel capacity $C(\boldsymbol{H})$ [83]:

$$P_{\text{outage}} = \Pr\{C(\boldsymbol{H}) < R\}. \tag{4.113}$$

The maximum rate that can be supported by this channel, for the outage probability P_{outage}, is called the *outage capacity*. More formally, the *p-percent outage capacity* is defined as the data capacity that is guaranteed in $(1 - p) \times 100\%$ of channel realizations [84]:

$$\Pr\left\{C(\boldsymbol{H}) \le C_{\text{outage},p}\right\} = p. \tag{4.114}$$

Given that $C_{\text{outage},p}$ occurs in $(1 - p)\%$ of channel uses, the *throughput* of the system associated with, denoted as R_p, can be determined as [84]

$$R_p = (1 - p)C_{\text{outage},p}. \tag{4.115}$$

An interesting special case of non-ergodic channels is so-called *block fading* channel model. In this channel model, the channel matrix is selected randomly and kept fixed of duration of N_s symbols. After N_s symbols a new channel matrix realization is drawn. The *B-block fading channel model* can be described as

$$\boldsymbol{y}_b[n] = \boldsymbol{H}_b\boldsymbol{x}_b[n] + \boldsymbol{z}_b[n];\ b = 1, \ldots, B;\ n = 1, \ldots, N_s, \tag{4.116}$$

where b is the index of the block and n is the index of the symbol within that block. Other assumptions for input, output, and noise vectors are the same as before. Again, we can perform the SVD of channel matrix $\boldsymbol{H}_b$ we obtain

$$\boldsymbol{H}_b = \boldsymbol{U}_b\boldsymbol{\Sigma}_b\boldsymbol{V}_b^\dagger, \tag{4.117}$$

and by pre-multiplying (4.117) by $\boldsymbol{U}_b^\dagger$ the output–input relationship becomes

$$\begin{aligned}\boldsymbol{U}_b^\dagger \boldsymbol{y}_b[n] &= \boldsymbol{U}_b^\dagger \underbrace{\boldsymbol{U}_b \boldsymbol{\Sigma}_b \boldsymbol{V}_b^\dagger}_{\boldsymbol{H}_b} \boldsymbol{x}_b[n] + \boldsymbol{U}_b^\dagger \boldsymbol{z}_b[n] \\ &= \boldsymbol{\Sigma}_b \boldsymbol{V}_b^\dagger \boldsymbol{x}_b[n] + \boldsymbol{U}_b^\dagger \boldsymbol{z}_b[n]. \end{aligned} \tag{4.118}$$

When transmitted vector $\boldsymbol{x}_b[n]$ get preprocessed on transmitter side as follows $\boldsymbol{V}_b \boldsymbol{x}_b[n]$, the output–input relationship (4.118) becomes free of spatial interference. As the number of blocks tends to plus infinity, the block fading model becomes ergodic and for CSIR case, the channel capacity can be evaluated as

$$C = \left\langle \sum\nolimits_i \log_2\left(1 + \frac{P/\sigma^2}{M_{Tx}} \sigma_i^2\right)\right\rangle. \tag{4.119}$$

A reader interested to learn more about the MIMO channels is referred to [32].

4.8.2 PLS for Wireless MIMO Channels

The MIMO wiretap channel consists of Alice's M_{Tx} antennas, Bob's M_{Rx} antennas, and Eve's M_{E} receive antennas. The matrix representation of the main (Alice-to-Bob) channel is given by

$$\boldsymbol{y}_B = \boldsymbol{H}_B \boldsymbol{x}_A + \boldsymbol{z}_B, \tag{4.120}$$

while the corresponding matrix representation of Alice-to-Eve's channel is

$$\boldsymbol{y}_E = \boldsymbol{H}_E \boldsymbol{x}_A + \boldsymbol{z}_E, \tag{4.121}$$

where $\boldsymbol{x}_{\mathrm{A}}$ is Alice transmitted vector, with covariance $\boldsymbol{R}_x = E\left(\boldsymbol{x}_A \boldsymbol{x}_A^\dagger\right)$, $\boldsymbol{y}_{\mathrm{B}}$ is Bob's received vector, $\boldsymbol{H}_B$ is Alice-to-Bob channel matrix of size $M_{\mathrm{Rx}} \times M_{\mathrm{Tx}}$, and $\boldsymbol{H}_E$ is Alice-to-Eve channel matrix of size $M_{\mathrm{E}} \times M_{\mathrm{Tx}}$. When the additive noise is Gaussian, the optimum source to achieve secrecy capacity is Gaussian, and the corresponding secrecy capacity expression under the average power constraint $\mathrm{Tr}(\boldsymbol{R}_x) \le P$ will be [85]

$$C_s = \max_{\boldsymbol{R}_x, \mathrm{Tr}(\boldsymbol{R}_x) \le P} [I(\boldsymbol{X}_A, \boldsymbol{Y}_B) - I(\boldsymbol{X}_A, \boldsymbol{Y}_E)]. \tag{4.122}$$

For Gaussian source, based on Eq. (4.109), if the noise is spatially white with unit variance, we can write

$$C_s = \max_{R_x, \mathrm{Tr}(R_x) \le P} \left[\log_2 \det\left(I + H_B R_X H_B^\dagger \right) - \log_2 \det\left(I + H_E R_X H_E^\dagger \right) \right]. \tag{4.123}$$

From equation above, we conclude that we still need to determine the transmit covariance matrix $\boldsymbol{R}_x$ maximizing the secrecy rate. Now we consider two scenarios, when the CSI is perfectly and partially known.

Assuming that transmitter has the *perfect CSI* for both Bob and Eve, for the multiple-input single-output multiple-Eve (MISOME) case, the optimal solution is transmit beamforming [85]:

$$\boldsymbol{R}_x = P|\lambda_m\rangle\langle\lambda_m|, \ \langle\lambda_m| = (|\lambda_m\rangle)^\dagger, \tag{4.124}$$

where the unit-norm column-eigenvector $|\lambda_m\rangle$ corresponding to the eigenvalue λ_m is obtained as the solution to eigenvalue equation [85]:

$$\left(I + P h_B^\dagger h_B \right)|\lambda_m\rangle = \lambda_m \left(I + P H_E^\dagger H_E \right)|\lambda_m\rangle, \tag{4.125}$$

where $\boldsymbol{h}_B$ is Bob's channel vector. The solution for the secrecy rate maximization problem of Eq. (4.123) is given by [86]

$$C_S = \frac{1}{2}\log_2\left[\lambda\max\left(I + P h_B^\dagger h_B, I + P H_E^\dagger H_E \right) \right], \tag{4.126}$$

where $\lambda_{\max}$ is the largest generalized eigenvalue of the following two matrices $\boldsymbol{I} + P\boldsymbol{h}_B^\dagger \boldsymbol{h}_B$ and $\boldsymbol{I} + P\boldsymbol{H}_E^\dagger \boldsymbol{H}_E$. Under the matrix power covariance constraint $\boldsymbol{R}_x \le \boldsymbol{S}$, where "$\le$" denotes less or equal to in the positive semidefinite partial ordering sense of the symmetric matrices, the MIMO secrecy capacity is given by [87]

$$C_S(S) = \sum_i \log_2 \lambda_i \left(I + S^{1/2} H_B^\dagger H_B S^{1/2}, I + S^{1/2} H_E^\dagger H_E S^{1/2} \right), \tag{4.127}$$

where λ_i are greater than one eigenvalue of matrices $\boldsymbol{I} + \boldsymbol{S}^{1/2}\boldsymbol{H}_B^\dagger \boldsymbol{H}_B \boldsymbol{S}^{1/2}$ and $\boldsymbol{I} + \boldsymbol{S}^{1/2}\boldsymbol{H}_E^\dagger \boldsymbol{H}_E \boldsymbol{S}^{1/2}$. The *average power constraint* is much less restrictive than the matrix power constraint, and the secrecy capacity of the MIMO wiretap channel under the average power constraint can be found by the exhaustive search over the set $\{S|S \ge \mathbf{0}), \mathrm{Tr}(S) \le P\}$; in other words, we can write [85, 88]

$$C_S(P_t) = \max_{S \ge 0, \mathrm{Tr}(S) \le P_t} C_S(S). \tag{4.128}$$

The practical schemes, due to limited feedback capabilities, we must deal with the imperfect CSI or *partial CSI* [89, 90]. A realistic model will be to assume that actual channel realizations are not known to Alice and Bob but are known to lie in an uncertainty set of possible channels [89, 90], and these channels are known as *compound channels* [89–92]. Now we have to perform the reliability and security studies over all possible realizations in the uncertainty sets. The uncertainty sets for Bob and Eve channel matrices' estimates can be defined, respectively, as [85, 90]

$$\mathcal{H}_B = \{\boldsymbol{H}_B | \boldsymbol{H}_B = \boldsymbol{H} + \Delta\boldsymbol{H},\ \|\Delta\boldsymbol{H}\|_F \leq \varepsilon\}, \tag{4.129}$$

$$\mathcal{H}_E = \{\boldsymbol{H}_E | \|\Delta\boldsymbol{H}_E\|_F \leq \varepsilon\}, \tag{4.130}$$

where $\|\cdot\|_F$ is the Frobenius norm, $\boldsymbol{H}$ is the estimate of Bob's channel matrix, and $\Delta\boldsymbol{H}$ is the estimation error not larger than ε. Clearly, we assume that the estimate of Eve's channel matrix is not known, only the maximum estimation error. This scenario is applicable when Eve cannot approach the transmitter beyond certain protection distance. In other words, this compound model is applicable when Eve is outside of the exclusion zone. Interestingly enough, the secrecy capacity for such compound model can be determined for the nondegraded case for Gaussian MIMO channels [90]. Based on (4.123), (4.129), and (4.130), we conclude that the compound MIMO wiretap secrecy capacity can be determined by

$$C_s = \max_{\mathrm{Tr}(\boldsymbol{R}_X) \leq P} \left[\min_{\boldsymbol{H}_B \in \mathcal{H}_B} \log_2 \det\left(\boldsymbol{I} + \boldsymbol{H}_B \boldsymbol{R}_X \boldsymbol{H}_B^{\dagger}\right) - \max_{\boldsymbol{H}_E \in \mathcal{H}_E} \log_2 \det\left(\boldsymbol{I} + \boldsymbol{H}_E \boldsymbol{R}_X \boldsymbol{H}_E^{\dagger}\right)\right]. \tag{4.131}$$

Clearly, the maximum secrecy rate is limited by the worst channel to Bob and the best channel to Eve.

The concept of *broadcast channel with confidential messages* has been already introduced in Sect. 4.4.1. In broadcast scenario, one user sends the message to multiple receivers, and this scenario is very common in downlink phase of cellular communication, in which the base station sends information to multiple mobile users. In a broadcast channel with confidential messages, Alice sends confidential message to Bob, which should be kept as secret as possible from other users/eavesdroppers. The broadcast channel with parallel independent subchannels is considered in [93] and the corresponding optimal source allocation, achieving the boundary of secrecy capacity region is determined. The transmission of two confidential messages over discrete memoryless broadcast channel is studied in [94], wherein each receiver serves as an eavesdropper of the other. The general MIMO Gaussian case over the matrix power constraint is studied in [95]. For two-user broadcast channel, with each user receiving the corresponding confidential message, it has been shown that the secrecy capacity region (R_1, R_2) is rectangular for the matrix power constraint $\boldsymbol{C}_x \leq \boldsymbol{S}$. Assuming so-called secret dirty-paper coding [95], the corner point rate (R_1^*, R_2^*) for secrecy capacity is given by [85, 95]

$$\begin{aligned} R_1^* &= \sum_i \log_2 \lambda_i,\ \lambda_i = \mathrm{eigenvalues}(\boldsymbol{I} + \boldsymbol{S}^{1/2} \boldsymbol{H}_B^{\dagger} \boldsymbol{H}_B \boldsymbol{S}^{1/2}) \\ R_2^* &= -\sum_j \log_2 \lambda_j',\ \lambda_j' = \mathrm{eigenvalues}(\boldsymbol{I} + \boldsymbol{S}^{1/2} \boldsymbol{H}_E^{\dagger} \boldsymbol{H}_E \boldsymbol{S}^{1/2}), \end{aligned} \tag{4.132}$$

wherein λ_i are larger than one eigenvalue of $\boldsymbol{I} + \boldsymbol{S}^{1/2}\boldsymbol{H}_B^{\dagger}\boldsymbol{H}_B\boldsymbol{S}^{1/2}$, while λ'_j are smaller than one eigenvalue of $\boldsymbol{I} + \boldsymbol{S}^{1/2}\boldsymbol{H}_E^{\dagger}\boldsymbol{H}_E\boldsymbol{S}^{1/2}$. Some more recent results on MIMO broadcast channels with confidential messages can be found in [96].

In a *multiaccess channel*, multiple users send information toward the same receiver, and this situation occurs in cellular communications when multiple mobile users transmit data toward the base station. *In multiaccess channel with confidential messages*, multiple senders Alice 1, Alice 2, … transmit confidential messages $\boldsymbol{M}_1$, $\boldsymbol{M}_2$, … to a single receiver Bob. Each confidential message should be decodable by Bob but without any leakage to other transmitters. This problem is considered in [97], wherein the inner and outer bounds for the region of secret rates have been determined, while the secrecy capacity region has been left unsolved.

The interference channel describes a scenario in which multiple transmitter–receiver communication links are simultaneously active introducing the crosstalk to each other. The *interference channel with confidential messages* represents the situation in which multiple transmitters want to transmit confidential message to the respective receivers in such a way to keep those messages secure from the counterpart receivers [90, 94, 98, 99]. As an illustration, in Fig. 4.29, the system model for two-user MIMO interference channel with confidential messages is provided. The $\boldsymbol{H}_i$ is used to denote the channel matrix of the *i*th ($i = 1, 2$) direct channel, while $\boldsymbol{H}_{ji}$ to denote the channel matrix of the crosstalk channel *j*. The figure of merit suitable to study the scaling behavior of the aggregate secrecy rate $R_{\oplus}$ in multiuser networks with signal-to-noise ratio at transmitter side, denoted as ρ, is the *secret multiplexing gain*, denoted as *SMG*, and can be defined as [85, 98]

$$SMG = \lim_{\rho \to \infty} \frac{R_{\oplus}}{\log \rho}. \tag{4.133}$$

The SMG, also known as the *secure degree of freedom* (DoF), has been studied in [98, 99] for the *K*-users Gaussian interference channel, where $K \geq 3$, assuming that multiuser network is composed of *K* transmitter–receiver pairs, employing *F* frequency bands, and it has been found that SMG = $(K - 2)/(2K - 2)$ secure DoF per frequency–time slot per user are almost surely achievable [98]. On the other hand, when an external eavesdropper is present, each user can achieve SMG = $(K - 1)/(2K)$ secure DoFs per frequency–time slot [98].

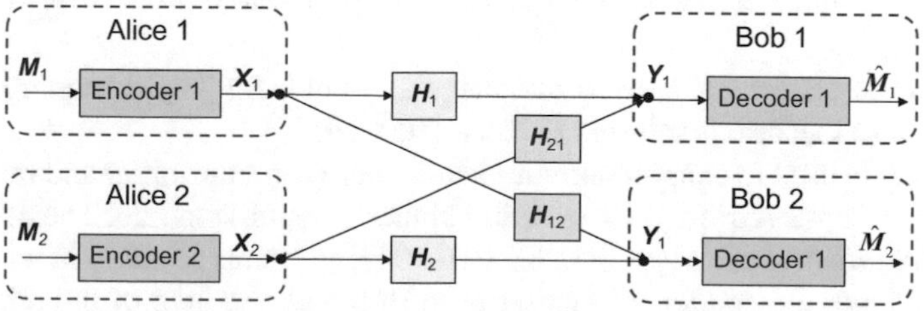

Fig. 4.29 The model for two-user MIMO interference channel with confidential messages

The two-user MIMO interference channel, with each node having arbitrary number of antennas, has been studied in [100, 101], and the corresponding interference channel model illustrated in Fig. 4.29 can be represented as follows:

$$\begin{aligned} \boldsymbol{y}_1 &= \boldsymbol{H}_1\boldsymbol{x}_1 + \boldsymbol{H}_{21}\boldsymbol{x}_2 + \boldsymbol{z}_1, \\ \boldsymbol{y}_2 &= \boldsymbol{H}_2\boldsymbol{x}_2 + \boldsymbol{H}_{12}\boldsymbol{x}_1 + \boldsymbol{z}_2. \end{aligned} \tag{4.134}$$

The achievable secrecy regions have been derived in [100, 101] under different CSI assumptions and various noncooperative and cooperative scenarios. In noncooperative scenarios, it is assumed that both transmitters have perfect CSI of both the direct and crosstalk channels. The generalized singular value decomposition (GSVD) can be employed, as explained in [100, 101]. Given the direct and crosstalk channel matrices $\boldsymbol{H}_1$ and $\boldsymbol{H}_{12}$, the GSVD procedure gives unitary matrices $\boldsymbol{U}_{\mathrm{Rx1}}$ and $\boldsymbol{U}_{\mathrm{E1}}$, positive semidefinite diagonal matrices $\boldsymbol{\Lambda}_1$ and $\boldsymbol{D}_1$, and $\boldsymbol{A}_1$ matrix of size $M_{\mathrm{Rx1}} \times d$, where $d = \min(M_{\mathrm{Rx1}}, M_{\mathrm{Tx1}} + M_{\mathrm{Tx2}})$, such that the following relationships are valid [100, 101]:

$$\begin{aligned} \boldsymbol{H}_1\boldsymbol{A}_1 &= \boldsymbol{U}_{Rx_1}\boldsymbol{\Lambda}_1, \\ \boldsymbol{H}_{12}\boldsymbol{A}_1 &= \boldsymbol{U}_{E_1}\boldsymbol{D}_1. \end{aligned} \tag{4.135}$$

The nonzero elements in $\boldsymbol{\Lambda}_1$ are ordered in ascending order, while in decreasing order in $\boldsymbol{D}_1$, and these two diagonal matrices are related by $\boldsymbol{\Lambda}_1^T\boldsymbol{\Lambda}_1 + \boldsymbol{D}_1^T\boldsymbol{D}_1 = \boldsymbol{I}$. The transmitter Tx_1 then precodes $\boldsymbol{x}_1$ as follows:

$$\boldsymbol{x}_1 = \boldsymbol{A}_1\boldsymbol{s}_1, \quad \boldsymbol{s}_1 \sim \mathcal{CN}(\boldsymbol{0}, \boldsymbol{\Sigma}_1), \tag{4.136}$$

where $\mathcal{CN}$ is the multivariate complex circular symmetric zero-mean Gaussian (normal) distribution with covariance matrix $\boldsymbol{\Sigma}_1$ being positive semidefinite diagonal matrix. The diagonal elements in $\boldsymbol{\Sigma}_1$ represent the powers allocated to different antennas. The transmitter Tx_2 applies an equivalent procedure. The achievable secrecy rate for Tx_1 is given by [100, 101]

$$\begin{aligned} 0 \le R_1 \le \log\left|\boldsymbol{I} + \boldsymbol{U}_{Rx_1}\boldsymbol{\Lambda}_1\boldsymbol{\Sigma}_1\boldsymbol{\Lambda}_1^\dagger\boldsymbol{U}_{Rx_1}^\dagger + \boldsymbol{U}_{E_2}\boldsymbol{D}_2\boldsymbol{\Sigma}_2\boldsymbol{D}_2^\dagger\boldsymbol{U}_{E_2}^\dagger\right| \\ - \log\left|\boldsymbol{I} + \boldsymbol{U}_{E_2}\boldsymbol{D}_2\boldsymbol{\Sigma}_2\boldsymbol{D}_2^\dagger\boldsymbol{U}_{E_2}^\dagger\right| - \log\left|\boldsymbol{I} + \boldsymbol{U}_{E_1}\boldsymbol{D}_1\boldsymbol{\Sigma}_1\boldsymbol{D}_1^\dagger\boldsymbol{U}_{E_1}^\dagger\right|. \end{aligned} \tag{4.137}$$

The similar achievable secrecy rate inequality holds for Tx_2. A reader interested in cooperative scenario is referred to Refs. [100, 101].

The relay is used to support communication between transmitter and receiver, by extending the range/coverage or enabling higher transmission rate. The *relay channel with confidential messages* [85, 90, 102–104] represents scenario in which Alice wants to deliver the confidential message to Bob with the help of the relay. Given that relay can be trusted on untrusted, it is possible to classify the various MIMO wiretap networks into two broad categories: *trusted* and *untrusted* relay wiretap net-

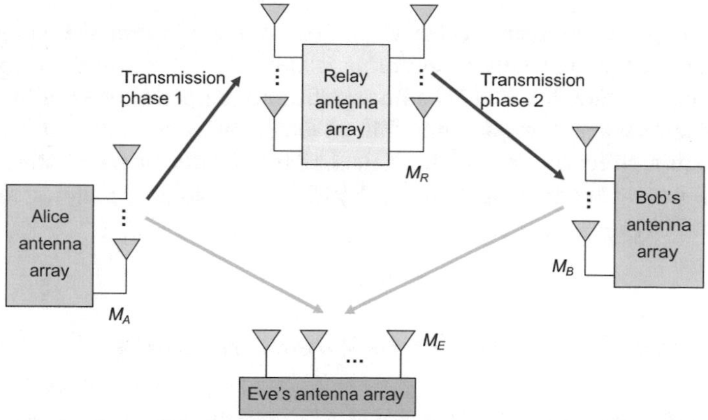

Fig. 4.30 The two-hop MIMO relay system architecture with an external eavesdropper

works. The cooperative jamming strategies enabling secure communication without employing the external helpers have been discussed in [104], with corresponding two-hop MIMO relay network architecture provided in Fig. 4.30. A number of antennas available to Alice, Bob, relay, and Eve are denoted, respectively, as M_A, M_B, M_R, and M_E. In cooperative jamming strategies, Alice and Bob transmit jamming signals in stages when they do not typically transmit any data. The relay retransmits the signal received from Alice. Eve has access to both Alice and Bob's transmissions over wireless channel. The signals received by relay and Eve in transmission phase 1 can be represented as [104]

$$\begin{aligned} \boldsymbol{y}_R &= \boldsymbol{H}_{AR}\left(\boldsymbol{T}_A\boldsymbol{x}_A + \boldsymbol{T}'_A z_A\right) + H_{BR}\boldsymbol{T}'_B z_B + z_R, \\ \boldsymbol{y}_{E_1} &= \boldsymbol{H}_{AE}\left(\boldsymbol{T}_A\boldsymbol{x}_A + \boldsymbol{T}'_A z_A\right) + H_{BE}\boldsymbol{T}'_B z_B + z_{E_1}, \end{aligned} \tag{4.138}$$

where $\boldsymbol{x}_A$ is the data signal transmitted by Alice toward the relay, while z_A and z_B are the jamming signals transmitted by Alice and Bob, respectively. (The relay and Eve's antennas noise sources are denoted by z_R and $\mathbf{z}_{E1}$, respectively.) We use $\boldsymbol{T}_A$ and $\boldsymbol{T}'_B$ to denote the beamformers corresponding to Alice and Bob. On the other hand, the signals received by Bob and Eve in transmission phase 2 can be represented as [104]

$$\begin{aligned} \boldsymbol{y}_R &= \boldsymbol{H}_{RB}\left(\boldsymbol{T}_R\boldsymbol{x}_R + \boldsymbol{T}'_R z_R\right) + \boldsymbol{H}_{AB}\boldsymbol{T}'_{A2} z_{A2} + z_B, \\ \boldsymbol{y}_{E_1} &= \boldsymbol{H}_{AE}\left(\boldsymbol{T}_R\boldsymbol{x}_R + \boldsymbol{T}'_R z_R\right) + \boldsymbol{H}_{AE}\boldsymbol{T}'_{A2} z_{A2} + z_{E_1}, \end{aligned} \tag{4.139}$$

where $\boldsymbol{x}_{\mathrm{R}}$ is the data signal transmitted by relay and $\boldsymbol{T}_{\mathrm{R}}$ is the corresponding beamformer. We use the z_{R} and z_{A2} to denote the jamming signals transmitted by relay and Alice, respectively. ($\boldsymbol{T}'_{\mathrm{R}}$ and $\boldsymbol{T}'_{\mathrm{A2}}$ are corresponding jammers' beamformers.) When both Alice and Bob's jammer signals are different from zero, that is, $z_{\mathrm{A}} \neq 0$ and $z_{\mathrm{B}} \neq 0$ in transmit phase 1, the corresponding scheme is called the *fully cooper-*

ating jamming (FCJ) scheme. Otherwise, if one of them is zero, the corresponding scheme is called the *partially cooperating jamming* (PCJ) scheme. As expected, in terms of secrecy rate, the FCJ scheme significantly outperforms both PCJ scheme and the scheme without the jamming. Moreover, the FCJ does not exhibit the secrete rate saturation effect for very high SNRs. For full details on cooperative jamming schemes for secure communication in relay MIMO networks, an interested reader is referred to [104].

4.8.3 Secret-Key Generation in Wireless Networks

The secret-key generation (SKG) protocols have already been described in Sect. 4.5, where source- and channel-type SKG protocols are introduced and analyzed. Here, we concentrate on SKG protocols in context of wireless networks. The wireless channels themselves can serve as a source of common randomness as discussed in [90, 105–108]. The SKG protocols for wireless channels are similar to those considered in Sect. 4.5, namely, Alice and Bob want to generate the secret with as little correlation with Eve as possible. To do so, they perform transmission over wireless fading channel and employ authenticated public channel for information reconciliation and privacy amplification steps. The key idea is to exploit the *fading channel coefficients reciprocity* between Alice and Bob, that is, $h_{AB} = h_{BA} = h_m$, for the channel coherence time T_c. Even if the reciprocity is not perfect, the occasional errors can be corrected for by Slepian–Wolf-like coding. In training phase, Alice sends symbol sequence $\boldsymbol{x}_A$ of duration T_A, followed by Bob's transmission of symbol sequence $\boldsymbol{x}_B$ of duration $T_c - T_A$. During Bob's transmission, Bob's and Eve's received vectors, denoted, respectively, as $\boldsymbol{y}_B$ and $\boldsymbol{y}_E$, can be represented as

$$\begin{aligned} \boldsymbol{y}_B &= h_m \boldsymbol{x}_A + \boldsymbol{z}_B, \\ \boldsymbol{y}_E &= h_{AE} \boldsymbol{x}_A + \boldsymbol{z}_E, \end{aligned} \tag{4.140}$$

with $\boldsymbol{z}_B$ and $\boldsymbol{z}_E$ being the multivariate complex circular symmetric zero-mean Gaussian (normal) distributions with covariance matrix $\boldsymbol{\Sigma}$ being $\sigma^2 \boldsymbol{I}$. On the other hand, when Bob transmits the symbol sequence $\boldsymbol{x}_B$, Alice and Eve's receive vectors can be written as

$$\begin{aligned} \boldsymbol{y}_A &= h_m \boldsymbol{x}_B + \boldsymbol{z}_A, \\ \boldsymbol{y}_E &= h_{BE} \boldsymbol{x}_B + \boldsymbol{z}_E. \end{aligned} \tag{4.141}$$

Let us assume that the main (Alice-to-Bob) channel coefficient is generated from a zero-mean Gaussian random variable of variance σ_m^2. Alice and Bob's channel coefficient estimates will be

$$h_m^{(A)} = h_m + \frac{\boldsymbol{x}_B^\dagger}{\|\boldsymbol{x}_B\|^2} z_A,$$
$$h_m^{(B)} = h_m + \frac{\boldsymbol{x}_A^\dagger}{\|\boldsymbol{x}_A\|^2} z_B. \tag{4.142}$$

Clearly, Alice (Bob) channel coefficient is a zero-mean Gaussian random variable with variance $(\sigma_m^{(A)})^2 = \sigma_m^2 + \sigma^2 / \|\boldsymbol{x}_B^2\|$ $[(\sigma_m^{(B)})^2 = \sigma_m^2 + \sigma^2 / \|\boldsymbol{x}_A^2\|]$. Assuming that both Alice and Bob's average transmit powers are the same and equal to P, we have that $\|\boldsymbol{x}_A^2\| = T_A P$, $\|\boldsymbol{x}_B^2\| = (T_c - T_A)P$. The secret-key rate (SKR) can be estimated by [107]

$$SKR = \frac{1}{T_c} I\left(h_m^{(A)}, h_m^{(B)}\right) = \frac{1}{T_c} \log_2\left[\frac{\left(\sigma^2 + \sigma_m^2 T_A P\right)\left(\sigma^2 + \sigma_m^2 (T_c - T_A)P\right)}{\sigma^4 + \sigma^2 \sigma_m^2 T_c P}\right]. \tag{4.143}$$

Obviously, the optimum frame duration for Alice is $T_A = T_c/2$, so that the previous expression simplifies to

$$SKR = \frac{1}{T_c} \log_2\left[1 + \frac{\sigma_m^2 (T_c P)^2}{4\left(\sigma^4 + \sigma^2 \sigma_m^2 T_c P\right)}\right]. \tag{4.144}$$

Form Eq. (4.144), we conclude that SKR increases as the average power P increases and decreases as the coherence time increases. Therefore, the fast fading is beneficial, while slow-varying fading results in low SKR.

To improve further the SKRs, the *joint source–channel secret-key generation* was proposed in [108]. In this scheme, the key-generation protocol, illustrated in Fig. 4.31, is composed of two phases: (i) the source-type training phase of duration T_s, which is used to generate the source-type key and estimate the main (Alice–Bob) channel coefficient at the same time and (ii) the transmission of duration $T_c - T_s$ that is used to generate the channel-type secret key. The total SKR will be then a summation of source- and channel-type SKRs. The fraction $0 < f < 1$ of training interval T_s is used by Alice to transmit the training sequence $\boldsymbol{x}_A$, while the portion of training interval $(1 - f)T_s$ is used by Bob to transmit the training symbol sequence $\boldsymbol{x}_B$. Therefore, this training source-type phase is very similar to the fading channel reciprocity-based protocol discussed above. The source-type SKR can be then calculated using an

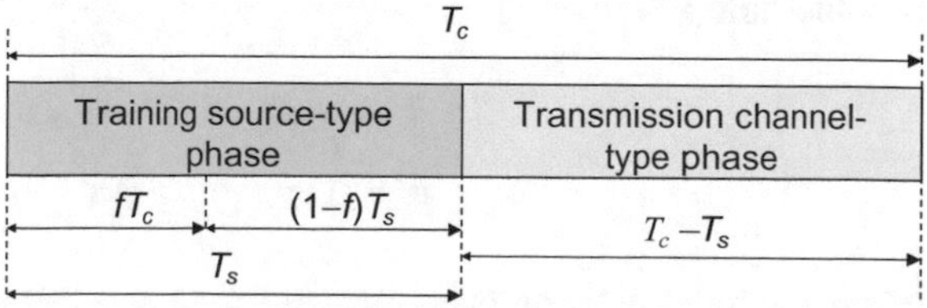

Fig. 4.31 The illustration of joint source–channel key generation scheme

expression similar to Eq. (4.143) as follows:

$$SKR_s = \frac{1}{T_c} I\left(h_m^{(A)}, h_m^{(B)}\right) = \frac{1}{T_c}\log_2\left[\frac{\left(\sigma^2 + \sigma_m^2 f T_s P_s\right)\left(\sigma^2 + \sigma_m^2 (1-f) T_s P_s\right)}{\sigma^4 + \sigma^2\sigma_m^2 T_c P}\right], \tag{4.145}$$

where P_s is the average power used during the training source-type phase. Given that the training interval T_s is shorter than coherence time, it is a good idea to use the minimum mean square error (MMSE) channel estimation approach [32, 109] instead of the least square method discussed above. Bob's MMSE estimate of the main channel coefficient will be [108]

$$\tilde{h}_m^{(B)} = \frac{\sigma_m^2}{\sigma^2 + f P_s T_s \sigma_m^2} \boldsymbol{x}_A^\dagger \boldsymbol{y}_B. \tag{4.146}$$

The true channel coefficient value can be represented as

$$h_m = \tilde{h}_m^{(B)} + \Delta h_m, \tag{4.147}$$

where the Δh_m is the estimation error, which is a zero-mean Gaussian random variable of variance $\sigma_m^2/(\sigma^2 + f P_s T_s \sigma_m^2)$.

In second phase, authors in [108] consider scheme in which Alice does not perform both power and rate adaptation to decouple the first and the second phases, since Eve can learn Alice channel coefficient estimate for the phase 1. So Alice transmits the symbol sequence to Bob with constant power, and the SKR for channel-type phase can be determined by [109]

$$\begin{aligned} SKR_{ch} &= \frac{T_c - T_s}{T_c}\left[I\left(X_A, Y_B|\tilde{h}_m^{(B)}\right) - I(X_A, Y_E|h_{AE})\right]^+ \\ &= \frac{T_c - T_s}{T_c}\left\{E\left[\log_2\left(1 + \frac{(\tilde{h}_m^{(A)})^2 P_{ch}}{\sigma^2 + \frac{\sigma_m^2 P_{ch}}{\sigma^2 + f P_s T_s \sigma_m^2}}\right)\right] - \log_2\left(1 + \frac{h_{AE}^2 P_{ch}}{\sigma^2}\right)\right\}, \end{aligned} \tag{4.148}$$

where the second term is an upper bound for Eve's mutual information, while for the first term we need to perform the averaging for different channel coefficient estimates. The following overall SKR for this two-phase-based secret-key generation protocol is achievable [107]:

$$\begin{aligned} SKR &= \max_{f, P_s, T_s} (SKR_s + SKR_{ch}) \\ s.t.\ & P_s T_s + P_{ch}(T_c - T_s) \le T_c P, \end{aligned} \tag{4.149}$$

where the maximization is performed for all possible parameters f, P_s, and T_s.

For additional details on joint source–channel key generation protocols and *relay-assisted key generation* an interested reader is referred to [107] and references therein.

4.9 Optical Channels' Physical-Layer Security

Thanks to its flexibility, security, immunity to interference, high-beam directivity, and energy-efficiency, the free-space optical (FSO) technology represents an excellent candidate for high-performance secure communications. Despite these advantages, large-scale deployment of FSO systems has so far been hampered by reliability and availability issues due to atmospheric turbulence in clear weather (*scintillation*), low visibility in foggy conditions, and high sensitivity to misalignment [32, 110]. Because of high directivity of optical beams, the FSO systems are much more challenging to intercept compared to RF systems. Nevertheless, the eavesdropper can still apply the beam splitter on transmitter side, the blocking attack, or exploit beam divergence at the receiver side. The research on FSO physical-layer security is getting momentum, which can be judged based on increased number of recent papers related to this topic [111, 112, 113–116]. Most of the papers on the physical-layer security for FSO communications are based on direct detection and employ wiretap channel approach introduced by Wyner [28].

In our recent papers [111, 112, 115, 116], we introduced a different strategy. It is well known that we can associate with a photon both spin angular momentum (SAM), related to polarization, and OAM, related to azimuthal dependence of the complex electric field [117–119]. Given that OAM eigenstates are orthogonal, these additional degrees of freedom can be utilized to improve both spectral efficiency and the physical-layer security in optical networks [11, 32, 73, 111, 112, 117–119]. Given that the spatial modes in spatial division multiplexing (SDM) fibers, such as few-mode fibers (FMFs), few-core fibers (FCFs), and few-mode-few-core fibers (FMFCFs), can be decomposed in terms of OAM eigenkets, the OAM can be used to enable the physical-layer security in both FSO and fiber-optic-based optical networks. Because OAM states provide an infinite basis state, while SAM states are two-dimensional only, the OAM can also be used to increase the security of QKD as we describe in [120]. Therefore, the OAM eigenkets can be employed to provide physical-layer security on classical, semiclassical, and quantum levels.

4.9.1 SDM-Fiber-Based Physical-Layer Security

Three types of physical-layer security schemes based on spatial modes are possible: classical, semiclassical, and QKD schemes; depending on the desired level of security. It is well known that classical protocols rely on the computational difficulty of reversing the one-way functions and, in principle, cannot provide any indication of Eve's presence at any point in the communication process. However, the optical communication links can be operated at a desired margin from the receiver sensitivity, and for known channel conditions, Eve's beam-splitting attack can be detected as it will cause sudden decrease in *secrecy capacity* C_S, defined as

$$C_S = C_{AB} - C_{AE}, \tag{4.150}$$

where C_{AB} is the instantaneous capacity of Alice-to-Bob channel and C_{AE} is the instantaneous capacity of Alice-to-Eve channel. Another relevant probabilistic measure that can be used to characterize the security of optical communication link is the *probability of strictly positive secrecy capacity*, defined as

$$P_S^+ = \Pr(C_S > 0). \tag{4.151}$$

Our assumption for DWDM applications is that we are concerned with the physical-layer security of a particular WDM channel and that we cannot manipulate other WDM channels. From recent studies of spatial division multiplexing (SDM) systems, such as [121–123], we have learned that channel capacity can be increased linearly with number of spatial modes N, rather than logarithmically with signal-to-noise ratio for conventional 2-D schemes. These observations motivate us to employ the spatial modes to dramatically improve secrecy capacity when compared to conventional 2-D schemes. The use of SDM schemes to increase the secret-key rates is always sensitive to the crosstalk among spatial modes and potential eavesdropper can compromise the security by relying on spatial coupling, without being detected by Alice and Bob. To solve this problem, in addition to compensating for spatial modes coupling effects, it is possible to employ the multidimensional signaling. In multidimensional signaling, the spatial modes are used as bases functions, and by detecting the signal in any particular spatial mode, Eve will not be able to compromise security as only a single coordinate will be detected. Since the multidimensional signaling based on spatial modes has been already described in [32, 73, 121–123], here we just briefly describe the corresponding multidimensional scheme to be used for raw key transmission, which is shown in Fig. 4.32.

The configurations of spatial-mode-based multidimensional modulator and mode-demultiplexer are already provided in [122, 123]. Alice generates the binary sequence randomly. The multidimensional mapper can be implemented as a look-up-table (LUT). For signal constellation size M, the $\log_2 M$ bits are used to find the coordinates of multidimensional signal constellation, obtained as described in [123]. The multidimensional coordinates are used as the inputs to corresponding Mach–Zehnder

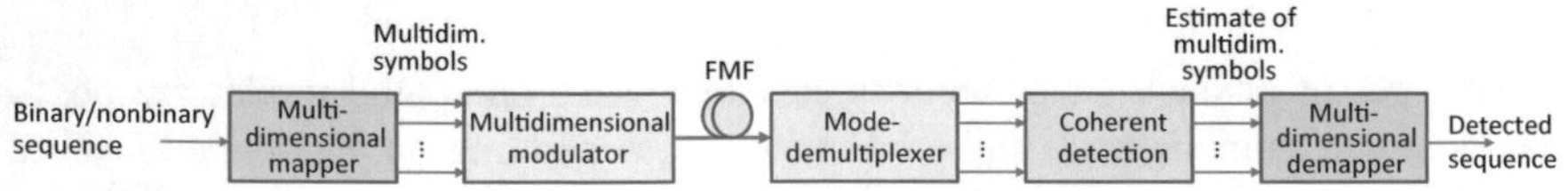

Fig. 4.32 The physical-layer security scheme employing mode-multiplexing-based multidimensional signaling

modulators (MZMs) of multidimensional modulator. After the mode-multiplexing, the signal is transmitted over SDM system of interest. On the receiver side, after mode-demultiplexing and coherent detection, the estimated multidimensional coordinates are used as inputs of multidimensional a posteriori probability (APP) demapper, which provides the most probably symbol being transmitted, and the detected sequence is delivered to Bob. After that, information reconciliation, based on systematic LDPC coding, is performed in similar fashion as already proposed for QKD applications [120]. To distil from the generated key a smaller set of bits whose correlation with Eve's string falls below the desired threshold, the privacy amplification is performed with the help of the *universal hash functions*.

In this rest of this subsection, we describe the corresponding *semiclassical physical-layer security* scheme. The simplest version, based on spatial position modulation like (SPML) approach, is illustrated in Fig. 4.33.

Alice generates a random binary sequence, which is with the help MZM converted into optical domain. The MZM's output drives the optical switch (OS) with ns switching speed, such as one introduced in [124]. The OS output is randomly selected. The OS outputs are used as inputs of spatial-domain multiplexer, whose configuration is provided in [122, 123]. Such encrypted signal is sent over FMF system to Bob. On the receiver side, in the simplest (direct detection) version, Bob applies spatial-domain demultiplexer, whose each output branch drives an avalanche photodiode (APD). In decision circuit, the largest output is selected as transmitted symbol. For M_s spatial modes, $\log_2 M_s$ bits of raw key are transmitted. In this simple scheme, MZM is used for framing purpose only. If coherent detection is used instead, and MZM is replaced by I/Q modulator (with two RF inputs), we can transmit $\log_2 M + \log_2 M_s$ bits of raw key per channel use, where M is the size of I/Q constellation. The polarization state (not shown in Figure) has not been used for raw key transmission, but to detect the presence of Eve, since the semiclassical encryption scheme is

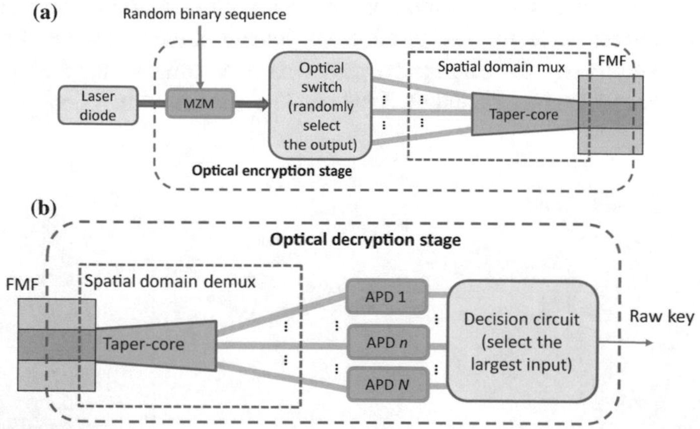

Fig. 4.33 SPML-semiclassical physical-layer security scheme: **a** optical encryption stage and **b** optical decryption stage. APD: avalanche photodiode

operated close to the quantum limit. After raw key transmission is completed, Alice encodes the raw key by employing multilevel nonbinary LDPC-coded modulation (ML-NB-LDPC-CM) [73] based on information reconciliation. The privacy amplification is then performed to distil for the shorter key with negligible correlation with Eve. This key is then used for secure communication, based on one-time pad or any symmetric cipher.

4.9.2 FSO Physical-Layer Security

For classical OAM-based physical-layer security scheme, the N raw key-carrying TEM_{00} modes are shone on a series of computer-generated holograms (CGHs), implemented, for instance, with the help of the spatial light modulators (SLMs), each programmed to one out of N OAM modes in use, as illustrated in Fig. 4.34. The corresponding diffraction angles are properly adjusted so that the coaxial propagation of outgoing OAM beams is obtained, and the resulting superposition beam is expanded by an expanding telescope. To impose N coordinates for multidimensional signaling, a series of MZMs is used on a transmitter side. The number of bits required to select a point from multidimensional signal constellation of size M is given by $\log_2 M$. Therefore, $\log_2 M$ bits are used to select a point from multidimensional signal constellation, whose coordinates are stored in N-dimensional mapper, which can be implemented with the help of an arbitrary waveform generator (AWG). On the receiver side, after compressing telescope, we pass the signal to adaptive optics subsystem to compensate for atmospheric turbulence effects. After that, a series of conjugate volume holograms recorded on SLMs are used to determine the projections along corresponding OAM modes. These OAM projections (in optical domain) are used as inputs of corresponding coherent detectors to estimate the coordinates of transmitted multidimensional signal. For efficient implementation, one local laser is used for all coherent detectors. After coherent detection, the corresponding analog-to-digital converters (ADCs) outputs are passed to an N-dimensional APP demapper, in which symbol LLRs are calculated. After raw key transmission is completed, Alice

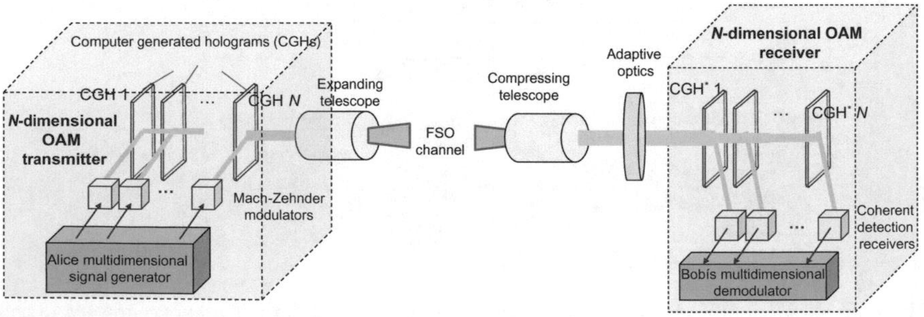

Fig. 4.34 The OAM-based FSO physical-layer security scheme

encodes the raw key by employing ML-NB-LDPC-CM-based information reconciliation. The privacy amplification is further performed, to distil for the shorter key with negligible correlation with Eve. This key is then used for secure FSO communication.

To demonstrate high potential of the OAM-based physical-layer security scheme, the propagation of OAM modes at 1550 nm with the azimuthal index l from −20 to 20 and radial index p set to 0 is simulated in [111] by performing split-step propagation method [110]. The FSO link of length 1 km is observed. To model the strong turbulence effects, the use of 11 random phase screens is sufficient [111]. The phase power spectral density (PSD) used to generate random phase screens is a modified version of the Kolmogorov spectrum which includes both inner and outer scales and is given as [110]

$$\Phi_n(\kappa) = 0.033C_n^2\left[1 + 1.802(\kappa/\kappa_l) - 0.254(\kappa/\kappa_l)^{7/6}\right]\frac{e^{-\kappa^2/\kappa_l^2}}{\left(\kappa^2 + \kappa_0^2\right)^{11/6}}, \quad (4.152)$$

where κ is the spatial frequency, $\kappa_l = 3.3/l_0$, $\kappa_0 = 1/L_0$, and C_n^2 is the refractive index structure parameter that indicates the turbulence strength; l_0 and L_0 are the inner and outer scales of the turbulence, respectively. The propagation of Laguerre–Gaussian (LG) beams with $n \in S = \{-20, \ldots, -1, 0, 1, \ldots, 20\}$ of $N = 41$ OAM modes in total is simulated [111] for three OAM subsets: $S' = \{-18, \ldots, -3, 0, 3, \ldots, 18\}$, $S'' = \{-20, \ldots, -5, 0, 5, \ldots, 20\}$, and $S''' = \{-14, -7, 0, 7, 14\}$ with spacing 2, 4, and 6 between selected modes, respectively, following by the calculation of the aggregate secrecy rate. The results of calculation are summarized in Fig. 4.35 [111]. The worst-case scenario, when Eve is located on a transmitter side and taps the portion r_e of the transmitted optical power, is considered. As we can see, in the weak turbulence regime, the total secrecy capacity can be improved to close to two orders of magnitude by using OAM multiplexed beam. However, as the refractive index structure parameter C_n^2 increases, the secrecy capacity decreases due to orthogonality loss caused by turbulence effects. Comparing with the secrecy capacity of

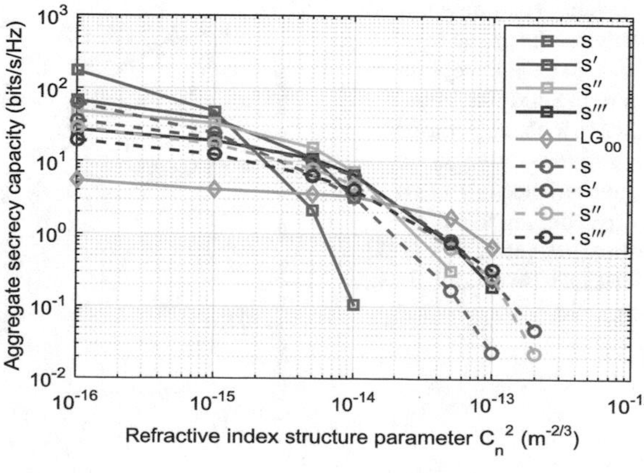

Fig. 4.35 The aggregate secrecy capacity versus refractive index structure parameter. Solid lines: equal power per channel. Dashed lines: fixed system power (fixed transmitted power equally divided among OAM channels). The portion of transmitted power taped by Eve is set to 0.01. Signal-to-noise ratio is set to 20 dB (after Ref. [111]; © IEEE 2016; reprinted with permission)

using single channel (denoted as LG_{00} mode), the use of OAM multiplexing is not beneficial in very strong turbulence regime without adaptive optics.

We can also use the Rytov variance $\sigma_R^2 = 1.23 C_n^2 k^{7/6} L^{11/6}$ to represent the turbulence strength, which takes the propagation distance L, the operating wavelength, and the refractive structure parameter into account. For our FSO system under study (a path of 1 km and wavelength at 1550 nm), the range of σ_R^2 spans from 0.002 to 10 for C_n^2 from 10^{-16} to 5×10^{-13}. As an illustration, the *probability of strictly positive secrecy capacity* P_S^+ by Eq. (4.151) is calculated, which represents a probabilistic metric for characterization of the secrecy for the communication. The results for P_S^+ corresponding to strong turbulence regime are summarized in Fig. 4.36 [111].

In an ideal homogeneous medium, the vorticity of OAM modes is preserved as they propagate and their wavefronts remain orthogonal even after undergoing diffraction. However, in an atmospheric turbulence channel, their orthogonality is no longer preserved due to fading and time-varying refractive index fluctuations that cause intensity scintillations in the beam profiles. This results in crosstalk as the power gets transferred from one particular mode to its neighbors. An intriguing possibility for FSO links that has been considered for quite some time has been to communicate using Bessel beams. In addition to forming orthogonal solutions to the free-space Helmholtz equation, these beams also have the property of being non-diffracting for extremely long distances and self-regenerating if partially blocked by obstructions along their propagation axis [125]. However, theoretical Bessel beams contain an infinite number of nearly equal energy rings and by definition they have an infinite energy. Thus, the physically realizable analog to the Bessel beam is the Bessel–Gaussian (BS) beam that can also non-diffractively propagate but only over a finite depth of focus distance. An experimental demonstration of this promising property being harnessed for use in transmitting data over an FSO channel was reported in [126]. However, the robustness of BS beams remains tenuous and very much depends on the nature of the obstructions that the beams encounter. While it is known that these beams can still regenerate when clipped by a discrete point

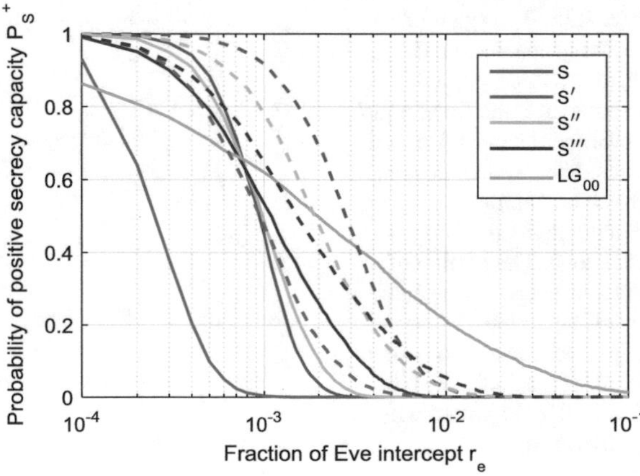

Fig. 4.36 The probability of positive secrecy capacity versus Eve's interception fraction. Solid lines: equal power per channel, dashed lines: fixed system power. Rytov variance is set to 4 (corresponding to strong turbulence) (after Ref. [111]; © IEEE 2016; reprinted with permission)

object blocking a small section of their wavefront, simulation studies have indicated that they cannot overcome strong turbulence-induced phase changes that distort their entire profile [127]. Nonetheless, under certain scenarios such as a weak to medium turbulence channel of distance on the order of several kilometers, it is possible to create BS beams that can propagate intact, as demonstrated in [112].

The BS and LG beams of orders from -15 to $+15$ have been individually propagated in consecutive manner over a 1 km turbulence channel of specified strength, with details provided in [112]. Monte Carlo trials are done for each C_n^2 value with the particular values being intentionally chosen to facilitate the plotting of the C_n^2 points on a semi-logarithmic horizontal scale. Figure 4.37 shows the simulated intensity profiles for several of the OAM modes after traveling through a strong turbulence channel. The complex electric field values at the final step are stored and from these data the channel crosstalk matrices are generated by pairwise computation of overlap integrals between the modal electric fields. An identity matrix would correspond to the ideal case where no channel crosstalk occurs. On the other hand, the turbulence would cause power leakage from each of the diagonal cells in the matrix to neighboring cells in the same row. Figure 4.38 contains crosstalk matrix results arranged in increasing turbulence strength for several C_n^2 values. A visual inspection shows slightly more power leakage among the LG modes than among the BS modes when the C_n^2 value is below 10^{-15} m$^{-2/3}$. To quantify these effects, the aggregate secrecy capacity is calculated over the entire set of modes and corresponding crosstalk matrices using the information-theoretic security formulas, with the results being summarized in Fig. 4.39 [112]. Two curves, one for the case of Eve near the transmitter and one for Eve near the receiver, are shown. As expected, the aggregate C_S values decrease with increasing turbulence strength. For reference, the maximum

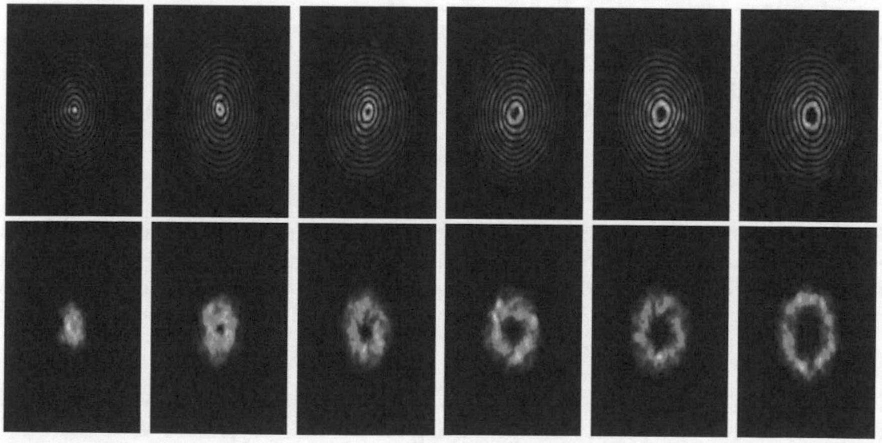

Fig. 4.37 The simulated OAM modes from order 0 to order +5 after 1 km of propagation distance through $C_n^2 = 10^{-14}$ turbulence. The top row contains sample beam profiles for BS modes and the bottom row contains sample beam profiles for LG modes (after Ref. [112]; © IEEE 2016; reprinted with permission)

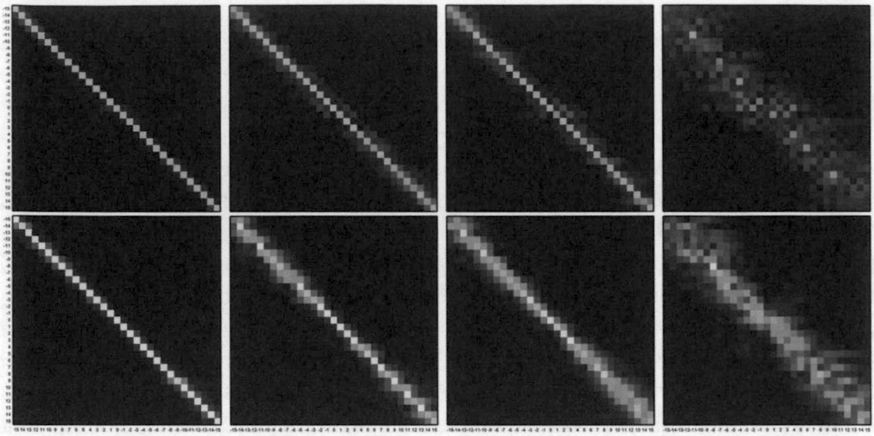

Fig. 4.38 The simulation crosstalk matrices for BS modes (top) and LG modes (bottom) for $C_n^2 = 10^{-17}, 10^{-16}, 1.75 \times 10^{-16}, 1.75 \times 10^{-15}$ (going left to right). Transmit orders are labeled along the vertical axis and the orders used to receive them are labeled along the horizontal axis (after Ref. [112]; © IEEE 2016; reprinted with permission)

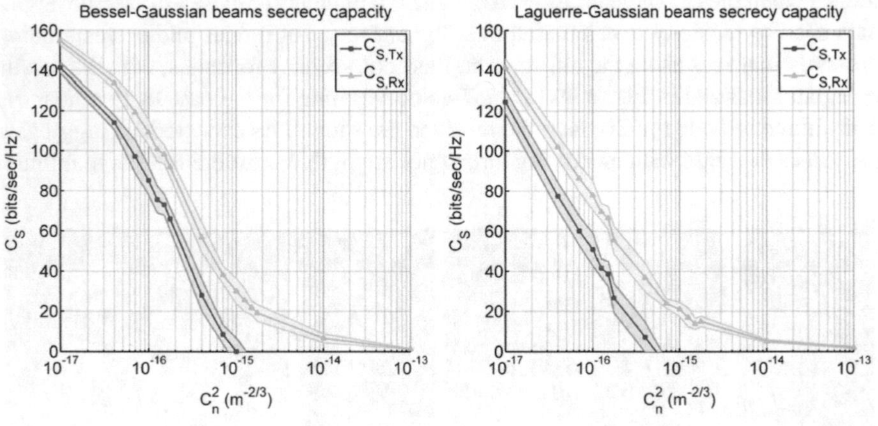

Fig. 4.39 The plots of aggregate secrecy capacity calculated from crosstalk matrices over orders −15 to +15 as a function of turbulence strength. The shaded bands surrounding the data points represent error bars over the Monte Carlo runs (after Ref. [112]; © IEEE 2016; reprinted with permission)

achievable aggregate secrecy capacity is 175 bits/s/Hz when using consecutive orders from −15 to +15 in a perfect channel having no crosstalk regardless of whether Eve is on the transmit or receive sides.

Comparing the data points in the BS and LG curves, we find that the C_S values are typically higher by 10–30 bits/s/Hz when using BS beams, which indicates better PLS performance than when using LG beams. However, this improvement is mostly negated in strong turbulence, when the C_n^2 value is above 10^{-15} $m^{-2/3}$. Under these channel conditions, we observe the capacity approach 0 bits/s/Hz if Eve is located near the receiver and become negative if Eve is located near transmitter. For corresponding experimental results, an interested reader is referred to [116].

4.10 Concluding Remarks

This chapter has been devoted to the physical-layer security fundamentals. The chapter starts with discussion on security issues (Sect. 4.1), followed by Sect. 4.2 where the information-theoretic security is introduced, and comparison against the computational security has been performed. In the same section, various information-theoretic security measures have been introduced, including strong secrecy and weak secrecy conditions. After that, in Sect. 4.3, the Wyner's wiretap channel model, also known as the degraded wiretap channel model, has been introduced. In the same section, the concept of secrecy capacity has been introduced as well as the nested wiretap coding. Further, in Sect. 4.4, the broadcast channel with confidential messages has been introduced, and the secrecy capacity definition has been generalized. The focus has been then moved to the secret-key generation (agreement) protocols in Sect. 4.5, where the source and channel-type models have been introduced, and corresponding secret-key generation protocols have been described. Section 4.6 has been devoted to the coding for the physical-layer security systems, including both coding for weak and strong secrecy systems. Regarding the coding for weak secrecy systems, described in Sect. 4.6.1, the special attention has been paid to two-edge type LDPC coding, punctured LDPC coding, and polar coding. Regarding the coding for strong secrecy systems, described in Sect. 4.6.2, the focus has been on coset coding with dual of LDPC codes and hash functions/extractor-based coding. In Sect. 4.6.3, the information reconciliation has been introduced and described. In Sect. 4.7, the concept of privacy amplification has been introduced and described. In wireless channels physical-layer security section, Sect. 4.8, the following topics have been covered: MIMO fundamentals (Sect. 4.8.1), wireless MIMO PLS (Sect. 4.8.2), and secret-key generation in wireless networks (Sect. 4.8.3). In Sect. 4.9, related to the optical channels PLS, both PLS for spatial division multiplexing (SDM)-fiber-based systems (Sect. 4.9.1) and FSO systems (Sect. 4.9.2) have been discussed.

References

1. X.800: Security architecture for open systems interconnection for CCITT applications, recommendation X.800 (03/91). https://www.itu.int/rec/T-REC-X.800-199103-I
2. Bloch M (2008) Physical-layer security. PhD dissertation, School of Electrical and Computer Engineering, Georgia Institute of Technology
3. Bloch M, Barros J (2011) Physical-layer security: from information theory to security engineering. Cambridge University Press, Cambridge

4. Bloch M (2014) Fundamentals of physical layer security. In: Zhou X, Song L, Zhang Y (eds) Physical layer security in wireless communications. CRC Press, Boca Raton, London, New York, pp 1–16
5. Chorti A et al (2016) Physical layer security: a paradigm shift in data confidentiality. In: Physical and data-link security techniques for future communications systems. Lecture notes in electrical engineering, vol 358. Springer, pp 1–15
6. Bloch M, Barros J, Rodrigues MRD, McLaughlin SW (2008) Wireless information-theoretic security. IEEE Trans Inform Theory 54(6):2515–2534
7. Bennet CH, Brassard G (1984) Quantum cryptography: public key distribution and coin tossing. In: Proceedings of the IEEE international conference on computers, systems, and signal processing, Bangalore, India, pp 175–179
8. Bennett CH (1992) Quantum cryptography: uncertainty in the service of privacy. Science 257:752–753
9. Djordjevic IB (2012) Quantum information processing and quantum error correction: an engineering approach. Elsevier/Academic Press, Amsterdam-Boston
10. Sun X, Djordjevic IB, Neifeld MA (2016) Multiple spatial modes based QKD over marine free-space optical channels in the presence of atmospheric turbulence. Opt Express 24(24):27663–27673
11. Qu Z, Djordjevic IB (2017) Four-dimensionally multiplexed eight-state continuous-variable quantum key distribution over turbulent channels. IEEE Photon J 9(6):7600408-1–7600408-8
12. Djordjevic IB (2018) FBG-based weak coherent state and entanglement assisted multidimensional QKD. IEEE Photon J 10(4):7600512-1–7600512-12
13. Neilsen MA, Chuang IL (2000) Quantum computation and quantum information. Cambridge University Press, Cambridge
14. Van Assche G (2006) Quantum cryptography and secrete-key distillation. Cambridge University Press, Cambridge-New York
15. Bennett CH, Brassard G, Crepeau C, Maurer U (1995) Generalized privacy amplification. IEEE Inform Theory 41(6):1915–1923
16. Shannon CE (1949) Communication theory of secrecy systems. Bell Syst Tech J 28(4):656–715
17. Vernam GS (1926) Cipher printing telegraph systems for secret wire and radio telegraphic communications. Trans Am Inst Electr Eng 1:295–301
18. Kahn D (1967) The codebreakers: the story of secret writing. Macmillan Publishing Co., New York
19. Forney GD (2003) On the role of MMSE estimation in approaching the information theoretic limits of linear Gaussian channels: Shannon meets Wiener. In: Proceedings of the 41st annual allerton conference on communication, control, and computing, Monticello, IL, pp 430–439
20. Schneier B (2015) Applied cryptography, second edition: protocols, algorithms, and source code in C. Wiley, Indianapolis, IN
21. Katz J, Lindell Y (2015) Introduction to modern cryptography, 2nd edn. CRC Press, Boca Raton, FL
22. Aumasson J-P (2018) Serious cryptography: a practical introduction to modern encryption. No Starch Press, San Francisco, CA
23. Sebbery J, Pieprzyk J (1989) Cryptography: an introduction to computer security. Prentice Hall, New York
24. Delfs H, Knebl H (2015) Introduction to cryptography: principles and applications (information security and cryptography), 3rd edn. Springer, Heidelberg, New York
25. Bloch M, Hayashi M, Thangaraj A (2015) Error-control coding for physical-layer secrecy. Proc IEEE 103(10):1725–1746
26. Cover TM, Thomas JA (1991) Elements of information theory. Wiley, New York
27. Bloch M, Laneman JN (2008) On the secrecy capacity of arbitrary wiretap channels. In: Proceedings of the 46th annual allerton conference on communication, control, and computing. Monticello, IL, pp 818–825
28. Wyner AD (1975) The wire-tap channel. Bell Syst Tech J 54(8):1355–1387

29. Lin F, Oggier F (2014) Coding for wiretap channels. In: Zhou X, Song L, Zhang Y (eds) Physical layer security in wireless communications. CRC Press, Boca Raton, London, New York, pp 17–32
30. Leung-Yan-Cheong SK (1976) Multi-user and wiretap channels including feedback. PhD dissertation, Stanford University
31. Csiszár I, Körner J (1978) Broadcast channels with confidential messages. IEEE Trans Inf Theory 24(3):339–348
32. Djordjevic IB (2017) Advanced optical and wireless communications systems. Springer International Publishing AG, Cham, Switzerland
33. Goldsmith A (2005) Wireless communications. Cambridge University Press, Cambridge
34. Tse D, Viswanath P (2005) Fundamentals of wireless communication. Cambridge University Press, Cambridge
35. Barros J, Rodrigues MRD (2006) Secrecy capacity of wireless channels. In: Proceedings of the IEEE international symposium on information theory 2006 (ISIT 2006), Seattle, WA, pp 356–360
36. Barros J, Bloch M (2008) Strong secrecy for wireless channels. In: Information-theoretic security. Lecture notes in computer science, vol 5155. Springer, Berlin, Germany, pp 40–53
37. Maurer UM (1993) Secret key agreement by public discussion from common information. IEEE Trans Inf Theory 39(3):733–742
38. Ahlswede R, Csiszár I (1993) Common randomness in information theory and cryptography-Part I: secret sharing. IEEE Trans Inf Theory 39(4):1121–1132
39. Maurer UM, Wolf S (1999) Unconditionally secure key agreement and intrinsic conditional information. IEEE Trans Inf Theory 45(2):499–514
40. Csiszár I, Narayan P (2004) Secrecy capacities for multiple terminals. IEEE Trans Inf Theory 50(12):3047–3061
41. Zhou X, Song L, Zhang Y (eds) (2014) Physical layer security in wireless communications. CRC Press, Boca Raton, London, New York
42. Ozarow LH, Wyner AD (1984) Wire-tap channel II. AT&T Bell Lab Tech J 63(10):2135–2157
43. Zamir R, Shamai S, Erez U (2002) Nested linear/lattice codes for structured multiterminal binning. IEEE Trans Inf Theory 48(6):1250–1276
44. Thangaraj A, Dihidar S, Calderbank AR, McLaughlin SW, Merolla J (2007) Applications of LDPC codes to the wiretap channel. IEEE Trans Inform Theory 53(8):2933–2945
45. Andersson M (2011) Coding for wiretap channels. Thesis, School of Electrical Engineering, Royal Institute of Technology (KTH), Sweden
46. Rathi V, Urbanke R, Andersson M, Skoglund M (2011) Rate-equivocation optimal spatially coupled LDPC codes for the BEC wiretap channel. In: Proceedings of the 2011 IEEE international symposium on information theory proceedings, St. Petersburg, pp 2393–2397
47. Tanner RM (1981) A recursive approach to low complexity codes. IEEE Tans Inf Theory 27:533–547
48. Boutros J, Pothier O, Zemor G (1999) Generalized low density (Tanner) codes. In: Proceedings of the IEEE international conference on communications (ICC'99), pp 441–445
49. Lentmaier M, Zigangirov K (1999) On generalized low-density parity-check codes based on Hamming component codes. IEEE Commun Lett 3(8):248–250
50. Djordjevic IB, Milenkovic O, Vasic B (2005) Generalized low-density parity-check codes for optical communication systems. J Lightw Technol 23(5):1939–1946
51. Djordjevic IB, Xu L, Wang T, Cvijetic M (2008) GLDPC codes with Reed-Muller component codes suitable for optical communications. IEEE Commun Lett 12(9):684–686
52. Djordjevic IB, Wang T (2014) Multiple component codes based generalized LDPC codes for high-speed optical transport. Opt Express 22(14):16694–16705
53. Zou D, Djordjevic IB (2016) An FPGA design of generalized low-density parity-check codes for rate-adaptive optical transport networks. In: Proceedings of the SPIE PhotonicsWest 2016, optical metro networks and short-haul systems VIII, SPIE vol 9773, 13–18 February 2016, San Francisco, California United States, pp 97730M-1–97730M-6

54. Wong CW, Wong T, Shea J (2011) Secret-sharing LDPC codes for the BPSK constrained Gaussian wiretap channel. IEEE Trans Inform Forensics Sec 6(3):551–564
55. Klinc D, Ha J, McLaughlin SW, Barros J, Kwak B (2011) LDPC codes for the Gaussian wiretap channel. IEEE Trans Inform Forens Sec 6(3):532–540
56. Arikan E (2009) Channel polarization: a method for constructing capacity-achieving codes for symmetric binary-input memoryless channels. IEEE Trans Inf Theory 55(7):3051–3073
57. Mahdavifar H, Vardy A (2011) Achieving the secrecy capacity of wiretap channels using polar codes. IEEE Trans Inf Theory 57(10):6428–6443
58. Koyluoglu OO, El Gamal H (2012) Polar coding for secure transmission and key agreement. IEEE Trans Inform Forens Sec 7(5):1472–1483
59. Liu R, Liang Y, Poor HV, Spasojevic P (2007) Secure nested codes for type II wiretap channels. In: Proceedings of the 2007 IEEE information theory workshop, Tahoe City, CA, pp 337–342
60. Belfiore J, Oggier F (2010) Secrecy gain: a wiretap lattice code design. In: Proceedings of the 2010 international symposium on information theory and its applications, Taichung, pp 174–178
61. Oggier F, Solé P, Belfiore J (2016) Lattice codes for the wiretap Gaussian channel: construction and analysis. IEEE Trans Inf Theory 62(10):5690–5708
62. Thangaraj A, Dihidar S, Calderbank AR, McLaughlin SW, Merolla J-M (2007) Applications of LDPC codes to the wiretap channel. IEEE Trans Inf Theory 53(8):2933–2945
63. Subramanian A, Thangaraj A, Bloch M, McLaughlin SW (2011) Strong secrecy on the binary erasure wiretap channel using large-girth LDPC codes. IEEE Trans Inf Forens Sec 6(3):585–594
64. Suresh AT, Subramanian A, Thangaraj A, Bloch M, McLaughlin SW (2010) Strong secrecy for erasure wiretap channels. In: Proceedings of the IEEE information theory workshop 2010 (ITW2010), Dublin, 30 Aug–3 Sept 2010
65. Thangaraj A (2014) Coding for wiretap channels: channel resolvability and semantic security. In: Proceedings of the 2014 IEEE information theory workshop (ITW 2014), Hobart, TAS, pp 232–236
66. Cohen G, Zémor G (2006) Syndrome-coding for the wiretap channel revisited. In: Proceedings of the IEEE information theory workshop 2006, Chengdu, China, pp 33–36
67. Hayashi M, Matsumoto R (2010) Construction of wiretap codes from ordinary channel codes. In: Proceedings of the IEEE international symposium on information theory (ISIT) 2010, Austin, TX, pp 2538–2542
68. Bellare M, Tessaro S, Vardy A (2012) Semantic security for the wiretap channel. In: Advances in cryptology—CRYPTO 2012. Lecture notes in computer science, vol 7417. Springer, Berlin, Heidelberg, Germany, pp 294–311
69. Cheraghchi M, Didier F, Shokrollahi A (2012) Invertible extractors and wiretap protocols. IEEE Trans Inf Theory 58(2):1254–1274
70. Harrison WK, Almeida J, Bloch MR, McLaughlin SW, Barros J (2013) Coding for secrecy: an overview of error-control coding techniques for physical-layer security. IEEE Signal Proc Mag 30(5):41–50
71. Slepian D, Wolf JK (1973) Noiseless coding of correlated information sources. IEEE Trans Inf Theory 19(4):471–480
72. Liveris AD, Xiong Z, Georghiades CN (2002) Compression of binary sources with side information using low-density parity-check codes. In Proceedings of the IEEE global telecommunications conference 2002 (GLOBECOM'02), Taipei, Taiwan, vol 2, pp 1300–1304
73. Djordjevic IB (2016) On advanced FEC and coded modulation for ultra-high-speed optical transmission. IEEE Commun Surv Tutor 18(3):1920–1951. https://doi.org/10.1109/COMST.2016.2536726
74. Cachin C, Maurer UM (1997) Linking information reconciliation and privacy amplification. J Cryptol 10:97–110
75. Carter JL, Wegman MN (1979) Universal classes of hash functions. J Comput Syst Sci 18(2):143–154

76. Wegman MN, Carter J (1981) New hash functions and their use in authentication and set equality. J Comput Sci Syst 22:265–279
77. Rényi A (1961) On measures of entropy and information. In: Proceedings of the 4th Berkeley symposium on mathematical statistics and probability, University Claifornia Press, vol 1, pp 547–561
78. Ilic I, Djordjevic IB, Stankovic M (2017) On a general definition of conditional Rényi entropies. In: Proceedings of the 4th international electronic conference on entropy and its applications, 21 November–1 December 2017. Sciforum electronic conference series, vol 4. https://doi.org/10.3390/ecea-4-05030
79. Teixeira A, Matos A, Antunes L (2012) Conditional Rényi entropies. IEEE Trans Inf Theory 58(7):4273–4277
80. Cachin C (1997) Entropy measures and unconditional security in cryptography. PhD dissertation, ETH Zurich, Hartung-Gorre Verlag, Konstanz
81. Maurer U, Wolf S (2000) Information-theoretic key agreement: from weak to strong secrecy for free. In: Preneel B (eds) Advances in cryptology—EUROCRYPT 2000. EUROCRYPT 2000. Lecture notes in computer science, vol 1807. Springer, Berlin, Heidelberg
82. Telatar E (1999) Capacity of multi-antenna Gaussian channels. Eur Trans Telecommun 10:585–595
83. Biglieri E (2005) Coding for wireless channels. Springer Science + Business Media Inc, New York, USA
84. Hampton JR (2014) Introduction to MIMO communications. Cambridge University Press, Cambridge, UK
85. Mukherjee A, Fakoorian SAA, Huang J, Swindlehurst L (2014) MIMO signal processing algorithms for enhanced physical layer security. In: Zhou X, Song L, Zhang Y (eds) Physical layer security in wireless communications. CRC Press, Boca Raton, London, New York, pp 93–114
86. Khisti A, Wornell GW (2010) Secure transmission with multiple antennas I: the MISOME wiretap channel/Part II: the MIMOME wiretap channel. IEEE Trans Inf Theory 56(7):3088–3104/5515–5532
87. Bustin R, Liu R, Poor HV, Shamai(Shitz) S (2009) An MMSE approach to the secrecy capacity of the MIMO Gaussian wiretap channel. EURASIP J Wirel Commun Netw 2009:370970. https://doi.org/10.1155/2009/370970
88. Narayan P, Tyagi H (2016) Multiterminal secrecy by public discussion. Found Trends Commun Inf Theory 13:129–275
89. Schaefer RF, Boche H, Poor HV (2015) Secure communication under channel uncertainty and adversarial attacks. Proc IEEE 103(10):1796–1813
90. Poor HV, Schaefer RF (2017) Wireless physical layer security. PNAS 114(1):19–26
91. Bjelakovic I, Boche H, Sommerfeld J (2013) Secrecy results for compound wiretap channels. Probl Inf Trans 49(1):73–98
92. Schaefer RF, Loyka S (2015) The secrecy capacity of compound MIMO Gaussian channels. IEEE Trans Inf Theory 61(10):5535–5552
93. Ly HD, Liu T, Liang Y (2010) Multiple-input multiple-output Gaussian broadcast channels with common and confidential messages. IEEE Trans Inf Theory 56(11):5477–5487
94. Liu R, Maric I, Spasojevic P, Yates RD (2008) Discrete memoryless interference and broadcast channels with confidential messages: secrecy rate regions. IEEE Trans Inf Theory 54(6):2493–2507
95. Liu R, Liu T, Poor HV, Shamai(Shitz) S (2010) Multiple-input multiple-output Gaussian broadcast channels with confidential messages. IEEE Trans Inf Theory 56(9):4215–4227
96. Liu R, Liu T, Poor HV, Shamai(Shitz) S (2013) New results on multiple-input multiple-output broadcast channels with confidential messages. IEEE Trans Inf Theory 59(3):1346–1359
97. Liang Y, Poor HV (2008) Multiple-access channels with confidential messages. IEEE Trans Inf Theory 54(3):972–1002
98. Koyluoglu OO, El Gamal H, Lai L, Poor HV (2011) Interference alignment for secrecy. IEEE Trans Inf Theory 57(6):3323–3332

99. He X, Aylin Yener A (2009) K-user interference channels: achievable secrecy rate and degrees of freedom. In: Proceedings of the 2009 IEEE information theory workshop on networking and information theory, 12–10 June 2009, Volos, Greece, pp 336–340
100. Fakoorian SAA, Swindlehurst AL (2010) MIMO interference channel with confidential messages: game theoretic beamforming designs. In 2010 Conference record of the forty fourth Asilomar conference on signals, systems and computers, 7–10 Nov. 2010, Pacific Grove, CA, USA
101. Fakoorian SAA, Swindlehurst AL (2011) MIMO interference channel with confidential messages: achievable secrecy rates and precoder design. IEEE Trans Inf Forens Sec 6(3):640–649
102. Oohama Y (2007) Capacity theorems for relay channels with confidential messages. In: Proceedings of the 2007 IEEE international symposium on information theory, 24–29 June 2007, Nice, France, pp 926–930
103. He X, Yener A (2010) Cooperation with an untrusted relay: a secrecy perspective. IEEE Trans Inf Theory 56(8):3801–3827
104. Huang J, Swindlehurst AL (2011) Cooperative jamming for secure communications in MIMO relay networks. IEEE Trans Signal Process 59(10):4871–4884
105. Wilson R, Tse D, Scholtz RA (2007) Channel identification: secret sharing using reciprocity in ultrawideband channels. IEEE Trans Inf Forens Sec 2(3):364–375
106. Ye C, Mathur S, Reznik A, Trappe W, Mandayam N (2010) Information theoretic key generation from wireless channels. IEEE Trans Inf Forens Sec 5(2):240–254
107. Lai L, Liang Y, Poor HV, Du W (2014) Key generation from wireless channels. In: Zhou X, Song L, Zhang Y (eds) Physical layer security in wireless communications. CRC Press, Boca Raton, London, New York, pp 47–68
108. Lai L, Liang Y, Poor HV (2012) A unified framework for key agreement over wireless fading channels. IEEE Trans Inf Forens Sec 7(2):480–490
109. Gopala PK, Lai L, El Gamal H (2008) On the secrecy capacity of fading channels. IEEE Trans Inf Theory 54(10):4687–4698
110. Andrews LC, Philips RL (2005) Laser beam propagation through random media. SPIE Press, Bellingham, WA
111. Sun X, Djordjevic IB (2016) Physical-layer security in orbital angular momentum multiplexing free-space optical communications. IEEE Photon J 8(1): paper 7901110
112. Wang T-L, Gariano J, Djordjevic IB (2018) Employing Bessel-Gaussian beams to improve physical-layer security in free-space optical communications. IEEE Photon J 10(5): paper 7907113. https://doi.org/10.1109/jphot.2018.2867173
113. Sidorovich VG (2004) Optical countermeasures and security of free-space optical communication links. In: Proceedings of the European symposium optics and photonics for defence and security, 2004, pp 97–108
114. Lopez-Martinez FJ, Gomez G, Garrido-Balsells JM (2015) Physical-layer security in free-space optical communications. IEEE Photon J 7(2): paper 7901014
115. Djordjevic IB, Sun X (2016) Spatial modes-based physical-layer security. In: Proceedings of the IEEE ICTON 2016, Paper Mo.C1.3, Trento, Italy, 10–14 July 2016. (Invited Paper.)
116. Wang T-L, Djordjevic IB (2018) Physical-layer security of a binary data sequence transmitted with Bessel-Gaussian beams over an optical wiretap channel. IEEE Photon J 10(6):7908611
117. Djordjevic IB, Arabaci M (2010) LDPC-coded orbital angular momentum (OAM) modulation for free-space optical communication. Opt Express 18(24):24722–24728
118. Djordjevic IB, Anguita A, Vasic B (2012) Error-correction coded orbital-angular-momentum modulation for FSO channels affected by turbulence. IEEE/OSA J Lightw Technol 30(17):2846–2852
119. Djordjevic IB, Qu Z (2016) Coded orbital-angular-momentum-based free-space optical transmission. In: Webster JG (ed) Wiley encyclopedia of electrical and electronics engineering, pp 1–12. https://doi.org/10.1002/047134608X.W8291
120. Djordjevic IB (2016) Integrated optics modules based proposal for quantum information processing, teleportation, QKD, and quantum error correction employing photon angular momentum. IEEE Photon J 8(1): paper 6600212. https://ieeexplore.ieee.org/document/7393447

121. Djordjevic IB (2012) Spatial-domain-based hybrid multidimensional coded-modulation schemes enabling multi-Tb/s optical transport. IEEE/OSA J Lightw Technol 30(14):2315–2328. (Invited Paper.)
122. Djordjevic IB, Cvijetic M, Lin C (2014) Multidimensional signaling and coding enabling multi-Tb/s optical transport and networking. IEEE Sig Proc Mag 31(2):104–117
123. Djordjevic IB, Jovanovic A, Peric ZH, Wang T (2014) Multidimensional optical transport based on optimized vector-quantization-inspired signal constellation design. IEEE Trans Commun 62(9):3262–3273
124. Djordjevic IB, Varrazza R, Hill M, Yu S (2004) Packet switching performance at 10 Gb/s across a 4 × 4 optical crosspoint switch matrix. IEEE Photon Technol Lett 16(1):102–104
125. McGloin D, Dholakia K (2005) Bessel beams: diffraction in a new light. Contemp Phys 46(1):15–28
126. Chen S, Li S, Zhao Y, Liu J, Zhu L, Wang A, Du J, Shen L, Wang J (2016) Demonstration of 20-Gbit/s high-speed Bessel beam encoding/decoding link with adaptive turbulence compensation. Opt Lett 41(20):4680–4683
127. Nelson W, Palastro JP, Davis CC, Sprangle P (2014) Propagation of Bessel and Airy beams through atmospheric turbulence. J Opt Soc Am A 31(3):603–609

Chapter 5
Quantum Information Theory and Quantum Information Processing Fundamentals

Abstract In this chapter, we provide the basic concepts of quantum information processing, quantum information theory, and quantum error correction. The following topics from quantum information processing will be covered: state vectors, operators, density operators, measurements, dynamics of a quantum system, superposition principle, quantum parallelism, no-cloning theorem, and entanglement. The following concepts from quantum information theory will be provided: Holevo information, accessible information, Holevo bound, Shannon Entropy and von Neumann Entropy, Schumacher's noiseless quantum coding theorem, and Holevo–Schumacher–Westmoreland theorem. The basic concepts of quantum error correction are introduced as well.

5.1 State Vectors, Operators, Projection Operators, and Density Operators

In quantum mechanics, the primitive undefined concepts are *physical system*, *observable* and *state* [1–15]. A physical system is any sufficiently isolated quantum object, say an electron, a photon, or a molecule. An observable will be associated with a measurable property of a physical system, say energy or z-component of the spin. The state of a physical system is a trickier concept in quantum mechanics compared to classical mechanics. The problem arises when considering composite physical systems. In particular, states exist, known as *entangled states*, for bipartite physical systems in which neither of the subsystems is in definite states. Even in cases where physical systems can be described as being in a state, two classes of states are possible: pure and mixed.

5.1.1 Sate Vectors and Operators

The condition of a quantum-mechanical system is completely specified by its *state vector* $|\psi\rangle$ in a Hilbert space H (a vector space on which a positive-definite scalar

I. B. Djordjevic, *Physical-Layer Security and Quantum Key Distribution*,
https://doi.org/10.1007/978-3-030-27565-5_5

product is defined) over the field of complex numbers [16]. Any state vector $|\alpha\rangle$, also known as a **ket**, can be expressed in terms of basis vectors $|\phi_n\rangle$ by

$$|\alpha\rangle = \sum_{n=1}^{\infty} a_n |\phi_n\rangle. \tag{5.1}$$

An **observable**, such as momentum and spin, can be represented by an **operator**, such as A, in the vector space of question. Quite generally, an operator acts on a ket from the left: $(A) \cdot |\alpha\rangle = A\,|\alpha\rangle$, which results in another ket. An operator A is said to *Hermitian* if

$$A^{\dagger} = A, \quad A^{\dagger} = \left(A^{\mathrm{T}}\right)^{*}. \tag{5.2}$$

Suppose that the Hermitian operator A has a discrete set of eigenvalues $a^{(1)}$, …, $a^{(n)}$, …. The associated eigenvectors (eigenkets) $|a^{(1)}\rangle$, …, $|a^{(n)}\rangle$, … can be obtained from

$$A\left|a^{(n)}\right\rangle = a^{(n)}\left|a^{(n)}\right\rangle. \tag{5.3}$$

The Hermitian conjugate of a ket $|\alpha\rangle$ is denoted by $\langle\alpha|$ and called the "bra". The space dual to ket space is known as **bra** space. There exists a one-to-one correspondence, dual-correspondence (D.C.), between a ket space and a bra space:

$$\begin{aligned}
&|\alpha\rangle \overset{\text{D.C.}}{\leftrightarrow} \langle\alpha| \\
&\left|a^{(1)}\right\rangle, \left|a^{(2)}\right\rangle, \ldots \overset{\text{D.C.}}{\leftrightarrow} \left\langle a^{(1)}\right|, \left\langle a^{(2)}\right|, \ldots \\
&|\alpha\rangle + |\beta\rangle \overset{\text{D.C.}}{\leftrightarrow} \langle\alpha| + \langle\beta| \\
&c_{\alpha}|\alpha\rangle + c_{b}|\beta\rangle \overset{\text{D.C.}}{\leftrightarrow} c_{\alpha}^{*}\langle\alpha| + c_{\beta}^{*}\langle\beta|.
\end{aligned} \tag{5.4}$$

The *scalar (inner) product* of two-state vectors $|\phi\rangle = \sum_n a_n|\phi_n\rangle$ and $|\psi\rangle = \sum_n b_n|\phi_n\rangle$ is defined by

$$\langle\beta \mid \alpha\rangle = \sum_{n=1}^{\infty} a_n b_n^{*}. \tag{5.5}$$

5.1.2 Projection Operators

The eigenkets $\{|\xi^{(n)}\rangle\}$ of operator Ξ form the basis so that arbitrary ket $|\psi\rangle$ can be expressed in terms of eigenkets by [17–26]

$$|\psi\rangle = \sum_{n=1}^{\infty} c_n |\xi^{(n)}\rangle. \tag{5.6}$$

By multiplying (5.6) with $\langle\xi^{(n)}|$ from the left, we obtain

$$\langle\xi^{(n)} \mid \psi\rangle = \sum_{j=1}^{\infty} c_j \langle\xi^{(n)} \mid \xi^{(j)}\rangle = c_n \langle\xi^{(n)} \mid \xi^{(n)}\rangle + \sum_{j=1, j\neq n}^{\infty} c_j \langle\xi^{(n)} \mid \xi^{(j)}\rangle. \tag{5.7}$$

Since the eigenkets $\{|\xi^{(n)}\rangle\}$ form the basis, the principle of orthonormality is satisfied $\langle\xi^{(n)} \mid \xi^{(j)}\rangle = \delta_{nj}, \quad \delta_{nj} = \begin{cases} 1, & n = j \\ 0, & n \neq j \end{cases}$ so that the Eq. (5.7) becomes

$$c_n = \langle\xi^{(n)} \mid \psi\rangle. \tag{5.8}$$

By substituting (5.8) into (5.6), we obtain

$$|\psi\rangle = \sum_{n=1}^{\infty} \langle\xi^{(n)} \mid \psi\rangle |\xi^{(n)}\rangle = \sum_{n=1}^{\infty} |\xi^{(n)}\rangle\langle\xi^{(n)} \mid \psi\rangle. \tag{5.9}$$

Because $|\psi\rangle = I|\psi\rangle$ from (5.9) is clear that

$$\sum_{n=1}^{\infty} |\xi^{(n)}\rangle\langle\xi^{(n)}| = I, \tag{5.10}$$

and the relation above is known *as completeness relation*. The operators under summation in (5.10) are known as *projection* operators P_n:

$$P_n = |\xi^{(n)}\rangle\langle\xi^{(n)}|, \tag{5.11}$$

which satisfy the relationship $\sum_{n=1}^{\infty} P_n = I$. It is easy to show that the ket (5.6) with c_n determined with (5.8) is of unit length:

$$\langle\psi|\psi\rangle = \sum_{n=1}^{\infty} \langle\psi|\xi^{(n)}\rangle\langle\xi^{(n)}|\psi\rangle = \sum_{n=1}^{\infty} |\langle\psi|\xi^{(n)}\rangle|^2 = 1. \tag{5.12}$$

The following theorem is an important theorem that will be used a lot throughout the chapter.

It can be shown that the eigenvalues of a Hermitian operator A are real, and the eigenkets are orthogonal:

$$\langle a^{(m)} \mid a^{(n)}\rangle = \delta_{nm}. \tag{5.13}$$

For the proof of this claim, an interested reader is referred to [1].

5.1.3 Photon, Spin ½ Systems, and Hadamard Gate

Photon. The x- and y-polarizations of the photon can be represented by

$$|E_x\rangle = \begin{pmatrix}1\\0\end{pmatrix} \quad |E_y\rangle = \begin{pmatrix}0\\1\end{pmatrix}.$$

On the other hand, the right- and left-circular polarizations can be represented by

$$|E_R\rangle = \frac{1}{\sqrt{2}}\begin{pmatrix}1\\j\end{pmatrix} \quad |E_L\rangle = \frac{1}{\sqrt{2}}\begin{pmatrix}1\\-j\end{pmatrix}.$$

The 45° polarization ket can be represented as follows:

$$|E_{45^\circ}\rangle = \cos\left(\frac{\pi}{4}\right)|E_x\rangle + \sin\left(\frac{\pi}{4}\right)|E_y\rangle = \frac{1}{\sqrt{2}}(|E_x\rangle + |E_y\rangle) = \frac{1}{\sqrt{2}}\begin{pmatrix}1\\1\end{pmatrix}.$$

The bras corresponding to the left- and right-polarization can be written by

$$\langle E_R| = \frac{1}{\sqrt{2}}(1 \;\; -j) \quad \langle E_L| = \frac{1}{\sqrt{2}}(1 \quad j).$$

It can be easily verified that the left- and right-states are orthogonal and that the right-polarization state is of unit length:

$$\langle E_R|E_L\rangle = \frac{1}{2}(1 \;\; -j)\begin{pmatrix}1\\-j\end{pmatrix} = 0 \quad \langle E_R|E_R\rangle = \frac{1}{2}(1 \;\; -j)\begin{pmatrix}1\\j\end{pmatrix} = 1.$$

The completeness relation is clearly satisfied because

$$|E_x\rangle\langle E_x| + |E_y\rangle\langle E_y| = \begin{pmatrix}1\\0\end{pmatrix}(1\;0) + \begin{pmatrix}0\\1\end{pmatrix}(0 \quad 1) = \begin{pmatrix}1&0\\0&1\end{pmatrix} = I_2.$$

An arbitrary polarization state can be represented by

$$|E\rangle = |E_R\rangle\langle E_R|E\rangle + |E_L\rangle\langle E_L|E\rangle.$$

For example, for $E = E_x$ we obtain

$$|E_x\rangle = |E_R\rangle\langle E_R|E_x\rangle + |E_L\rangle\langle E_L|E_x\rangle.$$

For the photon spin operator S matrix representation, we have to solve the following eigenvalue equation:

$$S|\psi\rangle = \lambda|\psi\rangle.$$

The photon spin operator satisfies $S^2 = I$ so that we can write

$$|\psi\rangle = S^2|\psi\rangle = S(S|\psi\rangle) = S(\lambda|\psi\rangle) = \lambda S|\psi\rangle = \lambda^2|\psi\rangle.$$

It is clear from previous equation that $\lambda^2 = 1$ so that the corresponding eigenvalues are $\lambda = \pm 1$. By substituting the eigenvalues into eigenvalue equation, we obtain that corresponding eigenkets are the left- and the right-polarization states:

$$S|E_R\rangle = |E_R\rangle \qquad S|E_L\rangle = -|E_L\rangle.$$

The photon spin represented in $\{|E_x\rangle, |E_y\rangle\}$-basis can be obtained by

$$S \doteq \begin{pmatrix} S_{xx} & S_{xy} \\ S_{yx} & S_{yy} \end{pmatrix} = \begin{pmatrix} \langle E_x|S|E_x\rangle & \langle E_x|S|E_y\rangle \\ \langle E_y|S|E_x\rangle & \langle E_y|S|E_y\rangle \end{pmatrix} = \begin{pmatrix} 0 & -j \\ j & 0 \end{pmatrix}.$$

Spin-1/2 systems. The S_z-basis in spin-1/2 systems can be written as $\{|E_z; +\rangle, |E_z; -\rangle\}$, where the corresponding basis kets represent the spin-up and spin-down states. The eigenvalues are $\{\hbar/2, -\hbar/2\}$, and the corresponding eigenket–eigenvalue relation is

$S_z|S_z; \pm\rangle = \pm\frac{\hbar}{2}|S_z; \pm\rangle$, where S_z is the spin operator that can be represented in basis above as follows:

$$S_z = \sum_{i=+,-}\sum_{j=+,-} |i\rangle\langle j|\underbrace{S_z|i\rangle}_{i\frac{\hbar}{2}|i\rangle}\langle j| = \sum_{i=+,-} i\frac{\hbar}{2}|i\rangle\langle i| = \frac{\hbar}{2}(|+\rangle\langle +| - |-\rangle\langle -|).$$

The matrix representation of spin-½ systems is obtained by

$$|S_z; +\rangle = \begin{pmatrix} \langle S_z; +|S_z; +\rangle \\ \langle S_z; -|S_z; +\rangle \end{pmatrix} \doteq \begin{pmatrix} 1 \\ 0 \end{pmatrix} \qquad |S_z; -\rangle \doteq \begin{pmatrix} 0 \\ 1 \end{pmatrix}$$

$$S_z \doteq \begin{pmatrix} \langle S_z; +|S_z|S_z; +\rangle & \langle S_z; +|S_z|S_z; -\rangle \\ \langle S_z; -|S_z|S_z; +\rangle & \langle S_z; -|S_z|S_z; -\rangle \end{pmatrix} = \frac{\hbar}{2}\begin{pmatrix} 1 & 0 \\ 0 & -1 \end{pmatrix}.$$

Hadamard Gate. The matrix representation of Hadamard operator (gate) is given by

$$H = \frac{1}{\sqrt{2}}\begin{bmatrix} 1 & 1 \\ 1 & -1 \end{bmatrix}.$$

It can easily be shown that the Hadamard gate is Hermitian and unitary as follows:

$$H^{\dagger} = \frac{1}{\sqrt{2}}\begin{bmatrix} 1 & 1 \\ 1 & -1 \end{bmatrix} = H$$

$$H^{\dagger}H = \frac{1}{\sqrt{2}}\begin{bmatrix} 1 & 1 \\ 1 & -1 \end{bmatrix}\frac{1}{\sqrt{2}}\begin{bmatrix} 1 & 1 \\ 1 & -1 \end{bmatrix} = \begin{bmatrix} 1 & 0 \\ 0 & 1 \end{bmatrix} = I.$$

The eigenvalues for Hadamard gate can be obtained from $\det(H - \lambda I) = 0$ to be $\lambda_{1,2} = \pm 1$. By substituting the eigenvalues into eigenvalue equation, namely, $H|\Psi_{1,2}\rangle = \pm|\Psi_{1,2}\rangle$, the corresponding eigenkets are obtained as follows:

$$|\Psi_1\rangle = \begin{bmatrix} \frac{1}{\sqrt{4-2\sqrt{2}}} \\ \frac{1}{\sqrt{2\sqrt{2}}} \end{bmatrix} \qquad |\Psi_2\rangle = \begin{bmatrix} \frac{1}{\sqrt{4+2\sqrt{2}}} \\ -\frac{1}{\sqrt{2\sqrt{2}}} \end{bmatrix}.$$

5.1.4 Density Operators

Let the large number of quantum systems of the same kind be prepared, each in one of a set of orthonormal states $|\phi_n\rangle$, and let the fraction of the system being in state $|\phi_n\rangle$ be denoted by probability P_n $(n = 1, 2, \ldots)$:

$$\langle \phi_m \mid \phi_n \rangle = \delta_{mn}, \qquad \sum_n P_n = 1. \tag{5.14}$$

Therefore, this ensemble of quantum states represents a classical *statistical mixture* of kets. The probability of obtaining ξ_n from the measurement of Ξ will be

$$\Pr(\xi_k) = \sum_{n=1}^{\infty} P_n |\langle \xi_k \mid \phi_n \rangle|^2 = \sum_{n=1}^{\infty} P_n \langle \xi_k \mid \phi_n \rangle \langle \phi_n \mid \xi_k \rangle = \langle \xi_k | \rho | \xi_k \rangle, \tag{5.15}$$

where the operator ρ is known as a *density operator* and it is defined by

$$\rho = \sum_{n=1}^{\infty} P_n |\phi_n\rangle\langle \phi_n|. \tag{5.16}$$

The expected value of the operator Ξ is given by

$$\langle \Xi \rangle = \sum_{k=1}^{\infty} \xi_k \Pr(\xi_k) = \sum_{k=1}^{\infty} \xi_k \langle \xi_k | \rho | \xi_k \rangle = \sum_{k=1}^{\infty} \langle \xi_k | \rho \Xi | \xi_k \rangle = \mathrm{Tr}(\rho \Xi). \tag{5.17}$$

The density operator *properties* can be summarized as follows [1–12, 17–26]:

1. The density operator is Hermitian ($\rho^+ = \rho$), with the set of orthonormal eigenkets $|\phi_n\rangle$ corresponding to the nonnegative eigenvalues P_n and $\mathrm{Tr}(\rho) = 1$.
2. Any Hermitian operator with nonnegative eigenvalues and trace 1 may be considered as a density operator.
3. The density operator is positive-definite: $\langle \psi | \rho | \psi \rangle \geq 0$ for all $|\psi\rangle$.
4. The density operator has the property $\mathrm{Tr}(\rho^2) \leq 1$, with equality if one of the prior probabilities is 1, and all the rest 0: $\rho = |\phi_n\rangle\langle\phi_n|$, and the density operator is then a projection operator.
5. The eigenvalues of a density operator satisfy $0 \leq \lambda_i \leq 1$.

The proof these properties is quite straightforward, and the proof is left as a homework problem. When ρ is the projection operator, we say that it represents the system in a *pure state*; otherwise with $\mathrm{Tr}(\rho^2) < 1$, it represents a *mixed state*. A mixed state in which all eigenkets occur with the same probability is known as a *completely mixed state* and can be represented by

$$\rho = \sum_{k=1}^{\infty} \frac{1}{n} |\phi_n\rangle\langle\phi_n| = \frac{1}{n} I \Rightarrow \mathrm{Tr}\left(\rho^2\right) = \frac{1}{n} \Rightarrow \frac{1}{n} \leq \mathrm{Tr}\left(\rho^2\right) \leq 1. \tag{5.18}$$

If the density matrix has off-diagonal elements different from zero, we say that it exhibits the *quantum interference*, which means that state term can interfere with each other. Let us observe the following pure state:

$$\begin{aligned} |\psi\rangle = \sum_{i=1}^{n} \alpha_i |\xi_i\rangle \Rightarrow \rho = |\psi\rangle\langle\psi| &= \sum_{i=1}^{n} |\alpha_i|^2 |a_i\rangle\langle a_i| + \sum_{i=1}^{n} \sum_{j=1, j\neq i}^{n} \alpha_i \alpha_j^* |\xi_i\rangle\langle\xi_j| \\ &= \sum_{i=1}^{n} \langle \xi_i | \rho | \xi_i \rangle |\xi_i\rangle\langle\xi_i| + \sum_{i=1}^{n} \sum_{j=1, j\neq i}^{n} \langle \xi_i | \rho | \xi_j \rangle |\xi_i\rangle\langle\xi_j|. \end{aligned} \tag{5.19}$$

The first term in (5.19) is related to the probability of the system being in state $|a_i\rangle$, and the second term is related to the quantum interference. It appears that the off-diagonal elements of a mixed state will be zero, while these of pure state will be nonzero. Notice that the existence of off-diagonal elements is base dependent; therefore, to check for purity it is a good idea to compute $\mathrm{Tr}(\rho^2)$ instead.

In quantum information theory, the density matrix can be used to determine the amount of information conveyed by the quantum state, i.e., to compute the von Neumann entropy:

$$S = \mathrm{Tr}(\rho\ \log\ \rho) = -\sum_i \lambda_i \log_2 \lambda_i, \tag{5.20}$$

where λ_i are the eigenvalues of the density matrix. The corresponding Shannon entropy can be calculated by [27, 28]

$$H = -\sum_i p_i \log_2 p_i. \tag{5.21}$$

Suppose now that S is a *bipartite composite system* with component subsystems A and B. For example, the subsystem A can represent the quantum register Q and subsystem B the environment E. The composite system can be represented by $AB = A \otimes B$, where $\otimes$ stands for the tensor product. If the dimensionality of Hilbert space H_A is m and the dimensionality Hilbert space H_B is n, then the dimensionality of Hilbert space H_{AB} will be mn. Let $|\alpha\rangle \in A$ and $|\beta\rangle \in B$, then $|\alpha\rangle|\beta\rangle = |\alpha\rangle \otimes |\beta\rangle \in AB$. If the operator A acts on kets from H_A and the operator B on kets from H_B, then the action of AB on $|\alpha\rangle|\beta\rangle$ can be described as follows:

$$(AB)|\alpha\rangle|\beta\rangle = (A|\alpha\rangle)(B|\beta\rangle). \tag{5.22}$$

The norm of state $|\psi\rangle = |\alpha\rangle|\beta\rangle \in AB$ is determined by

$$\langle\psi|\psi\rangle = \langle\alpha|\alpha\rangle\langle\beta|\beta\rangle. \tag{5.23}$$

Let $\{|\alpha_i\rangle\}$ ($\{|\beta_i\rangle\}$) be a basis for the Hilbert space H_A (H_B) and let E be an ensemble of physical systems S described by the density operator ρ. The *reduced density operator* ρ_A for subsystem A is defined to be the partial trace of ρ over B:

$$\rho_A = \mathrm{Tr}_B(\rho) = \sum_j \langle\beta_j|\rho|\beta_j\rangle. \tag{5.24}$$

Similarly, the *reduced density operator* ρ_B for subsystems B is defined to be the partial trace of ρ over A:

$$\rho_B = \mathrm{Tr}_A(\rho) = \sum_i \langle\alpha_j|\rho|\alpha_j\rangle. \tag{5.25}$$

5.2 Measurements, Uncertainty Relations, and Dynamics of a Quantum System

5.2.1 Measurements

Each measurable physical quantity—observable (such as position, momentum or angular momentum)—is associated with a Hermitian operator that has a complete set of eigenkets. According to P. A. Dirac "A measurement always causes the system to jump into an eigenstate of the dynamical variable that is being measured [11]." The Dirac's statement can be formulated as the following *postulate*: an exact measurement of an observable with operator A always yields as a result one of the eigenvalues $a^{(n)}$ of A. Thus, the measurement changes the state, with the measurement system is "thrown into" one of its eigenstates, which can be represented by $|\alpha\rangle \underrightarrow{A \text{ measurement}} |a^{(j)}\rangle$. If before measurement the system was in state $|\alpha\rangle$, the probability that the result of measurement will be the eigenvalue $a^{(i)}$ is given by

$$\Pr\left(a^{(i)}\right) = \left|< a^{(i)}|\alpha\rangle\right|^2. \tag{5.26}$$

Since at least one of the eigenvalues must occur as the result of the measurements, these probabilities satisfy

$$\sum_i \Pr\left(a^{(i)}\right) = \sum_i \left|< a^{(i)}|\alpha\rangle\right|^2 = 1. \tag{5.27}$$

The expected value of the outcome of the measurement of A is given by

$$\langle A\rangle = \sum_i a^{(i)} \Pr\left(a^{(i)}\right) = \sum_i a^{(i)} \left|\left\langle a^{(i)} \mid \alpha\right\rangle\right|^2 = \sum_i a^{(i)} \left\langle \alpha \mid a^{(i)}\right\rangle\left\langle a^{(i)} \mid \alpha\right\rangle. \tag{5.28}$$

By applying the eigenvalue equation $a^{(i)}\left|a^{(i)}\right\rangle = A\left|a^{(i)}\right\rangle$, Eq. (5.28) becomes

$$\langle A\rangle = \sum_i \langle\alpha|A\left|a^{(i)}\right\rangle\left\langle a^{(i)} \mid \alpha\right\rangle. \tag{5.29}$$

By using further the completeness relation $\sum_i \left|a^{(i)}\right\rangle\left\langle a^{(i)}\right| = I$, we obtain the expected value of the measurement of A to be simply

$$\langle A\rangle = \langle\alpha|A|\alpha\rangle. \tag{5.30}$$

In various situations, like initial state preparations for quantum information processing applications, we need to select one particular outcome of the measurement. This procedure is known as the *selective measurement* (or filtration) and it can be conducted as shown in Fig. 5.1.

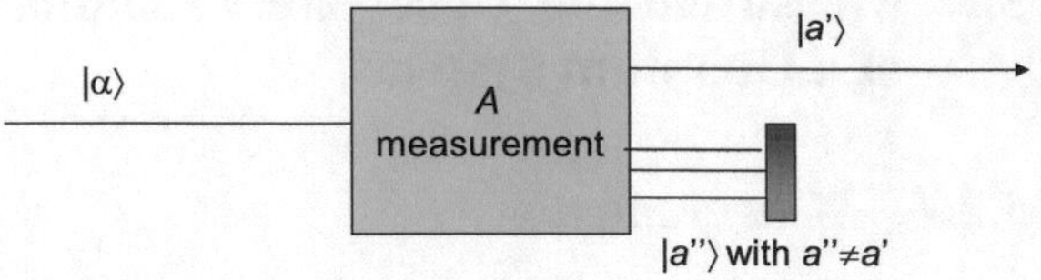

Fig. 5.1 The illustration of the concept of a selective measurement (filtration)

The result of the selective measurement can be interpreted as applying the *projection operator* $P_{a'}$ to $|\alpha\rangle$ to obtain

$$P_{a'}|\alpha\rangle = |a'\rangle\langle a'|\alpha\rangle. \tag{5.31}$$

The probability that the outcome of the measurement of observable Ξ with eigenvalues $\xi^{(n)}$ lies between (a, b) is given by

$$\Pr(\xi \in R(a,b)) = \sum_{\xi^{(n)} \in R(a,b)} \left|\left\langle \xi^{(n)} \mid \alpha \right\rangle\right|^2 = \sum_{\xi^{(n)} \in R(a,b)} \left\langle \alpha \mid \xi^{(n)} \right\rangle\left\langle \xi^{(n)} \mid \alpha \right\rangle = \langle \alpha|P_{ab}|\alpha\rangle = \langle P_{ab}\rangle, \tag{5.32}$$

where with P_{ab} we denoted the following projection operator:

$$P_{ab} = \sum_{\xi^{(n)} \in R(a,b)} \left|\xi^{(n)}\right\rangle\left\langle\xi^{(n)}\right|. \tag{5.33}$$

It can straightforwardly be shown that the projection operator P_{ab} satisfies

$$P_{ab}^2 = P_{ab} \Leftrightarrow P_{ab}(P_{ab} - I) = \mathbf{0}. \tag{5.34}$$

Therefore, the eigenvalues of projection operator P_{ab} are either 0 (corresponding to the "false proposition") or 1 (corresponding to the "true proposition"), and it is of high importance in *quantum detection theory* [2].

In terms of projection operators, the state of the system after the measurement is given by

$$|\alpha\rangle \underrightarrow{A \text{ measurement}} \frac{1}{\sqrt{\langle \alpha|P_j|\alpha\rangle}} P_j|\alpha\rangle,\ P_j = \left|a^{(j)}\right\rangle\left\langle a^{(j)}\right|. \tag{5.35}$$

In case operator A has the same eigenvalue a_i for the following eigenkets $\left\{\left|a_i^{(j)}\right\rangle\right\}_{j=1}^{d_i}$, with the corresponding characteristic equation:

$$A\left|a_i^{(j)}\right\rangle = a_i\left|a_i^{(j)}\right\rangle;\ j = 1, \ldots, d_i, \tag{5.36}$$

we say that eigenvalue a_i is *degenerate* of order d_i. The corresponding probability of obtaining the measurement result a_i can be found by

$$\Pr(a_i) = \sum_{j=1}^{d_i} \left|\left\langle a_i^{(j)} \mid \alpha\right\rangle\right|^2. \tag{5.37}$$

The projective measurements can be generalized as follows. Let the set of measurement operators be given by $\{M_m\}$, where index m stands for possible measurement result, satisfying the property $\sum_m M_m^\dagger M_m = I$. The probability of finding the measurement result m, given the state $|\psi\rangle$, is given by

$$\Pr(m) = \langle\psi|M_m^\dagger M_m|\psi\rangle. \tag{5.38}$$

After the measurement, the system will be left in the following state:

$$|\psi_f\rangle = \frac{M_m|\psi\rangle}{\sqrt{\langle\psi|M_m^\dagger M_m|\psi\rangle}}. \tag{5.39}$$

For projective measurements, clearly $M_m = P_m = |a^{(m)}\rangle\langle a^{(m)}|$, and from property above we obtain

$$\sum_m M_m^\dagger M_m = \sum_m |a^{(m)}\rangle \underbrace{\langle a^{(m)}|a^{(m)}\rangle}_{1}\langle a^{(m)}| = \sum_m |a^{(m)}\rangle\langle a^{(m)}| = \sum_m P_m = I, \tag{5.40}$$

which is the completeness relationship. The probability of obtaining the mth result of measurement will be then

$$\Pr(m) = \mathrm{Tr}\left(P_m^\dagger P_m \rho\right) = \mathrm{Tr}(P_m \rho) = \mathrm{Tr}\left(|a^{(m)}\rangle\langle a^{(m)}|\rho\right) = \langle a^{(m)}|\rho|a^{(m)}\rangle. \tag{5.41}$$

Another important type of measurement is known as a *positive operator-valued measure* (POVM). A POVM consists of the set of operators $\{E_m\}$, where each operator E_m is positive semidefinite, i.e., $\langle\psi|E_m|\psi\rangle \geq 0$, satisfying the relationship

$$\sum_m E_m = I. \tag{5.42}$$

The POVM can be constructed from generalized measurement operators $\{M_m\}$ by setting $E_m = M_m^\dagger M_m$. The probability of obtaining the mth result of measurements is given by $\mathrm{Tr}(E_m \rho)$. The POVM concept is in particular suitable to situations when the measurements are not repeatable. For instance, by performing the measurement on a photon, it can be destroyed so that the repeated measurements are not possible.

5.2.2 Uncertainty Principle

Let A and B be two operators, which in general do not commute, i.e., $AB \neq BA$. The quantity $[A, B] = AB - BA$ is called the *commutator* of A and B, while the quantity $\{A, B\} = AB + BA$ is called the *anticommutator*. Two observables A and B are said to be **compatible** when their corresponding operators commute: $[A, B] = 0$. Two observables A and B are said to be *incompatible* when $[A, B] \neq 0$. If in the set of operators $\{A, B, C, \ldots\}$ all operators commute in pairs, namely, $[A, B] = [A, C] = [B, C] = \ldots = 0$, we say the set is a *complete set of commuting observables* (CSCO).

If two observables, say A and B, are to be measured simultaneously and exactly on the same system; the system after the measurement must be left in the state $|a^{(n)}; b^{(n)}\rangle$ that is an eigenstate of both observables:

$$A\left|a^{(n)}; b^{(n)}\right\rangle = a^{(n)}\left|a^{(n)}; b^{(n)}\right\rangle,\ B\left|a^{(n)}; b^{(n)}\right\rangle = b^{(n)}\left|a^{(n)}; b^{(n)}\right\rangle. \tag{5.43}$$

This will be true only if $AB = BA$ or equivalently the commutator $[A, B] = AB - BA = 0$, that is, when two operators *commute* as shown below:

$$\begin{aligned} &AB\left|a^{(n)}; b^{(n)}\right\rangle = A\left(B\left|a^{(n)}; b^{(n)}\right\rangle\right) = Ab^{(n)}\left|a^{(n)}; b^{(n)}\right\rangle = b^{(n)} \cdot A\left|a^{(n)}; b^{(n)}\right\rangle = a^{(n)}b^{(n)}\left|a^{(n)}; b^{(n)}\right\rangle \\ &BA\left|a^{(n)}; b^{(n)}\right\rangle = a^{(n)}b^{(n)}\left|a^{(n)}; b^{(n)}\right\rangle \Rightarrow AB = BA. \end{aligned} \tag{5.44}$$

When two operators do not commute, they cannot be simultaneously measured with the complete precision. Given an observable A, we define the operator $\Delta A = A - \langle A\rangle$, and the corresponding expectation value of $(\Delta A)^2$ that is known as *dispersion* of A:

$$\left\langle(\Delta A)^2\right\rangle = \left\langle A^2 - 2A\langle A\rangle + \langle A\rangle^2\right\rangle = \left\langle A^2\right\rangle - \langle A\rangle^2. \tag{5.45}$$

Then for any state the following inequality is valid:

$$\left\langle(\Delta A)^2\right\rangle\left\langle(\Delta B)^2\right\rangle \geq \frac{1}{4}|\langle[\Delta A, \Delta B]\rangle|^2, \tag{5.46}$$

which is known as **the Heisenberg uncertainty principle**.

Example. The commutation relation for coordinate X and momentum P observables is $[X, P] = j\hbar$, as shown above. By substituting this commutation relation into (5.46), we obtain

$$\left\langle X^2\right\rangle\left\langle P^2\right\rangle \geq \frac{\hbar^2}{4}.$$

If we observe a large ensemble of N independent systems, all of them being in the state $|\psi\rangle$. On some systems X is measured, and on some systems P is measured. The uncertainty principle asserts that for none state the product of dispersions (variances) cannot be less than $\hbar^2/4$.

5.2.3 Time Evolution—Schrödinger Equation

Time evolution operator $U(t, t_0)$ operator transforms the initial ket at time instance t_0, $|\alpha, t_0\rangle$ into the final ket at the time instance t by

$$|\alpha, t_0; t\rangle = U(t, t_0)|\alpha, t_0\rangle. \tag{5.47}$$

This time evolution operator must satisfy the following two properties:

1. *Unitary property*: $U^{+}(t, t_0)\, U(t, t_0) = I$;
2. *Composition property*: $U(t_2, t_0) = U(t_2, t_1)\, U(t_1, t_0)$, $t_2 > t_1 > t_0$.

Following Eq. (5.47), the action of infinitesimal time evolution operator $U(t_0 + dt, t_0)$ can be described by

$$|\alpha, t_0; t_0 + dt\rangle = U(t_0 + dt, t_0)|\alpha, t_0\rangle. \tag{5.48}$$

The following operator satisfies all propositions above, when $dt \to 0$:

$$U(t_0 + dt, t_0) = 1 - j\Omega\, dt, \qquad \Omega^{\dagger} = \Omega, \tag{5.49}$$

where the operator Ω is related to the Hamiltonian H by $H = \hbar\Omega$, and the Hamiltonian eigenvalue s correspond to the energy $E = \hbar\omega$. For the infinitesimal time evolution operator $U(t_0 + dt, t_0)$, we can derive the time evolution equation as follows. The starting point in derivation is the composition property:

$$U(t + dt, t_0) = U(t + dt, t)U(t, t_0) = \left(1 - \frac{j}{\hbar}H dt\right)U(t, t_0). \tag{5.50}$$

Equation (5.50) can be rewritten into following form:

$$\lim_{dt\to 0} \frac{U(t + dt, t_0) - U(t, t_0)}{dt} = -\frac{j}{\hbar}HU(t, t_0), \tag{5.51}$$

which by taking the partial derivative definition into account the (5.51) becomes

$$j\hbar\frac{\partial}{\partial t}U(t, t_0) = HU(t, t_0), \tag{5.52}$$

and this equation is known as **Schrödinger equation for time evolution operator**.

The Schrödinger equation for a state ket can be obtained by applying the time evolution operator on initial ket:

$$j\hbar\frac{\partial}{\partial t}U(t,t_0)|\alpha,t_0\rangle = HU(t,t_0)|\alpha,t_0\rangle, \tag{5.53}$$

which based on (5.52) can be rewritten as

$$j\hbar\frac{\partial}{\partial t}|\alpha,t_0;t\rangle = H|\alpha,t_0;t\rangle. \tag{5.54}$$

For *conservative systems*, for which the Hamiltonian is time-invariant, we can easily solve Eq. (5.54) to obtain

$$U(t,t_0) = e^{-\frac{j}{\hbar}H(t-t_0)}. \tag{5.56}$$

The time evolution of kets in conservative systems can therefore be described by applying (5.56) in (5.47), which yields to

$$|\alpha(t)\rangle = e^{-\frac{j}{\hbar}H(t-t_0)}|\alpha(t_0)\rangle. \tag{5.57}$$

Therefore, the operators do not explicitly depend on time and this concept is known as the *Schrödinger picture.*

In *Heisenberg picture*, on the other hand, the state vector is independent of time, but operators depend on time:

$$A(t) = e^{\frac{j}{\hbar}H(t-t_0)}Ae^{-\frac{j}{\hbar}H(t-t_0)}. \tag{5.58}$$

The time evolution equation in Heisenberg picture is given by

$$j\hbar\frac{dA(t)}{dt} = [A(t),H] + j\hbar\frac{\partial A(t)}{\partial t}. \tag{5.59}$$

The density operator ρ, representing the statistical mixture of states, is independent on time in Heisenberg picture. The expectation value of a measurement of an observable Ξ(t) at time instance t is given by

$$\begin{aligned} &E_t[\Xi] = \mathrm{Tr}[\rho\Xi(t)] \\ &E_t[\Xi] = \mathrm{Tr}[\rho(t)\Xi],\ \rho(t) = e^{\frac{j}{\hbar}H(t-t_0)}\rho e^{-\frac{j}{\hbar}H(t-t_0)}. \end{aligned} \tag{5.60}$$

Example: The Hamiltonian for a two-state system is given by

$$H = \begin{bmatrix} \omega_1 & \omega_2 \\ \omega_2 & \omega_1 \end{bmatrix}.$$

The basis for this system is given by $\{|0\rangle = [1\ 0]^{\mathrm{T}}, |1\rangle = [0\ 1]^{\mathrm{T}}\}$.

(a) Determine the eigenvalues and eigenkets of H, and express the eigenkets in terms of basis.
(b) Determine the time evolution of the system described by the Schrödinger equation:

$$j\hbar\frac{\partial}{\partial t}|\psi\rangle = H|\psi\rangle, \ |\psi(0)\rangle = |0\rangle.$$

To determine the eigenkets of H, we start from the characteristic equation $\det(H - \lambda I) = 0$ and find that eigenvalues are $\lambda_{1,2} = \omega_1 \pm \omega_2$. The corresponding eigenvectors are

$$|\lambda_1\rangle = \frac{1}{\sqrt{2}}\begin{bmatrix}1\\1\end{bmatrix} = \frac{1}{\sqrt{2}}(|0\rangle + |1\rangle) \quad |\lambda_1\rangle = \frac{1}{\sqrt{2}}\begin{bmatrix}1\\-1\end{bmatrix} = \frac{1}{\sqrt{2}}(|0\rangle - |1\rangle).$$

We now have to determine the time evolution of arbitrary ket $|\psi(t)\rangle = [\alpha(t)\beta(t)]^{\mathrm{T}}$. The starting point is the Schrödinger equation:

$$j\hbar\frac{\partial}{\partial t}|\psi\rangle = j\hbar\begin{bmatrix}\dot{\alpha}(t)\\ \dot{\beta}(t)\end{bmatrix}, \qquad H|\psi\rangle = \begin{bmatrix}\omega_1 & \omega_2\\ \omega_2 & \omega_1\end{bmatrix}\begin{bmatrix}\alpha(t)\\ \beta(t)\end{bmatrix} = \begin{bmatrix}\omega_1\alpha(t) + \omega_2\beta(t)\\ \omega_2\alpha(t) + \omega_1\beta(t)\end{bmatrix} \Rightarrow$$

$$j\hbar\begin{bmatrix}\dot{\alpha}(t)\\ \dot{\beta}(t)\end{bmatrix} = \begin{bmatrix}\omega_1\alpha(t) + \omega_2\beta(t)\\ \omega_2\alpha(t) + \omega_1\beta(t)\end{bmatrix}.$$

By substitution $\alpha(t)+\beta(t) = \gamma(t)$ and $\alpha(t)-\beta(t) = \delta(t)$, we obtain the ordinary set of differential equations

$$j\hbar\frac{d\gamma(t)}{dt} = (\omega_1 + \omega_2)\gamma(t) \quad j\hbar\frac{d\delta(t)}{dt} = (\omega_1 - \omega_2)\delta(t),$$

whose solution is $\gamma(t) = C\exp\left(\frac{\omega_1+\omega_2}{j\hbar}\right)$ and $\delta(t) = D\exp\left(\frac{\omega_1-\omega_2}{j\hbar}t\right)$. From the initial state $|\psi(0)\rangle = |0\rangle = [1 \ \ 0]^{\mathrm{T}}$, we obtain the unknown constants $C = D = 1$ so that state time evolution is given by

$$|\psi(t)\rangle = \exp\left(-\frac{j}{\hbar}\omega_1 t\right)\begin{bmatrix}\cos\left(\frac{\omega_2 t}{\hbar}\right)\\ -j\sin\left(\frac{\omega_2 t}{\hbar}\right)\end{bmatrix}.$$

5.3 Quantum Information Processing (QIP) Fundamentals

Fundamental features of QIP are different from that of classical computing and can be summarized into three: (i) linear superposition, (ii) quantum parallelism, and (iii) entanglement. Below we provide some basic details of these features.

(i) Linear superposition. Contrary to the classical bit, a quantum bit or *qubit* can take not only two discrete values 0 and 1 but also *all* possible *linear combinations* of them. This is a consequence of a fundamental property of quantum states: it is possible to construct a *linear superposition* of quantum state $|0\rangle$ and quantum state $|1\rangle$.

(ii) Quantum parallelism. The *quantum parallelism* is a possibility to perform a large number of operations in parallel, which represents the key difference from classical computing. Namely, in classical computing, it is possible to know what is the internal status of the computer. On the other hand, because of no-cloning theorem, it is not possible to know the current state of quantum computer. This property has lead to the development of Shor factorization algorithm, which can be used to crack the Rivest–Shamir–Adleman (RSA) encryption protocol. Some other important quantum algorithms include Grover search algorithm, which is used to perform a search for an entry in an unstructured database; the quantum Fourier transform, which is a basis for a number of different algorithms; and Simon's algorithm. The quantum computer is able to encode all input strings of length N simultaneously into a single computation step. In other words, the quantum computer is able simultaneously to pursue 2^N classical paths, indicating that quantum computer is significantly more powerful than classical one.

(iii) Entanglement. At a quantum level, it appears that two quantum objects can form a single entity, even when they are well separated from each other. Any attempt to consider this entity as a combination of two independent quantum objects given by tensor product of quantum states fails, unless the possibility of signal propagation at superluminal speed is allowed. These quantum objects that cannot be decomposed into tensor product of individual independent quantum objects are called *entangled* quantum objects. Given the fact that arbitrary quantum states cannot be copied, which is the consequence of no-cloning theorem, the communication at superluminal speed is not possible, and as consequence the entangled quantum states cannot be written as tensor product of independent quantum states. Moreover, it can be shown that the amount of information contained in an entangled state of N-qubits grows exponentially instead of linearly, which is the case for classical bits.

In incoming subsections, we describe these fundamental features with more details.

5.3.1 Superposition Principle, Quantum Parallelism, Quantum Gates, and QIP Basics

We say that the allowable states $|\mu\rangle$ and $|\nu\rangle$ of the quantum system satisfy the <u>*superposition principle*</u> if their linear superposition $\alpha\ |\mu\rangle + \beta\ |\nu\rangle$, where α and β are the complex numbers ($\alpha, \beta \in C$), is also allowable quantum state. Without loss of generality, we typically observe the computational basis composed of the orthogonal canonical states $|0\rangle = \begin{bmatrix} 1 \\ 0 \end{bmatrix}, |1\rangle = \begin{bmatrix} 0 \\ 1 \end{bmatrix}$, so that quantum bit, also known as <u>qubit</u>, lies in a two-dimensional Hilbert space *H*, isomorphic to C^2-space, and can be represented as

$$|\psi\rangle = \alpha|0\rangle + \beta|1\rangle = \begin{pmatrix} \alpha \\ \beta \end{pmatrix}; \quad \alpha, \beta \in C; \quad |\alpha|^2 + |\beta|^2 = 1. \tag{5.61}$$

If we perform the measurement of a qubit, we will get $|0\rangle$ with probability $|\alpha|^2$ and $|1\rangle$ with probability of $|\beta|^2$. Measurement changes the state of a qubit from a superposition of $|0\rangle$ and $|1\rangle$ to the specific state consistent with the measurement result. If we parametrize the probability amplitudes α and β as follows,

$$\alpha = \cos\left(\frac{\theta}{2}\right), \quad \beta = e^{j\phi} \sin\left(\frac{\theta}{2}\right), \tag{5.62}$$

where θ is a polar angle, and ϕ is an azimuthal angle, we can geometrically represent the qubit by Bloch sphere (or the Poincaré sphere for the photon) as illustrated in Fig. 5.2. Bloch vector coordinates are given by (cosφ sinθ, sinφ sinθ, cosθ). This Bloch vector representation is related to computational basis (CB) by

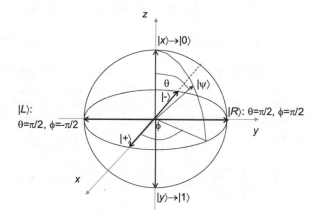

Fig. 5.2 Block (Poincaré) sphere representation of the single-qubit

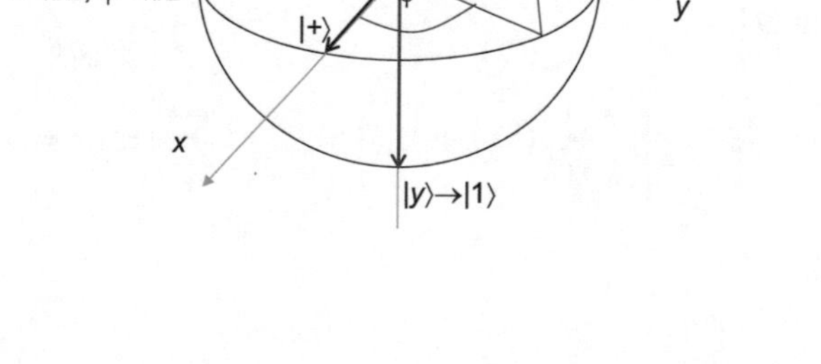

$$|\psi(\theta,\phi)\rangle = \cos(\theta/2)|0\rangle + e^{j\phi}\sin(\theta/2)|1\rangle \doteq \begin{pmatrix} \cos(\theta/2) \\ e^{j\phi}\sin(\theta/2) \end{pmatrix}, \tag{5.63}$$

where $0 \le \theta \le \pi$ and $0 \le \phi < 2\pi$. The north and south poles correspond to computational $|0\rangle$ ($|x\rangle$-polarization) and $|1\rangle$ ($|y\rangle$-polarization) basis kets, respectively. Other important bases are the *diagonal basis* $\{|+\rangle, |-\rangle\}$, very often denoted as $\{|\nearrow\rangle, |\searrow\rangle\}$, related to CB by

$$|+\rangle = |\nearrow\rangle = \frac{1}{\sqrt{2}}(|0\rangle + |1\rangle), \quad |-\rangle = |\searrow\rangle = \frac{1}{\sqrt{2}}(|0\rangle - |1\rangle), \tag{5.64}$$

and the *circular basis* $\{|R\rangle, |L\rangle\}$, related to the CB as follows:

$$|R\rangle = \frac{1}{\sqrt{2}}(|0\rangle + j|1\rangle), \qquad |L\rangle = \frac{1}{\sqrt{2}}(|0\rangle - j|1\rangle). \tag{5.65}$$

The pure qubit states lie on the Bloch sphere, while the mixed qubit states lie in the interior of the block sphere. The maximally mixed state $I/2$ (I denotes the identity operator) lies in the center of the Bloch sphere. The orthogonal states are antipodal. From Fig. 5.2, we see that CB, diagonal basis and circular bases are 90° apart from each other, and we often say that these three bases are mutually *conjugate bases*. These bases are used as three pairs of signal states for the six-state quantum-key distribution (QKD) protocol. Another important basis used in QKD and for eavesdropping is the *Breidbart basis* given by $\{\cos(\pi/8)|0\rangle + \sin(\pi/8)|1\rangle, -\sin(\pi/8)|0\rangle + \cos(\pi/8)|1\rangle\}$.

The superposition principle is the key property that makes quantum parallelism possible. To see this let us juxtapose n-qubits lying in n distinct two-dimensional Hilbert spaces H_0, H_1, …, H_{n-1} that are isomorphic to each other. In practice, this means that the qubits have been prepared separately, without any interaction, which can be mathematically described by the tensor product

$$|\boldsymbol{\psi}\rangle = |\psi_0\rangle \otimes |\psi_1\rangle \otimes \cdots \otimes |\psi_{n-1}\rangle \in H_0 \otimes H_1 \otimes \cdots \otimes H_{n-1}. \tag{5.66}$$

Any arbitrary basis can be selected as the computational basis for H_i, $i = 0, 1, \ldots, n-1$. However, for ease of exposition, we assume the computational basis to be $|0_i\rangle$ and $|1_i\rangle$. Consequently, we can represent the ith qubit as $|\psi_i\rangle = \alpha_i|0_i\rangle + \beta_i|1_i\rangle$. Introducing a further assumption, $\alpha_i = \beta_i = 2^{-1/2}$, without loss of generality, we now have

$$|\boldsymbol{\psi}\rangle = \prod_{i=0}^{n-1} \frac{1}{\sqrt{2}}(|0_i\rangle + |1_i\rangle) = 2^{-n/2}\sum_{\boldsymbol{x}} |\boldsymbol{x}\rangle, \boldsymbol{x} = x_0x_1\cdots x_{n-1}, x_j \in \{0, 1\}. \tag{5.67}$$

This composite quantum system is called the ***n*-qubit register** and as can be seen from equation above, it represents a superposition of 2^n quantum states that exist simultaneously! This is an example of quantum parallelism. In the classical realm, a linear increase in size corresponds roughly to a linear increase in processing power. In the quantum world, due to the power of quantum parallelism, a linear increase in size corresponds to an exponential increase in processing power. The downside, however, is the accessibility to this parallelism. Remember that superposition collapses the moment we attempt to measure it. The quantum circuit to create the superposition state above, in other words, Walsh–Hadamard transform, is shown in Fig. 5.3. Therefore, the Walsh–Hadamard transform on n ancilla qubits in state $|00 \ldots 0\rangle$ can be implemented by applying the Hadamard operators (gates) H, whose action is described in Fig. 5.3, on ancillary qubits.

More generally, a linear operator (gate) B can be expressed in terms of eigenkets $\{|a^{(n)}\rangle\}$ of a Hermitian operator A. The *operator B* is associated with a *square matrix* (albeit infinite in extent), whose elements are

$$B_{mn} = \langle a^{(m)} | B | a^{(n)} \rangle, \tag{5.68}$$

and can explicitly be written as

$$B \doteq \begin{pmatrix} \langle a^{(1)} | B | a^{(1)} \rangle & \langle a^{(1)} | B | a^{(2)} \rangle & \cdots \\ \langle a^{(2)} | B | a^{(1)} \rangle & \langle a^{(2)} | B | a^{(2)} \rangle & \cdots \\ \vdots & \vdots & \ddots \end{pmatrix}, \tag{5.69}$$

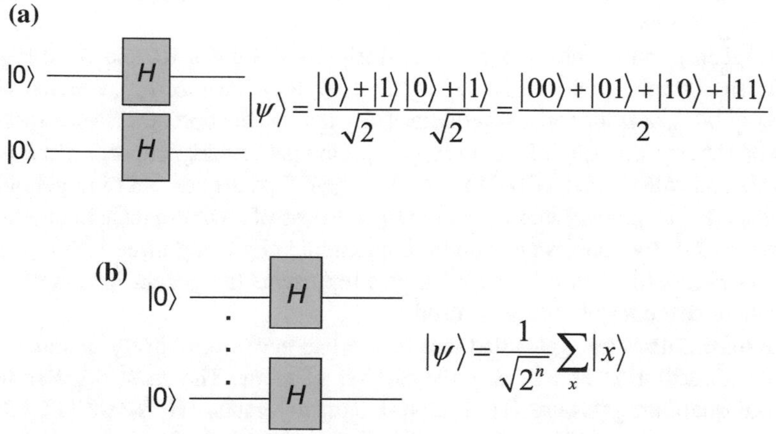

Fig. 5.3 The Walsh–Hadamard transform: **a** on two qubits, and **b** on n-qubits. The action of the Hadamard gate H on computational basis kets is given by $H|0\rangle = 2^{-1/2}(|0\rangle + |1\rangle)$ and $H|1\rangle = 2^{-1/2}(|0\rangle - |1\rangle)$

where we use the notation $\doteq$ to denote that operator B is represented by the matrix above. Very important single-qubit gates are Hadamard gate H, the phase shift gate S, the $\pi/8$ (or T) gate, controlled-NOT (or CNOT) gate, and Pauli operators X, Y, Z. The Hadamard gate H, phase shift gate, T gate, and CNOT gate have the following matrix representation in CB $\{|0\rangle, |1\rangle\}$:

$$H \doteq \frac{1}{\sqrt{2}}\begin{bmatrix} 1 & 1 \\ 1 & -1 \end{bmatrix}, S \doteq \begin{bmatrix} 1 & 0 \\ 0 & j \end{bmatrix}, T \doteq \begin{bmatrix} 1 & 0 \\ 0 & e^{j\pi/4} \end{bmatrix}, CNOT \doteq \begin{bmatrix} 1 & 0 & 0 & 0 \\ 0 & 1 & 0 & 0 \\ 0 & 0 & 0 & 1 \\ 0 & 0 & 1 & 0 \end{bmatrix}. \quad (5.70)$$

The Pauli operators, on the other hand, have the following matrix representation in CB:

$$X \doteq \begin{bmatrix} 0 & 1 \\ 1 & 0 \end{bmatrix}, Y \doteq \begin{bmatrix} 0 & -j \\ j & 0 \end{bmatrix}, Z \doteq \begin{bmatrix} 1 & 0 \\ 0 & -1 \end{bmatrix}. \quad (5.71)$$

The action of Pauli gates on an arbitrary qubit $|\psi\rangle = \alpha|0\rangle + \beta|1\rangle$ is given as follows:

$$X(\alpha|0\rangle + \beta|1\rangle) = \alpha|1\rangle + \beta|0\rangle, Y(\alpha|0\rangle + \beta|1\rangle) = j(\alpha|1\rangle - \beta|0\rangle),$$
$$Z(\alpha|0\rangle + \beta|1\rangle) = a|0\rangle - b|1\rangle. \quad (5.72)$$

So the action of X-gate is to introduce the bit-flip, the action of Z-gate is to introduce the phase-flip and the action of Y-gate is to simultaneously introduce the bit and phase flips.

Several important single-, double-, and triple-qubit gates are shown in Fig. 5.4. The action of single-qubit gate is to apply the operator U on qubit $|\psi\rangle$, which results in another qubit. Controlled-U gate conditionally applies the operator U on target qubit $|\psi\rangle$, when the control qubit $|c\rangle$ is in $|1\rangle$-state. One particularly important controlled-U-gate is controlled-NOT (CNOT) gate. This gate flips the content of target qubit $|t\rangle$ when the control qubit $|c\rangle$ is in $|1\rangle$-state. The purpose of SWAP-gate is to interchange the positions of two qubits and can be implemented by using three CNOT gates as shown in Fig. 5.4d. Finally, the Toffoli gate represents the generalization of CNOT gate, where two control qubits are used.

The minimum set of gates that can be used to perform arbitrary quantum computation algorithm is known as *universal set of gates*. The most popular sets of universal quantum gates are {H, S, CNOT, Toffoli} gates, {H, S, π/8 (T), CNOT} gates, Barenco gate, and Deutsch gate. By using these universal quantum gates, more complicated operations can be performed. As an illustration, Fig. 5.5 shows the Bell states [Einstein–Podolsky–Rosen (EPR) pairs] preparation circuit, which is of high importance in quantum teleportation and QKD applications.

So far single-, double-, and triple-qubit quantum gates have been considered. An arbitrary quantum state of K qubits has the form $\sum_s \alpha_s |\mathbf{s}\rangle$, where $\mathbf{s}$ runs over all binary

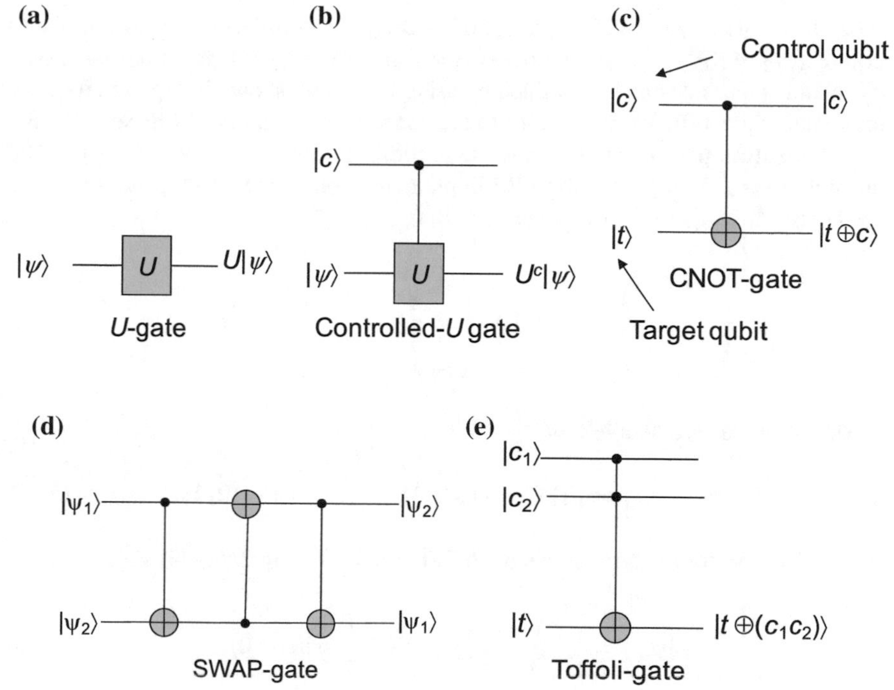

Fig. 5.4 Important quantum gates and their action: **a** single-qubit gate, **b** controlled-U gate, **c** CNOT gate, **d** SWAP-gate, and **e** Toffoli gate

Fig. 5.5 Bell states (EPR pairs) preparation circuit

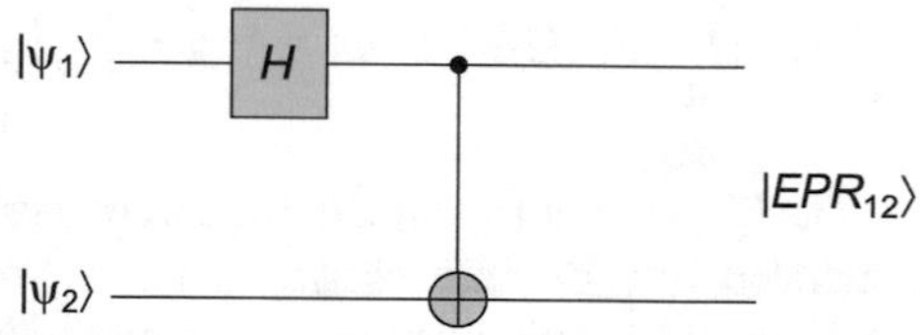

strings of length K. Therefore, there are 2^K complex coefficients, all independent except normalization constraint:

$$\sum_{s=00\cdots00}^{11\cdots11} |\alpha_s|^2 = 1. \tag{5.73}$$

For example, the state $\alpha_{00}|00\rangle + \alpha_{01}|01\rangle + \alpha_{10}|10\rangle + \alpha_{11}|11\rangle$ (with $|\alpha_{00}|_2 + |\alpha_{01}|^2 + |\alpha_{10}|^2 + |\alpha_{11}|^2 = 1$) is the general two-qubit state (we use $|00\rangle$ to denote the tensor product $|0\rangle \otimes |0\rangle$). The multiple qubits can be **entangled** so that they cannot be decomposed into two separate states. For example, the Bell state or EPR pair $(|00\rangle + |11\rangle)/\sqrt{2}$ cannot be written in terms of tensor product $|\psi 1\rangle\ |\psi 2\rangle = (\alpha_1|0\rangle + \beta_1|1\rangle)$

$\otimes (\alpha_2|0\rangle + \beta_2|1\rangle) = \alpha_1\alpha_2|00\rangle + \alpha_1\beta_2|01\rangle + \beta_1\alpha_2|10\rangle + \beta_1\beta_2|11\rangle$, because it would require $\alpha_1\alpha_2 = \beta_1\beta_2 = 1/\sqrt{2}$, while $\alpha_1\beta_2 = \beta_1\alpha_2 = 0$, which a priori has no reason to be valid. This state can be obtained by using the circuit shown in Fig. 5.5, for two-qubit input state $|00\rangle$. We will return to the concept of entanglement in Sect. 2.3.3.

The quantum parallelism can now be introduced more formally as follows. The quantum information processing *QIP* implemented on a quantum register maps the input string $i_1 \ldots i_N$ to the output string $O_1(i), \ldots, O_N(i)$:

$$\begin{pmatrix} O_1(i) \\ \vdots \\ O_N(i) \end{pmatrix} = U(QIP)\begin{pmatrix} i_1 \\ \vdots \\ i_N \end{pmatrix}; \quad (i)_{10} = (i_1 \cdots i_N)_2. \tag{5.74}$$

The CB states are denoted by

$$|i_1 \cdots i_N\rangle = |i_1\rangle \otimes \cdots \otimes |i_N\rangle; \quad i_1, \ldots, i_N \in \{0, 1\}. \tag{5.75}$$

The linear superposition allows us to form the following 2*N*-qubit state:

$$|\psi_{in}\rangle = \left[\frac{1}{\sqrt{2^N}}\sum_i |i_1 \cdots i_N\rangle\right] \otimes |0 \cdots 0\rangle, \tag{5.76}$$

and upon the application of quantum operation $U(QIP)$, the output can be represented by

$$|\psi_{out}\rangle = U(QIP)|\psi_{in}\rangle = \frac{1}{\sqrt{2^N}}\sum_i |i_1 \cdots i_N\rangle \otimes |O_1(i) \cdots O_N(i)\rangle. \tag{5.77}$$

The QIP circuit (or a quantum computer) has been able to encode all input strings generated by *QIP* into $|\psi_{\text{out}}\rangle$; in other words, it has simultaneously pursued 2^N classical paths. This ability of a QIP circuit to encode multiple computational results into a quantum state in a single quantum computational step is known as **quantum parallelism**, as mentioned earlier.

5.3.2 No-Cloning Theorem and Distinguishing the Quantum States

Just like in quantum parallelism, the quantum superposition is also the key concept behind our inability to clone arbitrary quantum states. To see this, let us think of a quantum copier that takes as input an arbitrary quantum state and outputs two copies of that state, resulting in a clone of the original state. For example, if the input state is $|\psi\rangle$, then the output of the copier is $|\psi\rangle|\psi\rangle$. For an arbitrary quantum state, such a copier raises a fundamental contradiction. Consider two arbitrary states $|\psi\rangle$ and $|\chi\rangle$ that are input to the copier. When they are inputted individually, we expect to get $|\psi\rangle|\psi\rangle$ and $|\chi\rangle|\chi\rangle$. Now consider a superposition of these two states given by

$$|\varphi\rangle = \alpha|\psi\rangle + \beta|\chi\rangle. \tag{5.78}$$

Based on the above description that the quantum copier clones the original state, we expect the output,

$$\begin{aligned}|\varphi\rangle|\varphi\rangle &= (\alpha|\psi\rangle + \beta|\chi\rangle)(\alpha|\psi\rangle + \beta|\chi\rangle)\\ &= \alpha^2|\psi\rangle|\psi\rangle + \alpha\beta|\psi\rangle|\chi\rangle + \alpha\beta|\chi\rangle|\psi\rangle + \beta^2|\chi\rangle|\chi\rangle.\end{aligned} \tag{5.79}$$

On the other hand, linearity of quantum mechanics, as evidenced by the Schrodinger wave equation hand, tells us that the quantum copier can be represented by a unitary operator that performs the cloning. If such a unitary operator were to act on the superposition state $|\varphi\rangle$, the output would be a superposition of $|\psi\rangle|\psi\rangle$ and $|\chi\rangle|\chi\rangle$, that is,

$$|\varphi'\rangle = \alpha|\psi\rangle|\psi\rangle + \beta|\chi\rangle|\chi\rangle. \tag{5.80}$$

As is clearly evident, the difference between previous two equations leads to contradiction mentioned above. As a consequence, there is no unitary operator that can clone $|\varphi\rangle$. We, therefore, formulate the no-cloning theorem as follows.

No-cloning Theorem: No quantum copier exists that can clone an arbitrary quantum state.

This result raises a related question: do there exist some specific states for which cloning is possible? The answer to this question is (surprisingly) yes. Remember, a key result of quantum mechanics is that unitary operators preserve probabilities. This implies that inner (dot) products $\langle\varphi \mid \varphi\rangle$ and $\langle\varphi' \mid \varphi'\rangle$ should be identical. The inner products $\langle\varphi \mid \varphi\rangle$ and $\langle\varphi' \mid \varphi'\rangle$ are, respectively, given by

$$\begin{aligned}\langle\varphi|\varphi\rangle &= \big(\langle\psi|\alpha^* + \langle\chi|\beta^*\big)(\alpha|\psi\rangle + \beta|\chi\rangle)\\ &= |\alpha|^2\langle\psi|\psi\rangle + |\beta|^2\langle\chi|\chi\rangle + \alpha^*\beta\langle\psi|\chi\rangle + \alpha\beta^*\langle\chi|\psi\rangle\\ \langle\varphi'|\varphi'\rangle &= \big(\langle\psi|\langle\psi|\alpha^* + \langle\chi|\langle\chi|\beta^*\big)(\alpha|\psi\rangle|\psi\rangle + \beta|\chi\rangle|\chi\rangle)\\ &= |\alpha|^2|\langle\psi|\psi\rangle|^2 + |\beta|^2|\langle\chi|\chi\rangle|^2 + \alpha^*\beta|\langle\psi|\chi\rangle|^2 + \alpha\beta^*|\langle\chi|\psi\rangle|^2.\end{aligned} \tag{5.81}$$

We know that $\langle\psi|\psi\rangle = \langle\chi|\chi\rangle = 1$. Therefore, the discrepancy lies in the cross terms. Specifically, to avoid the contradiction that resulted in the no-cloning theorem, we require that $|\langle\psi|\chi\rangle|^2 = \langle\psi|\chi\rangle$. This condition can only be satisfied when the states are orthogonal. Thus, cloning is possible only for mutually orthogonal states. It is, however, important to remember a subtle point here. Even if we have mutually orthogonal set of states, we need a quantum copier (or unitary operator) specifically for those states. If the unitary operator is specific to a different set of mutually orthogonal states, cloning would fail. It would seem that the no-cloning theorem would prevent us from exploiting the richness of quantum mechanics. It turns out that this is not the case. A key example is the QKD that with very high probability guarantees secure communication.

Not only non-orthogonal quantum states cannot be cloned, they also cannot be reliably distinguished. There is no a measurement device we can create that can reliably distinguish non-orthogonal states. This fundamental result plays an important role in quantum cryptography. Its proof is based on contradiction. Let assume that the measurement operator M is the Hermitian operator (with corresponding eigenvalues m_i and corresponding projection operators P_i) of an observable, $\mathfrak{M}$ which allows unambiguously to distinguish between two non-orthogonal states $|\psi_1\rangle$ and $|\psi_2\rangle$. The eigenvalue m_1 (m_2) unambiguously identifies the state $|\psi_1\rangle$ ($|\psi_2\rangle$) as the pre-measurement state. We know that for projection operators, the following properties are valid:

$$\begin{aligned} \langle\psi_1|P_1|\psi_1\rangle = 1 \ \langle\psi_2|P_2|\psi_2\rangle = 1 \\ \langle\psi_1|P_2|\psi_1\rangle = 0 \ \langle\psi_2|P_1|\psi_2\rangle = 0 \end{aligned} \tag{5.82}$$

Since $|\psi_1\rangle$ and $|\psi_2\rangle$ are non-orthogonal states, then $\langle\psi_1|\psi_2\rangle \neq 0$, and $|\psi_2\rangle$ can be represented in terms of $|\psi_1\rangle$ and another state $|\chi\rangle$ that is orthogonal to $|\psi_1\rangle$ ($\langle\psi_1|\chi\rangle = 0$) as follows:

$$|\psi_2\rangle = \alpha|\psi_1\rangle + \beta|\chi\rangle. \tag{5.83}$$

From the projection operators properties, listed above, we can conclude the following:

$$\begin{aligned} &0 = \langle\psi_1|P_2|\psi_1\rangle \overset{P_2^2=P_2}{=} \langle\psi_1|P_2P_2|\psi_1\rangle = \|P_2|\psi_1\rangle\|^2 \Rightarrow P_2|\psi_1\rangle = 0 \\ &1 = \langle\psi_2|P_2|\psi_2\rangle = \langle\psi_2|P_2P_2|\psi_2\rangle = \left(\alpha^*\langle\psi_1| + \beta^*\langle\chi|\right)(\alpha P_2|\psi_1\rangle + \beta P_2|\chi\rangle) = |\beta|^2\langle\chi|P_2|\chi\rangle. \end{aligned} \tag{5.84}$$

Now we use the completeness relationship:

$$1 = \langle\chi|\chi\rangle = \langle\chi|\overbrace{\sum_i P_i}^{I}|\chi\rangle = \sum_i \langle\chi|P_i|\chi\rangle \overset{\langle\chi|P_i|\chi\rangle\geq 0}{\geq} \langle\chi|P_2|\chi\rangle. \tag{5.85}$$

By combining the previous two equations, we obtain

$$1 = \langle\psi_2|P_2|\psi_2\rangle \leq |\beta|^2 \Rightarrow |\beta|^2 = 1, \tag{5.86}$$

indicating that the probability of finding of $|\psi_2\rangle$ in $|\chi\rangle$ is 1. Therefore, we conclude that $|\psi_2\rangle = |\chi\rangle$, which is a contradiction. Therefore, indeed, it is impossible to unambiguously distinguish non-orthogonal quantum states.

5.3.3 *Quantum Entanglement*

Let $|\psi_0\rangle, \ldots, |\psi_{n-1}\rangle$ be *n*-qubits lying in the Hilbert spaces $H_0, \ldots, H_{n-1}$, respectively, and the let the state of the joint quantum system lying in $H_0 \otimes \ldots \otimes H_{n-1}$ be denoted by $|\boldsymbol{\psi}\rangle$. The qubit $|\boldsymbol{\psi}\rangle$ is then said to be entangled if it cannot be written in the product state form

$$|\psi\rangle = |\psi_0\rangle \otimes |\psi_1\rangle \otimes \cdots \otimes |\psi_{n-1}\rangle. \tag{5.87}$$

Important examples of two-qubit states are *Bell states*, also known as Einstein–Podolsky–Rosen (*EPR*) states (pairs):

$$\begin{aligned} |B_{00}\rangle &= \frac{1}{\sqrt{2}}(|00\rangle + |11\rangle), |B_{01}\rangle = \frac{1}{\sqrt{2}}(|01\rangle + |10\rangle), \\ |B_{10}\rangle &= \frac{1}{\sqrt{2}}(|00\rangle - |11\rangle), |B_{11}\rangle = \frac{1}{\sqrt{2}}(|01\rangle - |10\rangle). \end{aligned} \tag{5.88}$$

The n-qubit ($n > 2$) analogs of Bell states will be now briefly reviewed. One popular family of entangled multiqubit states is Greenberger–Horne–Zeilinger (GHZ) states:

$$|\mathrm{GHZ}\rangle = \frac{1}{\sqrt{2}}(|00\cdots 0\rangle \pm |11\cdots 1\rangle). \tag{5.89}$$

Another popular family of multiqubit entangled states is known as W-states:

$$|\mathrm{W}\rangle = \frac{1}{\sqrt{N}}(|00\cdots 01\rangle + |00\cdots 10\rangle + \cdots + |01\cdots 00\rangle + |10\cdots 00\rangle). \tag{5.90}$$

The W-state of *n*-qubits represents a superposition of single-weighted CB states, each occurring with probability amplitude of $N^{-1/2}$.

For a bipartite system, we can elegantly verify whether or not the qubit $|\psi\rangle$ is a product state or an entangled one, by Schmidt decomposition. The *Schmidt decomposition theorem* states that a pure state $|\psi\rangle$ of the composite system $H_A \otimes H_B$ can be represented as

$$|\psi\rangle = \sum_i c_i |i_A\rangle |i_B\rangle, \tag{5.91}$$

where $|i_A\rangle$ and $|i_B\rangle$ are orthonormal basis of the subsystems H_A and H_A, respectively, and $c_i \in \Re^+$ ($\Re^+$ is the set of nonnegative real numbers) are *Schmidt coefficients* that satisfy the following condition $\sum_i c_i^2 = 1$. For the proof of the theorem, please refer to Ref. [1]. The Schmidt coefficients can be calculated from the partial density matrix $\text{Trace}_B(|\psi\rangle\langle\psi|)$. A corollary of the Schmidt decomposition theorem is that a pure state in a composite system is a product state if and only the Schmidt rank is 1, and is an entangled state if and only if the Schmidt rank is greater than one.

As an illustration, let us verify if the Bell state $|B_{11}\rangle$ is entangled one. We first determine the density matrix:

$$\begin{aligned}\rho = |\psi\rangle\langle\psi| &= \frac{1}{\sqrt{2}}(|01\rangle - |10\rangle)\frac{1}{\sqrt{2}}(\langle 01| - \langle 10|) \\ &= \frac{1}{2}(|01\rangle\langle 01| - |01\rangle\langle 10| - |10\rangle\langle 01| + |10\rangle\langle 10|).\end{aligned}$$

By tracing out the subsystem B, we obtain

$$\begin{aligned}\rho_A &= \text{Trace}_B(|\psi\rangle\langle\psi|) = \langle 0_B|\psi\rangle\langle\psi|0_B\rangle + \langle 1_B|\psi\rangle\langle\psi|1_B\rangle \\ &= \frac{1}{2}(|1\rangle\langle 1| + |0\rangle\langle 0|) = \frac{1}{2}I.\end{aligned}$$

The eigenvalues are $c_1 = c_2 = 1/2$, and the Schmidt rank is 2 indicating that the Bell state $|B_{11}\rangle$ is an entangled state.

5.3.4 Operator-Sum Representation

Let the composite system C be composed of quantum register Q and environment E. This kind of system can be modeled as a closed quantum system. Because the composite system is closed, its dynamic is unitary, and final state is specified by a unitary operator U as follows: $U(\rho \otimes \varepsilon_0)U^\dagger$, where ρ is a density operator of initial state of quantum register Q, and ε_0 is the initial density operator of the environment E. The reduced density operator of Q upon interaction ρ_f can be obtained by tracing out the environment:

$$\rho_f = \text{Tr}_E\left[U(\rho \otimes \varepsilon_0)U^\dagger\right] \equiv \xi(\rho). \tag{5.92}$$

The transformation (mapping) of initial density operator ρ to the final density operator ρ_f, denoted as $\xi: \rho \to \rho_f$, given by Eq. (5.92), is often called the *superoperator* or *quantum operation*. The final density operator can be expressed in so-called *operator-sum representation* as follows:

$$\rho_f = \sum_k E_k \rho E_k^\dagger, \tag{5.93}$$

where E_k are the operation elements for the superoperator. Clearly, in the absence of environment, the superoperator becomes $U\rho U^\dagger$, which is nothing else but a conventional time evolution quantum operation.

The operator-sum representation can be used in classification of quantum operations into two categories: (i) *trace-preserving* when $\mathrm{Tr}\xi(\rho) = \mathrm{Tr}\ \rho = 1$ and (ii) *non-trace-preserving* when $\mathrm{Tr}\xi(\rho) < 1$. Starting from the trace-preserving condition:

$$\mathrm{Tr}\rho = \mathrm{Tr}\xi(\rho) = \mathrm{Tr}\left[\sum_k E_k \rho E_k^\dagger\right] = \mathrm{Tr}\left[\rho \sum_k E_k E_k^\dagger\right] = 1, \tag{5.94}$$

we obtain

$$\sum_k E_k E_k^\dagger = I. \tag{5.95}$$

For non-trace-preserving quantum operation the Eq. (5.94) is not satisfied, and informally we can write $\sum_k E_k E_k^\dagger < I$.

If the environment dimensionality is large enough, it can be found in pure state, $\varepsilon_0 = |\phi_0\rangle\langle\phi_0|$, and the corresponding superoperator becomes

$$\begin{aligned}
\xi(\rho) &= \mathrm{Tr}_E\left[U(\rho \otimes \varepsilon)U^\dagger\right] = \sum_k \langle\phi_k|\left(U\rho \otimes \varepsilon U^\dagger\right)|\phi_k\rangle \\
&= \sum_k \langle\phi_k|\left(U\rho \otimes \left(\underbrace{|\phi_0\rangle\langle\phi_0|}_{\varepsilon}\right)U^\dagger\right)|\phi_k\rangle \\
&= \sum_k \underbrace{\langle\phi_k|U|\phi_0\rangle}_{E_k} \rho \underbrace{\langle\phi_0|U^\dagger|\phi_k\rangle}_{E_k^\dagger} \\
&= \sum_k E_k \rho E_k^\dagger,\ E_k = \langle\phi_k|U|\phi_0\rangle.
\end{aligned} \tag{5.96}$$

The E_k operators in operator-sum representation are known as *Kraus operators*.

As an illustration, let us consider bit-flip and phase-flip channels. Let the composite system be given by $|\phi_E\rangle|\psi_Q\rangle$, wherein the initial state of environment is $|\phi_E\rangle = |0_E\rangle$. Let further the quantum subsystem Q interacts to the environment E by Pauli-X operator:

$$U = \sqrt{1-p}I \otimes I + \sqrt{p}X \otimes X,\ 0 \le p \le 1. \tag{5.97}$$

Therefore, with probability $1 - p$, we leave the quantum system untouched, while with probability p we apply Pauli-X operator to both quantum subsystem and the environment. By applying the operator U on environment state, we obtain

$$U|\phi_E\rangle = \sqrt{1-p}I \otimes I|0_E\rangle + \sqrt{p}X \otimes X|0_E\rangle = \sqrt{1-p}|0_E\rangle I + \sqrt{p}|1_E\rangle X. \tag{5.98}$$

The corresponding Kraus operators are given by

$$E_0 = \langle 0_E|U|\phi_E\rangle = \sqrt{1-p}I, \quad E_1 = \langle 1_E|U|\phi_E\rangle = \sqrt{p}X. \tag{5.99}$$

Finally, the operator-sum representation is given by

$$\xi(\rho) = E_0\rho E_0^\dagger + E_1\rho E_1^\dagger = (1-p)\rho + pX\rho X. \tag{5.100}$$

In similar fashion, the Kraus operators for the phase-flip channel are given by

$$E_0 = \langle 0_E|U|\phi_E\rangle = \sqrt{1-p}I, \quad E_1 = \langle 1_E|U|\phi_E\rangle = \sqrt{p}Z \tag{5.101}$$

and the corresponding operator-sum representation is

$$\xi(\rho) = E_0\rho E_0^\dagger + E_1\rho E_1^\dagger = (1-p)\rho + pZ\rho Z. \tag{5.102}$$

5.3.5 Decoherence Effects, Depolarization, and Amplitude Damping Channel Models

Quantum computation works by manipulating quantum interference effect. The quantum interference, a manifestation of coherent superposition of quantum states, is the cornerstone behind all quantum information tasks such as quantum computation and quantum communication. A major source of problem is our inability to prevent our quantum system of interest from interacting with the surrounding environment. This interaction results in an entanglement between the quantum system and the environment leading to decoherence. To understand this system–environment entanglement and decoherence better, let us consider a qubit described by density state (matrix) $\rho = \begin{bmatrix} a & b \\ c & d \end{bmatrix}$ interacting with the environment, described by the following three states: $|0_E\rangle$, $|1_E\rangle$, and $|2_E\rangle$. Without loss of generality, we assume that environment was initially in state $|0_E\rangle$. The unitary operator introducing the entanglement between the quantum system and the environment is defined as

$$U|0\rangle|0_E\rangle = \sqrt{1-p}|0\rangle|0_E\rangle + \sqrt{p}|0\rangle|1_E\rangle$$

$$U|0\rangle|0_E\rangle = \sqrt{1-p}|1\rangle|0_E\rangle + \sqrt{p}|1\rangle|2_E\rangle. \tag{5.103}$$

The corresponding Kraus operators are given by

$$E_0 = \sqrt{1-p}I, \ E_1 = \sqrt{p}|0\rangle\langle 0|, \ E_2 = \sqrt{p}|1\rangle\langle 1|. \tag{5.104}$$

The operator-sum representation is given by

$$\xi(\rho) = E_0\rho E_0^\dagger + E_1\rho E_1^\dagger + E_2\rho E_2^\dagger = \begin{bmatrix} a & (1-p)b \\ (1-p)c & d \end{bmatrix}. \tag{5.105}$$

By applying these quantum operation n-times, the corresponding final state would be

$$\rho_f = \begin{bmatrix} a & (1-p)^n b \\ (1-p)^n c & d \end{bmatrix}. \tag{5.106}$$

If the probability p is expressed as $p = \gamma\Delta t$, we can write $n = t/\Delta t$ and in limit we obtain

$$\lim_{\Delta t \to 0} (1-p)^n = (1-\gamma\Delta t)^{t/\Delta t} = e^{-\gamma t}. \tag{5.107}$$

Therefore, the corresponding operator-sum representation as n tends to plus infinity is given by

$$\xi(\rho) = \begin{bmatrix} a & e^{-\gamma t} b \\ e^{-\gamma t} c & d \end{bmatrix}. \tag{5.108}$$

Clearly, the terms b and c go to zero as t increases, indicating that the relative phase in the original state of the quantum system is lost, and the corresponding channel model is known as the *phase damping* channel model.

In the above example, we have considered the coupling between a single-qubit quantum system and the environment and discussed the resulting loss of interference or coherent superposition. In general, for multiple qubit systems decoherence also results in loss of coupling between the qubits. In fact, with increasing complexity and size of the quantum computer, the decoherence effect becomes worse. Additionally, the quantum system lies in some complex Hilbert space where there are infinite variations of errors that can cause decoherence.

A more general example of dephasing is the depolarization. The *depolarizing channel*, as shown in Fig. 5.6, with probability $1-p$ leaves the qubit as it is, while with probability p moves the initial state into $\rho_f = I/2$ that maximizes the von Neumann entropy $S(\rho) = -\text{Tr}\,\rho\log\rho = 1$. The properties describing the model can be summarized as follows:

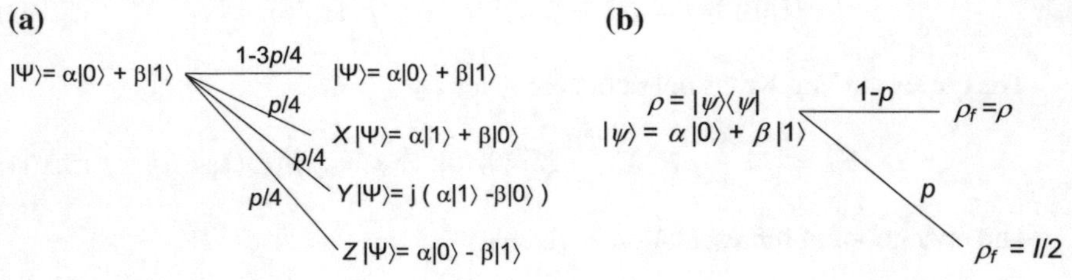

Fig. 5.6 Depolarizing channel model: **a** Pauli operator description and **b** density operator description

1. Qubit errors are independent,
2. Single-qubit errors (X, Y, Z) are equally likely, and
3. All qubits have the same single-error probability p/4.

The Kraus operators E_i of the channel should be selected as follows:

$$E_0 = \sqrt{1 - 3p/4I}; \; E_1 = \sqrt{p/4}X; \; E_1 = \sqrt{p/4}X; \; E_1 = \sqrt{p/4}Z. \tag{5.109}$$

The action of depolarizing channel is to perform the following mapping: $\rho \to \xi(\rho) = \sum_i E_i \rho E_i^\dagger$, where ρ is the initial density operator. Without loss of generality, we will assume that initial state was pure $|\psi\rangle = a|0\rangle + b|1\rangle$ so that

$$\begin{aligned} \rho &= |\psi\rangle\langle\psi| = (a|0\rangle + b|1\rangle)\big(\langle 0|a^* + \langle 1|b^*\big) \\ &= |a|^2 \begin{bmatrix} 1 & 0 \\ 0 & 0 \end{bmatrix} + ab^* \begin{bmatrix} 0 & 1 \\ 0 & 0 \end{bmatrix} + a^*b \begin{bmatrix} 0 & 0 \\ 1 & 0 \end{bmatrix} + |b|^2 \begin{bmatrix} 0 & 0 \\ 0 & 1 \end{bmatrix}. \\ &= \begin{bmatrix} |a|^2 & ab^* \\ a^*b & |b|^2 \end{bmatrix} \end{aligned} \tag{5.110}$$

The resulting quantum operation can be represented using operator-sum representation as follows:

$$\begin{aligned} \xi(\rho) &= \sum_i E_i \rho E_i^\dagger = \left(1 - \frac{3p}{4}\right)\rho + \frac{p}{4}(X\rho X + Y\rho Y + Z\rho Z) \\ &= \left(1 - \frac{3p}{4}\right)\begin{bmatrix} |a|^2 & ab^* \\ a^*b & |b|^2 \end{bmatrix} + \frac{p}{4}\begin{bmatrix} 0 & 1 \\ 1 & 0 \end{bmatrix}\begin{bmatrix} |a|^2 & ab^* \\ a^*b & |b|^2 \end{bmatrix}\begin{bmatrix} 0 & 1 \\ 1 & 0 \end{bmatrix} \\ &\quad + \frac{p}{4}\begin{bmatrix} 1 & 0 \\ 0 & -1 \end{bmatrix}\begin{bmatrix} |a|^2 & ab^* \\ a^*b & |b|^2 \end{bmatrix}\begin{bmatrix} 1 & 0 \\ 0 & -1 \end{bmatrix} + \frac{p}{4}\begin{bmatrix} 0 & -j \\ j & 0 \end{bmatrix}\begin{bmatrix} |a|^2 & ab^* \\ a^*b & |b|^2 \end{bmatrix}\begin{bmatrix} 0 & -j \\ j & 0 \end{bmatrix} \\ &= \left(1 - \frac{3p}{4}\right)\begin{bmatrix} |a|^2 & ab^* \\ a^*b & |b|^2 \end{bmatrix} + \frac{p}{4}\begin{bmatrix} |b|^2 & a^*b \\ ab^* & |a|^2 \end{bmatrix} + \frac{p}{4}\begin{bmatrix} |a|^2 & -ab^* \\ -a^*b & |b|^2 \end{bmatrix} + \frac{p}{4}\begin{bmatrix} |b|^2 & -a^*b \\ -ab^* & |a|^2 \end{bmatrix} \\ &= \begin{bmatrix} (1 - \frac{p}{2})|a|^2 + \frac{p}{2}|b|^2 & (1-p)ab^* \\ (1-p)a^*b & \frac{p}{2}|a|^2 + (1 - \frac{p}{2})|b|^2 \end{bmatrix} \end{aligned}$$

$$= \begin{bmatrix} (1-p)|a|^2 + \frac{p}{2}\left(|a|^2+|b|^2\right) & (1-p)ab^* \\ (1-p)a^*b & \frac{p}{2}\left(|a|^2+|b|^2\right) + (1-p)|b|^2 \end{bmatrix}. \quad (5.111.1)$$

Since $|a|^2 + |b|^2 = 1$, the operator-sum representation can be written as

$$\xi(\rho) = \sum_i E_i \rho E_i^\dagger = (1-p)\begin{bmatrix} |a|^2 & ab^* \\ a^*b & |b|^2 \end{bmatrix} + \frac{p}{2}I = (1-p)\rho + \frac{p}{2}I. \quad (5.111.2)$$

The first line in (5.111.1, 5.111.2) corresponds to model shown in Fig. 5.6a, and the last line corresponds to model shown in Fig. 5.6b. It is clear from (5.110) and (5.111.1, 5.111.2) that $\mathrm{Tr}\xi(\rho) = \mathrm{Tr}(\rho) = 1$ meaning that superoperator is trace-preserving. Notice that the depolarizing channel model in some other books/papers can slightly be different from one described here.

In the rest of this subsection, we describe the amplitude damping channel model. In certain quantum channels, the errors X, Y, and Z do not occur with the same probability. In amplitude damping channel, the operation elements are given by

$$E_0 = \begin{pmatrix} 1 & 0 \\ 0 & \sqrt{1-\varepsilon^2} \end{pmatrix}, \qquad E_1 = \begin{pmatrix} 0 & \varepsilon \\ 0 & 0 \end{pmatrix}. \quad (5.112)$$

The *spontaneous emission* is an example of a physical process that can be modeled using the amplitude damping channel model. If $|\psi\rangle = a\,|0\rangle + b\,|1\rangle$ is the initial qubit state $\left(\rho = \begin{bmatrix} |a|^2 & ab^* \\ a^*b & |b|^2 \end{bmatrix}\right)$, the effect of amplitude damping channel is to perform the following mapping:

$$\begin{aligned} \rho \to \xi(\rho) = E_0\rho E_0^\dagger + E_1\rho E_1^\dagger &= \begin{pmatrix} 1 & 0 \\ 0 & \sqrt{1-\varepsilon^2} \end{pmatrix}\begin{bmatrix} |a|^2 & ab^* \\ a^*b & |b|^2 \end{bmatrix}\begin{pmatrix} 1 & 0 \\ 0 & \sqrt{1-\varepsilon^2} \end{pmatrix} \\ &+ \begin{pmatrix} 0 & \varepsilon \\ 0 & 0 \end{pmatrix}\begin{bmatrix} |a|^2 & ab^* \\ a^*b & |b|^2 \end{bmatrix}\begin{pmatrix} 0 & 0 \\ \varepsilon & 0 \end{pmatrix} \\ &= \begin{pmatrix} |a|^2 & ab^*\sqrt{1-\varepsilon^2} \\ a^*b\sqrt{1-\varepsilon^2} & |b|^2\left(1-\varepsilon^2\right) \end{pmatrix} + \begin{pmatrix} |b|^2\varepsilon^2 & 0 \\ 0 & 0 \end{pmatrix} \\ &= \begin{pmatrix} |a|^2 + \varepsilon^2|b|^2 & ab^*\sqrt{1-\varepsilon^2} \\ a^*b\sqrt{1-\varepsilon^2} & |b|^2\left(1-\varepsilon^2\right) \end{pmatrix}. \end{aligned} \quad (5.113)$$

Probabilities $P(0)$ and $P(1)$ that E_0 and E_1 occur are given by

$$\begin{aligned} P(0) &= \mathrm{Tr}\left(E_0\rho E_0^\dagger\right) = \mathrm{Tr}\begin{pmatrix} |a|^2 & ab^*\sqrt{1-\varepsilon^2} \\ a^*b\sqrt{1-\varepsilon^2} & |b|^2\left(1-\varepsilon^2\right) \end{pmatrix} = 1 - \varepsilon^2|b|^2 \\ P(1) &= \mathrm{Tr}\left(E_1\rho E_1^\dagger\right) = \mathrm{Tr}\begin{pmatrix} |b|^2\varepsilon^2 & 0 \\ 0 & 0 \end{pmatrix} = \varepsilon^2|b|^2. \end{aligned} \quad (5.114)$$

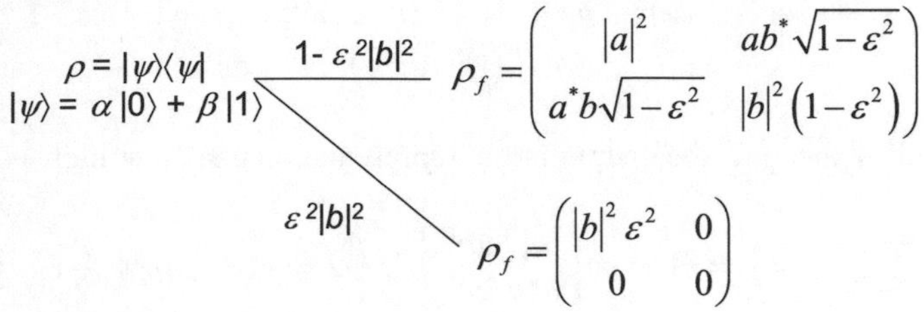

Fig. 5.7 Amplitude damping channel model

The corresponding amplitude damping channel model is shown in Fig. 5.7.

5.4 Classical (Shannon) and Quantum (von Neumann) Entropies

Let us observe a classical discrete memoryless source with the alphabet $X = \{x_1, x_2, \ldots, x_N\}$. The symbols from the alphabet are emitted by the source with probabilities $P(X = x_n) = p_n$, $n = 1, 2, \ldots N$. The amount of information carried out by the kth symbol is related to the uncertainty that is resolved when this symbol occurs, and it is defined as $I(x_n) = \log(1/p_n) = -\log(p_n)$, where the logarithm is to the base 2. The classical (Shannon) entropy is defined as the measure of the average amount of information per source symbol:

$$H(X) = E[I(x_n)] = \sum_{n=1}^{N} p_n I(x_n) = \sum_{n=1}^{N} p_n \log_2\left(\frac{1}{p_n}\right). \tag{5.115}$$

The Shannon entropy satisfies the following inequalities:

$$0 \leq H(X) \leq \log N = \log|X|. \tag{5.116}$$

Let Q be a quantum system with the state described by the density operator $\boldsymbol{\rho}_x$. The probability that the output $\boldsymbol{\rho}_x$ is obtained is given by $p_x = P(x)$. The quantum source is, therefore, described by the ensemble $\{\boldsymbol{\rho}_x, p_x\}$, characterized by the mixed density operator $\boldsymbol{\rho} = \sum_{x \in X} p_x \boldsymbol{\rho}_x$. Then, the *quantum (von Neumann) entropy* is defined as

$$S(\boldsymbol{\rho}) = -Tr(\boldsymbol{\rho} \log \boldsymbol{\rho}) = -\sum_{\lambda_i} \lambda_i \log \lambda_i, \tag{5.117}$$

where λ_i are eigenvalues of $\boldsymbol{\rho}$. When all quantum states are pure and mutually orthogonal, then von Neumann entropy equals the Shannon entropy as in that case

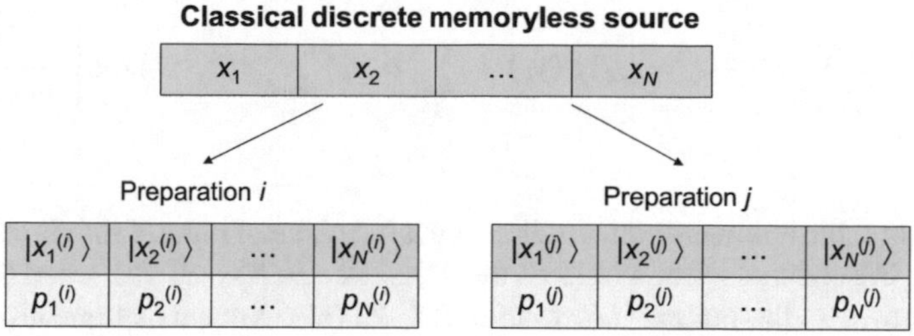

Fig. 5.8 The illustration of quantum representation of the classical information

$p_i = \lambda_i$. As an illustration, in Fig. 5.8, we provide an interpretation of quantum representation of the classical information.

Two different preparations $P^{(i)}$ and $P^{(j)}$, with $H(X^{(i)}) \neq H(X^{(j)})$ $(i \neq j)$, can generate the same $\boldsymbol{\rho}$ and hence have the same von Neumann entropy $S(\boldsymbol{\rho})$ because the states of the two preparations may not be physically distinguishable from each other.

5.5 Holevo Information, Accessible Information, and Holevo Bound

A classical discrete memoryless channel (DMC) is described by the set of transition probabilities

$$p(y_k|x_j) = P(Y = y_k|X = x_j); \;\; 0 \leq p(y_k|x_j) \leq 1 \tag{5.118}$$

satisfying the condition $\sum_k p\left(y_k|x_j\right) = 1$. The conditional entropy of X, given $Y = y_k$, is related to the uncertainty of the channel input X by observing the channel output y_k, and it is defined as

$$H(X|Y = y_k) = \sum_j p(x_j|y_k) \log\left[\frac{1}{p\left(x_j|y_k\right)}\right], \tag{5.119}$$

where

$$p(x_j|y_k) = p(y_k|x_j)p(x_j)/p(y_k), \;\; p(y_k) = \sum_j p(y_k|x_j)p(x_j). \tag{5.120}$$

The amount of uncertainty remaining about the channel input X after the observing channel output Y can be then determined by

$$H(X|Y) = \sum_k H(X|Y = y_k)p(y_k) = \sum_k p(y_k) \sum_j p(x_j|y_k) \log\left[\frac{1}{p(x_j|y_k)}\right]. \tag{5.121}$$

Therefore, the amount of uncertainty about the channel input X that is resolved by observing the channel input would be the difference $H(X) - H(X|Y)$, and this difference is known as the mutual information $I(X, Y)$ (also known as transinformation). In other words,

$$I(X, Y) = H(X) - H(X|Y) = \sum_j p(x_j) \sum_k p(y_k|x_j) \log\left[\frac{p(y_k|x_j)}{p(y_k)}\right]. \tag{5.122}$$

The mutual information can also be defined between any two random variables X and Y. In this case, the mutual information is related to the amount information that Y has about X in average, namely, $I(X;Y) = H(X) - H(X|Y)$. In other words, it represents the amount of uncertainty about X that is resolved given that we know Y.

In the quantum information theory, we can provide a similar interpretation of Holevo information. The *Holevo information* χ for the ensemble of states $\{\boldsymbol{\rho}_x, p_x\}$ corresponds to the average reduction in quantum entropy given that we know how $\boldsymbol{\rho}$ was prepared, namely, $\boldsymbol{\rho} = \Sigma_x\, p_x \boldsymbol{\rho}_x$, and it is defined as

$$\chi = S(\boldsymbol{\rho}) - \sum_x p_x S(\boldsymbol{\rho}_x). \tag{5.123}$$

Holevo information is upper bounded by Shannon entropy, namely, $\chi \leq H(X)$.

Maximum of the mutual information over all generalized POVM measurement schemes $\boldsymbol{M}_y$, denoted as $H(X{:}Y)$, is known as *accessible information*, and it is officially defined as

$$H(X : Y) = \max_{\boldsymbol{M}_y} I(X; Y). \tag{5.124}$$

If the quantum states are pure and mutually orthogonal, instead of POVM, we consider projective measurements such that $p(y|x) = \mathrm{Tr}(\boldsymbol{M}_y \boldsymbol{\rho}_x) = \mathrm{Tr}(\boldsymbol{P}_y \boldsymbol{\rho}_x) = 1$, iff $x = y$ and zero otherwise. In that case, $H(X{:}Y) = H(X)$ and Bob (receiver) is able accurately to estimate the information sent by Alice (transmitter). If the states are non-orthogonal, the accessible information is bounded by

$$H(X{:}Y) \leq S(\boldsymbol{\rho}) \leq H(X). \tag{5.125}$$

When $\boldsymbol{\rho}_x$ states are mixed states, the accessible information is bounded by the Holevo information:

$$H(X:Y) \leq \chi. \tag{5.126}$$

Since the Holevo information is upper bounded by the Shannon entropy, we can write

$$H(X:Y) \leq \chi \leq H(X). \tag{5.127}$$

Therefore, it is impossible for Bob to completely recover the classical information, characterized by $H(X)$, which Alice has sent him over the quantum channel!

5.6 Schumacher's Noiseless Quantum Coding Theorem and Holevo–Schumacher–Westmoreland (HSW) Theorem

5.6.1 *Schumacher's Noiseless Quantum Source Coding Theorem and Quantum Compression*

Consider a classical source X that generates symbols 0 and 1 with probabilities p and $1 - p$, respectively. The probability of output sequence $x_1, \ldots, x_N$ is given by

$$p(x_1, \ldots, x_N) = p^{\sum_i x_i}(1-p)^{\sum_i (1-x_i)}\overrightarrow{N \to \infty} p^{Np}(1-p)^{N(1-p)}. \tag{5.128}$$

By taking the logarithm of this probability, we obtain

$$\log p(x_1, \ldots, x_N) \approx Np \log p + N(1-p)\log(1-p) = -NH(X). \tag{5.129}$$

Therefore, the probability of occurrence of so-called typical sequence is $p(x_1, \ldots, x_N) \approx 2^{-NH(X)}$, and in average $NH(X)$ bits are needed to represent any typical sequence, which is illustrated in Fig. 5.3a. We denoted the typical set by $T(N, \varepsilon)$. Clearly, not much of the information will be lost if we consider $2^{NH(X)}$ typical sequences, instead of 2^N possible sequences, in particular, for large N. This observation can be used in data compression. Namely, with $NH(X)$ bits we can enumerate all typical sequences. If the source output is a typical sequence, it will be represented with $NH(X)$ bits. On the other hand, when atypical sequence get generated by the source, we will assign a fixed index to it resulting in compression loss. However, as $N \to \infty$, the probability of occurrence of atypical sequence will tend to zero resulting in arbitrary small information loss.

Let us now extend this concept to nonbinary sources. The law of large numbers applied to a nonbinary source with i.i.d. outputs $X_1, \ldots, X_N$ claims that as $N \to \infty$, the expected value of the source approaches the true value $E(X)$ in probability sense. In other words,

$$P\left(\left|\frac{1}{N}\sum_{i} X_i - E(X)\right| \le \varepsilon\right) > 1-\delta, \tag{5.130}$$

for sufficiently large N and $\varepsilon, \delta > 0$. Based on discussion for the typical set, we can write

$$-N[H(X)+\varepsilon] \le \log P(x_1,\ldots,x_N) \le -N[H(X)-\varepsilon]. \tag{5.131}$$

Since $P[T(N, \varepsilon)] > 1 - \delta$, the size of typical set, denoted as $|T(N, \varepsilon)|$, will be bounded by

$$(1-\varepsilon)2^{N[H(X)-\varepsilon]} \le |T(N,\varepsilon)| \le 2^{N[H(X)+\varepsilon]}. \tag{5.132}$$

Suppose now that a compression rate R is larger than $H(X)$, say $R > H(X) + \varepsilon$. Then based on the discussion above, the total number of typical sequences is bounded by

$$|T(N,\varepsilon)| \le 2^{N[H(X)+\varepsilon]} \le 2^{NR}, \tag{5.133}$$

and the probability of occurrence of typical sequence is lower bounded by $1 - \delta$, where δ is arbitrary small.

The source encoding strategy can be described as follows. Let us divide the set of all sequences generated by the source into typical and atypical sets as shown in Fig. 5.9a. From the equation above, it is clear that sequences in the typical set can be represented by at most NR bits. If an atypical sequence occurs, we assign to it a fixed index. As N tends to infinity, the probability for this event to occur tends to zero,

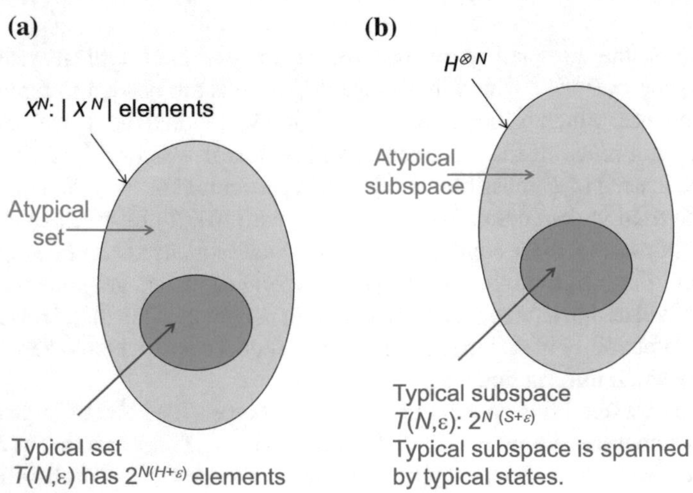

Fig. 5.9 The illustration of typical set (**a**) and typical subspace (**b**)

indicating that we can encode and decode typical sequences reliable. On the other hand, when $R < H(X)$, we can encode maximum 2^{NR} typical sequences, labeled as $T_R(N, \varepsilon)$. Since the probability of occurrence of typical sequence is upper bounded by $2^{-N[H(X)-\varepsilon]}$, the typical sequences in $T_R(N, \varepsilon)$ will occur with probability $2^{N[R-H(X)+\varepsilon]}$, which tends to zero as N tends to infinity, indicating that reliable compression scheme does not exist in this case. With this, we have just proved the Shannon's source coding theorem, which can be formulated as follows.

Shannon's Source Coding Theorem. For an i.i.d. source X, the reliable compression method exists for $R > H(X)$. If, on the other hand, $R < H(X)$ then the no reliable compression scheme exists [27, 28].

Consider now a quantum source emitting a pure state $|\psi_x\rangle$ with probability p_x, described in terms of the mixed density operator:

$$\rho = \sum_x p_x |\psi_x\rangle\langle\psi_x|. \tag{5.134}$$

The quantum message comprises N quantum source outputs, independent of each other, so that

$$\rho_{\otimes N} = \rho \otimes \cdots \otimes \rho. \tag{5.135}$$

The mixed density operator can be expressed in terms of eigenkets as

$$\rho = \sum_\alpha \lambda_\alpha |\lambda_\alpha\rangle\langle\lambda_\alpha|. \tag{5.136}$$

The *typical state* is a state $|\lambda_1\rangle \ldots |\lambda_N\rangle$ for which $\lambda_1 \ldots \lambda_N$ is a typical sequence satisfying the following inequality for sufficiently large N:

$$P\left(\left|\frac{1}{N}\log\left(\frac{1}{P(\lambda_1)P(\lambda_2)\cdots P(\lambda_N)}\right) - S(\rho)\right| \leq \varepsilon\right) > 1 - \delta. \tag{5.137}$$

The projector on a typical subspace is given by

$$P_T = \sum_\lambda |\lambda_1\rangle\langle\lambda_1| \otimes \cdots \otimes |\lambda_N\rangle\langle\lambda_N|. \tag{5.138}$$

The *projection* of the quantum message on the typical subspace is determined by

$$P_T\rho_{\otimes N}. \tag{5.139}$$

The probability of the projections is given by its trace. By using the projection operator, we can separate the total Hilbert space into typical and atypical subspaces, as illustrated in Fig. 5.9b. The probability that the state of the quantum message lies in the typical subspace is given by

$$P\left(\boldsymbol{P}_T \boldsymbol{\rho}_{\otimes N}\right) > 1 - \delta. \tag{5.140}$$

The bounds for the typical subspace $T_\lambda(N, \varepsilon)$, following similar methodology as for typical sequences, can be determined as

$$(1 - \varepsilon)2^{N[S(\rho) - \varepsilon]} \leq |T_\lambda(N, \varepsilon)| \leq 2^{N[S(\rho) + \varepsilon]}. \tag{5.141}$$

Now we are in a position to formulate the corresponding *compression procedure*. Define the projector on the typical subspace $\boldsymbol{P}_T$ and its complement projecting on orthogonal subspace $\boldsymbol{P}_T^\perp = \boldsymbol{I} - \boldsymbol{P}_T$ ($\boldsymbol{I}$ is the identity operator) with corresponding outcomes 0 and 1. For compression purposes, we perform the measurements using two orthogonal operators with outputs denoted as 1 and 0, respectively. If the outcome is 1, we know that the message is in typical state and we do nothing further. If, on the other hand, the outcome is 0, we know that it belongs to atypical subspace and numerate it as a *fixed* state from the typical subspace. Given that the probability of this to happen can be made as small as possible for large N, we can compress the quantum message without the loss of information. We shall now formulate the Schumacher's source coding theorem, the quantum information theory equivalent of the Shannon's source coding theorem. For derivation, an interested reader is referred to [1].

Schumacher's Source Coding Theorem. Let $\{|\psi_x\rangle, p_x\}$ be an i.i.d. quantum source. If $R > S(\rho)$ then there exists a reliable compression scheme of rate R for this source. Otherwise, if $R < S(\rho)$, then no reliable compression scheme of rate R exists.

In the formulation of the Schumacher's source coding theorem, we used the concept of reliable compression, without formally introducing it. We say that the compression is reliable if the corresponding *entanglement fidelity* tends to 1 for large N:

$$F\left(\boldsymbol{\rho}_{\otimes N}; D^N \circ C^N\right) \longrightarrow [N \to \infty]1; \boldsymbol{\rho}_{\otimes N} = |\psi_{\otimes N}\rangle\langle\psi_{\otimes N}|, |\psi_{\otimes N}\rangle = |\psi_{x_1}\rangle \otimes \ldots \otimes |\psi_{x_N}\rangle, \tag{5.142}$$

where we used D^N and C^N to denote the decompression and compression operations, respectively, defined as the following mappings:

$$D^N : H_c^N \to H^N, \quad C^N : H^N \to H_c^N, \tag{5.143}$$

where H_c^N is 2^{NR}-dimensional subspace of H^N. The entanglement fidelity F represents the measure of the preservation of entanglement before and after performing the trace-preserving quantum operation. There exist different definitions of fidelity. Let ρ and σ be two density operators. Then the fidelity can be defined as

$$F(\rho, \sigma) = Tr\left(\sqrt{\rho^{1/2}\sigma\rho^{1/2}}\right). \tag{5.144}$$

For two pure states with density operators $\rho = |\psi\rangle\langle\psi|,\ \sigma = |\phi\rangle\langle\phi|,$ the corresponding fidelity will be

$$\begin{aligned} F(\rho,\sigma) &= Tr\left(\sqrt{\rho^{1/2}\sigma\rho^{1/2}}\right) \overset{\rho^2=\rho}{=} Tr\left(\sqrt{|\psi\rangle\langle\psi|(|\phi\rangle\langle\phi|)|\psi\rangle\langle\psi|}\right) \\ &= Tr(|\psi\rangle \underbrace{\langle\psi|\phi\rangle\langle\phi|\psi\rangle}_{|\langle\phi|\psi\rangle|^2}\langle\psi|)^{1/2} = |\langle\phi|\psi\rangle| \underbrace{Tr(|\psi\rangle\langle\psi|)}_{1} = |\langle\phi|\psi\rangle|, \end{aligned} \tag{5.145}$$

where we used the property of the density operators for pure states $\rho^2 = \rho$ (or equivalently $\rho = \rho^{1/2}$). Clearly, for pure states, the fidelity corresponds to the square root of probability of finding the system in state $|\phi\rangle$ if it is known to be prepared in state $|\psi\rangle$ (and vice versa). Since fidelity is related to the probability, it ranges between 0 and 1, with 0 indicating there is no overlap and 1 meaning that the states are identical. The following *properties* of fidelity hold:

1. The symmetry property:

$$F(\rho,\sigma) = F(\sigma,\rho); \tag{5.146}$$

2. The fidelity is invariant under unitary operations:

$$F\left(U\rho U^{\dagger}, U\sigma U^{\dagger}\right) = F(\rho,\sigma); \tag{5.147}$$

3. If ρ and σ commute, the fidelity can be expressed in terms of eigenvalues of ρ, denoted as r_i, and σ, denoted as s_i, as follows:

$$F(\rho,\sigma) = \sum_i (r_i s_i)^{1/2}. \tag{5.148}$$

5.6.2 Holevo–Schumacher–Westmoreland (HSW) Theorem and Channel Coding

We have already introduced the concept of mutual information $I(X, Y)$ in Sect. 2.5, where we defined it as $I(X, Y) = H(X) - H(Y|X)$. $H(X)$ represents the uncertainty about the channel input X before observing the channel output Y, while $H(X|Y)$ denotes the conditional entropy or the amount of uncertainty remaining about the channel input after the channel output has been received. Therefore, the mutual information represents the amount of information (per symbol) that is conveyed by the channel. In other words, it represents the uncertainty about the channel input that is resolved by observing the channel output. The mutual information can be interpreted by means of Venn diagram shown in Fig. 5.10a. The left circle represents

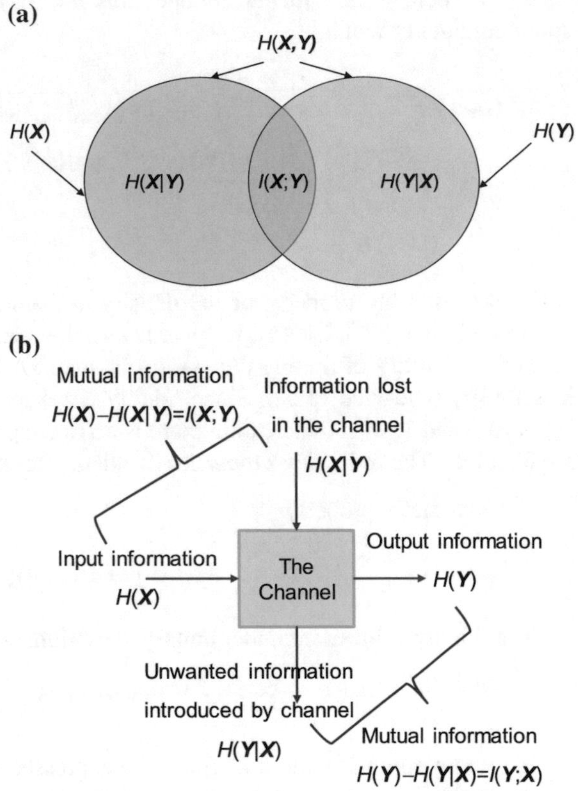

Fig. 5.10 Interpretation of the mutual information: **a** using Venn diagrams, and **b** using the approach due to Ingels

the entropy of channel input, the right circle represents the entropy of channel output, and the mutual information is obtained in intersection of these two circles. Another interpretation due to Ingels [28] is shown in Fig. 5.10b. The mutual information, i.e., the information conveyed by the channel, is obtained as the output information minus information introduced by the channel. One important figure of merit the classical channel is the *channel capacity*, which is obtained by maximization of mutual information $I(X, Y)$ over all possible input distributions:

$$C = \max_{\{p(x_i)\}} I(\boldsymbol{X};\boldsymbol{Y}). \tag{5.149}$$

The classical channel encoder accepts the message symbols and adds redundant symbols according to a corresponding prescribed rule. The channel coding is the act of transforming of a length K sequence into a length N codeword. The set of rules specifying this transformation are called the channel code, which can be represented as the following mapping:

$$C : M \to X, \tag{5.150}$$

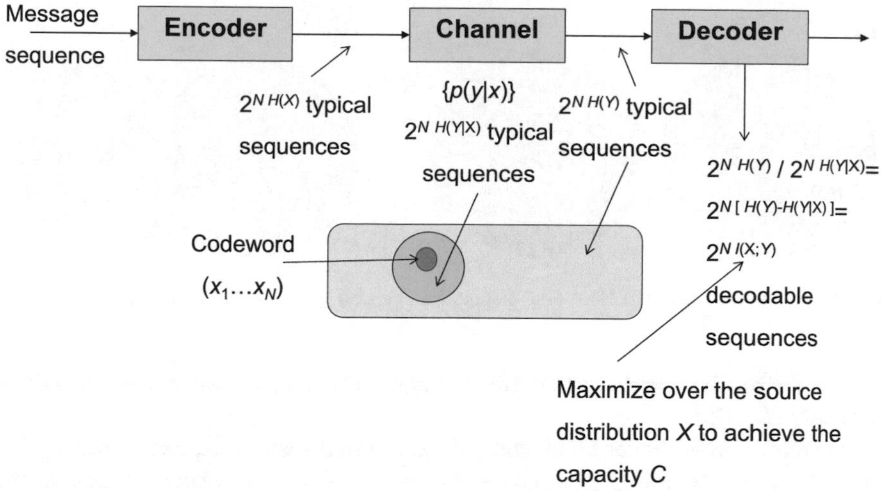

Fig. 5.11 The illustration of classical channel coding and Shannon's capacity theorem derivation

where C is the channel code, M is the set of information sequences of length K, and X is the set of codewords of length N. The decoder exploits these redundant symbols to determine which message symbol was actually transmitted. The concept of classical channel coding is introduced in Fig. 5.11. An important class of channel codes is the class of *block codes*. In an (N, K) *block code*, the channel encoder accepts information in successive K-symbol blocks and adds N-K redundant symbols that are algebraically related to the k message symbols, thus producing an overall encoded block of N symbols ($N > K$), known as a *codeword*. If the block code is *systematic*, the information symbols stay unchanged during the encoding operation, and the encoding operation may be considered as adding the N-K generalized parity checks to k information symbols. The code rate of the code is defined as $R = K/N$.

In order to determine the error correction capability of the linear block code, we have to introduce the concepts of Hamming distance and Hamming weight. The Hamming distance $d(\boldsymbol{x}_1, \boldsymbol{x}_2)$ between two codewords, $\boldsymbol{x}_1$ and $\boldsymbol{x}_2$, is defined as the number of locations in which these two vectors differ. The Hamming weight $wt(\boldsymbol{x})$ of a codeword vector $\boldsymbol{x}$ is defined as the number of nonzero elements in the vector. The minimum distance $d_{\min}$ of a linear block code is defined as the smallest Hamming distance between any pair of code vectors in the code space. Since the zero vector is also a codeword, the minimum distance of a linear block code can be defined as the smallest Hamming weight of the nonzero code vectors in the code. The codewords can be represented as points in n-dimensional space, as shown in Fig. 5.12. Decoding process can be visualized by creating the spheres of radius t around codeword points. The received word vector $\boldsymbol{r}$ in Fig. 5.12a will be decoded as a codeword $\boldsymbol{x}_i$ because its Hamming distance $d(\boldsymbol{x}_i, \boldsymbol{r}) \leq t$ is closest to the codeword $\boldsymbol{x}_i$. On the other hand, in example shown in Fig. 5.12b, the Hamming distance satisfies relation $d(\boldsymbol{x}_i, \boldsymbol{x}_j) \leq$

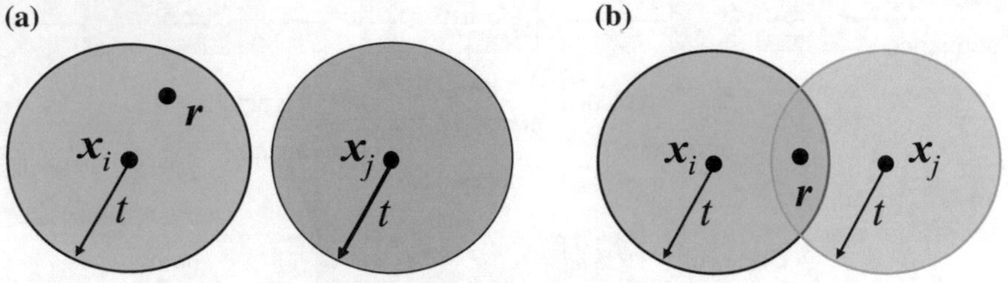

Fig. 5.12 The illustration of Hamming distance: **a** $d(\boldsymbol{x}_i, \boldsymbol{x}_j) \geq 2t + 1$, and **b** $d(\boldsymbol{x}_i, \boldsymbol{x}_j) < 2t + 1$

$2t$, and the received vector $\boldsymbol{r}$ that falls in intersection area of two spheres cannot be uniquely decoded.

Therefore, (n, k) linear block code of minimum distance $d_{\min}$ can correct up to t errors if, and only if, $t \leq \lfloor 1/2(d_{\min} - 1) \rfloor$ or $d_{\min} \geq 2t + 1$ (where $\lfloor \cdot \rfloor$ denotes the largest integer smaller or equal to the enclosed quantity). If we are only interested in detecting e_{d} errors, then the minimum distance should be $d_{\min} \geq e_{\mathrm{d}} + 1$. However, if we are interested in detecting e_{d} errors and correcting e_{c} errors, then the minimum distance should be $d_{\min} \geq e_{\mathrm{d}} + e_{\mathrm{c}} + 1$.

Shannon has shown that if $R < C$, we can construct 2^{NR} length N codewords that can be sent over the (classical) channel with maximum probability of error approaching zero for large N. In order to prove this claim, we introduce the concept of jointly typical sequences. Two length n sequences $\boldsymbol{x}$ and $\boldsymbol{y}$ are jointly typical sequences if they satisfy the following set of inequalities:

$$P\left(\left|\frac{1}{n}\log(\frac{1}{P(\boldsymbol{x})}) - H(X)\right| \leq \varepsilon\right) > 1 - \delta \tag{5.151}$$

$$P\left(\left|\frac{1}{n}\log(\frac{1}{P(\boldsymbol{y})}) - H(Y)\right| \leq \varepsilon\right) > 1 - \delta \tag{5.152}$$

$$P\left(\left|\frac{1}{n}\log(\frac{1}{P(\boldsymbol{x},\boldsymbol{y})}) - H(X, Y)\right| \leq \varepsilon\right) > 1 - \delta \tag{5.153}$$

where $P(\boldsymbol{x}, \boldsymbol{y})$ denotes the joint probability of the two sequences and $H(X, Y)$ is their joint entropy.

For an n-length input codeword randomly generated according to the probability distribution of a source X, the number of random input sequences is approximately $2^{nH(X)}$ and the number of output typical sequences is approximately $2^{nH(Y)}$. Furthermore, the total number of input and output sequences that are jointly typical is $2^{nH(X,Y)}$. Therefore, the total pairs of sequences that are simultaneously $\boldsymbol{x}$-typical, $\boldsymbol{y}$-typical, and also jointly typical are $2^{n[\underbrace{H(X) + H(Y) - H(X,Y)}_{I(X,Y)}]} = 2^{nI(X,Y)}$, where $I(X, Y)$ is the mutual information between X and Y, as illustrated in Fig. 5.11(bottom). These are the maximum number of codeword sequences that can be distinguished.

One way of seeing this is to consider a single codeword. For this codeword, the action of the channel, characterized by the conditional probability $P(\mathbf{y}|\mathbf{x})$, defines the Hamming sphere in which this codeword can lie after the action of the channel. The size of this Hamming sphere is approximately $2^{nH(Y|X)}$. Given that the total number of output typical sequences is approximately $2^{nH(Y)}$, if we desire to have no overlap between two Hamming spheres, the maximum number of codewords we can consider is given by $2^{nH(X)}/2^{nH(Y|X)} = 2^{nI(X,Y)}$. To increase this number, we need to maximize $I(X, Y)$ over the distribution of X, as we do not have a control over the channel. This maximal mutual information is referred to as the capacity of the channel. If we have a rate $R < C$, then Shannon's noisy channel coding theorem tells us that we can construct 2^{nR} length n codewords that can be sent over the channel with maximum probability of error approaching zero for large n. Now we can formally formulate Shannon channel coding theorem as follows.

Shannon's Channel Coding Theorem. Let us consider the transmission of $2^{N(C-\varepsilon)}$ equiprobable messages. Then there exists a classical channel coding scheme of rate $R < C$ in which the codewords are selected from all 2^N possible words such that decoding error probability can be made arbitrary small for sufficiently large N.

Now we consider the communication over the quantum channel, as illustrated in Fig. 5.13. The quantum encoder for each message m out of $M = 2^{NR}$ generates a product state codeword $\boldsymbol{\rho}^m$ drawn from the ensemble $\{p_x, \boldsymbol{\rho}_x\}$ as follows:

$$\boldsymbol{\rho}^m = \boldsymbol{\rho}_{m_1} \otimes \cdots \otimes \boldsymbol{\rho}_{m_N} = \boldsymbol{\rho}_{\otimes N}. \tag{5.154}$$

This quantum codeword is sent over the quantum channel described by the trace-preserving quantum operation ε, resulting in the received quantum word:

$$\boldsymbol{\sigma}_m = \varepsilon(\boldsymbol{\rho}_{m_1}) \otimes \varepsilon(\boldsymbol{\rho}_{m_2}) \otimes \cdots \otimes \varepsilon(\boldsymbol{\rho}_{m_N}). \tag{5.155}$$

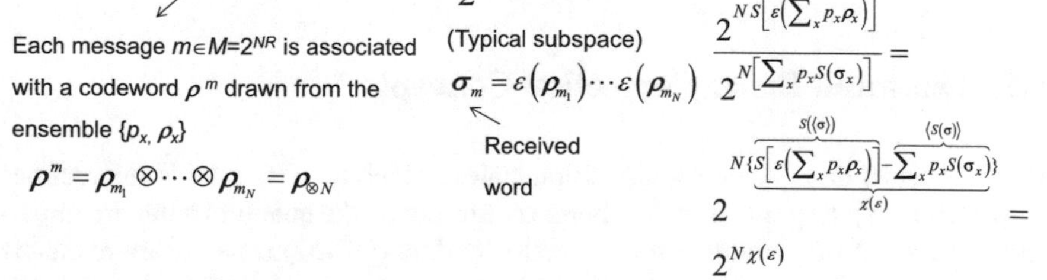

Fig. 5.13 The illustration of quantum channel coding and HSW theorem derivation

Bob performs the measurement on received state using the POVM measurement operators $\{\boldsymbol{M}_m\}$ in order to decode Alice's message. The probability of successful decoding is given by $p_m = \text{Trace}(\boldsymbol{\sigma}_m \boldsymbol{M}_m)$. The goal is to maximize the transmission rate over the quantum channel so that the probability of decoding error is arbitrarily small. The von Neumann entropy associated with the quantum encoder would be $S(\boldsymbol{\rho}) = S\left(\sum_x p_x \boldsymbol{\rho}_x\right)$, while the dimensionality of the typical subspace of quantum encoder is given by $2^{N\,S(\sum_x p_x \boldsymbol{\rho}_x)}$. On the other hand, the dimensionality of the quantum subspace characterizing the quantum channel is given by $2^{N[\sum_x p_x S(\boldsymbol{\sigma}_x)]}$. The quantum channel perturbs the quantum codeword transmitted over the quantum channel by performing the trace-preserving quantum operation so that the entropy at the channel output can be written as $S\left[\varepsilon\left(\sum_x p_x \boldsymbol{\rho}_x\right)\right]$, while the dimensionality of the corresponding subspace is given by $2^{N\,S[\varepsilon(\sum_x p_x \boldsymbol{\rho}_x)]}$. The number of decodable codewords would be then

$$\frac{2^{N\,S[\varepsilon(\sum_x p_x \boldsymbol{\rho}_x)]}}{2^{N[\sum_x p_x S(\boldsymbol{\sigma}_x)]}} = 2^{N\{\underbrace{\overbrace{S\left[\varepsilon\left(\sum_x p_x \boldsymbol{\rho}_x\right)\right]}^{S(\langle\sigma\rangle)} - \overbrace{\sum_x p_x S(\boldsymbol{\sigma}_x)}^{\langle S(\sigma)\rangle}}_{\chi(\varepsilon)}\}} = 2^{N\,\chi(\varepsilon)}. \tag{5.156}$$

In order to maximize the number of decodable codewords, we need to perform the optimization of $\chi(\varepsilon)$ over p_x and $\boldsymbol{\rho}_x$, which represents the simplified derivation of the HSW theorem, which can be formulated as follows.

Holevo–Schumacher–Westmoreland (HSW) Theorem. Let us consider the transmission of a codeword $\boldsymbol{\rho}^m$ drawn from the ensemble $\{p_x, \boldsymbol{\rho}_x\}$ over the quantum channel, characterized by the trace-preserving quantum operation ε, with the rate $R < C(\varepsilon)$, where $C(\varepsilon)$ is the product state capacity defined as

$$\begin{aligned} C(\varepsilon) &= \max_{\{p_x, \boldsymbol{\rho}_x\}} \chi(\varepsilon) = \max_{\{p_x, \boldsymbol{\rho}_x\}} [S(\langle\boldsymbol{\sigma}\rangle) - \langle S(\boldsymbol{\sigma})\rangle] \\ &= \max_{\{p_x, \boldsymbol{\rho}_x\}} \left[S\left[\varepsilon\left(\sum_x p_x \boldsymbol{\rho}_x\right)\right] - \sum_x p_x S\left(\varepsilon\left(\boldsymbol{\rho}_x\right)\right)\right], \end{aligned} \tag{5.157}$$

where the maximization of $\chi(\varepsilon)$ is performed over p_x and $\boldsymbol{\rho}_x$. Then, there exists a coding scheme that allows reliable error-free transmission over the quantum channel.

5.7 Quantum Error Correction Concepts

The QIP relies on delicate superposition states, which are sensitive to interactions with environment, resulting in decoherence. Moreover, the quantum gates are imperfect and the use of quantum error correction coding (QECC) is necessary to enable the fault-tolerant computing and to deal with quantum errors. QECC is also essential in quantum communication and quantum teleportation applications. The elements of

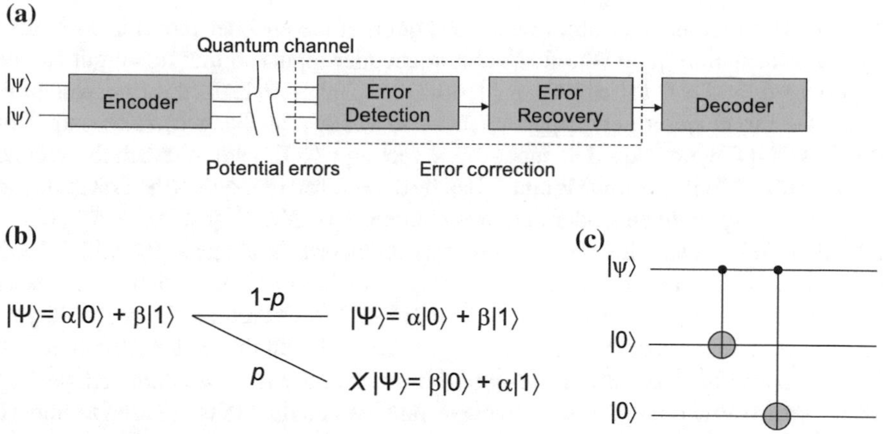

Fig. 5.14 **a** A quantum error correction principle. **b** Bit-flipping channel model. **c** Three-qubit flip-code encoder

quantum error correction codes are shown in Fig. 5.14a. The (N, K) QECC code performs encoding of the quantum state of K qubits, specified by 2^K complex coefficients α_s, into a quantum state of N-qubits, in such a way that errors can be detected and corrected, and all 2^K complex coefficients can be perfectly restored, up to the global phase shift. Namely, from quantum mechanics, we know that two states $|\psi\rangle$ and $e^{j\theta}|\psi\rangle$ are equal up to a *global phase shift* as the results of measurement on both states are the same. A quantum error correction consists of four major steps: encoding, error detection, error recovery, and decoding, as shown in Fig. 5.14a. The sender (Alice) encodes quantum information in state $|\psi\rangle$ with the help of local ancilla qubits $|0\rangle$ and then sends the encoded qubits over a noisy quantum channel (say free-space optical channel or optical fiber). The receiver (Bob) performs multiqubit measurement on all qubits to diagnose the channel error and performs a recovery unitary operation R to reverse the action of the channel. The quantum error correction is essentially more complicated than classical error correction. Difficulties for quantum error correction can be summarized as follows: (i) the no-cloning theorem indicates that it is impossible to make a copy of an arbitrary quantum state, (ii) quantum errors are continuous and a qubit can be in any superposition of the two bases states, and (iii) the measurements destroy the quantum information. The quantum error correction principles will be more evident after a simple example given below.

Assume we want to send a single qubit $|\psi\rangle = \alpha|0\rangle + \beta|1\rangle$ through the quantum channel in which during transmission the transmitted qubit can be flipped to $X|\psi\rangle = \beta|0\rangle + \alpha|1\rangle$ with probability p. Such a quantum channel is called a *bit-flip channel*, and it can be described as shown in Fig. 5.14b. Three-qubit flip code sends the same qubit three times, and therefore represents the *repetition code* equivalent. The corresponding codewords in this code are $|\bar{0}\rangle = |000\rangle$ and $|\bar{1}\rangle = |111\rangle$. The three-qubit flip-code encoder is shown in Fig. 5.14c. One input qubit and two ancillas are used at the input encoder, which can be represented by $|\psi_{123}\rangle = \alpha|000\rangle +$

$\beta|100\rangle$. The first ancilla qubit (the second qubit at the encoder input) is controlled by the information qubit (the first qubit at encoder input) so that its output can be represented by $CNOT_{12}(\alpha|000\rangle + \beta|100\rangle) = \alpha|000\rangle + \beta|110\rangle$ (if the control qubit is $|1\rangle$ the target qubit gets flipped, otherwise it stays unchanged). The output of the first CNOT gate is used as input to the second CNOT gate in which the second ancilla qubit (the third qubit) is controlled by the information qubit (the first qubit) so that the corresponding encoder output is obtained as $CNOT_{13}(\alpha|000\rangle + \beta|110\rangle) = \alpha|000\rangle + \beta|111\rangle$, which indicates that basis codewords are indeed $|\bar{0}\rangle$ and $|\bar{1}\rangle$. With this code, we are capable to correct a single-qubit flip error, which occurs with probability $(1-p)^3 + 3p(1-p)^2 = 1 - 3p^2 + 2p^3$. Therefore, the probability of an error remaining uncorrected or wrongly corrected with this code is $3p^2 - 2p^3$. It is clear from Fig. 5.14c that three-qubit bit-flip encoder is a *systematic encoder* in which the information qubit is unchanged, and the ancilla qubits are used to impose the encoding operation and create the parity qubits (the output qubits 2 and 3).

Let us assume that a qubit flip occurred on the first qubit leading to received quantum word $|\psi_r\rangle = \alpha|100\rangle + \beta|011\rangle$. In order to identify the error, it is needed to perform the measurements on the following observables Z_1Z_2 and Z_2Z_3, where the subscript denotes the index of qubit on which a given Pauli gate is applied. The result of measurement is the eigenvalue ± 1, and corresponding eigenvectors are two valid codewords, namely, $|000\rangle$ and $|111\rangle$. The observables can be represented as follows:

$$\begin{aligned} Z_1Z_2 &= (|00\rangle\langle 11| + |11\rangle\langle 11|) \otimes I - (|01\rangle\langle 01| + |10\rangle\langle 10|) \otimes I \\ Z_2Z_3 &= I \otimes (|00\rangle\langle 11| + |11\rangle\langle 11|) - I \otimes (|01\rangle\langle 01| + |10\rangle\langle 10|). \end{aligned} \tag{5.158}$$

It can be shown that $\langle\psi_r|Z_1Z_2|\psi_r\rangle = -1$, $\langle\psi_r|Z_2Z_3|\psi_r\rangle = +1$, indicating that an error occurred on either first or second qubit, but neither on second nor third qubit. The intersection reveals that the first qubit was in error. By using this approach, we can create the tree-qubit look-up table (LUT), given as Table 5.1.

Three-qubit flip-code error detection and error correction circuits are shown in Fig. 5.15. The results of measurements on ancillas (see Fig. 5.15a) will determine the error syndrome $[\pm 1 \pm 1]$, and based on LUT given by Table 5.1, we identify the error event and apply corresponding X_i gate on ith qubit being in error, and the error gets corrected since $X^2 = I$. The control logic operation is described in Table 5.1. For example, if both outputs at the measurements' circuits are -1, the operator X_2 is activated. The last step is to perform decoding as shown in Fig. 5.15b by simply reversing the order of elements in the corresponding encoder.

Table 5.1 The three-qubit flip-code LUT

Z_1Z_2	Z_2Z_3	Error
+1	+1	I
+1	−1	X_3
−1	+1	X_1
−1	−1	X_2

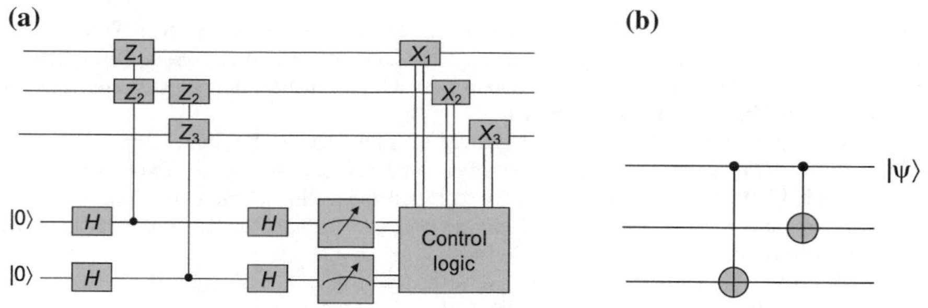

Fig. 5.15 **a** Three-qubit flip-code error detection and error correction circuit. **b** Decoder circuit configuration

5.8 Concluding Remarks

This chapter has provided an overview of the basic concepts of quantum information processing, quantum information theory, and quantum error correction. The following topics from quantum information processing have been described in Sects. 5.1, 5.2 and 5.3: state vectors, operators, density operators, measurements, dynamics of a quantum system, superposition principle, quantum parallelism, no-cloning theorem, and entanglement. The following concepts from quantum information theory have been described in Sects. 5.4, 5.5 and 5.6: Holevo information, accessible information, Holevo bound, Shannon Entropy and von Neumann Entropy, Schumacher's noiseless quantum coding theorem, and Holevo–Schumacher–Westmoreland theorem. The quantum error correction concepts have been introduced in Sect. 5.7.

References

1. Djordjevic IB (2012) Quantum information processing and quantum error correction: an engineering approach. Elsevier/Academic Press, Amsterdam, Boston
2. Helstrom CW (1976) Quantum detection and estimation theory. Academic, New York
3. Helstrom CW, Liu JWS, Gordon JP (1970) Quantum mechanical communication theory. Proc IEEE 58:1578–1598
4. Sakurai JJ (1994) Modern quantum mechanics. Addison-Wisley
5. Gaitan F (2008) Quantum error correction and fault tolerant quantum computing. CRC Press
6. Baym G (1990) Lectures on quantum mechanics. Westview Press
7. McMahon D (2007) Quantum computing explained. Wiley-IEEE Computer Society, Hoboken, NJ
8. McMahon D (2006) Quantum mechanics demystified. McGraw-Hill, New York
9. Neilsen MA, Chuang IL (2000) Quantum computation and quantum information. Cambridge University Press, Cambridge
10. Fleming S (2008) PHYS 570: Quantum mechanics (Lecture Notes). University of Arizona
11. Dirac PAM (1958) Quantum mechanics, 4th edn. Oxford University Press, London
12. Peleg Y, Pnini R, Zaarur E (1998) Schaum's outline of theory and problems of quantum mechanics. McGraw-Hill, New York

13. Fowles GR (1989) Introduction to modern optics. Dover Publications Inc, New York
14. Scully MO, Zubairy MS (1997) Quantum optics. Cambridge University Press, Cambridge
15. Le Bellac M (2006) A short introduction to quantum information and quantum computation. Cambridge University Press, Cambridge
16. Pinter CC (2010) A book of abstract algebra. Dover Publications Inc, New York
17. Griffiths D (1995) Introduction to quantum mechanics. Prentice-Hall, Englewood Cliffs
18. Zettilli N (2001) Quantum mechanics, concepts and applications. Wiley, New York
19. Von Neumann J (1955) Mathematical foundations of quantum mechanics. Princeton University Press, Princeton
20. Jauch JM (1968) Foundations of quantum mechanics. Addison-Wesley, Reading
21. Peres A (1995) Quantum theory: concepts and methods. Kluwer, Boston
22. Goswami A (1996) Quantum mechanics. McGraw-Hill, New York
23. Jammer M (1974) Philosophy of quantum mechanics. Wiley, New York
24. McWeeny R (2003) Quantum mechanics, principles and formalism. Dover Publications, Inc., Mineola, NY
25. Weyl H (1950) Theory of groups and quantum mechanics. Dover Publications Inc, New York
26. Liboff RL (1997) Introductory quantum mechanics. Addison-Wesley, Reading
27. Cover TM, Thomas JA (1991) Elements of information theory. Wiley, New York
28. Ingels FM (1971) Information and coding theory. Intext Educational Publishers, Scranton

Chapter 6
Quantum-Key Distribution (QKD) Fundamentals

Abstract This chapter is devoted to QKD fundamentals. The chapter starts with description of key differences between conventional cryptography, classical physical-layer security (PLS), and QKD. In section on QKD basics, after historical overview, we review different QKD types and describe briefly common postprocessing steps, namely, information reconciliation and privacy amplification steps. In the same section, we provide two fundamental theorems on which QKD relies on, no-cloning theorem and the theorem of inability to unambiguously distinguish non-orthogonal quantum states. In section on discrete variable (DV)-QKD systems, we describe in detail BB84 and B92 protocols as well as Ekert (E91) and EPR protocols. In the same section, the time-phase encoding protocol is also described. Regarding the BB84 protocols, different versions, suitable for different technologies, are described. In section on QKD security, the secret-key rate (SKR) is represented as thc product of raw key rate and fractional rate, followed by the description of different limitations to the raw key rate. After that, the generic expression for the fractional rate is provided, followed by description of different eavesdropping strategies including individual (independent or incoherent) attacks, collective attacks, coherent attacks, and the quantum hacking/side-channel attacks. For individual and coherent attacks, the corresponding secrete fraction expressions are described. The next section is devoted to various definitions of security, including the concept of ε-security. After that, the generic expressions for 2-D DV-QKD schemes are derived for both prepare-and-measure and decoy-state-based protocols. To facilitate the description of continuous variable (CV)-QKD protocols, the fundamentals of quantum optics are introduced first. In section on CV-QKD protocols, both squeezed state-based and coherent state-based protocols are described. Given that the coherent states are much easier to generate and manipulate, the coherent state-based protocols with both homodyne and heterodyne detections are described in detail. The secret fraction is derived for both direct- and reverse-reconciliation-based CV-QKD protocols. Furthermore, the details on practical aspects of GG02 protocol are provided. In the same section, the secret fraction calculation for collective attacks is discussed. After that, the basic concepts for measurement-device-independent (MDI)-QKD protocols are introduced. Then final section in the chapter provides some relevant concluding remarks.

I. B. Djordjevic, *Physical-Layer Security and Quantum Key Distribution*,
https://doi.org/10.1007/978-3-030-27565-5_6

6.1 From Conventional Cryptography to QKD

The basic key-based cryptographic system is provided in Fig. 6.1. The source emits the message (plaintext) $\boldsymbol{M}$ toward the encryption block, which with the help of key $\boldsymbol{K}$, obtained from key source, generates the cryptogram. On receiver side, the cryptogram transmitted over insecure channel gets processed by the decryption algorithm together with the key $\boldsymbol{K}$ obtained through the secure channel, which reconstructs the original plaintext to be delivered to the authenticated user. The encryption process can be mathematically described as $\boldsymbol{E}_{\boldsymbol{K}}(\boldsymbol{M}) = \boldsymbol{C}$, while the decryption process by $\boldsymbol{D}_{\boldsymbol{K}}(\boldsymbol{C}) = \boldsymbol{M}$. Similarly as before, the composition of decryption and encryption functions yields to identity mapping $\boldsymbol{D}_{\boldsymbol{K}}(\boldsymbol{E}_{\boldsymbol{K}}(\boldsymbol{M})) = \boldsymbol{M}$. The key source typically generates the key randomly from the *keyspace* (the range of possible key values) .

The key-based algorithms can be categorized into two broad categories:

- *Symmetric algorithms*, in which decryption key can be derived from encryption key and vice versa. Alternatively, the same key can be used for both encryption and decryption stages. Symmetric algorithms are also known as one-key (single-key) or secret-key algorithms. The well-known system employing this type of algorithms is digital encryption standard (DES) [1–3].
- *Asymmetric algorithms*, in which encryption and decryption keys are different. Moreover, the decryption key cannot be determined from encryption key, at least in any reasonable amount of time. Because of this fact, the encryption keys can be even made public, wherein the eavesdropper will not be able to determine the decryption key. The *public-key systems* [4] are based on this concept. In public-key systems, the encrypted keys have been made public, while the decryption key is known only to the intended user. The encryption key is then called the *public* key, while decryption is called the *secret (private)* key. The keys can be applied in arbitrary order to create the cryptogram from plaintext, and to reconstruct the plaintext from the cryptogram.

The simplest private key cryptosystem is the *Vernam cipher* (*one-time pad*). In *one-time pad* [5], a completely random sequence of characters, with the sequence length being equal to the message sequence length, is used as a key. When for each new message another random sequence is used as a key, the one-time pad scheme provides the perfect security, discussed in Chaps. 3 and 4. Namely, the brute-force

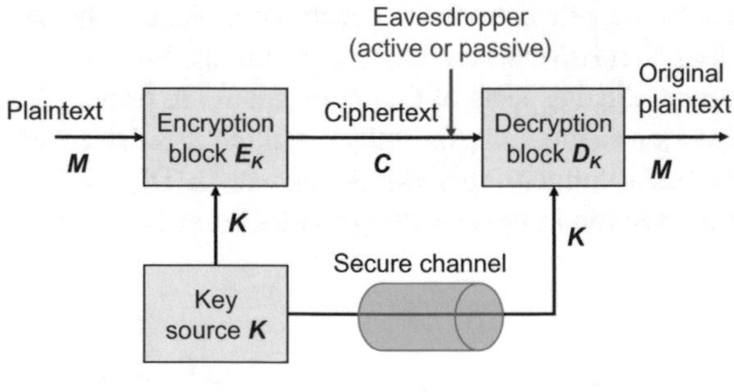

Fig. 6.1 The basic key-based cryptographic scheme

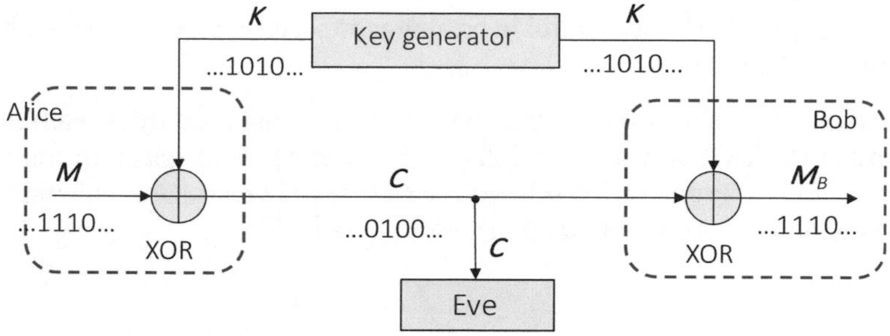

Fig. 6.2 The one-time pad encryption scheme

search approach would require to verify m^n possible keys, where m is the employed alphabet size and n is the length of intercepted cryptogram. In practice, in digital and computer communications, we typically operate on binary alphabet GF(2) = {0, 1}. To obtain the key, we need a special random generator and to encrypt using one-time pad scheme we simply perform addition mod 2, i.e., XOR operation, as illustrated in Fig. 6.2. Even though the one-time pad scheme offers perfect security, it has several drawbacks [6–8]: it requires the secure distribution of the key, the length of the key must be at least as long as the message, the key bits cannot be reused, the keys must be delivered in advance, securely stored until used, and destroyed after the use.

According to Shannon [9], the *perfect security*, also known as *unconditional security*, has been achieved when the messages $\boldsymbol{M}$ and cryptograms $\boldsymbol{C}$ are statistically independent so that the corresponding mutual information is equal to zero:

$$I(\boldsymbol{M}, \boldsymbol{C}) = H(\boldsymbol{M}) - H(\boldsymbol{M}|\boldsymbol{C}) = 0 \Leftrightarrow H(\boldsymbol{M}|\boldsymbol{C}) = H(\boldsymbol{M}). \tag{6.1}$$

The perfect secrecy condition can, therefore, be summarized as

$$H(\boldsymbol{M}) \leq H(\boldsymbol{K}). \tag{6.2}$$

In other words, the entropy (uncertainty) of the key cannot be lower than the entropy of the message, for an encryption scheme to be perfectly secure. Given that in Vernam cipher the length of the key is at least equal to the message length, it appears that one-time pad scheme is perfectly secure.

However, given that this condition is difficult to satisfy, in conventional cryptography instead of information-theoretic security the *computational security* is used [1, 3, 7, 10–12]. The computational security introduces two relaxations with respect to information-theoretic security [3]:

- Security is guaranteed against an efficient eavesdropper running the cryptanalytic attacks for certain limited amount of time. Of course, when eavesdropper has sufficient computational resources and/or sufficient time he/she will be able to break the security of the encryption scheme.

- Eavesdroppers can be successful in breaking the security protocols but with small success probability.

A reader interested to learn more about computational security is referred to an excellent book due to Katz and Lindell [3]. However, by using quantum computing, any conventional cryptographic scheme can be broken in reasonable amount of time by employing the Shor's factorization algorithm [13–17].

6.2 QKD Basics

Significant achievements have been recently made in quantum computing [6–8]. There are many companies currently working on development of the medium-scale quantum computers. Given that most of cryptosystems depend on the computational hardness assumption, the quantum computing represents a serious challenge to the modern cybersecurity systems. As an illustration, to break the Rivest–Shamir–Adleman (RSA) protocol [18], one needs to determine the period r of the function $f(x) = m^x \bmod n = f(x + r)$ ($r = 0, 1, \ldots, 2^l - 1$; m is an integer smaller than $n - 1$), as discussed in Chap. 3. This period is determined in one of the steps of the Shor's factorization algorithm [6–8, 15–17].

The QKD with symmetric encryption can be interpreted as one of the physical-layer security schemes that can provide the provable security against quantum computer-initiated attack [19]. The first QKD scheme was introduced by Bennett and Brassard, who proposed it in 1984 [13, 14], and it is now known as the BB84 protocol. The security of QKD is guaranteed by the quantum mechanics laws. Different photon degrees of freedom, such as polarization, time, frequency, phase, and orbital angular momentum (OAM) can be employed to implement various QKD protocols. Generally speaking, there are two generic QKD schemes, discrete variable (DV)-QKD and continuous variable (CV)-QKD, depending on strategy applied on Bob's side. In DV-QKD schemes, a single-photon detector (SPD) is applied on Bob's side, while in CV-QKD the field quadratures are measured with the help of homodyne/heterodyne detection. The DV-QKD scheme achieves the unconditional security by employing no-cloning theorem and theorem on indistinguishability of arbitrary quantum states, proved already in Chap. 5. The no-cloning theorem claims that arbitrary quantum states cannot be cloned, indicating that Eve cannot duplicate non-orthogonal quantum states even with the help of quantum computer. On the other hand, the second theorem claims that non-orthogonal states cannot be unambiguously distinguished. Namely, when Eve interacts with the transmitted quantum states, trying to get information on transmitted bits, she will inadvertently disturb the fidelity of the quantum states that will be detected by Bob. On the other hand, the CV-QKD employs the uncertainty principle claiming that both in-phase and quadrature components of a coherent states cannot be simultaneously measured with the complete precision. We can also classify different QKD schemes as either entanglement-assisted or prepare-and-measure types.

The research in QKD is getting momentum, in particular, after the first satellite-to-ground QKD demonstration [20]. Recently, the QKD over 404 km of ultralow-loss optical fiber is demonstrated; however, with ultralow secure-key rate (3.2×10^{-4} b/s). Given that quantum states cannot be amplified, the fiber attenuation limits the distance. On the other hand, the dead time (the time over which an SPD remains unresponsive to incoming photons due to long recovery time) of the SPDs, typically in 10–100 ns range, limits the baud rate and therefore the secure-key rate. The CV-QKD schemes, since they employ the homodyne/heterodyne detection, do not have dead time limitation; however, the typical distances are shorter.

By transmitting non-orthogonal qubit states between Alice and Bob and by checking for disturbance in transmitted state, caused by the channel or Eve's activity, they can establish an upper bound on noise/eavesdropping in their quantum communication channel [6]. The *threshold for maximum tolerable error rate* is dictated by the efficiency of the best information reconciliation and privacy amplification steps [6].

6.2.1 QKD System Types

The QKD protocols can be categorized into several general categories:

- **Device-dependent QKD**: In device-dependent QKD, typically, the quantum source is placed on Alice side and quantum detector at Bob's side. Popular classes include discrete variable QKD (DV-QKD), continuous variable QKD (CV-QKD) , entanglement-assisted (EA) QKD, distributed phase reference, etc. For EA QKD, the entangled source can be placed in the middle of the channel to extend the transmission distance.
- **Source-device-independent QKD**: In source-device-dependent QKD , the quantum source is placed at Charlie's (Eve's) side, while the quantum detectors at both Alice and Bob's sides.
- **Measurement-device-independent QKD (MDI-QKD)**: In measurement-device-independent QKD, the quantum detectors are placed at Charlie's (Eve's) side, while the quantum sources are placed at both Alice and Bob's sides. The quantum states get prepared at both Alice and Bob's sides and get transmitted toward Charlie's detectors. Charlie performs the partial Bell state measurements and announces when the desired partial Bell states are detected, with details to be provided later.

6.2.2 Information Reconciliation and Privacy Amplification Steps

The *classical postprocessing steps* are summarized in Fig. 6.3 [8].

The raw key is imperfect, and we need to perform the *information reconciliation* and *privacy amplification* to increase the correlation between sequences $\boldsymbol{X}$ (generated

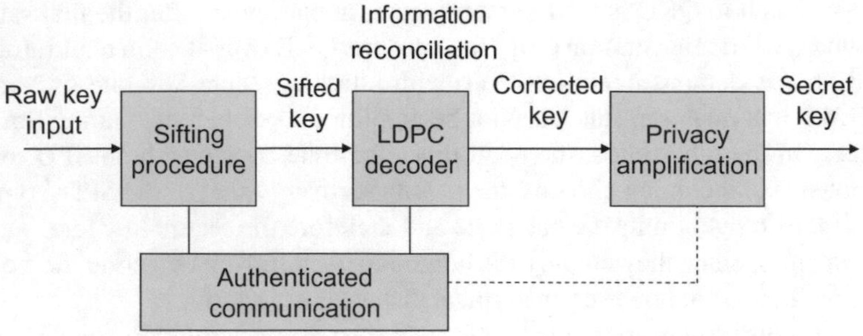

Fig. 6.3 Classical postprocessing steps

by Alice) and sequence ***Y*** (received by Bob) strings, while reducing eavesdropper Eve's mutual information about the result to a desired level of security.

Information reconciliation is nothing more than the error correction performed over a public channel, which reconciles errors between ***X*** and ***Y*** to obtain a shared bit string ***K*** while divulging the information as little as possible to the Eve.

Privacy amplification [21, 22] is used between Alice and Bob to distill from K a smaller set of bits S whose correlation with Eve's string ***Z*** is below a desired threshold. One way to accomplish privacy amplification is through the use of the *universal hash functions* G, which map the set of n-bit strings A to the set of m-bit strings B such that for any distinct $a_1, a_2 \in A$, when g is chosen uniformly at random from G, the probability of having $g(a_1) = g(a_2)$ is at most $1/|B|$.

6.2.3 No-Cloning Theorem and Distinguishing the Quantum States

No-Cloning Theorem: No quantum copier exists that can clone an arbitrary quantum state.

Proof If the input state is $|\psi\rangle$, then the output of the copier will be $|\psi\rangle|\psi\rangle$. For an arbitrary quantum state, such a copier raises a fundamental contradiction.

Consider two arbitrary states $|\psi\rangle$ and $|\chi\rangle$ that are input to the copier. When they are inputted individually we expect to get $|\psi\rangle|\psi\rangle$ and $|\chi\rangle|\chi\rangle$. Now consider a superposition of these two states given by

$$\begin{aligned}|\varphi\rangle = \alpha|\psi\rangle + \beta|\chi\rangle \Rightarrow |\varphi\rangle|\varphi\rangle &= (\alpha|\psi\rangle + \beta|\chi\rangle)(\alpha|\psi\rangle + \beta|\chi\rangle) \\ &= \alpha^2|\psi\rangle|\psi\rangle + \alpha\beta|\psi\rangle|\chi\rangle + \alpha\beta|\chi\rangle|\psi\rangle + \beta^2|\chi\rangle|\chi\rangle. \end{aligned} \tag{6.3}$$

On the other hand, linearity of quantum mechanics tells us that the quantum copier can be represented by a unitary operator that performs the cloning. If such unitary operator was to act on the superposition state $|\varphi\rangle$, the output would be a superposition of $|\psi\rangle|\psi\rangle$ and $|\chi\rangle|\chi\rangle$, that is,

$$|\varphi'\rangle = \alpha|\psi\rangle|\psi\rangle + \beta|\chi\rangle|\chi\rangle. \tag{6.4}$$

The difference between previous two equations leads to contradiction mentioned above. As a consequence, there is no unitary operator that can clone $|\varphi\rangle$.

This result raises a related question: do there exist some specific states for which cloning is possible? The answer to this question is (surprisingly) yes. The cloning is possible only for mutually orthogonal states.

Theorem *It is impossible to unambiguously distinguish non-orthogonal quantum states.*

In other words, there is no a measurement device we can create that can reliably distinguish non-orthogonal states. This fundamental result plays an important role in quantum cryptography.

Proof Its proof is based on contradiction. Let us assume that the measurement operator M is a Hermitian operator, with corresponding eigenvalues m_i and corresponding projection operators P_i, which allows us to unambiguously distinguish between two non-orthogonal states $|\psi_1\rangle$ and $|\psi_2\rangle$. The eigenvalue $m_1(m_2)$ unambiguously identifies the state $|\psi_1\rangle(|\psi_2\rangle$. We know that for projection operators, the following properties are valid:

$$\langle\psi_1|P_1|\psi_1\rangle = 1, \langle\psi_2|P_2|\psi_2\rangle = 1, \langle\psi_1|P_2|\psi_1\rangle = 0, \langle\psi_2|P_1|\psi_2\rangle = 0. \tag{6.5}$$

Given that $|\psi_2\rangle$ can be represented in terms of $|\psi_1\rangle$ and another state $|\chi\rangle$ that is orthogonal to $|\psi_1\rangle$ as follows:

$$|\psi_2\rangle = \alpha|\psi_1\rangle + \beta|\chi\rangle. \tag{6.6}$$

To satisfy Eq. (6.5), $|\psi_2\rangle$ must be equal to $|\chi\rangle$ representing the contradiction.

6.3 Discrete Variable (DV)-QKD Protocols

In this section, we describe some very popular DV-QKD protocols.

6.3.1 BB84 Protocol

This protocol was named after Bennett and Brassard, who proposed it in 1984 [13, 14]. Three key principles being employed are no-cloning theorem, state collapse during measurement, and irreversibility of the measurements [6]. The BB84 protocol can be implemented using different degrees of freedom (DOF) including the polarization DOF [23] or the phase of the photons [20]. Experimentally, the BB84 protocol has been demonstrated over both fiber optics and free-space optical (FSO) channels [23, 24]. The polarization-based BB84 protocol over fiber optics channel is affected by polarization mode dispersion (PDM), polarization-dependent loss (PDL), and fiber loss, which affect the transmission distance. To extend the transmission distance, the phase encoding is employed in [24]. Unfortunately, the secure-key rate over 405 km of ultralow-loss fiber is extremely low, only 6.5 b/s. In an FSO channel, the polarization effects are minimized; however, the atmospheric turbulence can introduce the wavefront distortion and random phase fluctuations. Previous QKD demonstrations include a satellite-to-ground FSO link demonstration and a demonstration over an FSO link between two locations in the Canary Islands [23, 24].

Two bases that are used in the BB84 protocol are the computational basis CB = $\{|0\rangle,\ |1\rangle\}$ and the diagonal basis:

$$DB = \left\{|+\rangle = (|0\rangle + |1\rangle)/\sqrt{2}, |-\rangle = (|0\rangle - |1\rangle)/\sqrt{2}\right\}. \tag{6.7}$$

These bases belong to the class of *mutually unbiased bases* (MUBs) [25–29]. Two orthonormal bases $\{|e_1\rangle, \ldots, |e_N\rangle\}$ and $\{|f_1\rangle, \ldots, |f_N\rangle\}$ in Hilbert space $\boldsymbol{C}^N$ are MUBs when the square of magnitude of the inner product between any two-basis states $\{|e_m\rangle\}$ and $\{|f_n\rangle\}$ is equal to the inverse of the dimension:

$$|\langle e_m|f_n\rangle|^2 = \frac{1}{N}\forall m, n \in \{1, \ldots, N\}. \tag{6.8}$$

The key word *unbiased* means that if a system is prepared in a state belonging to one of the basis, all outcomes of the measurement with respect to the other basis are equally likely.

Alice randomly selects the MUB, followed by random selection of the basis state. The logical 0 is represented by $|0\rangle,\ |+\rangle$, while logical one by $|1\rangle,\ |-\rangle$. Bob measures each qubit by randomly selecting the basis, computational or diagonal. In sifting procedure, Alice and Bob announce the bases being used for each qubit and keep only instances when they used the same basis.

6.3.1.1 BB84 Protocol: Formal Description

Alice generates two classical random sequences of bits: the data sequence $\boldsymbol{d}$ and the bases sequence $\boldsymbol{b}$ of length $N > 4n$ (n is the length of the sifted key) each, and encodes them as a block of N-qubits as follows [6]:

$$\begin{aligned}
&|\boldsymbol{\psi}\rangle = \bigotimes_{k=1}^{N} \left|\psi_{d_k b_k}\right\rangle \\
&|\psi_{00}\rangle = |0\rangle \qquad\qquad |\psi_{10}\rangle = |1\rangle \\
&|\psi_{01}\rangle = (|0\rangle + |1\rangle)/\sqrt{2}\ |\psi_{11}\rangle = (|0\rangle - |1\rangle)/\sqrt{2}.
\end{aligned} \tag{6.9}$$

Clearly, some states are not mutually orthogonal. The effect of this procedure is to encode the data sequence of bits $\boldsymbol{d}$ in either DB (X-basis) or CB (Z-basis) depending on sequence of bases $\boldsymbol{b}$. Bob receives $\xi(|\psi\rangle\langle\psi|)$, where ξ is the quantum operation describing the action of the quantum channel and eavesdropping. Bob measures each received qubit in either CB or DB depending on random bases sequence $\boldsymbol{b}'$ of length N, created by himself, to get the results of measurements denoted by $\boldsymbol{d}'$. Alice announces $\boldsymbol{b}$ and they both discard all bits in $\{\boldsymbol{d}', \boldsymbol{d}\}$ except those for which corresponding bits in $\boldsymbol{b}$ and $\boldsymbol{b}'$ are identical. Alice and Bob keep $2n$ bits. Alice randomly selects n out $2n$ and announces them. Bob and Alice compare the bits selected for parameter estimation and if more than t disagree (corresponding to the error correction capability of the FEC scheme) they abort protocol; otherwise, they perform information reconciliation and privacy amplification to get m acceptably secret shared key bits from the remaining n bits.

6.3.1.2 The BB84 Protocol Overview

The B84 protocol can be summarized as follows [6]:

- Alice randomly generates $N > 4n$ random data bits $\boldsymbol{d}$.
- Alice selects at random N-bit-long sequence of bases $\boldsymbol{b}$ and encodes the ith data bit d_i as $\{|0\rangle, |1\rangle\}$ if the corresponding bit in $\boldsymbol{b}$ is 0 or $\{|+\rangle, |-\rangle\}$ if corresponding bit in $\boldsymbol{b}$ is 1.
- The resulting state is sent to Bob by Alice.
- Once Bob receives the N-qubits, he announces this fact, and measures each qubit in either CB or DB, selected at random.
- Alice announces the sequence of bases $\boldsymbol{b}$ she used.
- Alice and Bob discard any bits where both used different basis. With high probability there are at least $2n$ bits left (otherwise abort protocol), and they use remaining $2n$ bits to continue with the protocol.
- Out of $2n$ remaining bits, Alice selects a subset of n bits to be used against Eve's interference and channel errors, and informs Bob which ones.
- Alice and Bob announce and compare the values of n bits used for quantum bit error rate estimation. If more than acceptable number of bits disagree, dictated by the error correction capability of the code, they abort protocol.
- Otherwise, Alice and Bob perform information reconciliation and privacy amplification on the remaining n bits to obtain m (wherein $m < n$) shared key bits.

6.3.1.3 Polarization-Based BB84 QKD Protocol

Now we describe the polarization-based BB84 QKD protocol, with corresponding four states being employed provided in Fig. 6.4. The bulky optics implementation of BB84 protocol is illustrated in Fig. 6.5.

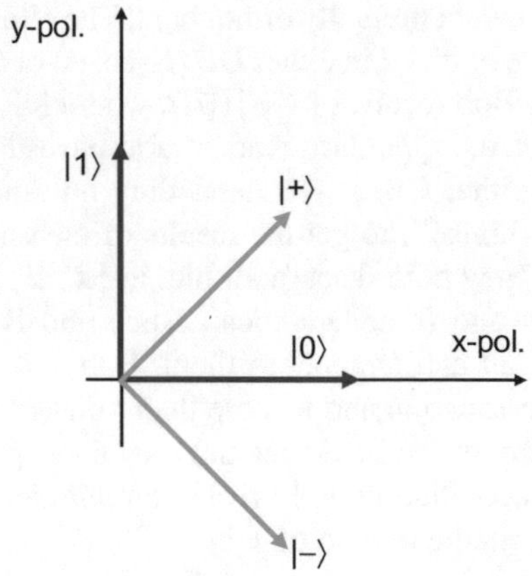

Fig. 6.4 The four states being employed in BB84 protocol

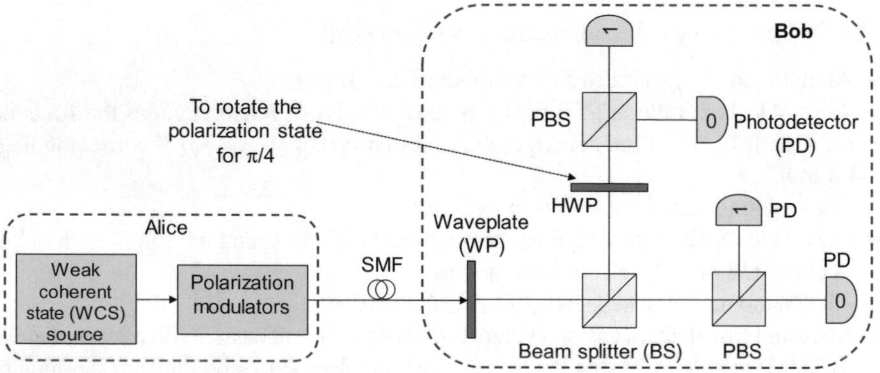

Fig. 6.5 Polarization-based BB84 protocol

Laser output state is known as the *coherent state* [29–34], representing the right eigenket of the *annihilation operator a*, namely, $a\,|\alpha\rangle = \alpha\,|\alpha\rangle$, where α is the complex eigenvalue. The coherent state vector $|\alpha\rangle$ can be represented in terms of orthonormal eigenkets $|n\rangle$ (the number or Fock state) of the number operator $N = a^\dagger a$ as follows:

$$|\alpha\rangle = \exp\left[-|\alpha|^2/2\right]\sum_{n=0}^{+\infty}(n!)^{-1/2}\alpha|n\rangle. \tag{6.10}$$

The coherent states are not orthogonal since the following is valid:

$$|\langle\alpha|\beta\rangle|^2 = \left|e^{-\left(|\alpha|^2+|\beta|^2\right)/2}\sum_{n=0}^{+\infty}\sum_{m=0}^{+\infty}(n!)^{-1/2}(m!)^{-1/2}\alpha^n\left(\beta^*\right)^m\langle n|m\rangle\right|^2$$

$$= \left| e^{-(|\alpha|^2+|\beta|^2)/2} \sum_{n=0}^{+\infty} (n!)^{-1} \alpha^n (\beta^*)^n \right|^2 = \left| e^{-(|\alpha|^2+|\beta|^2-2\alpha\beta^*)/2} \right|^2$$
$$= e^{-|\alpha-\beta|^2}. \tag{6.11}$$

When sufficiently attenuated, the coherent state becomes the weak coherent state, which can ensure the coherent state not to have more than one photon with a high probability. For example, for $\alpha = 0.1$ we obtain $|0.1\rangle = \sqrt{0.90}|0\rangle + \sqrt{0.09}|1\rangle + \sqrt{0.002}|2\rangle + \cdots$. The number state $|n\rangle$ can be expressed in terms of the ground state ($|0\rangle$) with the help of the creation operator by

$$|n\rangle = \frac{(a^\dagger)^n |0\rangle}{(n!)^{1/2}}. \tag{6.12}$$

The density operator of a coherent state is given by

$$\rho = |\alpha\rangle\langle\alpha| = e^{-|\alpha|^2} \sum_{n=0}^{+\infty} \sum_{m=0}^{+\infty} (n!)^{-1/2} (m!)^{-1/2} \alpha^n (\alpha^*)^m |n\rangle\langle m|. \tag{6.13}$$

The diagonal terms of ρ are determined by

$$\langle n|\rho|n\rangle = e^{-|\alpha|^2} \frac{|\alpha|^{2n}}{n!}, \tag{6.14}$$

which is clearly a Poisson distribution for average number of photons $|\alpha|^2$. Therefore, the probability that n-photons can be found in coherent state $|\alpha\rangle$ follows the Poisson distribution:

$$P(n) = |\langle n|\alpha\rangle|^2 = e^{-\mu} \frac{\mu^n}{n!}, \tag{6.15}$$

where we use $\mu = |\alpha|^2$ to denote the average number of photons. Given that the probability that multiple photons are transmitted is now nonzero, Eve can exploit this fact to gain information from the channel known as the *photon number splitting (PNS) attack*. Assuming that Eve is capable of detecting the number of photons that reach her without performing a measurement on the system, when multiple photons are detected Eve will take a single photon and places it into quantum memory until Alice and Bob perform time sifting and basis reconciliation. By learning Alice and Bob's measurement basis from the discussion over the authenticated public channel, the photon stored in quantum memory will be measured in the correct basis providing Eve with all the information contained in the photon.

To overcome the PNS attack, the use of *decoy-state*-based quantum key distribution was proposed in [35]. In decoy-state QKD, the average number of photons transmitted is increased during random timeslots, allowing Alice and Bob to detect if Eve is stealing photons when multiple photons are transmitted.

6.3.1.4 Phase-Encoding-Based BB84 QKD

Now we describe the phase-encoding-based BB84 QKD protocol, whose bulky optics version is illustrated in Fig. 6.6.

The input two-mode Fock basis state $|01\rangle_{n1n2}$ after the first beam splitter (BS) gets transformed to

$$|01\rangle_{n_1n_2} \rightarrow 2^{-1/2}\left(j|01\rangle_{n_3n_4} + |10\rangle_{n_3n_4}\right). \tag{6.16}$$

After the phase shifts in the Mach–Zehnder (MZ) branches get introduced, we can write

$$2^{-1/2}\left(j|01\rangle_{n_3n_4} + |10\rangle_{n_3n_4}\right) \rightarrow 2^{-1/2}\left(je^{-j\phi/2}|01\rangle_{n_3n_4} + e^{j\phi/2}|10\rangle_{n_3n_4}\right), \phi = \phi_A - \phi_B, \tag{6.17}$$

where ϕ_A (ϕ_B) corresponds to the phase shift introduced in the upper (lower) branch belonging to Alice (Bob). After the second BS, the corresponding state will be

$$|01\rangle_{n_3n_4} \rightarrow 2^{-1/2}\left(j|01\rangle_{n_5n_6} + |10\rangle_{n_5n_6}\right), |10\rangle_{n_3n_4} \rightarrow 2^{-1/2}\left(j|10\rangle_{n_5n_6} + |01\rangle_{n_5n_6}\right). \tag{6.18}$$

The overall output state can be represented by

$$|\psi(\phi)\rangle = j\left(\sin(\phi/2)|01\rangle_{n_5n_6} + \cos(\phi/2)|10\rangle_{n_5n_6}\right). \tag{6.19}$$

By setting different phases ϕ we can define the states for BB84 protocol as follows:

$$\begin{aligned} &|\psi(0)\rangle = j|10\rangle_{n_5n_6} \doteq |0\rangle, \qquad |\psi(\pi)\rangle = j|01\rangle_{n_5n_6} \doteq |1\rangle \\ &|\psi(\pi/2)\rangle = j2^{-1/2}\left(|10\rangle_{n_5n_6} + |01\rangle_{n_5n_6}\right) \doteq |+\rangle, |\psi(-\pi/2)\rangle \\ &\qquad = j2^{-1/2}\left(|10\rangle_{n_5n_6} - |01\rangle_{n_5n_6}\right) \doteq |-\rangle. \end{aligned} \tag{6.20}$$

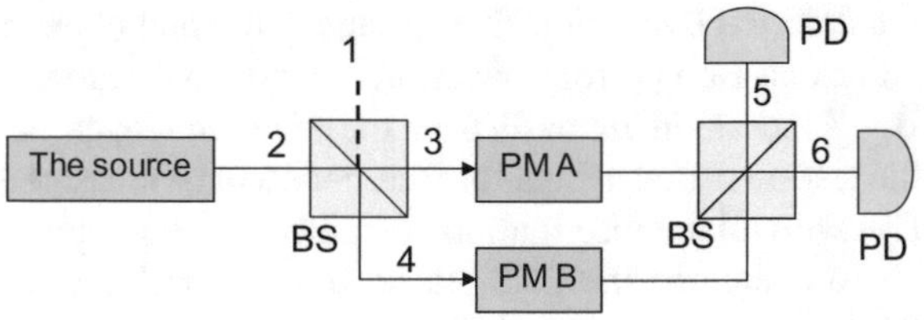

Fig. 6.6 Phase-encoding-based BB84 protocol

Alice randomly sets the control voltage to introduce the phase shift $\phi_A \in \{-\pi/2, 0, \pi/2, \pi\}$ and therefore randomly selects the state. Bob randomly selects the measurement basis by randomly choosing $\phi_B \in \{0, \pi/2\}$. Conclusive measurement occurs when either: (i) $\phi_B = 0$ and $\phi_A \in \{0, \pi\}$ or (ii) $\phi_B = \pi/2$ and $\phi_A \in \{-\pi/2, \pi/2\}$.

6.3.2 B92 Protocol

The B92 protocol, introduced by Bennet in 1992 [36], employs only two non-orthogonal states, as illustrated in Fig. 6.7. Alice randomly generates a classical bit *d*, and depending on its value (0 or 1) she sends the following state to Bob [6]:

$$|\psi\rangle = \begin{cases} |0\rangle, d = 0 \\ |+\rangle = (|0\rangle + |1\rangle)/\sqrt{2}, d = 1 \end{cases}. \tag{6.21}$$

Bob randomly generates a classical bit d' and he subsequently measures the received qubits in the CB $= \{|0\rangle, |1\rangle\}$ if $d' = 0$ or DB $= \{|+\rangle, |-\rangle\}$ if $d' = 1$, with the result of measurement being $r = 0$ or 1, corresponding to -1 and $+1$ eigenstates of X and Z observables, which is illustrated in Table 6.1. Clearly, when the result of a measurement is $r = 0$ the possible states sent by Alice are $|0\rangle$ and $|+\rangle$ and

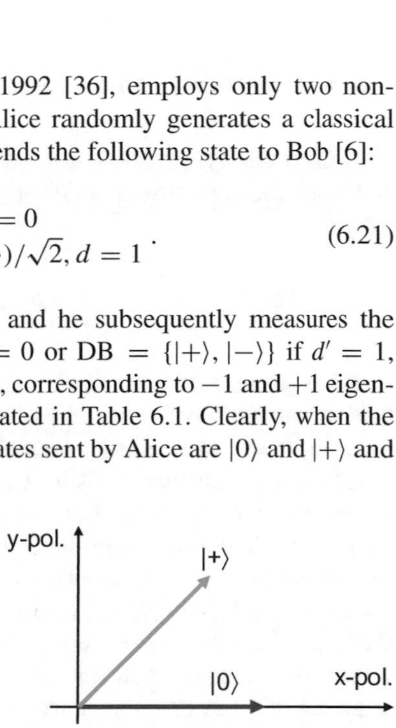

Fig. 6.7 The two states being employed in B92 protocol

Table 6.1 Explanation of the B92 protocol

Bits generated by Alice	$d = 0$		$d = 1$	
States sent by Alice	$\|0\rangle$		$\|+\rangle$	
Bits generated by Bob	$d' = 0$	$d' = 1$	$d' = 0$	$d' = 1$
Bob's measurement base	CB: $\{\|0\rangle, \|1\rangle\}$	DB: $\{\|+\rangle, \|-\rangle\}$	CB: $\{\|0\rangle, \|1\rangle\}$	DB: $\{\|+\rangle, \|-\rangle\}$
The possible resulting states to be detected by Bob	$\|0\rangle$ $\|1\rangle$	$\|+\rangle$ $\|-\rangle$	$\|0\rangle$ $\|1\rangle$	$\|+\rangle$ $\|-\rangle$
The probability of Bob measuring a given state	1 0	1/2 1/2	1/2 1/2	1 0
Result of Bob's measurement *r*	0 –	0 1	0 1	0 –

Bob does not know which state Alice has sent. On the other hand, when the result of the measurement is $r = 1$ the possible states detected by Bob are $|1\rangle$ and $|-\rangle$, and Bob's detected bit is a complement of Alice bit.

Bob publicly announces r (and of course keeps d' secret). Alice and Bob keep only those pairs $\{d, d'\}$ for which the result of measurement was $r = 1$. The *final key bit* is d for Alice and $1 - d'$ for Bob. After that, the information reconciliation and privacy amplification stages take place.

6.3.3 Ekert (E91) and EPR Protocols

The Ekert (E91) protocol was introduced by Ekert in 1991 [37] (see also [6]). Suppose Alice and Bob share n entangled pairs of qubits in the Bell state (EPR pair) $|B_{00}\rangle$:

$$|B_{00}\rangle = \frac{|00\rangle + |11\rangle}{\sqrt{2}}. \tag{6.22}$$

Alice generates a random classical bit b to determine the base, and measures her half on the Bell state in either CB $= \{|0\rangle, |1\rangle\}$ basis for $b = 0$ or DB $= \{|+\rangle, |-\rangle\}$ basis for $b = 1$ to obtain the data bit d. On the other hand, Bob in similar fashion randomly generates the basis bit b', and performs the measurement in either CB or DB to obtain d'. Alice and Bob compare the bases being used, namely, b and b', and keep for their raw key only those instances of $\{d, d'\}$ for which the basis was the same $b = b'$. Clearly, the raw key such generated is *truly random*, it is undetermined until Alice and Bob perform measurements on their Bell state half. For this reason, QKD is essentially a *secret-key generation* method, as it is dictated by the entangled source. If instead the Bell state $|B_{01}\rangle = (|01\rangle + |10\rangle)/\sqrt{2}$ was used, Bob's raw key sequence will be complementary to the Alice raw key sequence.

The *EPR protocol* is a three-state protocol that employs the Bell's inequality to detect the presence of the Eve as a hidden random variable, as described in [38]. Let $|\theta\rangle$ denote the polarization state of a photon that is linearly polarized at an angle θ. Three possible polarization states for EPR pair include [38]

$$\begin{aligned}
|B_0\rangle &= \frac{|0\rangle|3\pi/6\rangle + |3\pi/6\rangle|0\rangle}{\sqrt{2}}, \\
|B_1\rangle &= \frac{|\pi/6\rangle|4\pi/6\rangle + |4\pi/6\rangle|\pi/6\rangle}{\sqrt{2}}, \\
|B_2\rangle &= \frac{|2\pi/6\rangle|5\pi/6\rangle + |5\pi/6\rangle|2\pi/6\rangle}{\sqrt{2}}.
\end{aligned} \tag{6.23}$$

For each of these EPR pairs, we apply the following encoding rules [38]:

$$|\psi\rangle_1 = \begin{cases} |0\rangle, & d = 0 \\ |3\pi/6\rangle, & d = 1 \end{cases}$$

$$|\psi\rangle_2 = \begin{cases} |\pi/6\rangle, & d = 0 \\ |4\pi/6\rangle, & d = 1 \end{cases}$$

$$|\psi\rangle_3 = \begin{cases} |2\pi/6\rangle, & d = 0 \\ |5\pi/6\rangle, & d = 1 \end{cases}. \tag{6.24}$$

The corresponding measurement operators will be [38]

$$M_0 = |0\rangle\langle 0|, \; M_1 = |\pi/6\rangle\langle\pi/6|, \; M_2 = |2\pi/6\rangle\langle 2\pi/6|. \tag{6.25}$$

The EPR protocol can be now described as follows. The EPR state $|B_i\rangle$ ($i = 0$, 1, 2) is randomly selected. The first photon in the EPR pair is sent to Alice and the second to Bob. Alice and Bob select at random, independently and with equal probability, one of the measurement operators M_0, M_1, and M_2. They subsequently measure their respective photon (qubit), and the results of measurement determine the corresponding bit for the sifted key. Clearly, the selection of EPR pairs as in Eq. (6.23) results in complementary raw keys. Alice and Bob then communicate over an authenticated public channel to determine instances when they used the same measurement operator and keep those instances. The corresponding shared key represents the *sifted key*. The instances when they used different measurement operators represent the *rejected key*, and these are used to detect Eve's presence by employing the Bell's inequality on rejected key. Alice and Bob then perform information reconciliation and privacy amplification steps.

Let us now describe how the rejected key can be used to detect Eve's presence [38]. Let $P(\neq|i, j)$ denote the probability that Alice and Bob's bits in rejected key are different given that Alice and Bob's measurement operators are either M_i and M_j or M_j and M_i, respectively. Let further $P(=|i, j) = 1 - P(\neq|i, j)$. Let us denote the difference of these two probabilities by $\Delta P(i, j) = P(\neq|i, j) - P(=|i, j)$. The Bell's parameter can be defined by [38]

$$\beta = 1 + \Delta P(1, 2) - |\Delta P(0, 1) - \Delta P(0, 2)|, \tag{6.26}$$

which for measurement operators above reduces to $\beta \geq 0$. However, the quantum mechanics prediction gives $\beta = -1/2$, which is clearly in violation of the Bell's inequality.

6.3.4 Time-Phase Encoding

The BB84 protocol can also be implemented using different degrees of freedom, in addition to polarization states, such as time-phase encoding [39–41] and orbital angular momentum [42, 43]. The time-phasing encoding states for BB84 protocol are provided in Fig. 6.8, which is a different version of phase-encoding scheme compared to Fig. 6.6. The time-basis corresponds to the computational basis, while the phase-basis to the diagonal basis. The pulse is localized within the time bin of duration $\tau = T/2$. The time-basis is like the pulse-position modulation (PPM). The state in which the photon is placed in the first time bin is denoted by $|t_0\rangle$, while the state in which the photons are placed in the second time bin is denoted by $|t_1\rangle$. The phase MUB states are defined by

$$|f_0\rangle = \frac{1}{\sqrt{2}}(|t_0\rangle + |t_1\rangle), \ |f_1\rangle = \frac{1}{\sqrt{2}}(|t_0\rangle - |t_1\rangle). \tag{6.27}$$

Alice randomly selects either the time-basis or the phase-basis, followed by random selection of the basis state. The logical 0 is represented by $|t_0\rangle$, $|f_0\rangle$, while logical one by $|t_1\rangle$, $|f_1\rangle$. Bob measures each qubit by randomly selecting the basis, the time-basis or the phase-basis. In sifting procedure, Alice and Bob announce the bases being used for each qubit and keep only instances when they used the same basis.

To implement the time-phase encoder either the electro-optical polar modulator, composed of concatenation of an amplitude modulator and phase modulator, or I/Q modulator can be used [44]. The amplitude modulator (Mach–Zehnder modulator) is used for pulse shaping and the phase modulator to introduce the desired phase shift, as illustrated in Fig. 6.9. The arbitrary waveform generator (AWG) is used to generate the corresponding RF waveforms as needed. The variable optical attenuator (VOA) is used to attenuate the modulated beam down to single-photon level.

On receiver side, the time-basis states can be detected with the help of single-photon counter and properly designed electronics. On the other hand, to detect the phase-basis states, the time-delay interferometer can be used as described in [45], see Fig. 6.10. The difference in the path between two arms is [45] $\Delta L = \Delta L_0 + \delta L$, where ΔL_0 is the path difference equal to $c\tau$ (c is the speed of light) and $\delta L \ll \Delta L_0$ is small path difference used to adjust the phase since $\phi = k\delta L$ (k is the wave number). When the phase-state $|f_0\rangle$ is incident to the time-phase decoder, the outputs of second 50:50 beam splitter (BS) occupy three-time slots, and the + output

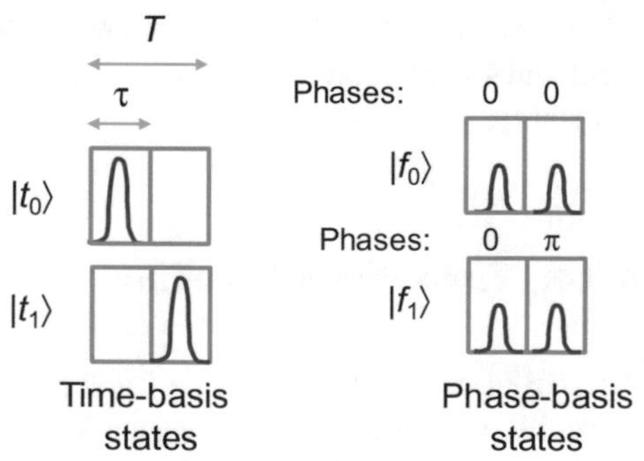

Fig. 6.8 The time-phase encoding states to be used in BB84 protocol

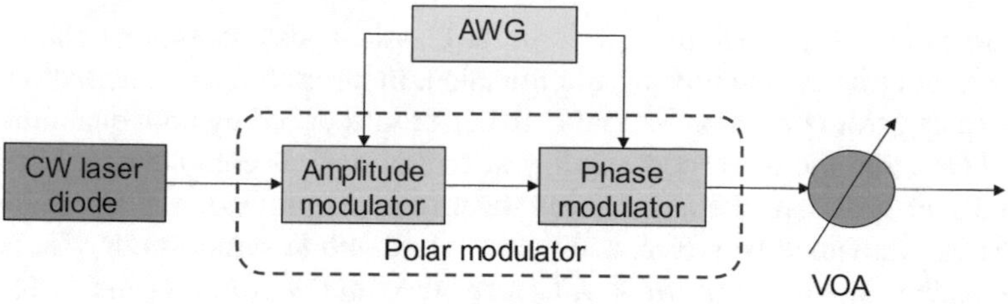

Fig. 6.9 The time-phase encoder for BB84 QKD protocol. AWG: arbitrary waveform generator, VOA: variable optical attenuator

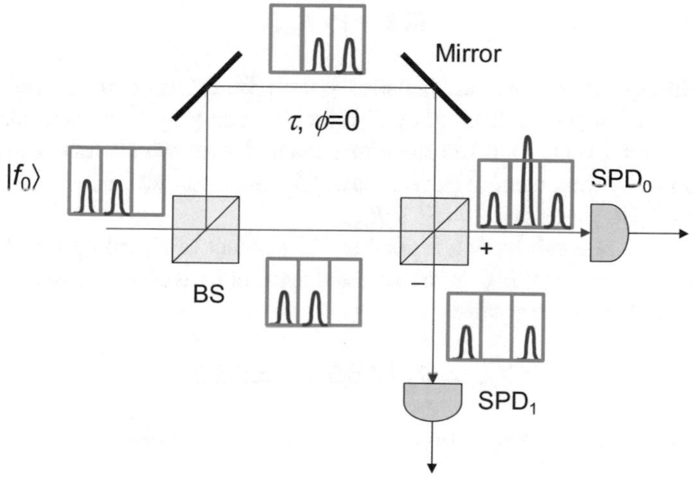

Fig. 6.10 The time-phase decoder for BB84 QKD protocol. BS: 50:50 beam splitter, SPD: single-photon detector

denotes the case when the interferometer outputs interfere constructively, while the − output denotes that corresponding interferometer outputs interfere destructively. For constructive interference, the middle pulse gets doubled, while for destructive interference middle pulses cancel each other. Therefore, the click of SPD in the middle slot at the + output identifies the $|f_0\rangle$-state, while the corresponding click at the − output identifies the $|f_1\rangle$-state.

6.4 Security Issues of QKD Systems

As discussed in BB84 protocol overview, after raw key transmission phase is completed, Alice and Bob perform sifting procedure and parameter estimation, so that they are left with n symbols out of N transmitted, and corresponding key is known as the *sifted key*. After that, they perform the classical postprocessing of such information reconciliation and privacy amplification. In information reconciliation stage, they employ error correction to correct the error introduced by both quantum channel and Eve, and the corresponding key after this stage is completed is commonly referred to as the *corrected key*. During the privacy amplification, they remove correlation that remained with Eve, and they are left with m symbols out of n, and the corresponding key is the *secure key*. Clearly, the *fractional rate* of the secret key r can be determined as $r = m/n$ (as $N \to \infty$), and it is often referred as the *secret fraction* [46]. The *secret-key rate* (SKR) is then determined as the product of the secret fraction and the raw key rate, denoted R_{raw}, and we can write [46]

$$SKR = r \cdot R_{\text{raw}}. \tag{6.28}$$

Interestingly enough, in many papers, the term SKR refers to the secret fraction r. Given that raw key rate is dictated by devices being employed, in particular, single-photon detector (SPD) and the quantum channel over which transmission takes place, and not by the quantum physics laws, the fraction rate can also be called the normalized SKR, given that $r = SKR/R_{\text{raw}}$.

The raw key rate can be determined as the product of signaling rate R_s and the probability of Bob accepting the transmitted symbol in average, denoted as Pr(Bob accepting), so that we can write

$$R_{\text{raw}} = R_s \cdot \Pr(\text{Bob accepting}). \tag{6.29}$$

On the other hand, the signaling rate is determined by [46]

$$R_s = \min\left(R_{s,\max}, 1/T_{\text{dc}}, \frac{\mu T T_B \eta}{\tau_{\text{dead}}}\right), \tag{6.30}$$

where $R_{s,\max}$ is the maximum signaling rate allowed by the source, T_{dc} is the duty cycle, and τ_{dead} is the dead time of the SPD (the time required for SPD to reset after a photon gets detected). The term $\mu T T_{\text{B}} \eta$ corresponds to Bob's probability of detection, with μ being the average number of photons generated by the source, η being the SPD efficiency, T is the transmissivity of the quantum channel, and T_B denotes the losses in Bob's receiver. The quantum channel is typically either fiber optics channel or free-space optical channel. The attenuation in fiber optics channel is determined by $T = 10^{-\alpha L/10}$, where α is the attenuation coefficient in dB/km, while L is the transmission distance in km. The typical attenuation coefficient in standard single-mode fiber (SSMF) is 0.2 dB/km at 1550 nm. Regarding, the FSO link, if we ignore the atmospheric turbulence effects, for the line-of-site link of transmission distance L, we can estimate the FSO link loss by [47] $T = [d_{\text{Rx}}/(d_{\text{Tx}} + DL)]^2 \times 10^{-\alpha L/10}$, where d_{Tx} and d_{Rx} represent diameters of transmit and receive telescopes' apertures, while D is the divergence. Therefore, the first term represents the geometric losses, while the second term is the average attenuation due to scattering, with scattering attenuation coefficient being <0.1 dB/km in clear sky condition at 1520–1600 nm.

In BB84 protocol, the probability that Bob accepts the transmitted symbol is related to the probability he used the same basis, denoted as p_{sift}, so that the signaling rate becomes $R_s p_{\text{sift}}$, and raw key rate when Alice sends n-photons to Bob will be [46]

$$R_{\text{raw},n} = R_s p_{\text{sift}} p_{A,n} p_n, \tag{6.31}$$

where $p_{A,n}$ is the probability that pulse sent by Alice contains n-photons (see section below) and p_n is the probability that Eve sends Bob a pulse with n-photons. The overall total raw key rate will be $R_{\text{raw}} = \sum_n R_{\text{raw},n}$.

What remains to be determined is the fraction rate. The QKD can achieve the unconditional security, which means that its security can be verified without imposing any restrictions on either Eve's computational power or eavesdropping strategy. The bounds on the fraction rate are dependent on the classical postprocessing steps. The most common is the one-way postprocessing, in which either Alice or Bob holds the reference key and sends the classical information to the other party through the public channel, while the other party performs certain procedure on data without providing the feedback. Similarly to the physical-layer security (PLS) schemes described in Chap. 4, the most common one-way processing consists of two steps, the information reconciliation and privacy amplification. The expression for *secret fraction*, obtained by *one-way postprocessing,* is very similar to that for the classical PLS schemes [46]:

$$r = I(A; B) - \min_{\text{Eve's strategies}} (I_{EA}, I_{EB}), \tag{6.32}$$

where $I(A; B)$ is the mutual information between Alice and Bob, while the second term corresponds to Eve's information I_E about Alice or Bob's raw key, where minimization is performed over all possible eavesdropping strategies. Alice and Bob will decide to employ either direct or reverse reconciliation so that they can minimize Eve's information.

We now describe different eavesdropping strategies that Eve may employ, which determine Eve's information I_E.

6.4.1 *The Eavesdropping Strategies and Corresponding Secret Fractions*

Here we discuss different levels of security.

6.4.1.1 Independent (Individual) or Incoherent Attacks

This is the most constrained family of attacks, in which Eve attacks each qubit independently, and interacts with each qubit by applying the same strategy. Moreover, she measures the quantum states before the classical postprocessing takes place. The security bound for incoherent attacks is the same as that for classical PLS, as described in Chap. 4, and it is determined by Csiszár–Körner bound [48], given by Eq. (6.32), wherein the mutual information between Alice and Eve is given by

$$I_{EA} = \max_{\text{Eve's strategies}} I(A;E), \tag{6.33}$$

where the maximization is performed over all possible incoherent eavesdropping strategies. The similar definition holds for I_{BE}.

An important family of incoherent attacks is the *intercept-resend* (*IR*) attack, in which Eve intercepts the quantum signal sent by Alice, performs the measurement on it, and based on the measurement result she prepares new quantum signal (in the same MUB as the measured quantum state) and sends such prepared quantum signal to Bob. In BB84 protocol, Eve's basis choice will match Alice's basis 50% of the time. When she uses the wrong basis there is still 50% chance to guess the correct bit. When Bob does the measurement, there is 50% chance that Bob will use the same basis as Eve. However, these bits will correlate with Eve's sequence not Alice's one, and overall probability of error will be 1/4, while the mutual information between Alice and Eve will be $I(A;E) = 1/2$. Given high probability of error introduced by the IR attack, Alice and Bob can easily identify Eve's activity. However, Eve might choose to apply the IR attack with probability of p_{IR}. In that case the probability of error will be $q = p_{\mathrm{IR}}/4$. The mutual information between Alice and Eve will be then $I(A;E) = h(p_{\mathrm{IR}}/4)$, where $h(\cdot)$ is the binary entropy function $h(q) = -q\log_2 q - (1 - q)\log_2(1 - q)$. If all errors get contributed to Eve, the secret fraction can be estimated as

$$r = \underbrace{1 - h(q)}_{I(A;B)} - h(q) = 1 - 2h(q). \tag{6.34}$$

Another important family of incoherent attacks is the *photon number splitting* (PNS) attack [49], in which Eve acts on multi-photons quantum states. In a weak coherent state-based QKD system, the quantum states are generated by modulating the beam from the coherent light source, which is then attenuated so that in average one photon per pulse gets transmitted by Alice. However, as discussed above, the coherent source emits the photons based on Poisson distribution so that the probability of emitting n-photons in a state generated with a mean photon number μ is determined by the following equation:

$$p_{A,n} = p(n|\mu) = e^{-\mu}\frac{\mu^n}{n!}. \tag{6.35}$$

The Poisson distribution for three different mean photon numbers is shown in Fig. 6.11.

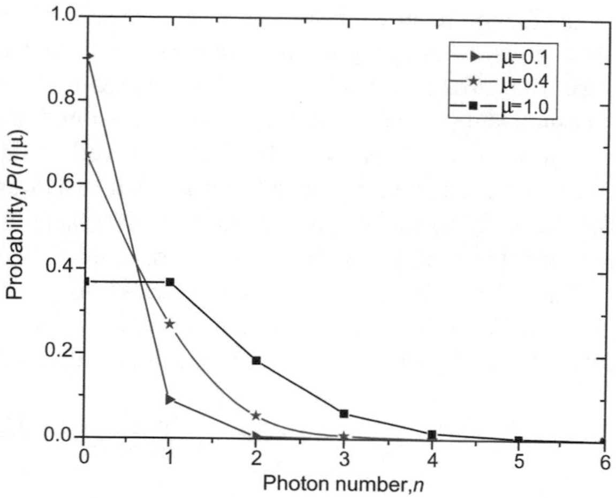

Fig. 6.11 The photon number (Poisson) distribution versus photon number (to see the trend the distribution is shown as continuous even though it is discrete)

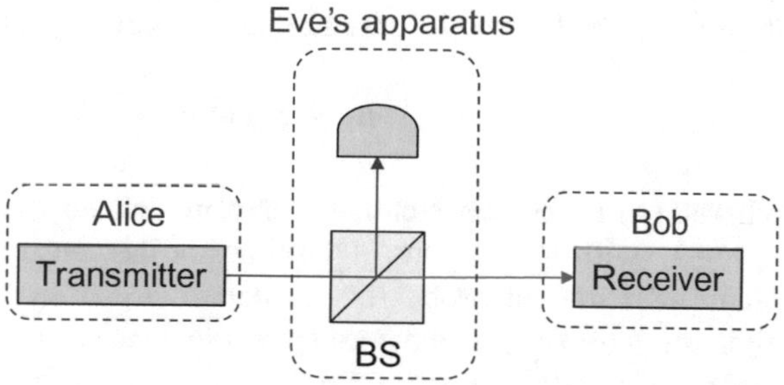

Fig. 6.12 The illustration of Eve's realization of the PNS attack. BS: the beam splitter

As the parameter μ increases, the probability of getting more than one photon increases as well. To perform the PNS attack, Eve can employ the beam splitter to take one of the photons from the multi-photon state, and pass the rest to Bob, as illustrated in Fig. 6.12. She can measure the photon by randomly selecting the basis. Eve can even replace the quantum link with ultralow-loss fiber so that Alice and Bob cannot figure that the transmitted signal gets attenuated. To solve for the PNS attack, Alice and Bob can employ decoy-state-based protocol [35], in which Alice transmits the quantum states with different mean photon numbers, representing signal and decoy states. Eve cannot distinguish between decoy state and signal state, and given that Alice and Bob know the decoy signal level they can identify the PNS attack.

6.4.1.2 Collective Attacks

The collective attacks represent generalization of the incoherent attacks given that Eve's interaction with each qubit is also independent and identically distributed (i.i.d). However, in these attacks, Eve can store her ancilla qubits in a quantum memory until the end of classical postprocessing steps. For instance, given that information reconciliation requires the exchange of the parity bits over an authenticated classical channel, Eve can apply the best known classical attacks to learn the content of the parity bits. Based on all information available to her, Eve can perform the best measurement strategy on her ancilla qubits (stored in the quantum memory).

The PNS attack is stronger when Eve applies the quantum memory. Namely, from sifting procedure, Eve can learn which basis was used by Alice and can apply the correct basis on her photon stored in the quantum memory and thus double the number of identified bits compared to the case when quantum memory is not used.

The security bound for collective attacks, assuming one-way postprocessing, is given by Eq. (6.32), wherein Eve's information about Alice sequence is determined from Holevo information as follows [50] (see also [46]):

$$I_{EA} = \max_{\text{Eve's strategies}} \chi(A; E), \tag{6.36}$$

where maximization is performed over all possible collective eavesdropping strategies. The similar definition holds for I_{BE}. This bound is also known as Devetak–Winter bound. The Holevo information, introduced in [51], is defined here as

$$\chi(A; E) = S(\rho_E) - \sum_a S(\rho_{E|a})p(a), \tag{6.37}$$

where $S(\rho)$ is the von Neumann entropy defined as $S(\rho) = -\text{Tr}(\log(\rho)) = -\sum_i \lambda_i \log \lambda_i$, with λ_i being the eigenvalues of the density operator (state) ρ. In (6.37), $p(a)$ represents the probability of occurrence of symbol a from Alice's classical alphabet, while $\rho_{E|a}$ is the corresponding density operator of Eve's ancilla. Finally, ρ_E is Eve's partial density state defined by $\rho_E = \sum_a p(a)\rho_{E|a}$. In other words, the Holevo information corresponds to the average reduction in von Neumann entropy given that we know how ρ_E get prepared.

6.4.1.3 Coherent Attacks

The coherent attacks represent the most general and the most flexible strategies that Eve can apply on quantum states. She can adapt the eavesdropping strategy on the fly, based on previous and current measurements. She might further entangle as many quantum states as she wants and stores them in the quantum memory. Therefore, the minimization of mutual information between Alice/Bob and Eve is impossible. Nevertheless, the bounds have been determined in many cases, and these are very similar to those obtained for the collective attacks. Namely, the states sent by Alice $|\psi_i\rangle$ to Bob are independent and can be represented by the tensor product $|\psi_1\rangle \otimes \cdots \otimes |\psi_n\rangle \otimes \cdots$ and the quantum channel typically does not correlate them. So, Eve will not get any advantage to introduce the artificial correlations. However, the correlations get introduced later during the classical postprocessing, and Eve should exploit those correlations, rather than guessing the transmitted symbol in the raw key. The final key is determined by the relations among the symbols in the raw key, and by employing the entanglement-based approaches Eve can learn some of these relationships.

6.4.1.4 Quantum Hacking Attacks and/or Side-Channel Attacks

Quantum hacking attacks exploit the weakness of the particular QKD system implementation. Many of hacking attacks are feasible with current technology. In *Trojan horse attack* [52], Eve sends the bright laser beam toward Alice's encoder and measures the reflected photons to gain the information about the secret. This attack can be avoided by using an optical isolator. It was also noticed that silicon-based photon counters employing the avalanche photodiodes (APDs) emit some light at different wavelengths when they detect a single photon, which can be exploited by Eve [53]. In time-shifting attack [54], Eve exploits the efficiency mismatch of Bob's single-photon detectors to estimate Bob's basis selection. In *blinding attack* [55], Eve's exploits the APDs-based single-photon counterproperties to force Bob to pick up the same basis as Eve does. The APDs, for photon counting, often operate in gated mode so that the APD is active only when the photon arrival is expected. When in off mode the, APD operates in a linear regime, meaning that the photocurrent is proportional to the received optical power, and therefore it behaves as the classical photodetector. To exploit this, Eve can send the weak classical light beam, sufficiently strong to generate the click when Bob employs the same basis, and not to register the click otherwise. In such scenario, 50% of events will be recorded as inconclusive. Fortunately, the common property of quantum hacking attacks is that they are preventable once the attack is known. However, the threat from an unknown quantum hacking attack is still present.

Many of *side-channel attacks* are at the same time zero-error attacks [46]. Namely, Eve can exploit the quantum channel loss to hide her side-channel attack. The beam-splitting attack exploits the losses of the quantum channel, and Eve can place the beam splitter just after Alice's transmitter, take the portion of the beam, and pass the rest to Bob. Since Alice does not modify the optical mode, this attack will not be detected by Bob. Interestingly enough the Poisson distribution is preserved during the PNS attack [56]. In realistic implementations, Eve can perform *unambiguous state discrimination* followed by signal resending, when Alice employs the linearly independent signal states and achieves better than beam-splitting efficiency as shown in [57]. Fortunately, the existence of side-channel attacks does not compromise the security as long as the corresponding side-channel attacks are taken into account during the privacy amplification stage [46].

6.4.2 Security Definitions

Here we are interested in the security against the collective attacks. Let the *perfect key* represent a sequence of perfectly correlated symbols between Alice and Bob, for which Eve does not possess any information. (Different secure keys from the set of secure keys must appear uniformly.) Then the key $\mathcal{K}$ that deviates from the perfect key by ε is said to be *ε-secure*. Early definitions of security of QKD were defined in analogy with corresponding classical definition. Eve, possessing the density state ρ_E, performs the measurement $\mathcal{M}$ that maximizes her mutual information with the $\mathcal{K}$, denoted as $I(\mathcal{K}; E)$, which yields the accessible information, and we can write

$$H(\mathcal{K} : E) = \max_{\mathcal{M}(\rho_E)=E} I(\mathcal{K} : E), \tag{6.38}$$

where maximization is performed over all possible Eve's strategies and $H(\mathcal{K}; E)$ denotes the accessible information. Then the ε-security can be defined as

$$H(\mathcal{K} : E) \leq \varepsilon. \tag{6.39}$$

Given that Eve can wait to perform the measurement on her system until she learns the portion of the key, this inequality does not guarantee the security of the key.

6.4.2.1 Trace-Distance-Based ε-Security

One of the key requirements that the definition of security must satisfy is the *composability* property, indicating that the security of the key is guaranteed regardless of the application [46]. In other words, if the ε-secure key is used in an ε'-secure task the whole procedure must be at least $(\varepsilon + \varepsilon')$-secure. It has been shown by Ben-Or et al. [58] that ε-security defined by Eq. (6.39) holds when two-universal hashing is used in privacy amplification step. On the other hand, it has been shown by König et al. [59] that the accessible information is not composable for arbitrary ε.

To deal with this problem, Ben-Or et al. in [58] as well as Renner and König in [60] introduced the concept of trace-distance-based ε-security (see also [29, 43, 61–63]). We say that the key is *ε-secure* with respect to an eavesdropper $\mathcal{E}$ if the trace distance between the joint state of key $\mathcal{K}$ and Eve $\mathcal{E}$, denoted as $\rho_{\mathcal{K},\mathcal{E}}$, and the product state of completely mixed state (CMS) with respect to the set $\mathcal{K}$ of all possible secure keys, denoted as ρ_{CMS}, and Eve's state, denoted as $\rho_{\mathcal{E}}$, is smaller than or equal to ε. In other words, we can write

$$D\big(\rho_{\mathcal{K},\mathcal{E}}, \rho_{CMS} \otimes \rho_{\mathcal{E}}\big) = \frac{1}{2}\big\|\rho_{\mathcal{K},\mathcal{E}} - \rho_{CMS} \otimes \rho_{\mathcal{E}}\big\|_1 = \frac{1}{2}\mathrm{Tr}\big|\rho_{\mathcal{K},\mathcal{E}} - \rho_{CMS} \otimes \rho_{\mathcal{E}}\big| \leq \varepsilon, \tag{6.40}$$

where $|\rho| = \sqrt{\rho^\dagger \rho}$. With this definition, the parameter ε has a quite clear interpretation, and it represents the maximum tolerable failure probability for the key extraction process. Given that the composability of this definition is satisfied, we can decompose the security of the final key ε in terms of securities for error correction (EC) ε_{EC}, privacy amplification (PA) ε_{PA}, and parameter estimation (PE) ε_{PE} steps as well as the Renyi entropy estimate's failure probability [58], denoted as ε_R. In other words, we can write $\varepsilon = \varepsilon_{EC} + \varepsilon_{PA} + \varepsilon_{PE} + \varepsilon_R$.

Now the main question is how to relate the trace-distance-based ε-security definition to the length m of the secret key being extracted. To do so one needs to prove the following inequality representing the generalization of the Chernoff bound [46]:

$$\Pr\left(D\left(\rho_{\mathcal{K},\mathcal{E}}, \rho_{CMS} \otimes \rho_{\mathcal{E}}\right) > \varepsilon\right) \le e^{m-f\left(\rho_{\mathcal{K},\mathcal{E}},\varepsilon\right)}, \tag{6.41}$$

wherein the constant factors are omitted. Clearly, the key will be insecure with very small probability provided that the following condition is satisfied:

$$m \le f\left(\rho_{\mathcal{K},\mathcal{E}}, \varepsilon\right). \tag{6.42}$$

In other words, as long as the key length is shorter than Eve's uncertainty about the input to the privacy amplification step, the key will be secure.

The security bounds discussed in Sect. 6.4.1 are valid asymptotically for infinitely long keys, when $\varepsilon \to 0$.

6.4.2.2 Fidelity-Based ε-Security

In addition to the trace-distance-based ε-security, the concept of *fidelity-based* ε*-security* can also be used [58]. Let us assume that Alice possesses n pairs of entangled $|B_{00}\rangle = (|00\rangle + |1\rangle)\sqrt{2}$ qubits, denoted as $|B_{00}\rangle^{\otimes n}$. The *fidelity of two density states* ρ and ρ', as shown in Chap. 5, is defined to be

$$F\left(\rho, \rho'\right) = Tr\left(\sqrt{\rho^{1/2}\rho'\rho^{1/2}}\right). \tag{6.43}$$

When $\rho = |\psi\rangle\langle\psi|$ and $\rho' = |\psi'\rangle\langle\psi'|$ are pure states, given that for pure states $\rho^2 = \rho$, the fidelity becomes simply

$$F\left(\rho, \rho'\right) = Tr\left(\sqrt{|\psi\rangle\langle\psi|\psi'\rangle\langle\psi'|\psi\rangle\langle\psi|}\right) = \left|\langle\psi|\psi'\rangle\right|. \tag{6.44}$$

The fidelity between the pure state $|\psi\rangle$ and arbitrary density state ρ is given by

$$F(\rho, |\psi\rangle) = \mathrm{Tr}\left(\sqrt{|\psi\rangle\langle\psi|\rho|\psi\rangle\langle\psi|}\right) = \mathrm{Tr}\left(\sqrt{\langle\psi|\rho|\psi\rangle|\psi\rangle\langle\psi|}\right) = \sqrt{\langle\psi|\rho|\psi\rangle}. \tag{6.45}$$

An important theorem related to the fidelity is the *Uhlmann theorem* [64], claiming that for two density operators ρ and ρ' and for purification of ρ being $|\psi\rangle\langle\psi|$, the following is valid:

$$F\left(\rho, \rho'\right) = \max_{|\psi'\rangle\langle\psi'|} F\left(|\psi\rangle\langle\psi|, |\psi'\rangle\langle\psi'|\right), \tag{6.46}$$

where the maximization is performed over all purifications $|\psi'\rangle\langle\psi'|$ of ρ'.

By applying Eq. (6.45) on $|B_{00}\rangle^{\otimes n}$ we obtain

$$F^2\left(\rho_{AB}^n, |B_{00}\rangle^{\otimes n}\right) = {}^{\otimes n}\langle B_{00}|\rho_{AB}^n|B_{00}\rangle^{\otimes n}, \tag{6.47}$$

where ρ_{AB}^n is the shared density state between Alice and Bob. The final keys for Alice and Bob, denoted by $\mathcal{K}_A$ and $\mathcal{K}_B$, respectively, are of length $m < n$, and in the absence of Eve, the ρ_{AB}^m will be $|B_{00}\rangle^{\otimes m}$. So, we can define the fidelity-based ε-security as follows:

$$1 - F\left(\rho_{AB}^m, |B_{00}\rangle^{\otimes m}\right) \leq \varepsilon''. \tag{6.48}$$

Therefore, in the absence of Eve $F\left(\rho_{AB}^m, |B_{00}\rangle^{\otimes m}\right) = 1$ and clearly $\varepsilon'' = 0$ in that case. Let $|\psi_1^m\rangle\langle\psi_1^m|$ be the purification of ρ_{ABE}^m on systems A, B, and E. According to the Uhlmann's theorem, there exists a purification $|\psi_1^m\rangle\langle\psi_1^m|$ such that

$$F\left(|\psi_1^m\rangle\langle\psi_1^m|, |\psi_2^m\rangle\langle\psi_2^m|\right) = F\left(\rho_{AB}^m, |B_{00}\rangle^{\otimes m}\right). \tag{6.49}$$

By properly constructing the purification states $|\psi_1^m\rangle$ and $|\psi_2^m\rangle$, and performing the measurements on Alice and Bob's system, while tracing out Eve's system we obtain that

$$F\left(\rho_{\text{QKD}}^m, \rho_{\text{perfect}}^m\right) > F\left(\rho_{AB}^m, |B_{00}\rangle^{\otimes m}\right), \tag{6.50}$$

where ρ_{perfect}^m is the density operator corresponding to the perfect key, the key for which the distribution over two m-strings is [58] $p_{\text{perfect}}^{(m)}(l, l') = \delta(l, l')2^{-m}$. In (6.50), ρ_{QKD}^m denotes the density operator corresponding to the QKD scheme. In other words, by performing the purification of two density states, the fidelity improves. Given that the trace distance and the fidelity are related by [6, 61]

$$\left\| \rho_{\text{QKD}}^m - \rho_{\text{perfect}}^m \right\|_1 \leq 2\sqrt{1 - F\left(\rho_{\text{QKD}}^m, \rho_{\text{perfect}}^m\right)}, \tag{6.51}$$

we conclude that the fidelity-based ε-security, given by Eq. (6.48), is related to the trace-distance-based ε-security since

$$\frac{1}{2}\left\| \rho_{\text{QKD}}^m - \rho_{\text{perfect}}^m \right\|_1 \leq \sqrt{\varepsilon''}, \tag{6.52}$$

and the fidelity-based ε-security definition is, therefore, also composable.

If fidelity is high, then there exists a positive integer s such that $F^2\left(\rho_{AB}^m, |B_{00}\rangle^{\otimes m}\right) = {}^{\otimes m}\langle B_{00}|\rho_{AB}^m|B_{00}\rangle^{\otimes m} > 1 - 2^{-s}$. Clearly, the largest eigenvalue of ρ_{AB}^m must be larger than $1 - 2^{-s}$. The von Neumann entropy of ρ_{AB}^m is upper bounded by the entropy of the density operator [6]:

$$\rho_{\max} = \mathrm{diag}\left(1 - 2^{-s}, \underbrace{2^{-s}/\left(2^{2m} - 1\right), \cdots, 2^{-s}/\left(2^{2m} - 1\right)}_{2^{2m}-1 \text{ entries}}\right) \tag{6.53}$$

given by

$$S_{\max} = S(\rho_{\max}) = -\left(1 - 2^{-s}\right)\log\left(1 - 2^{-s}\right) - \left(2^{2n} - 1\right)\frac{2^{-s}}{2^{2m} - 1}\log\left(\frac{2^{-s}}{2^{2m} - 1}\right). \tag{6.54}$$

By using the approximation $\ln(1 + x) \approx x$, when x is small, we obtain that

$$S_{\max} \le \left(1 - 2^{-s}\right)2^{-s}/\ln 2 + 2^{-s}s + 2^{-s}2m = (2m + s + 1/\ln 2)2^{-s} - \underbrace{2^{-2s}/\ln 2}_{O(2^{-2s})}. \tag{6.55}$$

In conclusion, the high fidelity implies low von Neuman entropy since

$$S\left(\rho_{AB}^{m}\right) < S_{\max} \le (2m + s + 1/\ln 2)2^{-s} - O\left(2^{-2s}\right). \tag{6.56}$$

Therefore, by ensuring that Alice and Bob possess the EPR pairs of fidelity at least $1 - 2^{-s}$, the QKD protocol is secure [6].

6.4.3 Secure-Key Rates for 2-D DV-QKD Systems

The bound for two-dimensional (2-D) DV-QKD schemes, assuming individual attacks, is given by Eqs. (6.32) and (6.33). The mutual information between Alice and Bob can be determined by employing the binary symmetric channel (BSC), in which the probability of error q, in DV-QKD commonly referred to as the quantum bit error rate (QBER), the mutual information is determined by $I(A;B) = 1 - h(q)$. To account for the imperfect information reconciliation step, or equivalent error control coding (ECC), the ECC information leakage to Eve, denoted by $\mathrm{leakage}_{\mathrm{ECC}}(q)$ [46], is clearly lower bounded by $h(q)$, and we can rewrite the SKR expression as follows:

$$SKR = R_{\mathrm{raw}}\left[1 - \mathrm{leakage}_{\mathrm{ECC}}(q) - I_E\right],\ I_E = \min_{\text{Eve's strategies}}(I_{EA}, I_{EB}). \tag{6.57}$$

The expression similar to this has been derived in [65–68]. We assume that Eve can gain information only for non-empty time slots, provided that Bob detects what Eve has resent to him. Given that Alice's source can be represented as sending a pulse containing n-photons with probability $p_{A,n}$, we can apply the de Finetti theorem [61] and we can decompose Eve's information I_E in terms of I_E for each photon number as follows:

$$I_E = \min(I_{EA}, I_{EB}) = \min_{\text{Eve's strategies}} \left\{ \max\left(\sum\nolimits_n I_{EA,n} y_n\right), \max\left(\sum\nolimits_n I_{EB,n} y_n\right) \right\}, \tag{6.58}$$

where $y_n = R_{\text{raw},n}/R_{\text{raw}}$ and represents the fraction of the total raw key rate when Alice used n-photons per pulse.

6.4.3.1 Prepare-and-Measure BB84 Protocol

In prepare-and-measure BB84 protocol, $I_{EA} = I_{EB}$, and expression (6.58) simplifies to $I_E = \max\left(\sum_n I_{EA,n} y_n\right)$. In zero-photon case, clearly $I_{EA,0} = 0$. For single-photon case, the only way Eve can gain some information is by employing the intercept-resent (IR) attack, which introduces the error rate of q_1, and corresponding Eve's information will be $I_{EA,1} = h(q_1)$. For number of photons per pulse being two and above, Eve employs the PNS attack, and her information in this case is equal to 1, that is, $I_{EA,2} = 1$, $n \geq 2$. Clearly, based on this discussion, we conclude that Eve's information will be

$$\begin{aligned} I_E &= \max_{\text{Eve's strategies}} \left(\sum\nolimits_n I_{EA,n} y_n\right) = \max_{\text{Eve's strategies}} \left[y_1 h(q_1) + y_2\right] \\ &= \max_{\text{Eve's strategies}} \left[y_1 h(q_1) + 1 - y_0 - y_1\right] = 1 - \min_{\text{Eve's strategies}} \left\{y_0 + y_1\left[1 - h(q_1)\right]\right\}. \end{aligned} \tag{6.59}$$

In prepare-and-measure 2-D QKD schemes, such as BB84, the only parameters being measured are raw key rate $R_{\text{raw}} = \sum_n R_{\text{raw},n}$ and the average quantum bit error rate (QBER) $q = \sum_n q_n y_n$. In that case, we need to assume that $q_n = 0$, $n \geq 2$, so that $q_1 = q/y_1$. Further, in Eq. (6.31) it makes sense to set $p_0 = 0$ and $p_n = 0$, $n \geq 2$ so we can write

$$R_{\text{raw}} = R_{\text{raw},1} + R_{\text{raw},2} = R_{\text{raw},1} + R_s p_{\text{sift}} p_{A,n\geq 2}, \tag{6.60}$$

and after dividing both sides of equation with R_{raw} we obtain the following solution for $y_1 = R_{\text{raw},1}/R_{\text{raw}}$ [68]:

$$y_1 = 1 - (R_s/R_{\text{raw}}) p_{\text{sift}} p_{A,n\geq 2}. \tag{6.61}$$

For this y_1, Eq. (6.59) becomes

$$I_E = 1 - y_1\left[1 - h(q_1)\right] = 1 - y_1\left[1 - h(q/y_1)\right]. \tag{6.62}$$

Finally, after substituting such derived I_E into Eq. (6.57), we obtain the following equation for the SKR:

$$SKR = R_{\text{raw}}\left[y_1\left[1 - h(q/y_1)\right] - \text{leakage}_{\text{ECC}}(q)\right]. \tag{6.63}$$

6.4.3.2 Decoy-State-Based Protocol

In the rest of this section, we study the *2-D QKD employing the decoy states* [35, 69–76]. In decoy-state-based protocols, Alice randomly changes the nature of the quantum signal, such as the intensity of the laser. At the end of transmission process, she reveals which intensities she used so that Eve cannot adapt her attack on the fly. In the postprocessing stage, Alice and Bob use this information for parameters' estimation. Alice and Bob can maximize the secret fraction by optimizing over both signal levels and probability of occurrence of each level. In practice, even in the presence of atmospheric turbulence effects, three levels are sufficient as shown in [77]. Let p be some parameters of the source that can be adjusted, such as the mean photon number μ (or the intensity) of the laser source. Alice randomly changes the value of parameter p, from pulse to pulse, and at the end of raw key transmission phase she reveals the list of values of $p \in \mathcal{P}$. Alice and Bob sort out the available data and perform the parameter estimation. Alice and Bob measure $2|\mathcal{P}|$ parameters $R_{\text{raw}}^{(p)}$ and $q^{(p)}$ $(p = 1, \ldots, |\mathcal{P}|)$. The set of parameters $\mathcal{P}$ becomes publicly known; nevertheless, for $|\mathcal{P}| > 1$, Eve cannot adapt her measurement strategy based on p since she does not know it in advance. Evidently, p_n and q_n are independent on p. The raw key rate when Alice sends a pulse with n-photons becomes $R_{\text{raw}}^{(p)} = R_s p_{\text{sift}} P_{A,n}(p) p_n$. The measured parameters then become

$$R_{\text{raw}}^{(p)} = \sum_{n\geq 0} R_{\text{raw},n}^{(p)}, \quad q^{(p)} = \sum_{n\geq 0} q_n^{(p)} y_n^{(p)}; \; p = 1, 2, \cdots, |\mathcal{P}| \tag{6.64}$$

The equation above introduces a linear system of $2|\mathcal{P}|$ equations for p_n and q_n. As mentioned above, in practice, the decoy-state protocol with $|\mathcal{P}| = 3$ performs comparably to ideal one [76, 77]. For each p, Eve's information is given by

$$I_E^{(p)} = 1 - y_0^{(p)} - y_1^{(p)}\left[1 - h(q_1)\right]; \; y_n^{(p)} = R_{\text{raw},n}^{(p)}/R_{\text{raw}}^{(p)}, \; n = 0, 1. \tag{6.65}$$

The total SKR can be found by summing up the SKRs for different values of p, that is, $SKR = \sum_p SKR^{(p)}$. When the classical postprocessing has been applied on the whole raw key, the SKR simplifies to

$$SKR = R_{\text{raw}}\left[1 - \text{leakage}_{\text{ECC}}(q)\right] - \sum_p R_{\text{raw}}^{(p)} I_E^{(p)}. \tag{6.66}$$

6.5 Quantum Optics Fundamentals

In this section, we briefly describe the quantum optics fundamentals [7, 30, 78–80] to facilitate the description of CV-QKD schemes. Monochromatic plane wave can be represented as

$$\boldsymbol{E}(\boldsymbol{r}, t) = \boldsymbol{e}|\boldsymbol{E}|\big[I \cos(\boldsymbol{kr} - \omega t) - Q \sin(\boldsymbol{kr} - \omega t)\big], \tag{6.67}$$

wherein the first component represents in-phase, while the second one represents the quadrature component. The vector $\boldsymbol{e}$ denotes the polarization orientation, $\boldsymbol{k}$ is the wave vector, and $\boldsymbol{r}$ is the position vector.

6.5.1 Quadrature Operators, Creation and Annihilation Operators, Uncertainty Principle

In quantum optics, the quadrature components I and Q are replaced with corresponding operators $\hat{I}$ and $\hat{Q}$, equivalent to the position and momentum operators, satisfying a similar commutation relationship:

$$\left[\hat{I}, \hat{Q}\right] = 2jN_0, \tag{6.68}$$

where N_0 represents the variance of vacuum fluctuation (shot noise). (Very often in quantum optics, a different notation is used for quadrature operators, namely, $\hat{x}$ and $\hat{p}$ are used instead.) The eigenstates of $\hat{I}$ can be denoted by $\langle I|$ since $\hat{I}|I\rangle = I|I\rangle$. Annihilation a and creation $a^\dagger$ operators are defined as

$$a = \frac{\hat{I} + j\hat{Q}}{2\sqrt{N_0}}, \quad a^\dagger = \frac{\hat{I} - j\hat{Q}}{2\sqrt{N_0}}. \tag{6.69}$$

In other words, we can express the quadrature operators in terms of annihilation and creation operators as follows:

$$\hat{I} = \left(a^\dagger + a\right)\sqrt{N_0}, \; \hat{Q} = j\left(a^\dagger - a\right)\sqrt{N_0}. \tag{6.70}$$

The *photon number operator* was introduced earlier by

$$\begin{aligned} N = a^\dagger a = \frac{\hat{I} + j\hat{Q}}{2\sqrt{N_0}} \frac{\hat{I} - j\hat{Q}}{2\sqrt{N_0}} &= \frac{1}{4N_0}\left[\left(\hat{I}^2 + \hat{Q}^2\right) - j\underbrace{(\hat{I}\hat{Q} - \hat{Q}\hat{I})}_{\left[\hat{I},\hat{Q}\right]=2N_0 j}\right] \\ &= \frac{1}{4N_0}\left[\left(\hat{I}^2 + \hat{Q}^2\right) + 2N_0\right]. \end{aligned} \tag{6.71}$$

Fock (photon number) states are eigenkets of the photon number operator:

$$N|n\rangle = n|n\rangle, \tag{6.72}$$

where n represents the number of photons.

The *vacuum state*, denoted as $|0\rangle$, is an eigenket of the annihilation operator:

$$a|0\rangle = 0 \;\Rightarrow\; \langle 0|a^{\dagger}a|0\rangle = 0. \tag{6.73}$$

A coherent state $|\alpha\rangle$, introduced earlier, is the right eigenket of the *annihilation operator* a, namely, $a\,|\alpha\rangle = \alpha\,|\alpha\rangle$, where α is the complex eigenvalue. The expected values of quadrature operators with respect to $|\alpha\rangle$ are given by

$$\begin{aligned}\left\langle \hat{I}\right\rangle_{\alpha} &= \langle\alpha|\hat{I}|\alpha\rangle = \langle\alpha|a + a^{\dagger}|\alpha\rangle\sqrt{N_0} = \langle\alpha|\alpha + \alpha^{*}|\alpha\rangle\sqrt{N_0} = 2\mathrm{Re}\{\alpha\}\sqrt{N_0},\\ \left\langle \hat{Q}\right\rangle_{\alpha} &= \langle\alpha|\hat{Q}|\alpha\rangle = j\langle\alpha|a^{\dagger} - a|\alpha\rangle\sqrt{N_0} = j\langle\alpha|\alpha^{*} - \alpha|\alpha\rangle\sqrt{N_0} = 2\mathrm{Im}\{\alpha\}\sqrt{N_0}.\end{aligned} \tag{6.74}$$

So if we apply the scaling factor $1/(2\sqrt{N_0})$ on α, the expected values of in-phase and quadrature operators will represent the real and imaginary parts of α. For the vacuum state, the expected values of quadrature operators are equal to zero, and therefore Eq. (6.74) defines the displacement. The expected values of $\hat{I}^2$ and $\hat{Q}^2$ are given by

$$\begin{aligned}\left\langle \hat{I}^2\right\rangle_{\alpha} &= \langle\alpha|\hat{I}^2|\alpha\rangle = \langle\alpha|(a + a^{\dagger})(a + a^{\dagger})|\alpha\rangle N_0 = \left[(\alpha + \alpha^{*})^2 + 1\right]N_0,\\ \left\langle \hat{Q}^2\right\rangle_{\alpha} &= \langle\alpha|\hat{Q}^2|\alpha\rangle = j^2\langle\alpha|(a^{\dagger} - a)(a^{\dagger} - a)|\alpha\rangle N_0 = \left[1 - (\alpha^{*} - \alpha)^2\right]N_0.\end{aligned} \tag{6.75}$$

The corresponding variances for the quadrature operators will be then

$$\begin{aligned}\left\langle \Delta\hat{I}^2\right\rangle_{\alpha} &= \left\langle \hat{I}^2\right\rangle_{\alpha} - \left[\left\langle \hat{I}\right\rangle_{\alpha}\right]^2 = \left[(\alpha + \alpha^{*})^2 + 1\right]N_0 - (\alpha + \alpha^{*})^2 N_0 = N_0,\\ \left\langle \Delta\hat{Q}^2\right\rangle_{\alpha} &= \left\langle \hat{Q}^2\right\rangle_{\alpha} - \left[\left\langle \hat{Q}\right\rangle_{\alpha}\right]^2 = \left[1 - (\alpha^{*} - \alpha)^2\right]N_0 + (\alpha^{*} - \alpha)^2 N_0 = N_0,\end{aligned} \tag{6.76}$$

indicating the variances of fluctuations for both quadratures are the same. By setting $\alpha = 0$, the same conclusion applies to the vacuum state. Interestingly enough, the product of variances of quadratures is given by

$$\left\langle \Delta\hat{I}^2\right\rangle_{\alpha}\left\langle \Delta\hat{Q}^2\right\rangle_{\alpha} = N_0^2, \tag{6.77}$$

indicating that the Heisenberg uncertainty principle is satisfied with equality sign. In other words, the coherent state satisfies the minimum uncertainty relation. Uncertainty principle is employed in CV-QKD systems. Very often, the uncertainty principle is expressed in terms of the *shot-noise unit* (SNU), where N_0 is normalized to 1, as follows $\left\langle \Delta\hat{I}^2\right\rangle\left\langle \Delta\hat{Q}^2\right\rangle \geq 1$.

6.5.2 Coherent States, Gaussian State, and Squeezed States

The *Gaussian state* is coherent state for which the projections along eigenkets of $\hat{I}$ have the Gaussian shape:

$$\langle I|\alpha\rangle = \frac{1}{(2\pi N_0)^{1/4}} e^{-\frac{\left(I-\langle\hat{I}\rangle\right)^2 + j\langle\hat{I}\rangle\langle\hat{Q}\rangle}{4N_0} + \frac{jI\langle\hat{Q}\rangle}{2N_0}}. \tag{6.78}$$

A *squeezed state* is a particular Gaussian state which has unequal fluctuations in quadratures:

$$\begin{aligned} \langle I|\alpha, s\rangle &= \frac{1}{(2\pi N_0 s)^{1/4}} e^{-\frac{s^{-1}\left(I-\langle\hat{I}\rangle\right)^2 + j\langle\hat{I}\rangle\langle\hat{Q}\rangle}{4N_0} + \frac{jI\langle\hat{Q}\rangle}{2N_0}}, \quad s > 0 \\ V\left(\hat{I}\right) &= \mathrm{Var}\left(\hat{I}\right) = sN_0, \quad V\left(\hat{Q}\right) = \mathrm{Var}\left(\hat{Q}\right) = s^{-1}N_0, \end{aligned} \tag{6.79}$$

where s is the squeezing parameter.

To visualize the quantum states in quantum optics, the Wigner function is used, which is for density state ρ defined as [7]

$$W(I,Q) = \frac{1}{4\pi N_0}\int e^{j\frac{\xi Q}{2N_0}}\langle I-\xi/2|\rho|I+\xi/2\rangle d\xi. \tag{6.80}$$

If someone measures the in-phase component, the result will follow the probability density function (PDF) $f(I)$ obtained by averaging the Wigner function over the quadrature component, that is, $f(I) = \int_Q W(I,Q)dQ$. On the other hand, if someone measures the quadrature component, the result will follow the following PDF $f(Q) = \int_I W(I,Q)dI$. Therefore, we can interpret the Wigner function of the coherent state $|\alpha\rangle$ as the 2-D Gaussian distribution centered at (Re{α}, Im{α}) with variance being N_0 (and zero covariance). Therefore, in the (I, Q)-plane, the coherent state is represented by the circle, with radius related to the standard deviation (uncertainty), which is illustrated in Fig. 6.13. For the squeezed states, the I-coordinate is scaled by factor $s^{1/2}$, while the Q-coordinate by factor $s^{-1/2}$.

Let us now represent the coherent state in terms of number states. By using the completeness relationship, we can write

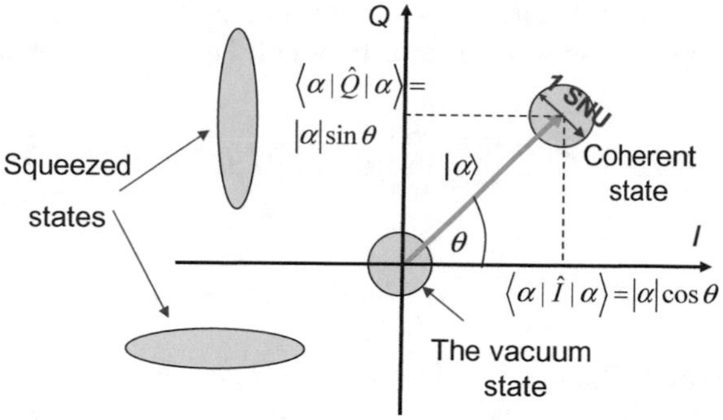

Fig. 6.13 The illustration of coherent and squeezed states

$$|\alpha\rangle = \sum\nolimits_n (|n\rangle\langle n|)|\alpha\rangle = \sum\nolimits_n |n\rangle\langle n|\alpha\rangle. \tag{6.81}$$

Since the number state $|n\rangle$ can be represented in terms of ground state by [81]

$$|n\rangle = \left(a^\dagger\right)^n |0\rangle/\sqrt{n!}, \tag{6.82}$$

the projection coefficient can be determined as

$$\langle n|\alpha\rangle = \langle 0| \underbrace{a^n|\alpha\rangle}_{\alpha^n|\alpha\rangle} /\sqrt{n!} = \langle 0|\alpha\rangle \alpha^n/\sqrt{n!}, \tag{6.83}$$

and the coherent state after substituting (6.83) into (6.81) becomes

$$|\alpha\rangle = \langle 0|\alpha\rangle \sum\nolimits_n \frac{\alpha^n}{\sqrt{n!}}|n\rangle. \tag{6.84}$$

The projection along the ground state $\langle 0|\alpha\rangle$ can be determined from the normalization condition:

$$1 = \sum\nolimits_n |\langle n|\alpha\rangle|^2 = \sum\nolimits_n \langle 0|\alpha\rangle \frac{\alpha^n}{\sqrt{n!}} \langle \alpha|0\rangle \frac{\alpha^{*n}}{\sqrt{n!}} = |\langle 0|\alpha\rangle|^2 \underbrace{\sum\nolimits_n \frac{\left(|\alpha|^2\right)^n}{n!}}_{e^{|\alpha|^2}} = |\langle 0|\alpha\rangle|^2 e^{|\alpha|^2}, \tag{6.85}$$

and by solving for $\langle 0|\alpha\rangle$ we obtain

$$\langle 0|\alpha\rangle = e^{-|\alpha|^2/2}, \tag{6.86}$$

and after substituting (6.86) into (6.84), the coherent state gets represented as

$$|\alpha\rangle = e^{-|\alpha|^2/2} \sum\nolimits_n \frac{\alpha^n}{\sqrt{n!}}|n\rangle. \tag{6.87}$$

Another convenient form can be obtained by expressing the number operator in terms of the ground state, based on Eq. (6.82), to obtain

$$|\alpha\rangle = e^{-|\alpha|^2/2}\sum\nolimits_n \frac{\alpha^n}{\sqrt{n!}}\left(a^\dagger\right)^n|0\rangle/\sqrt{n!} = e^{-|\alpha|^2/2}\underbrace{\sum\nolimits_n \frac{\left(\alpha a^\dagger\right)^n}{n!}}_{\exp(\alpha a^\dagger)}|0\rangle = e^{-\frac{|\alpha|^2}{2}+\alpha a^\dagger}|0\rangle$$

$$= e^{\alpha a^\dagger - \alpha^* a}|0\rangle = D(\alpha)|0\rangle,\ D(\alpha) = e^{\alpha a^\dagger - \alpha^* a}, \tag{6.88}$$

where $D(\alpha)$ is the *displacement operator*. The displacement operator is used to describe one coherent state as the displacement of another coherent state in the I–Q diagram.

The squeezed coherent state $|\alpha, s\rangle$ can be now represented in terms of the ground state as follows:

$$|\alpha, s\rangle = D(\alpha)S(s)|0\rangle,\ S(s) = e^{-\frac{1}{2}\left(sa^{\dagger 2} - s^* a^2\right)}, \tag{6.89}$$

where $S(s)$ is the *squeezing operator*. The squeezing parameter s in quantum optics is typically a real number, so that squeezing operator gets simplified to $S(s) = \exp\left[-s(a^{\dagger 2} - a^2)/2\right]$. By applying the squeezing operator on the vacuum state, we obtain the *squeezed vacuum state* [80, 82]:

$$|0, s\rangle = (\cosh s)^{-1/2}\sum_{n=0}^{\infty}\frac{\sqrt{(2n)!}}{2^n n!}\tanh s^n|2n\rangle. \tag{6.90}$$

When a nonlinear crystal is pumped with the bright laser beam of frequency 2ω, the pump protons will be split into pairs of photons of frequency ω. Whenever the phase-matching condition for the degenerate optical parametric amplifier (OPA) is satisfied, the output will be a superposition of even number states $|2n\rangle$ [78, 79, 80], which is consistent with the equation above. In the Heisenberg picture, the annihilation operator gets transformed by the linear unitary Bogoliubov transformation as follows $a \to b = (\cosh s)a - (\sinh s)a^\dagger$. On the other hand, the quadrature operators $\hat{\boldsymbol{x}} = \left(\hat{x}_1\, \hat{x}_2\right)^{\mathrm{T}} = \left(\hat{I}\ \hat{Q}\right)^{\mathrm{T}}$ get transformed to [80]

$$\hat{\boldsymbol{x}} \to \hat{\boldsymbol{x}}_{\text{out}} = \underbrace{\begin{bmatrix} e^{-s} & 0 \\ 0 & e^s \end{bmatrix}}_{\boldsymbol{S}(s)}\hat{\boldsymbol{x}} = \boldsymbol{S}(s)\hat{\boldsymbol{x}},\ \ \boldsymbol{S}(s) = \begin{bmatrix} e^{-s} & 0 \\ 0 & e^s \end{bmatrix}. \tag{6.91}$$

For the quadrature operators, we can define the *covariance matrix* $\boldsymbol{\Sigma} = \left(\Sigma_{ij}\right)_{2\times 2}$ elements as follows [80]:

$$\Sigma_{ij} = \frac{1}{2}\left\langle\left\{\Delta\hat{x}_i, \Delta\hat{x}_j\right\}\right\rangle = \frac{1}{2}\left(\left\langle\hat{x}_i\hat{x}_j\right\rangle + \left\langle\hat{x}_j\hat{x}_i\right\rangle\right) - \left\langle\hat{x}_i\right\rangle\left\langle\hat{x}_j\right\rangle,\ \ \Delta\hat{x}_i = \hat{x}_i - \left\langle\hat{x}_i\right\rangle, \tag{6.92}$$

where $\{\cdot,\cdot\}$ denotes the anticommutator. The diagonal elements of the covariance matrix are clearly the variances of the quadratures:

$$\Sigma_{ii} = V(\hat{x}_i) = \left\langle (\Delta \hat{x}_i)^2 \right\rangle = \langle \hat{x}_i^2 \rangle - \langle \hat{x}_i \rangle^2. \tag{6.93}$$

The covariance matrix for arbitrary number of Gaussian states can easily be generalized. The covariance matrix for the squeezed vacuum state is given by $\boldsymbol{V} = \boldsymbol{S}(s)\boldsymbol{S}(s)^T = \boldsymbol{S}(2s)$ and has different variances for the quadratures, with one of the variances being squeezed below the quantum shot noise and the other one being antisqueezed above the quantum shot noise.

6.5.3 EPR State and Manipulation of Photon States

When we pump a nonlinear crystal in the nondegenerate regime, the crystal will generate pairs of photons in two different modes, commonly referred to as the signal and the idler. This process is known as the *spontaneous parametric down-conversion (SPDC)* [8, 78–80]. By sending the photons at frequency ω_p (pump signal) into a nonlinear crystal, such KH_2PO_4, beta-barium borate (BBO), or periodically poled LiNbO3 (PPLN) crystal, we generate the photon pairs at frequencies ω_s (signal) and ω_i (idler) satisfying the energy conservation principle $\hbar(\omega_s + \omega_i) = \hbar\omega_p$ and momentum conservation principle: $\boldsymbol{k}_s + \boldsymbol{k}_i = \boldsymbol{k}_p$, which is illustrated in Fig. 6.14. The nonlinear crystal is used to split photons into pairs of photons that satisfy the law of conservation of energy, have combined energies and momenta equal to the energy and momentum of the original photon, phase matched in the frequency domain, and have correlated polarizations. If the photons have the same polarization they belong to Type I correlation; otherwise, they have perpendicular polarizations and belong to Type II. As the SPDC is stimulated by random vacuum fluctuations, the photon pairs are created at random time instances. The output of a Type I down-converter is known as a squeezed vacuum and it contains only even number of photon number. The output of the Type II down-converter is known as a two-mode squeezed vacuum (TMSV) state, which is described below. Alternatively, the highly nonlinear fiber (HNLF) can also be used as an SPDC source [83, 84]. HLNF employs the four-wave mixing (FWM) effect [44, 85]. The Hamiltonian describing this system will contain the bilinear term $a^\dagger b^\dagger$. The corresponding Gaussian unitary operator is known as the *two-mode squeezing operator*:

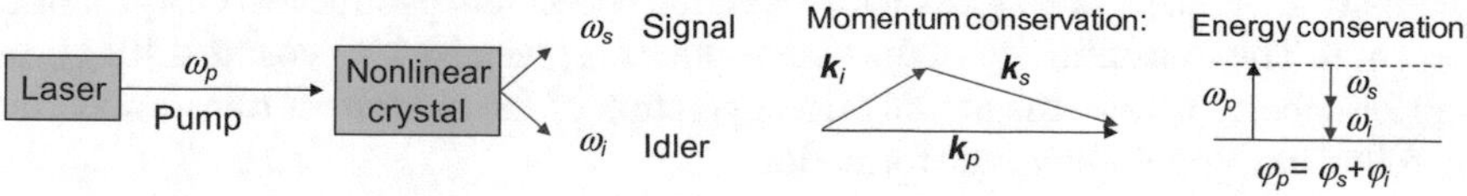

Fig. 6.14 The illustration of SPDC process

$$S_2(s) = \exp\left[-s\left(a^\dagger b^\dagger - ab\right)/2\right], \tag{6.94}$$

where s quantifies the two-mode squeezing.

In the Heisenberg picture, the quadrature operators $\hat{\boldsymbol{x}} = \left(\hat{x}_1\, \hat{x}_2\, \hat{x}_3\, \hat{x}_4\right)^{\mathrm{T}} = \left(\hat{I}_a\ \hat{Q}_a\ \hat{I}_b\ \hat{Q}_b\right)^{\mathrm{T}}$ get transformed to [80]

$$\hat{\boldsymbol{x}} \to \boldsymbol{S}_2(s)\hat{\boldsymbol{x}},\ \ \boldsymbol{S}_2(s) = \begin{bmatrix} \cosh s\,\mathbf{1} & \sinh s\,\boldsymbol{Z} \\ \sinh s\,\boldsymbol{Z} & \cosh s\,\mathbf{1} \end{bmatrix}, \tag{6.95}$$

where $\mathbf{1}$ is the identity matrix and $\boldsymbol{Z} = \mathrm{diag}(1, -1)$ is the Pauli Z-matrix. When we apply $S_2(s)$ to a couple of the vacuum states, we obtain the TMSV state, commonly known as the EPR state $\rho^{\mathrm{EPR}} = |s\rangle\langle s|_{\mathrm{EPR}}$, where [80]

$$|s\rangle_{EPR} = \sqrt{1-\lambda^2}\sum_{n=0}^{\infty}\lambda^n|n\rangle_a|n\rangle_b = \sqrt{\frac{2}{v+1}}\sum_{n=0}^{\infty}\left(\frac{v-1}{v+1}\right)^{n/2}|n\rangle_a|n\rangle_b,\ \lambda^2 = \frac{v-1}{v+1}, \tag{6.96}$$

with $\lambda = \tanh(s)$ and $v = \cosh(2s)$ representing the variance of the quadratures. The EPR state represent the Gaussian state with zero-mean and covariance matrix given by

$$\boldsymbol{V}_{\mathrm{EPR}} = \begin{bmatrix} v\mathbf{1} & \sqrt{v^2-1}\boldsymbol{Z} \\ \sqrt{v^2-1}\boldsymbol{Z} & v\mathbf{1} \end{bmatrix},\ \ v = \cosh 2s, \tag{6.97}$$

wherein the elements of the covariance matrix are determined by applying the definition Eqs. (6.92) and (6.93) to the EPR state. The EPR state (6.96) is directly applicable to the CV-QKD systems.

The most commonly used devices to manipulate the photon states are [6, 8]: (i) mirrors, which are used to change the direction propagation; (ii) phase shifters, which are used to introduce a given phase shift; and (iii) beam splitters, which are used to implement various quantum gates. The *phase shifter* is a slab of transparent medium, say borosilicate glass, with index of refraction n being higher than that of the air and it is used to perform the following operation:

$$|\psi_{\mathrm{out}}\rangle = \begin{bmatrix} e^{j\phi} & 0 \\ 0 & 1 \end{bmatrix}|\psi\rangle;\ \ \phi = kL,\ k = n\omega/c. \tag{6.98}$$

The corresponding equivalent scheme is shown in Fig. 6.15. The *beam splitter* is a partially silvered piece of glass with reflection coefficient parameterized as follows $R = \cos\theta$. The transmittivity of the beam splitter is given by $T = \cos^2\theta \in [0, 1]$. In the Heisenberg picture, the annihilation operators of the modes get transformed by the following Bogoliubov transformation:

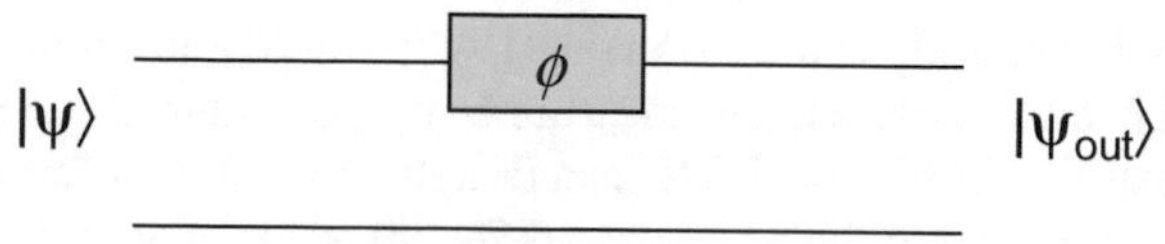

Fig. 6.15 The equivalent scheme of the phase shifter

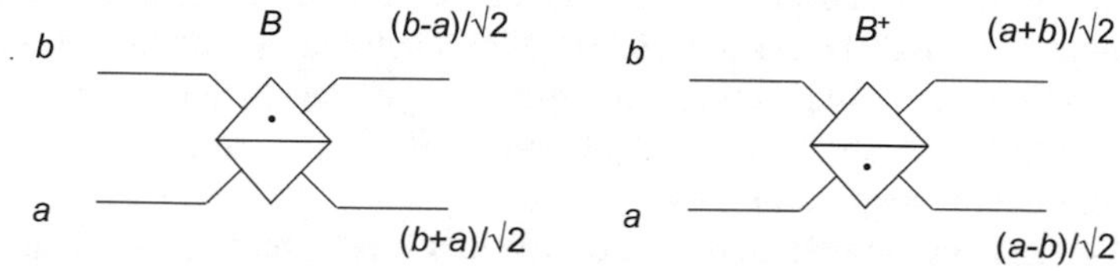

Fig. 6.16 Illustrating the operation principle of 50/50 beam splitter (B) and its Hermitian conjugate ($B^{\dagger}$)

$$\begin{bmatrix} a \\ b \end{bmatrix} \to \begin{bmatrix} a_{\text{out}} \\ b_{\text{out}} \end{bmatrix} = \begin{bmatrix} \sqrt{T} & \sqrt{1-T} \\ -\sqrt{1-T} & \sqrt{T} \end{bmatrix} \begin{bmatrix} a \\ b \end{bmatrix} = \begin{bmatrix} \cos\theta & \sin\theta \\ -\sin\theta & \cos\theta \end{bmatrix} \begin{bmatrix} a \\ b \end{bmatrix}. \tag{6.99}$$

The 50:50 beam splitter is obtained for $\theta = \pi/4$, and its implementation is shown in Fig. 6.16.

The beam splitter acts on two modes, which can be described by the creation (annihilation) operators a ($a^{\dagger}$) and b ($b^{\dagger}$). The corresponding Hamiltonian is given by

$$\hat{H}_{BS} = j2\theta\left(a^{\dagger}b - ab^{\dagger}\right). \tag{6.100}$$

The action of beam splitter can be represented by the evolution operation as

$$BS(\theta) = e^{-j\hat{H}_{BS}/2} = e^{\theta\left(a^{\dagger}b - ab^{\dagger}\right)}. \tag{6.101}$$

The quadrature operators of the modes $\hat{\boldsymbol{x}} = \left(\hat{x}_1\, \hat{x}_2\, \hat{x}_3\, \hat{x}_4\right)^{\mathrm{T}} = \left(\hat{I}_a\, \hat{Q}_a\, \hat{I}_b\, \hat{Q}_b\right)^{\mathrm{T}}$ get transformed by the following symplectic transformation:

$$\hat{\boldsymbol{x}} \to \hat{\boldsymbol{x}}_{\text{out}} = \begin{bmatrix} \sqrt{T}\mathbf{1} & \sqrt{1-T}\mathbf{1} \\ -\sqrt{1-T}\mathbf{1} & \sqrt{T}\mathbf{1} \end{bmatrix} \hat{\boldsymbol{x}}. \tag{6.102}$$

6.6 Continuous Variable (CV)-QKD Protocols

The CV-QKD can be implemented by employing either homodyne detection, where only one quadrature component is measured at a time (because of the uncertainty principle) , or with heterodyne detection (HD), where one beam splitter (BS) and two balanced photodetectors are used to measure both quadrature components simultaneously, as illustrated in Fig. 6.17. HD can double the mutual information between Alice and Bob compared to the homodyne detection scheme at the expense of additional 3 dB loss of the beam splitter (BS). In order to reduce the laser phase noise, the quantum signals are typically co-propagated together with the time-domain multiplexed high-power pilot tone (PT) to align Alice's and Bob's measurement bases. To implement CV-QKD, both squeezed states and coherent states can be employed. CV-QKD studies are getting momentum, by judging by increasing number of papers related to CV-QKD [7, 86, 87, 88–101], thanks to their compatibility with the state-of-the-art optical communication technologies.

The CV-QKD system experiences the 3 dB loss limitation in transmittance when the direct reconciliation is used. To avoid this problem, either reverse reconciliation [86] or the postelection [96] methods are used. Different eavesdropping strategies discussed in Sect. 6.4.1 are applicable to CV-QKD systems as well. It has been shown that for Gaussian modulation, Gaussian attack is optimum attack for both individual attacks [102] and collective attacks [103, 104].

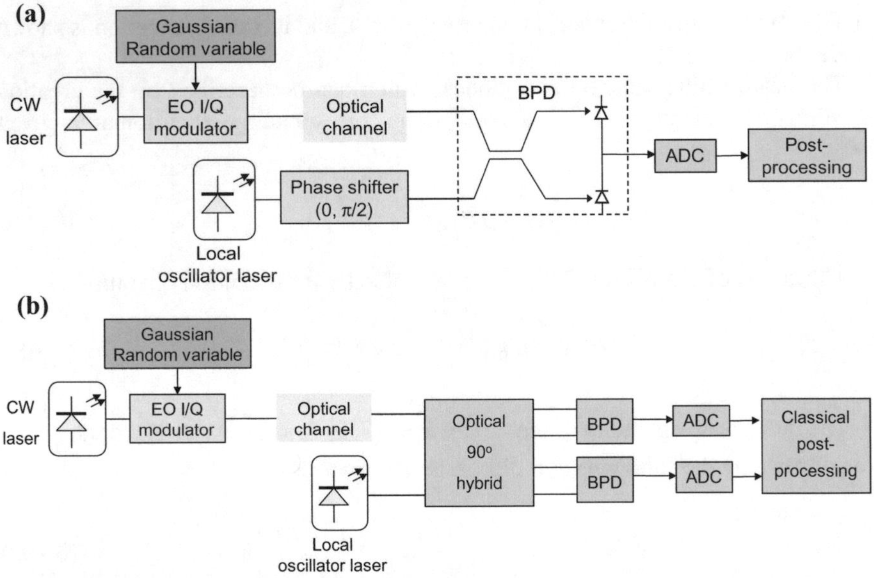

Fig. 6.17 Illustrating discrete Gaussian-modulation-based CV-QKD protocols: **a** homodyne-detection-based CV-QKD scheme, and **b** heterodyne-detection-based CV-QKD scheme. ADC: analog-to-digital conversion, BPD: balanced photodetection

6.6.1 *Squeezed State-Based Protocol*

Squeezed state-based protocols employ the Heisenberg uncertainty principle, claiming that it is impossible to measure both quadratures with complete precision. To impose the information, Alice randomly selects to use either in-phase or quadrature degree of freedom (DOF), as illustrated in Fig. 6.18(left). In Alice encoding rule I (when in-phase DOF is used), the squeezed state is imposed on the in-phase component (with squeezed parameter $s_I < 1$):

$$X_{A,I} \to \left|X_{A,I}, s_I\right\rangle, \ s_I < 1, \ X_{A,I} \sim \mathcal{N}\left(0, \sqrt{v_{A,I}N_0}\right) \tag{6.103}$$

The amplitude is selected from a zero-mean Gaussian source of variance $v_{A,I}N_0$. On the other hand, in Alice encoding rule Q (when the quadrature is used), the squeezed state is imposed on the quadrature (with squeezed parameter $s_Q > 1$) and we can write

$$X_{A,Q} \to \left|jX_{A,Q}, s_Q\right\rangle, \ s_Q > 1, \ X_{A,Q} \sim \mathcal{N}\left(0, \sqrt{v_{A,Q}N_0}\right) \tag{6.104}$$

The variances of squeezed states I and Q will be $V\left(\hat{I}\right) = \sigma_I^2 N_0 = s_I N_0$ and $V\left(\hat{Q}\right) = \sigma_Q^2 N_0 = N_0/s_Q$, respectively (as described above). On the receiver side, Bob randomly selects whether to measure either $\hat{I}$ or $\hat{Q}$. Alice and Bob exchange the encoding rules being used by them to measure the quadrature for every squeezed state and keep only instances when they measured the same quadrature in the sifting procedure. Therefore, this protocol is very similar to the BB84 protocol. After that, the information reconciliation takes place, followed by the privacy amplification.

From Eve's point of view, the corresponding mixed states when Alice employed either I or Q DOF will be [7]

$$\rho_I = \int f_N\left(0, \sqrt{v_{A,I}N_0}\right)|x, s_I\rangle\langle x, s_I|dx,$$

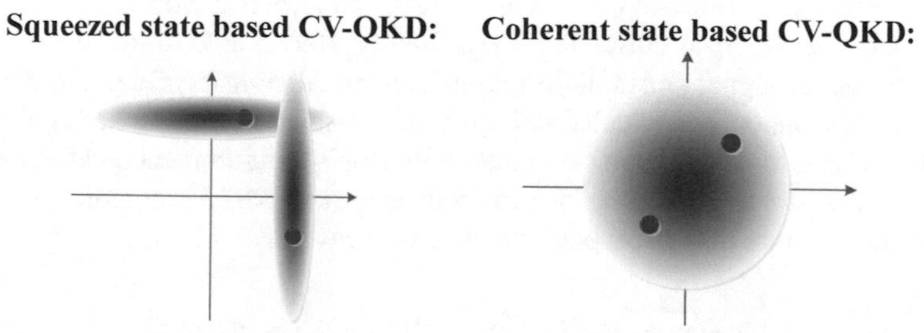

Fig. 6.18 Illustration of operation principles of squeezed (left) and coherent (right) state-based protocols

$$\rho_Q = \int f_N\left(0, \sqrt{v_{A,Q} N_0}\right) |jQ, s_Q\rangle\langle jQ, s_Q| dx, \tag{6.105}$$

where $f_N(0, \sigma)$ is zero-mean Gaussian distribution with variance σ^2. When the condition $\rho_I = \rho_Q$ is satisfied, Eve cannot distinguish whether Alice was imposing the information on the squeezed state I or Q. Based on Eq. (6.77), to ensure that Eve cannot distinguish between I and Q squeezed states, we require that

$$\left(v_{A,I} + \sigma_I^2\right)\sigma_Q^2 = 1, \left(v_{A,Q} + \sigma_Q^2\right)\sigma_I^2 = 1. \tag{6.106}$$

By dividing the left sides of both equations, we obtain

$$\frac{\left(v_{A,I} + \sigma_I^2\right)\sigma_Q^2}{\left(v_{A,Q} + \sigma_Q^2\right)\sigma_I^2} = 1 \Rightarrow 1 + \frac{v_{A,1}}{\sigma_I^2} = 1 + \frac{v_{A,Q}}{\sigma_Q^2}. \tag{6.107}$$

Given that squeezed states are much more difficult to generate than the coherent states, we focus our attention to the coherent state-based CV-QKD protocols.

6.6.2 Coherent State-Based Protocols

In coherent state-based protocols, there is only one encoding rule for Alice:

$$(X_{A,I}, X_{A,Q}) \to |X_{A,I} + jX_{A,Q}\rangle, \ X_{A,I} \sim \mathfrak{N}\left(0, \sqrt{v_A N_0}\right), \ X_{A,Q} \sim \mathfrak{N}\left(0, \sqrt{v_A N_0}\right) \tag{6.108}$$

In other words, Alice randomly selects a point in 2-D (I, Q) space from a zero-mean circular symmetric Gaussian distribution, as illustrated in Fig. 6.18(right). Clearly, both quadratures have the same uncertainty. Here we again employ the Heisenberg uncertainty, claiming that it is impossible to measure both quadratures with complete precision. On receiver side, Bob performs the random measurement on either $\hat{I}$ or $\hat{Q}$, and let the result of his measurement be denoted by Y_B. When Bob measures $\hat{I}$, his measurement Y_B is correlated $X_{A,I}$. On the other hand, when Bob measures $\hat{Q}$, his measurement result Y_B is correlated $X_{A,Q}$. Clearly, Bob is able to measure a single coordinate of the signal constellation point sent by Alice using Gaussian coherent state. Bob then announces which quadrature he measured in each signaling interval, and Alice selects the coordinate that agrees with Bob's measurement quadrature. The rest of the protocol is the same as for the squeezed state-based protocol.

From Eve's point of view, the mixed state will be

$$\rho = \int f_N\left(0, \sqrt{v_A N_0}\right) f_N\left(0, \sqrt{v_A N_0}\right) |I + jQ\rangle\langle I + jQ| dI dQ, \tag{6.109}$$

v_A is the Alice modulation variance and N_0 represents the variance of the coherent states fluctuations. For lossless transmission, the total variance of Bob's random variable will be

$$V(Y_B) = \mathrm{Var}(Y_B) = v_A N_0 + N_0 = (v_A + 1)N_0, \tag{6.110}$$

which is composed of Alice's modulation term $v_A N_0$ and the variance of intrinsic coherent state fluctuations N_0.

6.6.2.1 Lossy Channel Model and Mutual Information

For lossy transmission, the channel can be modeled as a beam splitter with attenuation $0 \leq T \leq 1$, as illustrated in Fig. 6.19. From this figure, we conclude that the output state can be represented in terms of input state $\boldsymbol{x}_{\mathrm{A}}$ and ground state $\boldsymbol{x}_0$ as follows:

$$x_B = \sqrt{T}x_A + \sqrt{1-T}x_0 = \sqrt{T}\left(x_A + \sqrt{\chi_{\mathrm{line}}}x_0\right), \; \chi_{\mathrm{line}} = (1-T)/T = 1/T - 1. \tag{6.111}$$

For lossy transmission when the Gaussian modulation of coherent states is employed instead of vacuum states, the previous expression needs to be modified as follows:

$$x_B = \sqrt{T}\left(x_A + \sqrt{\chi_{\mathrm{line}}}x_0\right), \; \chi_{\mathrm{line}} = (1-T)/T + \varepsilon = 1/T - 1 + \varepsilon, \, \varepsilon \geq 0. \tag{6.112}$$

The first term in χ_{line} is contributed to the channel attenuation, while the second term ε, commonly referred to as the excess noise, is due to other noisy factors such as the laser phase noise, imperfect orthogonality of quadratures, and so on. The total variance of Bob's random variable will be now composed of three terms

$$\begin{aligned} V(Y_B) &= \mathrm{Var}(Y_B) = T(v_A N_0 + N_0 + \chi_{\mathrm{line}} N_0) \\ &= T\Bigg(\underbrace{v_A + 1}_{v} + \chi_{\mathrm{line}}\Bigg)N_0 = T(v + \chi_{\mathrm{line}})N_0, \end{aligned}$$

where $v = v_{\mathrm{A}} + 1$.

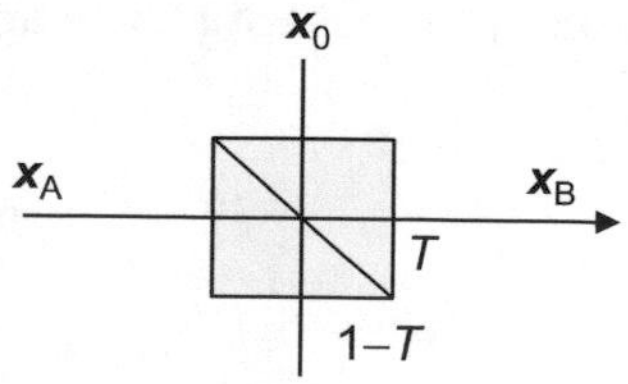

Fig. 6.19 The illustration of lossy transmission channel

Mutual information between Alice and Bob based on Shannon's formula can be calculated by

$$I(A;B) \doteq I(X_A;Y_B) = \frac{1}{2}\log\left(1 + \frac{T v_A N_0}{T(N_0 + \chi_{\text{line}} N_0)}\right)$$
$$= \frac{1}{2}\log\left(1 + \frac{v_A}{1 + \chi_{\text{line}}}\right). \tag{6.113}$$

The Eve's information, in CV-QKD schemes, is determined by reconciliation strategy.

6.6.2.2 Information Reconciliation

In *direct reconciliation*, the key is determined based on Alice random variable X_A (and Bob performs error correction on his sequence), so that the *secret fraction* is given by

$$r = I(X_A;Y_B) - I(X_A;Z). \tag{6.114}$$

Bob and Eve cannot both acquire all information about the key elements of the protocol; the variance of Bob on $\hat{I}$ and Eve on $\hat{Q}$ satisfies uncertainty relationship $\chi_{\text{line}}\chi_E \geq 1$ so that the mutual information between Alice and Eve is given by

$$I(A;E) = \frac{1}{2}\log(1 + \frac{v_A}{1 + \underbrace{\chi_E}_{1/\chi_{\text{line}}}}) = \frac{1}{2}\log\left(1 + \frac{v_A}{1 + 1/\chi_{\text{line}}}\right). \tag{6.115}$$

After substituting (6.113) and (6.115) into (6.114) we obtain

$$r = \frac{1}{2}\log\left(1 + \frac{v_A}{1 + \chi_{\text{line}}}\right) - \frac{1}{2}\log\left(1 + \frac{v_A}{1 + 1/\chi_{\text{line}}}\right) = \frac{1}{2}\log\left[\frac{1 + \chi_{\text{line}} + v_A}{1 + \chi_{\text{line}} + v_A\chi_{\text{line}}}\right]. \tag{6.116}$$

For $T = 1/2$, the $\chi_{\text{line}} = 1$ and the argument in logarithm becomes 1, while r becomes zero. Therefore, in direct reconciliations scheme, the channel attenuation must be smaller than 3 dB.

In *reverse reconciliation*, the key is determined from Bob's measurements (and Alice performs error correction on her sequence), so that the secret fraction is given by

$$r = I(X_B;Y_A) - I(X_B;Z), \tag{6.117}$$

while the mutual information between Bob and Alice is determined by Eq. (6.113), but rewritten in a different form:

$$I(X_B; Y_A) = \frac{1}{2}\log\left(\frac{\overbrace{v_A + 1}^{v} + \chi_{\text{line}}}{1 + \chi_{\text{line}}}\right) = \frac{1}{2}\log\left(\frac{v + \chi_{\text{line}}}{1 + \chi_{\text{line}}}\right), \quad v = v_A + 1. \quad (6.118)$$

The optimum Gaussian eavesdropping strategy is based on *entangling cloning machine*, as shown in Fig. 6.20. In Eve's apparatus, the EPR source is used with one half being kept by Eve in the quantum memory (QM) and the second half being injected into Alice–Bob channel by a beam splitter. The beam splitter also extracts the portion of the coherent state sent by Alice and Eve and stores into another QM, until Bob announces which quadrature he has measured at every signaling interval. Let the upper quadratures (kept by eve in QMD) be denoted by $\hat{I}_1,\ \hat{Q}_1$, while the quadratures inject into Alice–Bob channel by $\hat{I}_2,\ \hat{Q}_2$. Clearly, Eve can simultaneously correlate the quadratures sent to Bob with quadratures kept in her QM since $\left[\hat{I}_1 - \hat{I}_2, \hat{Q}_1 + \hat{Q}_2\right] = 0$. On the other hand, given that $\left[\hat{I}_1 - \hat{I}_2, \hat{Q}_2\right] = 2jN_0$, the correlation between two in-phase components cannot be perfect and from the uncertainty principle, we have that $\left\langle\left(\hat{I}_1 - \hat{I}_2\right)^2\right\rangle\left\langle\hat{Q}_2^2\right\rangle \geq N_0^2$, and similar relationship holds for correlations in quadrature $\hat{Q}$. To summarize, we can represent uncertainty principle (in shot-noise unit) by [46]

$$v_{B|E} v_{B|A} \geq 1, \quad (6.119)$$

where B stands for any quadrature of the Bob.

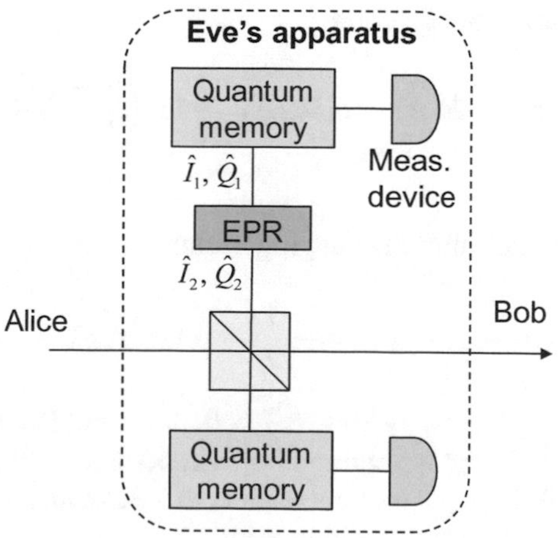

Fig. 6.20 The optimum Gaussian eavesdropping strategy based on entangling cloning machine

We also use the following definition of conditional variance $v_{X|Y}$ that quantifies the amount of uncertainty on X after the measurement on Y is performed [46, 105]:

$$v_{X|Y} = \langle x^2 \rangle - |\langle xy \rangle|^2 / \langle y^2 \rangle. \tag{6.120}$$

Alice employs the estimators (I_A, Q_A) for the field that she sends $(I_A + A_I, Q_A + A_Q)$, wherein $\langle A_I^2 \rangle = \left\langle A_Q^2 \right\rangle = sN_0$, where $s > 1/v$ is the parameter she applies to ensure the consistency with the uncertainty principle. By calculating $\langle Q_A^2 \rangle = (v - s)N_0$, $\langle Q_B^2 \rangle = T(v + \chi_{\text{line}})N_0$, and $\langle Q_A Q_B \rangle = T^{1/2} \langle Q_A^2 \rangle$, and substituting into (6.120), we obtain the following conditional variance [105]:

$$v_{B|A} = \langle Q_B^2 \rangle - |\langle Q_B Q_A \rangle|^2 / \langle Q_A^2 \rangle = T(s + \chi_{\text{line}})N_0. \tag{6.121}$$

The constrain $s > 1/v$ gives the following relationship for conditional variance $v_{B|A} \geq T(1/v + \chi_{\text{line}})N_0$, and from uncertainty principle (6.119), we obtain the following conditional variance of Bob given Eve's measurements:

$$v_{B|E} \geq N_0^2 / [T(1/v + \chi_{\text{line}})N_0]. \tag{6.122}$$

Now we can calculate the mutual information between Bob and Eve as follows [105]:

$$\begin{aligned} I(B;E) &= \frac{1}{2}\log\left(\frac{v_B}{v_{B|E}}\right) = \frac{1}{2}\log\left(\frac{T(v + \chi_{\text{line}})N_0}{\frac{N_0}{T(1/v + \chi_{\text{line}})}}\right) \\ &= \frac{1}{2}\log\left[T^2(v + \chi_{\text{line}})(1/v + \chi_{\text{line}})\right]. \end{aligned} \tag{6.123}$$

After substitution of (6.118) and (6.123) into (6.117), we obtain the expression for secret fraction:

$$r = I(B;A) - I(B;E) = \frac{1}{2}\log\left(\frac{v + \chi_{\text{line}}}{1 + \chi_{\text{line}}}\right) - \frac{1}{2}\log\left[T^2(v + \chi_{\text{line}})(1/v + \chi_{\text{line}})\right], \tag{6.124}$$

which after rearranging becomes

$$r = -\frac{1}{2}\log\left[(1 + \chi_{\text{line}})T^2(1/v + \chi_{\text{line}})\right] > 0, \; T > 0. \tag{6.125}$$

Clearly, as long as $T > 0$, the secret fraction is positive as well.

So far, we ignored Bob's imperfect detection and electrical noise. Referring now to Bob's input (or equivalently to the *channel output*), we can represent Bob's equivalent detector noise variance by

$$\chi_{\det} = \frac{d(1+v_{el})-\eta}{\eta}, \tag{6.126}$$

where $d = 1$ for homodyne detection (either in-phase or quadrature can be measured at given signaling interval in homodyne detection scheme), and $d = 2$ for heterodyne detection (both quadratures can be measured simultaneously). The parameter η denotes the detection efficiency, while v_{el} denotes the variance of equivalent electrical noise. The variance of the total noise can be determined by summing up the variance of the channel noise and detection noise, which can be expressed by referring to the *channel input* as follows:

$$\chi_{\text{total}} = \chi_{\text{line}} + \frac{\chi_{\det}}{T} = \underbrace{\frac{1-T}{T} + \varepsilon}_{\chi_{\text{line}}} + \frac{d(1+v_{el})-\eta}{T\eta}. \tag{6.127a}$$

When referring to the channel output, variance of the total noise at Bob's side can be determined by

$$\begin{aligned}\chi_{\text{total,Bob}} &= T\eta\left(\chi_{\text{line}} + \frac{\chi_{\det}}{T}\right) = T\eta\frac{1-T}{T} + T\eta\varepsilon + T\eta\frac{d(1+v_{el})-\eta}{T\eta} \\ &= \eta(1-T) + T\eta\varepsilon + d(1+v_{el}) - \eta.\end{aligned} \tag{6.127b}$$

The corresponding expression for the secret fraction (6.125) needs to be modified as follows:

$$r = -\frac{1}{2}\log\left[(1+\chi_{\text{total}})(T\eta)^2(1/v+\chi_{\text{total}})\right]. \tag{6.128}$$

6.6.3 GG02 Protocol Implementation

In this subsection, we describe a typical GG02 CV-QKD system configuration, named after the authors Grosshans and Grangier who proposed it in 2002 [89] (see also [7]). The corresponding scheme is depicted in Fig. 6.21.

Alice employs the weak coherent state (WCS) source, such as sufficiently attenuated laser beam, to get the Gaussian state, and further uses the balanced beam splitter (BBS) to split the signal into two parts. The upper output of BS is modulated by the I/Q modulator to place the constellation point in a desired location in I-Q space and to control Alice's variance v_A. The second output of BBS is used to send the local oscillator (LO) reference signal for Bob. The number of photons at receive BS outputs 1 and 2 can be expressed as:

$$N_1 = \frac{1}{2}\left(\boldsymbol{a}_{PM}^{\dagger} + \boldsymbol{a}_{Rx}^{\dagger}\right)(\boldsymbol{a}_{PM} + \boldsymbol{a}_{Rx}),\ N_2 = \frac{1}{2}\left(\boldsymbol{a}_{PM}^{\dagger} - \boldsymbol{a}_{Rx}^{\dagger}\right)(\boldsymbol{a}_{PM} - \boldsymbol{a}_{Rx}), \tag{6.129}$$

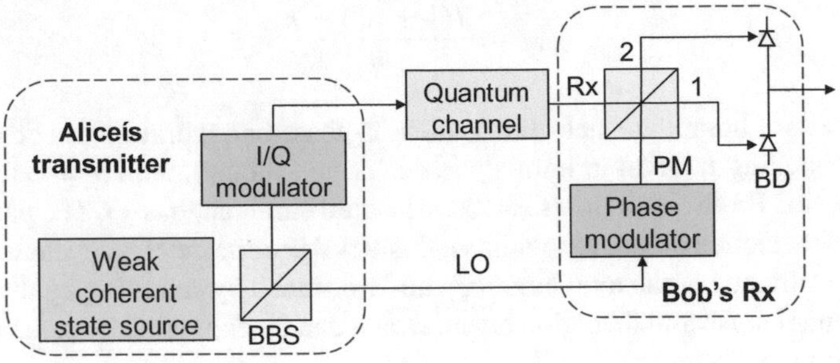

Fig. 6.21 The GG02 CV-QKD system configuration. LO: local oscillator (reference) signal, BBS: balanced beam splitter, and BD: balanced detector

where the creation (annihilation) operators at receive BS inputs Rx and PM are denoted by $\boldsymbol{a}_{Rx}^{\dagger}(\boldsymbol{a}_{Rx})$ and $\boldsymbol{a}_{PM}^{\dagger}(\boldsymbol{a}_{PM})$, respectively. The balanced detector (BD) output operator, assuming the photodiode responsivity of $P_{\mathrm{ph}} = 1$ A/W, is given by

$$\hat{i}_{BD} = 4N_0(N_1 - N_2) = 4N_0\left(\boldsymbol{a}_{PM}^{\dagger}\boldsymbol{a}_{Rx} + \boldsymbol{a}_{Rx}^{\dagger}\boldsymbol{a}_{PM}\right). \tag{6.130}$$

The phase modulation output annihilation operator can be represented by $\boldsymbol{a}_{PM} = |\alpha|e^{j\theta}\mathbf{1}/(2\sqrt{N_0})$, and after substitution into (6.130), we obtain

$$\hat{i}_{BD} = 4N_0\left(\frac{|\alpha|e^{-j\theta}}{2\sqrt{N_0}}\boldsymbol{a}_{Rx} + \frac{|\alpha|e^{-j\theta}}{2\sqrt{N_0}}\boldsymbol{a}_{Rx}^{\dagger}\right) = 2|\alpha|\sqrt{N_0}\left(e^{j\theta}\boldsymbol{a}_{Rx}^{\dagger} + e^{-j\theta}\boldsymbol{a}_{Rx}\right). \tag{6.131}$$

In the absence of Eve, the receive signal annihilation operator will be $\boldsymbol{a}_{Rx} = (I, Q)\mathbf{1}/(2\sqrt{N_0})$, which after substitution into (6.131) yields

$$\begin{aligned}\hat{i}_{BD} &= |\alpha|\left[(\cos\theta, \sin\theta)\mathbf{1}\cdot(I\hat{I}, -Q\hat{Q}) + (\cos\theta, \sin\theta)\mathbf{1}\cdot(I\hat{I}, Q\hat{Q})\right] \\ &= |\alpha|\left(\cos\theta\cdot I\hat{I} + \sin\theta\cdot Q\hat{Q}\right).\end{aligned} \tag{6.132}$$

Bob measures in-phase component $\hat{I}$ by selecting $\theta = 0$ to get $|\alpha|I$ and the quadrature component $\hat{Q}$ by setting $\theta = \pi/2$ to get $|\alpha|Q$.

The variance of Bob's measurement has the following components: (i) signal variance, (ii) intrinsic fluctuations N_0, (iii) channel noise $\chi_{\mathrm{line}}N_0$, (iv) the noise due to imperfect homodyne detection $\chi_{\mathrm{homodyne}}N_0 = (1-\eta)N_0/\eta$, and (v) the electronic noise $v_{\mathrm{el}}N_0$, so that we can write

$$\chi_{\mathrm{total}}N_0 = \left[\chi_{\mathrm{line}} + \frac{\overbrace{\chi_{\mathrm{homodyne}} + v_{\mathrm{el}}/\eta}^{\chi_{\mathrm{el}}}}{T_{\mathrm{line}}}\right]N_0. \tag{6.133}$$

From variances of Alice and Bob signals $v_A N_0 = \mathrm{Var}(X_A)$ and $\mathrm{Var}(X_B)$ as well as correlation coefficient $\gamma(X_A, X_B)$, we can determine [7]

$$v_A = \mathrm{Var}(X_A)/N_0,\ T_{\text{line}} = \gamma^2 \mathrm{Var}(Y_B)/\mathrm{Var}(X_A),$$
$$\chi_{\text{total}} + 1 = \mathrm{Var}(X_A)\left(\gamma^{-2} - 1\right)/N_0. \tag{6.134}$$

Mutual information for Alice-to-Eve channel is given by

$$I(A;E) = \frac{1}{2}\log\left(1 + \frac{v_A}{1 + 1/\chi_{BE}}\right), \tag{6.135}$$

which is essentially Eq. (6.115). On the other hand, the mutual information for Bob-to-Eve channel will be [7, 106, 107]

$$I(B;E) = \frac{1}{2}\log\left\{\frac{T_{\text{line}}T_{\text{extra}}(v + \chi_{\text{total}})}{\chi_{BB}T_{\text{line}} + 1/[T_{\text{line}}T_{\text{extra}}(1/v + \chi_{BE})]}\right\}, \tag{6.136}$$

which represents the generalization of Eq. (6.123). For a *paranoid approach* (when detection noises are controlled by Eve), we have to set the parameters from above equations as follows [7, 106, 107]:

$$T_{\text{extra}} = T_{\text{homodyne}},\ \chi_{BE} = \chi_{\text{total}},\ \chi_{BB} = 0. \tag{6.137}$$

On the other hand, for a *realistic approach* (when Eve cannot control the detection noises), we set the parameters in Eqs. (6.135) to be [7, 106, 107]

$$T_{\text{extra}} = 1,\ \chi_{BE} = \chi_{\text{line}},\ \chi_{BB} = (\chi_{\text{homodyne}} + \chi_{\text{el}})/T_{\text{line}}. \tag{6.138}$$

For direct reconciliation, we can calculate the secret fraction by $r = I(A;B) - I(A;E)$. On the other hand, for the reverse reconciliation, we can calculate the secret fraction by $r = I(B;A) - I(B;E)$.

6.6.4 Collective Attacks

In this subsection, we briefly describe how to calculate the secret fraction for coherent state homodyne-detection-based CV-QKD scheme when Eve employs the Gaussian collective attack [90, 103, 104], which is known to be optimum for Gaussian modulation. This attack is completely characterized by estimating the covariance matrix by Alice and Bob $\boldsymbol{V}_{\text{AB}}$, which is based on Eq. (6.97) and previously determined Bob's variance of the in-phase component $V(\hat{I}) = T\eta(v + \chi_{\text{total}})$ can be represented as

$$\boldsymbol{V}_{\mathrm{AB}} = \begin{bmatrix} v\boldsymbol{I} & \sqrt{T\eta}\sqrt{v^2-1}\boldsymbol{Z} \\ \sqrt{T\eta}\sqrt{v^2-1}\boldsymbol{Z} & T\eta(v+\chi_{\mathrm{total}})\boldsymbol{I} \end{bmatrix}, \quad v = v_A + 1. \tag{6.139}$$

As before, we assume that $v = v_A + 1$, where v_A is Alice's Gaussian modulation (normalized) variance, T is the channel transmittance, η detection efficiency, and χ_{total} is defined by Eq. (6.127a).

When the reverse reconciliation is used, the secret fraction is given by

$$r = \beta I(A;B) - \chi(B;E), \tag{6.140}$$

where β is the reconciliation efficiency and $I(A;B)$ is the mutual information between Alice and Bob determined in the same fashion as for individual attacks:

$$I(A;B) = \frac{d}{2}\log_2\left(\frac{v+\chi_{\mathrm{total}}}{1+\chi_{\mathrm{total}}}\right), \quad d = \begin{cases} 1, & \text{homodyne detection} \\ 2, & \text{heterodyne detection} \end{cases}. \tag{6.141}$$

The Holevo information between Bob and Eve, denoted as $\chi(B;E)$, is determined by [90, 103, 104]

$$\chi(B;E) = g\left(\frac{\lambda_1-1}{2}\right) + g\left(\frac{\lambda_2-1}{2}\right) - g\left(\frac{\lambda_3-1}{2}\right) - g\left(\frac{\lambda_4-1}{2}\right), \tag{6.142}$$

where $g(x) = (x+1)\log_2(x+1) - x\log_2 x$ is the entropy of a thermal state with the mean number of photons being x. The λ-parameters are defined as [90, 108]

$$\lambda_{1,2} = \sqrt{\frac{1}{2}\left(A \pm \sqrt{A^2-4B}\right)}, \quad \lambda_{3,4} = \sqrt{\frac{1}{2}\left(C \pm \sqrt{C^2-4D}\right)},$$
$$A = v^2(1-2T) + 2T + T^2(v+\chi_{\mathrm{line}})^2, \quad B = T^2(1+v\chi_{\mathrm{line}})^2. \tag{6.143}$$

The parameters C and D are different for homodyne and heterodyne detection [90, 108]:

$$C = \begin{cases} \frac{A\chi_{\mathrm{homodyne}} + v\sqrt{B} + T(v+\chi_{\mathrm{line}})}{T(v+\chi_{\mathrm{total}})}, & \text{homodyne detection} \\ \frac{A\chi_{\mathrm{het}}^2 + B + 1 + 2\chi_{\mathrm{het}}\left[v\sqrt{B} + T(v+\chi_{\mathrm{line}})\right] + 2T(v^2-1)}{T^2(v+\chi_{\mathrm{total}})^2}, & \text{heterodyne detection} \end{cases} \tag{6.144}$$

$$D = \begin{cases} \sqrt{B}\frac{v+\chi_{\mathrm{homodyne}}\sqrt{B}}{T(v+\chi_{\mathrm{total}})}, & \text{homodyne detection} \\ \frac{\left(v+\chi_{\mathrm{het}}\sqrt{B}\right)^2}{T^2(v+\chi_{\mathrm{total}})^2}, & \text{heterodyne detection} \end{cases}. \tag{6.145}$$

6.7 Measurement-Device-Independent (MDI) Protocols

In MDI protocols [109–114], Alice and Bob are connected to a third party, Charlie, through a quantum channel (such as an FSO or a fiber optics link). The single-photon detectors are located at Charlie's side, as illustrated in Fig. 6.22. The simplified version of this protocol is described in the rest of section.

Alice and Bob have single-photon sources available and randomly select one of four states $|0\rangle, |1\rangle, |+\rangle, |-\rangle$ to be sent to Charlie. Charlie performs the partial Bell state measurement (BSM) with the help of a beam splitter (BS) and announces the events for each of the measurement which resulted in either $|\psi^-\rangle = 2^{-1/2}(|0\rangle_A|1\rangle_B - |1\rangle_A|0\rangle_B)$ or $|\psi^+\rangle = 2^{-1/2}(|0\rangle_A|1\rangle_B + |1\rangle_A|0\rangle_B)$. A successful BSM corresponds to the observation when precisely two detectors (with orthogonal polarizations) are triggered as follows:

- The simultaneous clicks on detectors D_{10} and D_{21} (or D_{11} and D_{20}) identify the projection into the Bell state $|\psi^-\rangle$.
- On the other hand, the clicks on detectors D_{10} and D_{11} (D_{20} and D_{21}) identify the projection into the Bell state $|\psi^+\rangle$.

Other detection patterns are not considered successful and are discarded.

Alice and Bob then exchange the information, over an authenticated classical, noiseless, channel, about the bases being used: Z-basis spanned by $|0\rangle$ and $|1\rangle$ [computational basis (CB)] states, or X-basis spanned by $|+\rangle$ and $|-\rangle$ (diagonal basis) states. Alice and Bob keep only events in which they used the same basis, while at the same time Charlie announced the BSM states $|\psi^-\rangle$ or $|\psi^+\rangle$. One of them (say Bob) flips the sent bits except when both used diagonal basis and Charlie announces the state $|\psi^+\rangle$. The remaining events are discarded.

Further, the X-key is formed out of those bits when Alice and Bob prepared their photons in X-basis (diagonal basis). The error rate on these bits is used to bound the information obtained by Eve. Next, Alice and Bob form the Z-key out of those bits for which they both used the Z-basis (computational basis). Finally, they perform

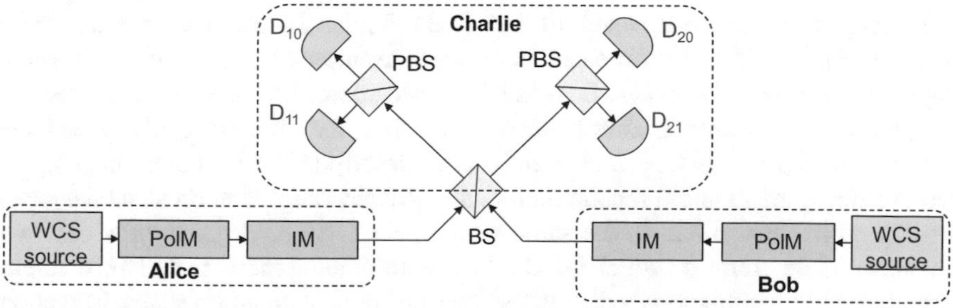

Fig. 6.22 The MDI-QKD system configuration. PolM: polarization modulator (used to generate one out of four polarization states $|0\rangle, |1\rangle, |+\rangle, |-\rangle$), IM: intensity modulator (used to generate the signal and decoy states of different intensities)

information reconciliation and privacy amplification on the Z-key to get the secret key.

The secret fraction of the MDI-QKD protocol can be estimated by [112]

$$r = q_{11}^{(Z)}\left[1 - h_2\left(e_{11}^{(X)}\right)\right] - q_{\mu\sigma}^{(Z)} f_e h_2\left(e_{\mu\sigma}^{(Z)}\right), \tag{6.146}$$

where we introduced the following notation:

- $q_{11}^{(Z)}$ denotes the probability that Charlie declares a successful result ("the gain") when both Alice and Bob sent single photon each in the Z-basis,
- $e_{11}^{(X)}$ denotes the phase-error rate of the single-photon signals, both sent in the Z-basis, and
- $q_{\mu\sigma}^{(Z)}$ denotes the gain in the Z-basis when Alice sent the weak coherent state (WCS) of intensity μ to Charlie, while Bob at the same time sent WCS of intensity σ.

The function $h_2(x)$ in equation above denotes the binary entropy function defined as $h_2(x) = -x\log_2(x) - (1-x)\log_2(1-x)$.

In the equation above, the information removed from the final key in privacy amplification step is described by the following term $q_{11}^{(Z)} h_2\left(e_{11}^{(X)}\right)$. The information revealed by Alice in the information reconciliation's step is described by the following term:

$$q_{\mu\sigma}^{(Z)} f_e h_2\left(e_{\mu\sigma}^{(Z)}\right),$$

with f_e being the error correction inefficiency ($f_e \geq 1$). The error rates $q_{11}^{(Z)}$ and $e_{11}^{(X)}$ cannot be measured directly but instead bounded by employing the decoy-state approach. Additional details about this protocol can be found in Chap. 9.

6.8 Concluding Remarks

This chapter has been devoted to the QKD fundamentals. The chapter starts with description of key differences between conventional cryptography, classical physical-layer security, and QKD, which have been described in Sect. 6.1. In section on QKD basics, Sect. 6.2, after historical overview, the different QKD types have been introduced (see Sect. 6.2.1), and the brief description of the common postprocessing steps, information reconciliation, and privacy amplification steps has been introduced in Sect. 6.2.2. In the same section, Sect. 6.2.3, we have provided two fundamental theorems on which QKD relies on no-cloning theorem and the theorem of inability to unambiguously distinguish non-orthogonal quantum states. In section on discrete variable (DV)-QKD systems, Sect. 6.3, we have described in detail BB84 protocols (Sect. 6.3.1) and B92 protocol (Sect. 6.3.2) as well as Ekert (E91) and EPR protocols (Sect. 6.3.3). In the same section, Sect. 6.3.4, the time-phase encoding protocol has also been described. Regarding the BB84 protocols, different versions, suitable for different technologies, have been described (see Sect. 6.3.1). In section on QKD security, Sect. 6.4, the secret key rate has been represented as the product of raw key rate and fractional rate, followed by the description of different limitations to the raw key rate. After that, the generic expression for the fractional rate has been provided followed by description in Sect. 6.4.1 of different eavesdropping strategies including individual (independent or incoherent) attacks, collective attacks, and coherent attacks as well as the quantum hacking/side-channel attacks. For individual and coherent attacks, the corresponding secrete fraction expressions have been described (see Sect. 6.4.1). The next Sect. 6.4.2 has been devoted to various definitions of security, including the concept of ε-security. After that, the generic expressions for 2-D DV-QKD schemes have been derived in Sect. 6.4.3 for both prepare-and-measure and decoy-state-based protocols. To facilitate the description of continuous variable (CV)-QKD protocols, the fundamentals of quantum optics have been introduced first in Sect. 6.5. In section on CV-QKD protocols, Sect. 6.6, both squeezed state-based protocols (Sect. 6.6.1) and coherent state-based protocols (Sect. 6.6.2) have been described. Given that the coherent states have been much easier to generate and manipulate, the coherent state-based protocols with both homodyne and heterodyne detections have been described in detail in Sect. 6.6.2. The secret fraction has been derived for both direct- and reverse-reconciliation-based CV-QKD protocols. Furthermore, the details on practical aspects of GG02 protocol have been provided in Sect. 6.6.3. In the same section, the secret fraction calculation for collective attacks has been discussed in Sect. 6.6.4. After that, the basic concepts of measurement-device-independent QKD protocols have been introduced in Sect. 6.7.

References

1. Schneier B (2015) Applied cryptography, second edition: protocols, algorithms, and source code in C. Wiley, Indianapolis, IN
2. Drajic D, Ivanis P (2009) Introduction to information theory and coding, 3rd edn. Akademska Misao, Belgrade, Serbia (in Serbian)
3. Katz J, Lindell Y (2015) Introduction to modern cryptography, 2nd edn. CRC Press, Boca Raton, FL
4. Diffie W, Hellman ME (1976) New direction in cryptography. IEEE Trans Inform Theory 22:644–654
5. Kahn D (1967) The codebreakers: the story of secret writing. Macmillan Publishing Co., Ney York
6. Neilsen MA, Chuang IL (2000) Quantum computation and quantum information. Cambridge University Press, Cambridge
7. Van Assche G (2006) Quantum cryptography and secrete-key distillation. Cambridge University Press, Cambridge-New York
8. Djordjevic IB (2012) Quantum information processing and quantum error correction: an engineering approach. Elsevier/Academic Press, Amsterdam-Boston
9. Shannon CE (1949) Communication theory of secrecy systems. Bell Syst Tech J 28:656–715
10. Aumasson J-P (2018) Serious cryptography: a practical introduction to modern encryption. No Starch Press, San Francisco, CA

11. Sebbery J, Pieprzyk J (1989) Cryptography: an introduction to computer security. Prentice Hall, New York
12. Delfs H, Knebl H (2015) Introduction to cryptography: principles and applications (information security and cryptography), 3rd edn. Springer, Heidelberg, New York
13. Bennet CH, Brassard G (1984) Quantum cryptography: public key distribution and coin tossing. In: Proceedings of the IEEE international conference on computers, systems, and signal processing, Bangalore, India, pp 175–179
14. Bennett CH (1992) Quantum cryptography: uncertainty in the service of privacy. Science 257:752–753
15. Le Bellac M (2006) An introduction to quantum information and quantum computation. Cambridge University Press
16. Shor PW (1997) Polynomial-time algorithms for prime number factorization and discrete logarithms on a quantum computer. SIAM J Comput 26(5):1484–1509
17. Ekert A, Josza R (1996) Quantum computation and Shor's factoring algorithm. Rev Modern Phys 68(3):733–753
18. Rivest RL, Shamir A, Adleman L (1978) A method for obtaining digital signatures and public-key cryptosystems. Commun ACM 21(2):120–126
19. Barnett SM (2009) Quantum information. Oxford University Press, Oxford
20. Liao S-K et al (2017) Satellite-to-ground quantum key distribution. Nature 549:43–47
21. Bennett CH, Brassard G, Crepeau C, Maurer U (1995) Generalized privacy amplification. IEEE Inform Theory 41(6):1915–1923
22. Chorti A et al (2016) Physical layer security: a paradigm shift in data confidentiality. In: Physical and data-link security techniques for future communications systems. Lecture notes in electrical engineering, vol 358. Springer, pp 1–15
23. Ursin R et al (2007) Entanglement-based quantum communication over 144 km. Nat Phys 3(7):481–486
24. Boaron A et al (2018) Secure quantum key distribution over 421 km of optical fiber. Phys Rev Lett 121:190502
25. Schwinger J (1960) Unitary operator bases. Proc Natl Acad Sci USA 46:570–579
26. Wootters WK, Fields BD (1989) Optimal state-determination by mutually unbiased measurements. Ann Phys 191:363–381
27. Ivanovic ID (1981) Geometrical description of quantal state determination. J Phys A 14:3241–3245
28. Bengtsson I (2007) Three ways to look at mutually unbiased bases. AIP Conf Proc 889(1):40
29. Djordjevic IB (2018) FBG-based weak coherent state and entanglement assisted multidimensional QKD. IEEE Photon J 10(4):7600512
30. Helstrom CW (1976) Quantum detection and estimation theory. Academic Press, New York
31. Helstrom CW, Liu JWS, Gordon JP (1970) Quantum-mechanical communication theory. Proc IEEE 58 (10):1578–1598
32. Vilnrotter C, Lau C-W (2001) Quantum detection theory for the free-space channel. IPN Progr Rep 42–146:1–34
33. Djordjevic IB (2007) LDPC-coded optical coherent state quantum communications. IEEE Photon Technol Lett 19(24):2006–2008
34. Djordjevic IB (2009) LDPC-coded *M*-ary PSK optical coherent state quantum communication. IEEE/OSA J Lightw Technol 27(5):494–499
35. Lo H-K, Ma X, Chen K (2005) Decoy state quantum key distribution. Phys Rev Lett 94:230504
36. Bennett CH (1992) Quantum cryptography using any two nonorthogonal states. Phys Rev Lett 68(21):3121–3124
37. Ekert AK (1991) Quantum cryptography based on Bell's theorem. Phys Rev Lett 67(6):661–663
38. Lomonaco SJ (1999) A quick glance at quantum cryptography. Cryptologia 23(1):1–41
39. Townsend PD, Rarity JG, Tapster PR (1993) Single photon interference in 10 km long optical fibre interferometer. Electron Lett 29(7):634–635

40. Marand C, Townsend PD (1995) Quantum key distribution over distances as long as 30 km. Opt Lett 20(16):1695–1697
41. Islam NT, Lim CCW, Cahall C, Kim J, Gauthier DJ (2017) Provably secure and high-rate quantum key distribution with time-bin qudits. Sci Adv 3(11):e1701491
42. Spedalieri FM (2006) Quantum key distribution without reference frame alignment: exploiting photon orbital angular momentum. Opt Commun 260(1):340–346
43. Djordjevic IB (2013) Multidimensional QKD based on combined orbital and spin angular momenta of photon. IEEE Photon J 5(6):7600112
44. Djordjevic IB (2017) Advanced optical and wireless communications systems. Springer International Publishing, Switzerland
45. Islam NT (2018) High-rate, high-dimensional quantum key distribution systems. PhD dissertation, Duke University
46. Scarani V, Bechmann-Pasquinucci H, Cerf NJ, Dušek M, Lütkenhaus N, Peev M (2009) The security of practical quantum key distribution. Rev Mod Phys 81:1301
47. Bloom S, Korevaar E, Schuster J, Willebrand H (2003) Understanding the performance of free-space optics [invited]. J Opt Netw 2(6):178–200
48. Csiszár I, Körner J (1978) Broadcast channels with confidential messages. IEEE Trans Inf Theory 24(3):339–348
49. Brassard G, Lütkenhaus N, Mor T, Sanders BC (2000) Limitations on practical quantum cryptography. Phys Rev Lett 85:1330
50. Devetak I, Winter A (2005) Distillation of secret key and entanglement from quantum states. Proc R Soc London Ser A 461(2053):207–235
51. Holevo AS (1973) Bounds for the quantity of information transmitted by a quantum communication channel. Probl Inf Trans 9(3):177–183
52. Lucamarini M, Choi I, Ward MB, Dynes JF, Yuan ZL, Shields AJ (2015) Practical security bounds against the trojan-horse attack in quantum key distribution. Phys Rev X 5:031030
53. Kurtsiefer C, Zarda P, Mayer S, Weinfurter H (2001) The breakdown flash of silicon avalanche photodiodes—back door for eavesdropper attacks? J Mod Opt 48(13):2039–2047
54. Qi B, Fung C-HF, Lo H-K, Ma X (2006) Time-shift attack in practical quantum cryptosystems. Quantum Inf Comput 7(1):73–82. https://arxiv.org/abs/quant-ph/0512080
55. Lydersen L, Wiechers C, Wittmann S, Elser D, Skaar J, Makarov V (2010) Hacking commercial quantum cryptography systems by tailored bright illumination. Nat Photon 4(10):686
56. Lütkenhaus N, Jahma M (2002) Quantum key distribution with realistic states: photon-number statistics in the photon-number splitting attack. New J Phys 4:44
57. Dušek M, Jahma M, Lütkenhaus N (2000) Unambiguous state discrimination in quantum cryptography with weak coherent states. Phys Rev A 62(2):022306
58. Ben-Or M, Horodecki M, Leung DW, Mayers D, Oppenheim J (2005) The universal composable security of quantum key distribution. In: Theory of cryptography: second theory of cryptography conference, TCC 2005. Lecture notes in computer science, vol 3378. Springer, Berlin, pp 386–406
59. König R, Renner R, Bariska A, Maurer U (2007) Small accessible quantum information does not imply security. Phys Rev Lett 98:140502
60. Renner R, König R (2005) Universally composable privacy amplification against quantum adversaries. In: Theory of cryptography: second theory of cryptography conference, TCC 2005. Lecture notes in computer science, vol 3378. Springer, Berlin, pp 407–425
61. Renner R (2005) Security of quantum key distribution. PhD dissertation, Swiss Federal Institute of Technology, Zurich
62. Scarani V, Renner R (2008) Quantum cryptography with finite resources: unconditional security bound for discrete-variable protocols with one-way postprocessing. Phys Rev Lett 100(20):200501
63. Djordjevic IB (2016) Integrated optics modules based proposal for quantum information processing, teleportation, QKD, and quantum error correction employing photon angular momentum. IEEE Photon J 8(1):6600212

64. Uhlmann A (1976) The transition probability in the state space of a *-algebra. Rep Math Phys 9:273–279
65. Gottesman D, Lo H-K, Lütkenhaus N, Preskill J (2004) Security of quantum key distribution with imperfect devices. Quantum Inf Comput 4(5):325–360
66. Fung C-HF, Tamaki K, Lo H-K (2006) Performance of two quantum-key-distribution protocols. Phys Rev A 73:012337
67. Kraus B, Branciard C, Renner R (2007) Security of quantum-key-distribution protocols using two-way classical communication or weak coherent pulses. Phys Rev A 75:012316
68. Inamori H, Lütkenhaus N, Mayers D (2007) Unconditional security of practical quantum key distribution. Eur Phys J D 41:599–627
69. Hwang W-Y (2003) Quantum key distribution with high loss: toward global secure communication. Phys Rev Lett 91:057901
70. Ma X, Fung C-HF, Dupuis F, Chen K, Tamaki K, Lo H-K (2006) Decoy-state quantum key distribution with two-way classical postprocessing. Phys Rev A 74:032330
71. Zhao Y, Qi B, Ma X, Lo H-K, Qian L (2006) Experimental quantum key distribution with decoy states. Phys Rev Lett 96:070502
72. Rosenberg D, Harrington JW, Rice PR, Hiskett PA, Peterson CG, Hughes RJ, Lita AE, Nam SW, Nordholt JE (2007) Long-distance decoy-state quantum key distribution in optical fiber. Phys Rev Lett 98(1):010503
73. Yuan ZL, Sharpe AW, Shields AJ (2007) Unconditionally secure one-way quantum key distribution using decoy pulses. Appl Phys Lett 90:011118
74. Hasegawa J, Hayashi M, Hiroshima T, Tanaka A, Tomita A (2007) Experimental decoy state quantum key distribution with unconditional security incorporating finite statistics. arXiv: 0705.3081
75. Tsurumaru T, Soujaeff A, Takeuchi S (2008) Exact minimum and maximum of yield with a finite number of decoy light intensities. Phys Rev A 77:022319
76. Hayashi M (2007) General theory for decoy-state quantum key distribution with an arbitrary number of intensities. New J Phys 9:284
77. Sun X, Djordjevic IB, Neifeld MA (2016) Secret key rates and optimization of BB84 and decoy state protocols over time-varying free-space optical channels. IEEE Photonics J 8(3):7904713-1–7904713-13
78. Mandel L, Wolf E (1995) Optical coherence and quantum optics. Cambridge University Press, Cambridge-New York-Melbourne
79. Scully MO, Zubairy MS (1997) Quantum optics. Cambridge University Press, Cambridge-New York-Melbourne
80. Weedbrook C, Pirandola S, García-Patrón R, Cerf NJ, Ralph TC, Shapiro JH, Seth Lloyd L (2012) Gaussian quantum information. Rev Mod Phys 84:621
81. Sakurai JJ (1994) Modern Quantum Mechanics. Addison-Wisley
82. Yuen HP (1976) Two-photon coherent states of the radiation field. Phys Rev A 13:2226
83. Radic S, McKinistrie CJ (2005) Optical amplification and signal processing in highly nonlinear optical fiber. IEICE Trans Electron E88-C:859–869
84. Jansen SL, Van den Borne D, Krummrich PM, Spälter S, Khoe G-D, De Waardt H (2006) Long-haul DWDM transmission systems employing optical phase conjugation. IEEE J Sel Top Quantum Electron 12:505–520
85. Cvijetic M, Djordjevic IB (2013) Advanced optical communications and networks. Artech House
86. Grosshans F, Grangier P (2002) Reverse reconciliation protocols for quantum cryptography with continuous variables. arXiv:quant-ph/0204127
87. Qu Z, Djordjevic IB (2017) RF-assisted coherent detection based continuous variable (CV) QKD with high secure key rates over atmospheric turbulence channels. In: Proceedings of the ICTON 2017, paper Tu.D2.2, Girona, Spain, 2–6 July 2017 (Invited Paper)
88. Ralph TC (1999) Continuous variable quantum cryptography. Phys Rev A 61:010303(R)
89. Grosshans F, Grangier P (2002) Continuous variable quantum cryptography using coherent states. Phys Rev Lett 88:057902

90. Grosshans F (2005) Collective attacks and unconditional security in continuous variable quantum key distribution. Phys Rev Lett 94:020504
91. Leverrier A, Grangier P (2009) Unconditional security proof of long-distance continuous-variable quantum key distribution with discrete modulation. Phys Rev Lett 102:180504
92. Becir A, El-Orany FAA, Wahiddin MRB (2012) Continuous-variable quantum key distribution protocols with eight-state discrete modulation. Int J Quantum Inform 10:1250004
93. Xuan Q, Zhang Z, Voss PL (2009) A 24 km fiber-based discretely signaled continuous variable quantum key distribution system. Opt Express 17(26):24244–24249
94. Huang D, Huang P, Lin D, Zeng G (2016) Long-distance continuous-variable quantum key distribution by controlling excess noise. Sci Rep 6:19201
95. Qu Z, Djordjevic IB (2017) High-speed free-space optical continuous-variable quantum key distribution enabled by three-dimensional multiplexing. Opt Express 25(7):7919–7928
96. Silberhorn C, Ralph TC, Lütkenhaus N, Leuchs G (2002) Continuous variable quantum cryptography: beating the 3 dB loss limit. Phys Rev A 89:167901
97. Patel KA, Dynes JF, Choi I, Sharpe AW, Dixon AR, Yuan ZL, Penty RV, Shields J (2012) Coexistence of high-bit-rate quantum key distribution and data on optical fiber. Phys Rev X 2:041010
98. Qu Z, Djordjevic IB (2017) Four-dimensionally multiplexed eight-state continuous-variable quantum key distribution over turbulent channels. IEEE Photon J 9(6):7600408
99. Qu Z, Djordjevic IB (2017) Approaching Gb/s secret key rates in a free-space optical CV-QKD system affected by atmospheric turbulence. In: Proceedings of the ECOC 2017, P2.SC6.32, Gothenburg, Sweden
100. Qu Z, Djordjevic IB (2018) High-speed free-space optical continuous variable-quantum key distribution based on Kramers-Kronig scheme. IEEE Photon J 10(6):7600807
101. Heid M, Lütkenhaus N (2006) Efficiency of coherent-state quantum cryptography in the presence of loss: influence of realistic error correction. Phys Rev A 73:052316
102. Grosshans F, Cerf NJ (2004) Continuous-variable quantum cryptography is secure against non-Gaussian attacks. Phys Rev Lett 92:047905
103. García-Patrón R, Cerf NJ (2006) Unconditional optimality of gaussian attacks against continuous-variable quantum key distribution. Phys Rev Lett 97:190503
104. Navascués M, Grosshans F, Acín A (2006) Optimality of Gaussian attacks in continuous-variable quantum cryptography. Phys Rev Lett 97:190502
105. Grosshans F, Van Assche G, Wenger J, Tualle-Brouri R, Cerf NJ, Grangier P (2003) Quantum key distribution using Gaussian-modulated coherent states. Nature 421:238
106. Grosshans F (2002) Communication et Cryptographi e Quantiques avec des Variables Continues. PhD dissertation, Université Paris XI
107. Wegner J (2004), Dispositifs Imulsionnels pour la Communication Quantique à Variables Contines. PhD dissertation, Université Paris XI
108. Qu Z (2018) Secure high-speed optical communication systems. PhD dissertation, University of Arizona
109. Masanes L, Pironio S, Acín A (2011) Secure device-independent quantum key distribution with causally independent measurement devices. Nat Commun 2:238
110. Lo H-K, Curty M, Qi B (2012) Measurement-device-independent quantum key distribution. Phys Rev Lett 108:130503
111. Xu F, Curty M, Qi B, Lo H-K (2015) Measurement-device-independent quantum cryptography. IEEE J Sel Top Quantum Electron 21(3):148–158
112. Chan P, Slater JA, Lucio-Martinez I, Rubenok A, Tittel W (2014) Modeling a measurement-device-independent quantum key distribution system. Opt Express 22(11):12716–12736
113. Curty M, Xu F, Lim CCW, Tamaki K, Lo H-K (2014) Finite-key analysis for measurement-device-independent quantum key distribution. Nat Commun 5:3732
114. Yin H-L et al (2016) Measurement-device-independent quantum key distribution over a 404 km optical fiber. Phys Rev Lett 117:190501

Chapter 7
Discrete Variable (DV) QKD

Abstract This chapter is devoted to the discrete variable (DV) QKD protocols and represents the continuation of Chap. 6. The chapter starts with the description of BB84 and decoy-state-based protocols, and evaluation of their secrecy fraction performance in terms of achievable distance. The next topic is related to the security of DV-QKD protocols when the resources are finite. We introduce the concept of composable ε-security and describe how it can be evaluated for both collective and coherent attacks. We also discuss how the concept of correctness and secrecy can be combined to come up with tight security bounds. After that, we evaluate the BB84 and decoy-state protocols for finite-key assumption over atmospheric turbulence effects. We also describe how to deal with time-varying free-space optical channel conditions. The focus is then moved to high-dimensional (HD) QKD protocols, starting with the description of mutually unbiased bases (MUBs) selection, followed by the introduction of the generalized Bell states. We then describe how to evaluate the security of HD QKD protocols for finite resources. We describe various HD QKD protocols, including time-phase encoding, time-energy encoding, OAM-based HD QKD, fiber Bragg grating (FBGs)-based HD QKD, and waveguide Bragg gratings (WBGs)-based HD QKD protocols.

7.1 BB84 and Decoy-State Protocols

In this section, we describe the BB84 and decoy-state-based protocols and discuss their performance.

7.1.1 The BB84 Protocol Revisited

In BB84 protocol [1–5], Alice randomly selects the computational basis (Z-basis) $\{|0\rangle, |1\rangle\}$ or diagonal basis (X-basis) $\{|+\rangle, |-\rangle\}$, followed by random selection of the basis state. The logical 0 is represented by $|0\rangle$, $|+\rangle$ states, while logical one by $|1\rangle$, $|-\rangle$ states. Bob measures each qubit by randomly selecting the basis, computa-

I. B. Djordjevic, *Physical-Layer Security and Quantum Key Distribution*,
https://doi.org/10.1007/978-3-030-27565-5_7

tional, or diagonal. In sifting procedure, Alice and Bob announce the bases being used for each qubit and keep only instances when they used the same basis. From remaining bits, Alice selects a subset of bits to be used against Eve's interference and channel errors and informs Bob which ones. Alice and Bob announce and compare the values of these bits used for quantum bit error (QBER) rate estimation. If more than acceptable number of bits disagree, dictated by the error correction capability of the code, they abort protocol. Otherwise, Alice and Bob perform information reconciliation and privacy amplification on the remaining bits to obtain shared key bits. The corresponding polarization-based BB84 and time-phase encoding-based BB84 protocols' implementations have already been described in Chap. 6, together with corresponding simplification B92 protocol [6]. Here we briefly describe the *secret fraction* calculation in *asymptotic case*, when the size of raw key tends to infinity. The corresponding *secret-key rate* (SKR) is then determined as the product of the secret fraction and the raw key rate. The expression for *secret fraction*, obtained by *one-way postprocessing*, is given by

$$r = I(A;B) - \min_{\text{Eve's strategies}} (I_{EA}, I_{EB}), \tag{7.1}$$

where $I(A; B)$ is the mutual information between Alice and Bob, while the second term corresponds to the Eve's information I_E about Alice or Bob's raw key, where minimization is performed over all possible eavesdropping strategies.

For *collective attacks*, Eve's information about Alice sequence is determined from Holevo information as follows [7, 8]:

$$I_{EA} = \max_{\text{Eve's strategies}} \chi(A; E), \tag{7.2}$$

where maximization is performed over all possible collective eavesdropping strategies. The similar definition holds for I_{BE}. This bound is also known as Devetak–Winter bound. The Holevo information, introduced in [9], is defined here as

$$\chi(A; E) = S(\rho_E) - \sum\nolimits_a S(\rho_{E|a}) p(a), \tag{7.3}$$

where $S(\rho)$ is the von Neumann entropy defined as $S(\rho) = -\text{Tr}(\log(\rho)) = -\sum_i \lambda_i \log \lambda_i$, with λ_i being the eigenvalues of the density operator (state) ρ. In (7.3), $p(a)$ represents the probability of occurrence of symbol a from Alice's classical alphabet, while $\rho_{E|a}$ is the corresponding density operator of Eve's ancilla. Finally, ρ_E is the Eve's partial density state defined by $\rho_E = \sum_a p(a)\rho_{E|a}$. In other words, the Holevo information corresponds to the average reduction in Eve's von Neumann entropy (uncertainty) given that we know how ρ_E get prepared.

The secret fraction that can be achieved with BB84 protocol is lower bounded by

$$r = q_1^{(Z)}\left[1 - h_2\left(e_1^{(X)}\right)\right] - q^{(Z)} f_e h_2\left(e^{(Z)}\right), \tag{7.4}$$

where $q_1^{(Z)}$ denotes the probability of declaring a successful result ("the gain") when Alice sent a single photon and Bob detected it in the Z-basis, f_e denotes the error correction inefficiency ($f_e \geq 1$), $e^{(X)}[e^{(Z)}]$ denotes the QBER in the X-basis (Z-basis), and $h_2(x)$ is the binary entropy function, which is defined as

$$h_2(x) = -x \log_2(x) - (1-x)\log_2(1-x). \tag{7.5}$$

The second term $q^{(Z)}h_2[e^{(X)}]$ denotes the amount of information Eve was able to learn during the raw key transmission, and this information is typically removed from the final key during the privacy amplification stage. The last term $q^{(Z)}f_e h_2[e^{(Z)}]$ denotes the amount of information revealed during the information reconciliation (error correction) stage, typically related to the parity bits exchanged over noiseless public channel (when systematic error correction is used).

7.1.2 The Decoy-State Protocols

When weak coherent state (WCS) source is used instead of single-photon source, this scheme is sensitive to the photon number splitting (PNS) attack, as discussed in Chap. 6. Given that the probability that multiple photons are transmitted from WCS is now nonzero, Eve can exploit this fact to gain information from the channel. To overcome the PNS attack, the use of *decoy-state* quantum-key distribution systems can be used [10–22]. In decoy-state QKD, the average number of photons transmitted is increased during random timeslots, allowing Alice and Bob to detect if Eve is stealing photons when multiple photons are transmitted. There exist different versions of decoy-state-based protocols. In one-decoy-state-based protocol, in addition to signal state with mean photon number μ, the decoy state with mean photon number $\nu < \mu$ is employed. Alice first decides on signal and decoy mean photon number levels, and then determines the optimal probabilities for these two levels be used, based on corresponding SKR expression. In weak-plus-vacuum decoy-state-based protocol, in addition to μ and ν levels, the vacuum state is also used as the decoy state. Both protocols have been evaluated in [19], both theoretically and experimentally. It was concluded that the use of one signal state and two decoy states is enough, which agrees with findings in [20, 21].

The probability that Alice can successfully detect the photon when Alice employs the WCS with mean photon number μ, denoted as q_μ, also known as the *gain*, can be determined by

$$q_\mu = \sum_{n=0}^{\infty} y_n e^{-\mu} \frac{\mu^n}{n!}, \tag{7.6}$$

where y_n, also known as the *yield*, is the probability of Bob's successful detection when Alice has sent n-photons. The similar expression holds for decoy states with mean photon number σ_i, that is,

$$q_{\sigma_i} = \sum_{n=0}^{\infty} y_n e^{-\sigma_i} \frac{\sigma_i^n}{n!}. \tag{7.7}$$

Let e_n denote the quantum bit error rate (QBER) corresponding to the case when Alice has sent n-photons, the average QBER for the coherent state $|\mu \exp(\mathrm{j}\,\phi)\rangle$ (where ϕ is the arbitrary phase) can be then estimated by

$$\bar{e}_\mu = \frac{1}{q_\mu} \sum_{n=0}^{\infty} y_n e_n e^{-\mu} \frac{\mu^n}{n!}. \tag{7.8}$$

If T is the overall transmission efficiency, composed of the channel transmissivity T_{ch}, photodetector efficiency η, and Bob's optics transmissivity T_{Bob}, the transmission efficiency of n-photon pulse will be given by

$$T_n = 1 - (1 - T)^n. \tag{7.9}$$

In the absence of Eve, for $n = 0$, the yield y_0 is affected by the background detection rate, denoted as p_d, so that we can write $y_0 = p_d$. On the other hand, the yield for $n \geq 1$ is contributed by both signal photons and background rate as follows:

$$y_n = T_n + y_0 - T_n y_0 = T_n + y_0(1 - T_n) = 1 - (1 - y_0)(1 - T)^n. \tag{7.10}$$

The error rate of n-photon states e_n can be estimated by [10, 22]

$$e_n = \frac{1}{y_n}\left(e_0 y_0 + e_d \underbrace{T_n}_{\cong y_n - y_0} \right) \cong e_d + \frac{y_0}{y_n}(e_0 - e_d), \tag{7.11}$$

where e_d is the probability of a photon hitting erroneous detector. For time-phase encoding, this represents the intrinsic misalignment error rate. In the absence of signal photons, assuming that two detectors have equal background rates, we can write $e_0 = 1/2$. In similar fashion, the average QBER can be estimated by

$$\bar{e}_\mu = e_d + \frac{y_0}{q_\mu}(e_0 - e_d). \tag{7.12}$$

The secrecy fraction can now be lower bounded by modifying the corresponding expression for BB84 protocols as follows:

$$r = q_1^{(Z)}\left[1 - h_2\left(e_1^{(X)}\right)\right] - q_\mu^{(Z)} f_e h_2\left(\bar{e}_\mu^{(Z)}\right), \tag{7.13}$$

where we used the subscript 1 to denote the single-photon pulses and μ to denote the pulse with the mean photon number μ.

As an illustration, in Fig. 7.1, we provide secrecy fraction results for decoy-state protocol and weak coherent state (WCS)-based, prepare-and-measure, BB84 protocol, assuming that the standard SMF with attenuation coefficient 0.21 dB/km is used.

Other parameters used in simulations are set as follows [10, 22–24]: the detector efficiency $\eta = 0.045$, the dark count rate $p_d = 1.7 \times 10^{-6}$, the probability of a photon hitting erroneous detector $e_d = 0.033$, and reconciliation inefficiency $f_e = 1.22$. For decoy-state-based BB84 protocol to calculate the secrecy fraction we use Eq. (7.13) for the optimum value of mean photon number per pulse μ, wherein q_μ is determined by $q_\mu = 1 - (1 - y_0)\exp(-\mu T)$, $\bar{e}_\mu$ is determined by Eq. (7.12), e_1 is determined by Eq. (7.11), and q_1 as the argument of (7.1) for $n = 1$, that is, by $q_1 = y_1 \mu \exp(-\mu)$. For WCS-based BB84 protocol, without the decoy state, we use Eq. (7.4), assuming that Eve controls the channel, performs intercept-resent attacks, and she is transparent to multi-photon states. In this lower bound scenario, the error from single-photon events dominates. Now by calculating $q^{(Z)}$ by $q_\mu = 1 - (1 - y_0)\exp(-\mu T)$, $e^{(Z)}$ by $E_\mu = e_\mathrm{d} + (e_0 - e_d)y_0/q_\mu$, and assuming suboptimum μ proportional to T, we have obtained the red curve in Fig. 7.1, which is consistent with Ref. [10]. When using the optimum μ instead (maximizing the secrecy fraction), we obtain the blue curve, which is consistent with Ref. [23]. For these two curves, we estimate the $q_1^{(Z)}$ from (7.6), by setting $y_n = 1$ for $n \geq 2$ (see [22]) as follows:

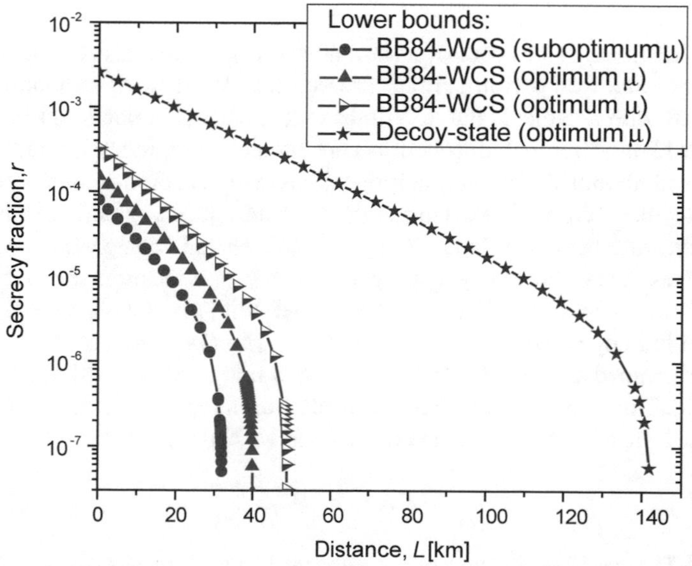

Fig. 7.1 Secrecy fraction versus transmission distance for decoy-state-based protocol and WCS-based BB84 protocol

$$q_1^{(X)} \geq q_\mu - \sum_{n=2}^{\infty} e^{-\mu} \frac{\mu^n}{n!}. \tag{7.14}$$

Given that multi-photon states do not introduce errors, that is, $e_n = 0$ for $n \geq 2$, we can bound the e_1 as follows:

$$e_1^{(X)} \leq q_\mu E_\mu / q_1^{(X)}. \tag{7.15}$$

More accurate expression for q_μ, for WCS-based BB84 protocol, will be the following:

$$q_\mu = y_0 e^{-\mu} + y_1 e^{-\mu} \mu + \sum_{n=2}^{\infty} e^{-\mu} \frac{\mu^n}{n!}, \tag{7.16}$$

which is obtained from (7.6) by setting $y_n = 1$ for $n \geq 2$. The corresponding curve is denoted in purple, which indicated that the distance of $L = 48.62$ km is achievable, when WCS-based BB84 protocol is used. In contrast, for decoy-state-based BB84 protocol and optimum μ, it is possible to achieve the distance close to 141.8 km, which agrees with Ref. [23].

7.2 Security of QKD Systems with Finite Resources

As discussed in BB84 protocol overview, after raw key transmission phase is completed, Alice and Bob perform sifting procedure and parameter estimation, so that they are left with n symbols out of N transmitted, and corresponding key is known as the sifted key. After that, they perform the classical postprocessing such as information reconciliation and privacy amplification. In information reconciliation stage, they employ error correction to correct the error introduced by both quantum channel and Eve, and the corresponding key after this stage is completed is commonly referred to as the corrected key. During the privacy amplification, they remove correlation that remained with Eve, and they are left with m symbols out of n, and the corresponding key is the secure key. Clearly, the *fractional rate* of the secret key r can be determined as $r = m/n$ (as $N \to \infty$), and it is often referred to as the *secret fraction* [8]. The *secret-key rate* (SKR) is then determined as the product of the secret fraction and the raw key rate, denoted R_{raw}, and we can write [8]

$$SKR = r \cdot R_{\text{raw}}. \tag{7.17}$$

The QKD can achieve the unconditional security, which means that its security can be verified without imposing any restrictions on either Eve's computational power or eavesdropping strategy. The bounds on the fraction rate are dependent on the classical postprocessing steps. The most common is the one-way postprocessing, in which either Alice or Bob holds the reference key and sends the classical information to the other party through the public channel, while the other party performs certain procedure on data without providing the feedback. Similarly to the physical-layer security (PLS) schemes described in Chap. 4, the most common one-way processing consists of two steps, the information reconciliation and privacy amplification. The expression for *secret fraction*, obtained by *one-way postprocessing*, is very similar to that for the classical PLS schemes [8]:

$$r = I(A;B) - \min_{\text{Eve's strategies}} (I_{EA}, I_{EB}), \tag{7.18}$$

where $I(A;B)$ is the mutual information between Alice and Bob, while the second term corresponds to the Eve's information I_E about Alice or Bob's raw key, where minimization is performed over all possible eavesdropping strategies. Alice and Bob will decide to employ either direct or reverse reconciliation so that they can minimize Eve's information. Here we are interested in the security against the collective attacks. In incoming section, we turn our attention to different security definitions.

7.2.1 *Finite-Length Secret-Key Fraction Rate*

Let the *perfect key* represent a sequence of perfectly correlated symbols between Alice and Bob, for which Eve does not possess any information. The key $\mathcal{K}$ that deviates from the perfect key by ε is said to be ε-*secure* key. One of the key requirements that the definition of security must satisfy is the *composability* property, indicating that the security of the key is guaranteed regardless of the application [8, 25–29]. In other words, if the ε-secure key is used in an ε'-secure task, the whole procedure must at least $(\varepsilon + \varepsilon')$-secure.

7.2.1.1 Trace-Distance-Based ε-Security Revisited

To deal with this problem, Ben-Or et al. in [25] as well as Renner and König in [27] introduced the concept of trace-distance-based ε-security (see also [28, 30–33]). We say that the key is ε-*secure* with respect to an eavesdropper $\mathfrak{E}$ if the trace distance between the joint state of key $\mathcal{K}$ and Eve $\mathfrak{E}$, denoted as $\rho_{\mathcal{K},\mathfrak{E}}$, and the product state of completely mixed state (CMS) with respect to the set $\mathcal{K}$ of all possible secure keys, denoted as ρ_{CMS}, and Eve's state, denoted as $\rho_{\mathfrak{E}}$, is smaller than or equal to ε. In other words, we can write

$$D\left(\rho_{\mathcal{K},\mathfrak{E}}, \rho_{CMS} \otimes \rho_{\mathfrak{E}}\right) = \frac{1}{2}\left\|\rho_{\mathcal{K},\mathfrak{E}} - \rho_{CMS} \otimes \rho_{\mathfrak{E}}\right\|_1 = \frac{1}{2}\text{Tr}\left|\rho_{\mathcal{K},\mathfrak{E}} - \rho_{CMS} \otimes \rho_{\mathfrak{E}}\right| \leq \varepsilon, \tag{7.19}$$

where $|\rho| = \sqrt{\rho^{\dagger}\rho}$. With this definition, the parameter ε has a quite clear interpretation, it represents the maximum tolerable failure probability for the key extraction process. Given that the composability of this definition is satisfied, we can decompose the security of the final key ε in terms of securities for error correction (EC) ε_{EC}, privacy amplification (PA) ε_{PA}, and parameter estimation (PE) ε_{PE} steps, as well as the Rényi entropies estimates failure probability [25], denoted as ε_R. In other words, we can write $\varepsilon = \varepsilon_{EC} + \varepsilon_{PA} + \varepsilon_{PE} + \varepsilon_R$. The security bounds discussed in previous section are valid only asymptotically for infinitely long keys, when $\varepsilon \to 0$. In addition to the trace-distance-based ε-security, the concept of *fidelity-based ε-security* can also be used [25]. However, as discussed in Chap. 6, these two approaches are equivalent to each other.

Now the main question is how to relate the trace-distance-based ε-security definition to the length m of the secret key being extracted. To do so one needs to prove the following inequality representing the generalization of the Chernoff bound [8]:

$$\Pr\left(D\left(\rho_{\mathcal{K},\mathcal{E}}, \rho_{CMS} \otimes \rho_{\mathcal{E}}\right) > \varepsilon\right) \le e^{m - f(\rho_{\mathcal{K},\mathcal{E}}, \varepsilon)}, \tag{7.20}$$

wherein the constant factors are omitted. Clearly, the key will be insecure with very small probability provided that the following condition is satisfied:

$$m \le f\left(\rho_{\mathcal{K},\mathcal{E}}, \varepsilon\right). \tag{7.21}$$

In other words, as long as the key length is shorter than Eve's uncertainty about the input to the privacy amplification step, the key will be secure.

7.2.1.2 Smooth Min-Entropy and Finite-Key Secret-Key Fraction Under Collective Attacks

In Chap. 4,we introduced the *Rényi entropy of X of order* α as follows [34]:

$$R_{\alpha}(X) = \frac{1}{1-\alpha}\log_2\left\{\sum\nolimits_x [P_X(x)]^{\alpha}\right\}, \tag{7.22}$$

where $P_X(x)$ is the distribution of random variable X. The *min-entropy* is defined as the special case of Rényi entropy in the limit when $\alpha \to \infty$, that is,

$$H_{\min}(X) = H_{\infty}(X) = \lim_{\alpha\to\infty} R_{\alpha}(X) = -\log_2 \max_{x\in X} P_X(x). \tag{7.23}$$

Renner introduced in [28] the concept of the *smooth min-entropy*, for the smoothness parameter $\varepsilon > 0$ as follows. Let $\mathcal{P}(\mathcal{H})$ denote the set of nonnegative operators on the Hilbert space $\mathcal{H}$. The density operator ρ on $\mathcal{H}$ is nonnegative if Hermitian conjugate with nonnegative eigenvalues. Let further $\rho_{\mathrm{AB}} \in \mathcal{P}(\mathcal{H}_{\mathrm{A}} \otimes \mathcal{H}_{\mathrm{B}})$ and $\sigma_{\mathrm{B}} \in \mathcal{P}(\mathcal{H}_{\mathrm{B}})$. The min-entropy of ρ_{AB} relative to σ_{B} is defined by [28]

$$H_{\min}(A|B) \doteq H_{\min}(\rho_{AB}|\sigma_B) = -\log\min\{\lambda|\lambda I_A \otimes \sigma_B - \rho_{AB} \geq 0\}, \qquad (7.24)$$

where I_A is the identity operator on subspace A. When $\mathcal{H}_B$ is the trivial space of complex numbers $\mathcal{C}$, the min-entropy of density operator ρ is simply

$$H_{\min}(\rho_A) = -\log\lambda_{\max}(\rho). \qquad (7.25)$$

The ε_R-*smooth min-entropy* of ρ_{AB} relative to σ_B is defined by [28]

$$H_{\min}^{\varepsilon_R}(\rho_{AB}|\sigma_B) = \sup_{\bar{\rho}_{AB}} H_{\min}(\bar{\rho}_{AB}|\sigma_B), \qquad (7.26)$$

where the supremum is taken over all operators $\bar{\rho}_{AB} \in \mathcal{P}(\mathcal{H}_A \otimes \mathcal{H}_B)$ such that $\mathrm{Tr}(\bar{\rho}_{AB}) \leq \mathrm{Tr}(\rho_{AB})$, $\|\bar{\rho}_{AB} - \rho_{AB}\| \leq \varepsilon_R \mathrm{Tr}(\rho_{AB})$. Let us now observe the QKD protocol in which Alice and Bob share N entangled pairs denoted by $\rho_{A^N B^N}$. Alice and Bob then apply the measurement on corresponding qubits in entangled pairs. For von Neumann measurements, Alice and Bob will then possess N highly correlated bits. In parameter estimation step, Alice and Bob select the subset of p correlated pairs to determine the statistics $\lambda_{(A,B)}$ of their data. In sifting phase, some of the data pairs are discarded. Alice and Bob are then left with $n \leq N - p$ bits, representing raw key, denoted, respectively, as X^n and Y^n. Let the sequence of bits Eve was able to get be denoted as E^n. In one-way postprocessing, Alice and Bob perform further information reconciliation (error correction) and privacy amplification to get the secure key of length $m \leq n$. The smooth min-entropy describes the number of uniform bits that can be extracted by the privacy amplification step [28, 29]. The *leftover hash lemma* from classical cryptography [35, 36] tells us that we can extract $m < H_{\min}(X)$ almost uniformly distributed bits by privacy amplification from a random variable X. More precisely, for universal$_2$ hashing function $h(S, X)$, where S is uniform over the set of keys $\mathbb{S}$, if the following is valid:

$$m \leq H_{\min}(X) - 2\log\left(\frac{1}{\varepsilon_{PA}}\right), \qquad (7.27)$$

then the statistical distance is bounded by

$$\delta(h(S, X), (U, S)) \leq \varepsilon_{PA}, \qquad (7.28)$$

where U is uniform over $(0, 1)^m$. The *statistical distance* for two random variables Y and Z is defined by

$$0 \leq \delta(Y, Z) = \frac{1}{2}\sum_v |\Pr(Y = v) - \Pr(Z = v)| \leq 1. \qquad (7.29)$$

The corresponding inequality for QKD will be just generalization of (7.27) as follows:

$$m \leq H_{\min}^{\varepsilon_R}(X^n|E^n) - 2\log\left(\frac{1}{\varepsilon_{PA}}\right). \tag{7.30}$$

For strict derivation of (7.30), an interested reader is referred to [28, 37]. The corresponding distance equivalent to (7.29) is given by

$$\Delta(A|B) = \min_{\sigma_B} \frac{1}{2}\|\rho_{AB} - I_A \otimes \sigma_B\|_1, \, \mathrm{Tr}(\sigma_B) = \mathrm{Tr}(\rho_B), \, \sigma_B \in \mathcal{P}(\mathcal{H}_B). \tag{7.31}$$

By rearranging the inequality (7.30), we obtain

$$2^{m-H_{\min}^{\varepsilon_R}(X^n|E^n)} \leq 2^{\log_2 \varepsilon_{PA}^2}, \tag{7.32}$$

which can be simplified as

$$\sqrt{2^{m-H_{\min}^{\varepsilon_R}(X^n|E^n)}} \leq \varepsilon_{PA}. \tag{7.33}$$

Now by comparing (7.33) and (7.28), we conclude that

$$\Delta(A|B) = \sqrt{2^{m-H_{\min}^{\varepsilon_R}(X^n|E^n)}} \leq \varepsilon_{PA}. \tag{7.34}$$

To account for the information leakage due to error correction, we need to modify the inequality (7.30) as follows:

$$m \leq H_{\min}^{\varepsilon_R}(X^n|E^n) - 2\log\left(\frac{1}{\varepsilon_{PA}}\right) - \mathrm{leakage}_{EC}. \tag{7.35}$$

In engagement-assisted protocols, Alice and Bob initially share the density state as follows $\rho_{A^N B^N} = (\sigma_{AB})^{\otimes N}$, where σ_{AB} is two-qubit Bell state. Given that different purifications performed by Eve are equivalent under local unitary transformation, we can write $\rho_{X^N E^N} = (\sigma_{XE})^{\otimes N}, \sigma_{XE} \in \Gamma$, where Γ is the set of density operators compatible with statistics $\lambda_{(A,B)}$, except for probability $\varepsilon'_{\mathrm{R}}$. It has been shown in [28–30] that the smooth min-entropy, for $\varepsilon_{\mathrm{R}} > \varepsilon'_{\mathrm{R}}$, is lower bounded by

$$H_{\min}^{\varepsilon_R}(X^n|E^n) \geq n\left(\min_{\sigma_{XE}} S(X|E) - (2D+3)\sqrt{\frac{\log(2/(\varepsilon_R - \varepsilon'_R))}{n}}\right), \tag{7.36}$$

where D is the dimensionality of the system and $S(\cdot|\cdot)$ denotes the conditional von Neumann entropy.

The corresponding expression for the finite-key *secret-key fraction under collective attacks* becomes

$$r_N = \frac{n}{N}\left[\min_{\sigma_{XE}\in\Gamma} S(X|E) - S(X|B) - \frac{1}{n}\log\left(\frac{2}{\varepsilon_{EC}}\right) - \frac{2}{n}\log\left(\frac{1}{\varepsilon_{PA}}\right) - (2D+3)\sqrt{\frac{\log(2/\varepsilon_R)}{n}}\right]. \tag{7.37}$$

The set of states Γ considers the fact that the parameter estimation is based on p samples. From the *law of large number*, the Theorem 12.2.1 in Ref. [38], we know that for samples $X_1, \ldots, X_p$ that are i.i.d. with distribution $P(x)$, the following is valid:

$$\Pr(\mathcal{D}(P_{x^n}||P) > \varepsilon_{PE}) \leq 2^{-p(\varepsilon_{PE}-D\ln(p+1)/p)}, \tag{7.38}$$

where $\mathcal{D}(\cdot)$ is the relative entropy. Further, from Lemma 12.6.1 in [38], we know that for two distributions P_1 and P_2, the following is valid:

$$\mathcal{D}(P_1||P_2) \geq \frac{1}{2\ln 2}\|P_1 - P_2\|_1^2. \tag{7.39}$$

When the statistics of λ_p is obtained by performing the POVM measurements with D outcomes on σ, then for every $\varepsilon_{\mathrm{PE}} > 0$, the state σ will be contained in the set [28, 29]:

$$\Gamma_\xi = \left\{\sigma \,|\, \left\|\lambda_p - \lambda_\infty(\sigma)\right\| \leq \xi\right\}, \xi = \sqrt{\frac{2\ln(1/\varepsilon_{PA}) + D\ln(p+1)}{p}}, \tag{7.40}$$

where $\lambda_\infty(\sigma)$ is the POVM-based probability distribution.

7.2.1.3 Finite-Key Secret-Key Fraction Under Coherent Attacks

Equation (7.37) is valid for collective attacks, but it can be extended to coherent attacks, as discussed in [30]. Let $\varepsilon_{\mathrm{coh}}$ denote the maximum tolerable failure probability under the coherent attacks. By introducing the following substitution [30],

$$\varepsilon = \varepsilon_{\mathrm{coh}}(N+1)^{-(D^4-1)}, \tag{7.41}$$

the corresponding secrecy fraction rate for coherent attacks becomes [30]

$$r_{N,\mathrm{coh}} = r_N - \frac{2\left(D^4-1\right)\log(N+1)}{N}. \tag{7.42}$$

Another approach to finite-key secret rates would be to apply *de Finetti representation theorem* [28, 30, 39]. The *de Finetti representation theorem* bounds the distance between the original N-partite state ρ_{AB} and a mixture of product states $\rho_{A^N B^N} = (\sigma_{AB})^{\otimes N}$, which are typically obtained when collective attack is applied by Eve [28, 30, 39], and are known as Finetti–Hilbert–Schmidt states. This theorem is tight when somebody wants to compare different QKD protocols on the quantum states' level. However, as shown in [30], it gives quite pessimistic results for finite-key secret-key rates. It has been shown in [40] that for the purpose of QKD, and some other quantum information processing applications, it is sufficient to study the distance between two permutation-invariant maps, implementing the QKD protocol. Let us assume that p out of $N_s < N$ sifted bits are used for parameter estimation, and k systems are traced out to come close to the product state to satisfy the conditions of the de Finetti theorem. The remaining $n = N_s - p - k$ bits represent the raw key. Let further $\rho_n^{|\theta\rangle}$ represent the rotationally invariant output of a QKD protocol, which in general is not in a product form (for any $|\theta\rangle$). It can be represented in the symmetric (permutation-invariant) subspace of $\mathcal{H}^{\otimes n}$ by $\rho_n^{|\theta\rangle} = \sum_\pi |\theta\rangle^{\otimes(n-t)}|\phi\rangle_t$, where the summation is over all possible permutation π, wherein $0 \le t \le p/2$. The parameter t is related to the distance between $\rho_n^{|\theta\rangle}$ and the perfect pure n-fold product state, and can be determined by [28, 30] $t = (N_s/k)[2\ln(2/\varepsilon_{\text{deF}}) + D^4 \ln k]$, where ε_{deF} is the error parametrizing t. The total error for coherent attack can be now decomposed by $\varepsilon = \varepsilon_{EC} + \varepsilon_{PA} + \varepsilon_R + \varepsilon_{PE} + \varepsilon_{\text{deF}}$. The maximum parameter estimation error for p samples now becomes [30]

$$\xi(p) = \frac{1}{D-1}\left[\frac{2\ln(1/\varepsilon_{PA}) + D\ln(p+1)}{p} + (1+\ln 2)h_2(t/p)\right]^{1/2}, \quad (7.43)$$

where the parameter k is determined in an optimum fashion, and $h_2(\cdot)$ is the binary entropy function. The corresponding expression for finite-length secret fraction rate, for the de Finetti theorem, under coherent attacks becomes [28, 30]

$$\begin{aligned} r_N = \frac{n}{N}\Bigg[&\min_{\sigma_{XE}\in\Gamma} S(X|E) - S(X|B) - \frac{1}{n}\log\left(\frac{2}{\varepsilon_{EC}}\right) - \frac{2}{n}\log\left(\frac{1}{\varepsilon_{PA}}\right) \\ &- \frac{4}{n}(p+k)\log(D) - \left(\frac{5}{2}D+4\right)\sqrt{\frac{\log(2/\varepsilon_R)}{n} + h_2\left(\frac{t}{n}\right)}\Bigg]. \end{aligned} \quad (7.44)$$

However, as shown in [30], this finite-length secret fraction rate results in too pessimistic SKR values.

7.2.2 Tight Finite-Key Analysis

The authors in [41] extended the security definitions to include both correctness of the key and ε-secrecy as introduced by Renner [28]. The QKD protocol outputs the key $\boldsymbol{K}$ on Alice side and key $\boldsymbol{K}'$ on Bob's side, both of which are of the same length, say l. We say that the key is *correct* if for any strategy that Eve applies they still end with the same key, that is, $\boldsymbol{K}' = \boldsymbol{K}$. We say that the QKD protocol is ε_c-*correct* if $\Pr[\boldsymbol{K}' \neq \boldsymbol{K}] \le \varepsilon_c$. On the other hand, the definition of security is a generalization to that given by Eq. (7.19). We say that the protocol is ε_s-*secret* if it is ε_s-indistinguishable from the perfectly secret protocol. The *condition for ε_s-secrecy* can be formulated as follows:

$$\begin{aligned}\min_{\rho_{\mathcal{E}}}(1-p_{\text{abort}})D\big(\rho_{\mathcal{K},\mathcal{E}},\rho_{CMS}\otimes\rho_{\mathcal{E}}\big) &= (1-p_{\text{abort}})\frac{1}{2}\min_{\rho_{\mathcal{E}}}\big\|\rho_{\mathcal{K},\mathcal{E}}-\rho_{CMS}\otimes\rho_{\mathcal{E}}\big\|_1 \\ &= (1-p_{\text{abort}})\frac{1}{2}\min_{\rho_{\mathcal{E}}}\operatorname{Tr}\big|\rho_{\mathcal{K},\mathcal{E}}-\rho_{CMS}\otimes\rho_{\mathcal{E}}\big| \le \varepsilon_s,\end{aligned} \tag{7.45}$$

p_{abort} is the probability for protocol to abort. The protocol for which $\varepsilon_s = 0$ is called here perfectly secret. We say that the *protocol is secure* if it is simultaneously correct and secret. To be more precise, we say that the *protocol is ε-secure*, if it simultaneously ε_c-correct and ε_s-secret, so that $\varepsilon_c + \varepsilon_s \le \varepsilon$. Finally, we say that the protocol is *robust* if it aborts with probability ε_r in the absence of Eve. As an illustration, a trivial protocol which always aborts is secure, but it does not result in any secure key. The ε-secure protocol results in a secure key of length l. To account for protocol robustness, we define the finite-key secrecy fraction rate as follows [41]:

$$r = (1-\varepsilon_r)\frac{l}{N(n,p)}, \tag{7.46}$$

where $N(n, p)$ denotes the number of qubits to be transmitted until n raw key bits and p bits to be used in parameter estimation are collected. By using this concept, for the quantum source of preparation quality q, the key is ε-secure if the following inequality is satisfied:

$$l \le n(1-h_2(Q+\Delta Q)) - \text{leak}_{EC} - \log_2\left(\frac{2}{\varepsilon_s^2\varepsilon_c}\right),\quad \Delta Q = \sqrt{\frac{n+p}{np}\frac{p+1}{p}\ln\left(\frac{2}{\varepsilon_s}\right)}, \tag{7.47}$$

where Q is the maximum tolerable QBER, and ΔQ corresponds to the correction to be added to account for statistical fluctuations. Evidently, this tight finite-key length estimate is applicable to two-dimensional protocols. However, it can straightforwardly be extended to two-base D-dimensional protocols. It has been shown that corresponding secrecy fraction rates agree very well with Renner definition-based expressions for sufficiently large N, but numerical results indicate that actual secrecy rates are much higher than predicted by Renner's theory for low and medium sifted key lengths N.

7.3 Finite-Key Analysis for BB84 and Decoy-State QKD Protocols Over Atmospheric Turbulence Channels

In this section, we provide finite-key analysis of secure fraction rates for weak coherent state (WCS)-based BB84 and decoy-state protocols in the presence of atmospheric turbulence effects [21]. (Notice that here we use the notation consistent with the chapter, rather than with paper [21].)

7.3.1 *BB84 Protocol Over Time-Varying FSO Channels*

The photon number of each signal pulse, for such a WCS source, follows a Poisson distribution with a source parameter μ, which represents the expected average photon number. The probability of weak coherent pulse to contain n-photons is given by $p_n = e^{-\mu}\mu^n/n!$, and the total probability that the pulses contain multi-photons is $p_{\text{multi}} = \sum_{n\geq 2} p_n$. In such cases, occasional multi-photon emission from the source gives an opportunity to Eve to perform the photon number splitting (PNS) attack [42] (see also Chap. 6). In order to ensure the security, source intensity (parameter μ) needs to be additionally attenuated as the channel transmittance decreases, thus greatly reducing the achievable secure fraction rate and maximum possible transmission distance. For a given channel loss, secure fraction rate strongly depends on the source intensity. Based on tight finite-key theory [41], briefly described in previous section, we represent the secure fraction rate as follows [21]:

$$r = \frac{1}{2}p_{\text{exp}}\left\{\frac{p_{\text{exp}} - p_{AE}p_{\text{multi}}}{p_{\text{exp}}}[1 - h(Q + \Delta Q)] - \frac{\text{leak}_{EC}}{N} - \frac{1}{N}\log\frac{2}{\varepsilon_s^2\varepsilon_c}\right\}, \tag{7.48}$$

where Q is the QBER, while $\Delta Q \cong \sqrt{1/N\ln(1/\varepsilon_s)}$ is introduced to account for statistical fluctuations due to finite-size key length. The expected detection probability is determined by $p_{\text{exp}} = 1 - \exp(-\mu T_B T)$, where T_{B} denotes the total detection efficiency of Bob's side, and it is determined as the product of receive aperture transmittance T_{Rx}, Bob's optics transmittance η_{B}, detector transmittance η_{det}, and the quantum efficiency of detector η_{Qeff}, that is, $T_B = T_{RX}\eta_B\eta_{\text{det}}\eta_{Qeff}$. We use $T = 10^{-aL/10}$ to denote the channel transmittance, wherein a is the attenuation coefficient measured in dB/km and L is the transmission distance in km. With $h_2(\cdot)$, as before, we denoted the binary entropy function. Further, the QBER can be decomposed as $Q = Q_{\text{mis}} + 1/2p_b$, where Q_{mis} is the probability of having an error per time slot due to misalignment and p_b denotes the background noise error probability. The misalignment error probability is determined by the probability that Eve performs the intercept-resend (IR) attack on single-photon pulses, denoted as p_{IR}, so that the misalignment error probability is determined by $Q_{\text{mis}} = 1/4\, p_{\text{IR}}\, p_1\, p_{\text{exp}}$. The leakage due to error correction is estimated by $\text{leak}_{EC} \approx f_{EC} h_2(Q)$, with f_{EC} being the error correction inefficiency which is bounded as $f_{EC}(Q) \geq 1$. The security parameters ε_c and ε_s denote ε_c-correctness and ε_s-secrecy, which are defined in previous section. The total ε-security of the protocol is determined by $\varepsilon \geq \varepsilon_c + \varepsilon_s$. Finally, p_{AE} in Eq. (7.48) denotes the detection probability at Eve's detectors, which is also related to Alice's optics transmittance T_{Alice} and Eve's location. The position of Eve will affect the optimal choice of source intensity, as shown below. Intuitively, the more Eve can access the more we need to attenuate the source brightness.

We calculate in the absence of turbulence the optimal source intensity μ for different channel losses and different Eve's positions with results being summarized in Fig. 7.2. As evident from the figure, both the secure fraction rate and optimal parameter μ decrease as the channel loss increases. As the distance between Alice and Eve, denoted as d_{AE}, decreases, which means Eve is closer to Alice and that the ratio between d_{AE} and the total distance d_{AB} decreases from 1 (Eve at Bob's side) to 0 (Eve at Alice's side), the optimal source intensity μ decreases, and hence the secure fraction rate decreases as well.

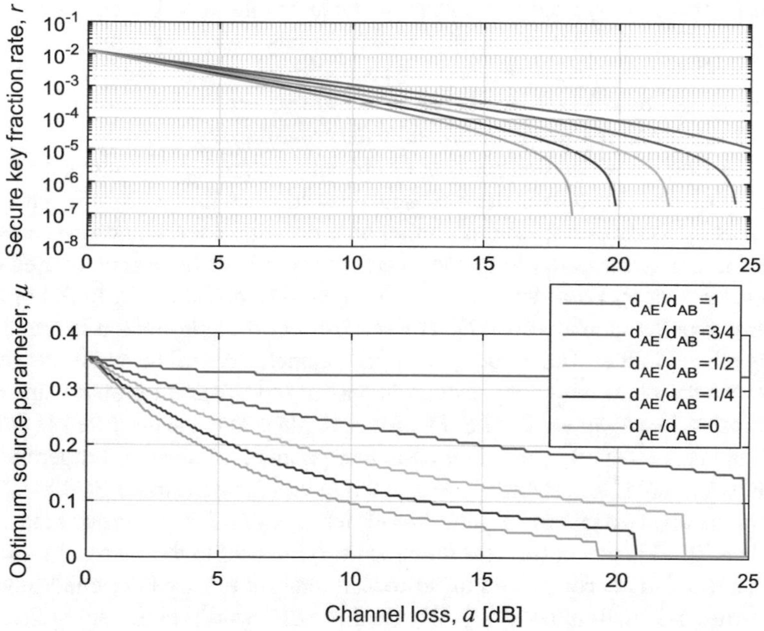

Fig. 7.2 (Top) Secret fraction rate against channel loss in dBs, with Eve located at different distance from Bob's side ($d_{\text{AE}}/d_{\text{AB}} = 1$) to Alice's side ($d_{\text{AE}}/d_{\text{AB}} = 0$). (Bottom) The optimal source intensity μ versus channel loss (also expressed in dBs). In calculations, the parameters are selected as follows: $T_{\text{Alice}} = 0.4725$, $T_{\text{Rx}} = 0.7884$, $\eta_{\text{B}} = 0.898$, $\eta_{\text{det}} = 0.4352$, $\eta_{\text{Qeff}} = 0.9$, $a = 0.192$ dB/km, $p_{\text{b}} = 10^{-5}$, $f_{\text{EC}} = 1.16$, $N = 10^8$, $\varepsilon_{\text{c}} = \varepsilon_{\text{s}} = 10^{-6}$. [Modified from [21]. © IEEE 2016; reprinted with permission.]

Now we describe the *time-varying FSO turbulent channel model* [21, 43]. The light beam propagating through a time-varying FSO channel experiences both phase and intensity perturbations due to the random distribution of the refractive index, which is known as turbulence [43–45]. The intensity or irradiance fluctuation is commonly referred to as scintillation. Scintillation effect in FSO channels has been extensively studied to derive a probability density function (PDF) of the random fluctuation of irradiance. The Gamma–Gamma (Γ–Γ) distribution serves as a good model for the optical scintillation ranging from weak to strong turbulence regimes, which is given as [43–45]

$$p(I) = \frac{2(\alpha\beta)^{(\alpha+\beta)/2}}{\Gamma(\alpha)\Gamma(\beta)} I^{(\alpha+\beta)/2-1} K_{\alpha-\beta}\left[2(\alpha\beta I)^{1/2}\right], I > 0, \tag{7.49}$$

where I is the normalized irradiance at the receiver, $\Gamma(\cdot)$ is the gamma function, and $K_{\mathrm{p}}(\cdot)$ is a modified Bessel function of the second kind and the order p. The parameters α and β represent the effective number of large-scale cells and small-scale cells in the scattering process, respectively, and are defined as functions of Rytov variance σ_R^2 which is commonly used as a metric for the atmospheric turbulence strength. For their relationship to Rytov variance for nonzero inner scale, an interested reader is referred to [21]. Weak turbulence is associated with $\sigma_R^2 < 1$, the moderate turbulence is characterized by $\sigma_R^2 \sim 1$, while the strong turbulence regime with $\sigma_R^2 > 1$. The so-called saturation regime is specified by the condition $\sigma_R^2 \to \infty$.

For the frequency characterization of optical scintillation, the intensity fluctuation is highly correlated over the atmospheric coherence time. The coherence time, denoted as τ_{c}, is defined as the minimum time duration over which two successive received intensity samples are uncorrelated. Many studies have been carried out to model the power spectral density (PSD) of optical scintillation for both horizontal and vertical paths such as [46, 47]. The spectrum of the scintillation is found to be directly related to the wind velocity over the channel. General formulas are given in [45] for the temporal frequency spectra of optical scintillation for both plane waves and Gaussian beam waves. The PSD varies from place to place for different environments, and it is affected by weather condition parameters, such as temperature and humidity. A simple Butterworth-type spectral model for terrestrial FSO links, which is determined by two parameters, the cutoff frequency and the spectral slope, is proposed in [47]. This power transfer function fits quite well to the shape of the optical scintillation's PSD derived from experimental data for a 1 km horizontal link in an urban terrestrial environment [47]. We use this Butterworth model of the first order with a cutoff frequency of 12 Hz, in FSO channel model, to achieve a channel coherence time of $\tau_{\mathrm{c}} \sim 10$ ms. To obtain the optical scintillation under strong turbulence, we first generate a sequence of uncorrelated normal distributed random samples, and then filter the sequence by the Butterworth filter. We then map each sample from the normal distribution to a Gamma–Gamma distribution given in Eq. (7.49) with the Rytov variance $\sigma_R^2 = 30$ by using the memoryless nonlinear transform (MNLT) technique [48], which maps data from the normal distribution to Gamma–Gamma distribution by equating their cumulative density functions (CDFs), wherein the cor-

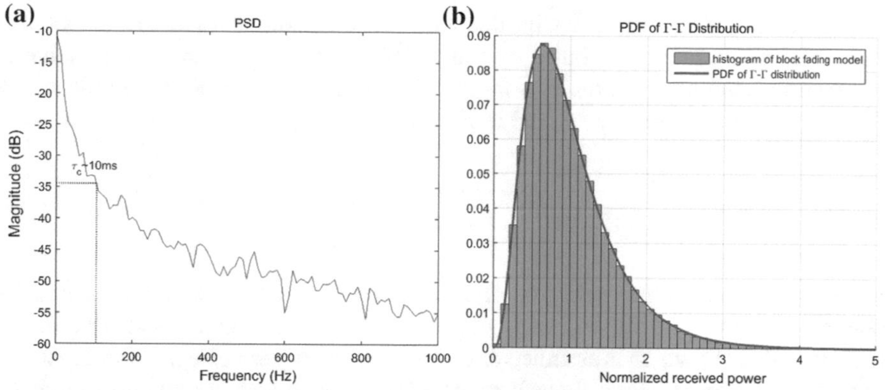

Fig. 7.3 Time-varying channel model: **a** PSD of generated block fading model **b** PDF of Gamma–Gamma distribution with Rytov variance of 30 and histogram of block fading model of normalized received irradiance. [After Ref. [21]. © IEEE 2016; reprinted with permission.]

relation in the sequence is preserved. The PSD and PDF of the time-varying FSO channel model are illustrated in Fig. 7.3a and b, respectively. The coherence time is approximately 10 ms as targeted. We further simplify the scintillation model to a time-varying block fading model by assuming that channel is constant within coherence time since the intensity samples are highly correlated. As it can be seen from Fig. 7.3b, the PDF of the block fading model fits very well the theoretical Gamma–Gamma distribution. We also assume that fluctuations in the received intensity are entirely due to fluctuations in the channel transmittance, so that the transmission efficiency of the channel, e.g., the channel transmittance T, is proportional to the received irradiance, which can be calculated as the product of average transmittance and normalized received power given by the block fading model.

The secure fraction rates performance of FSO-based QKD system can be significantly affected by the atmospheric turbulence effects. Typically, a fixed source intensity was often assumed, based on an average channel condition, while the time-varying nature of the FSO channel is commonly neglected. Therefore, in the presence of turbulence, the fixed transmit power is often mismatched with respect to real-time channel transmittance, and hence leading to a reduced security.

To deal with time-varying FSO channel conditions, we proposed an adaptation method in which we monitor the FSO channel with an auxiliary classical probe signal, and then predict channel states based on the previous and current channel states and adapt the source brightness according to the channel predictions to get improved secure fraction rates [21]. In a realistic FSO system, there might be some delay between measurement of the channel state and implementation of the adjustment to the source parameters, which could arise from the physical transmission latency through the path or due to some signal processing procedures in the system. In such case, predicting the channel states in a real time is necessary. In our adaptation method, an *autoregressive (AR) predictor* is used on Alice's side to predict channel state information, and the accuracy of the predictions determines how much improvement we can achieve with this adaptation scheme. An AR model can also be treated as an infinite impulse response (IIR) filter and the problem to solve is to determine the filter coefficients so that the mean square error is minimized. By definition, an autoregressive model is given as

$$y_t = \sum_{i=1}^{o} a_i y_{t-i} + \varepsilon_t, \tag{7.50}$$

where a_i are the autoregressive coefficients ($i = 1, \ldots, o$), y_t represent successive channel states of time-varying channel, o is the order of the model, and ε_t represent the residual error, which is assumed to be a zero-mean white noise process. By using the AR model above, each channel state y_t can be predicted from o previous states as follows:

$$\widehat{y}_t = \sum_{i=1}^{\rho} a_i y_{t-i}, \tag{7.51}$$

where $\widehat{y}_t$ denotes an estimate of the channel state y_t. The prediction error is defined as the difference between the estimated and actual channel states, that is, $\widehat{\varepsilon}_t = \widehat{y}_t - y_t$. As the order of the AR model increases, the performance of predictor gets improved. To evaluate the accuracy of the AR method, the root mean square (RMS) error between the predicted and the actual channel states' values is typically used, as illustrated in Fig. 7.4. We model an AR(o) predictor based on the block fading model described above and perform either single-step prediction or multi-step prediction with the order of up to 20. As shown in Fig. 7.4a, the RMS error drops very fast at order o = 3 and then decreases slowly as the filter order increases. The prediction length K affects the prediction accuracy, which is denoted as "K-steps-ahead" prediction in Fig. 7.4b. As expected, the RMS error increases as the prediction length increases.

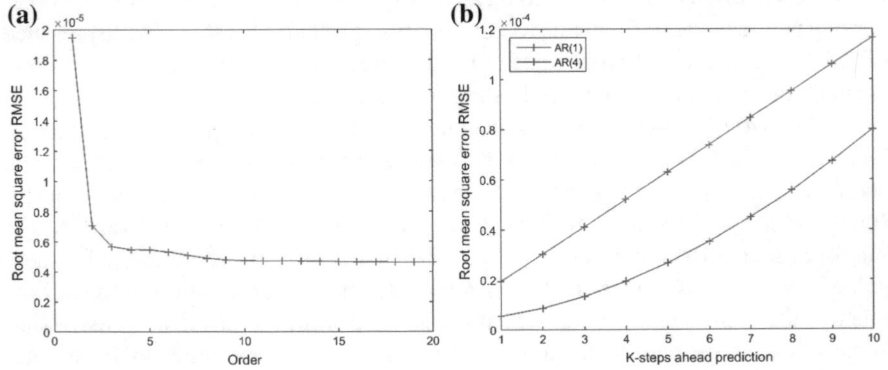

Fig. 7.4 **a** Root mean square error (RMSE) against the AR model order. **b** RMSE versus prediction length K. [Modified from [21]. © IEEE 2016; reprinted with permission.]

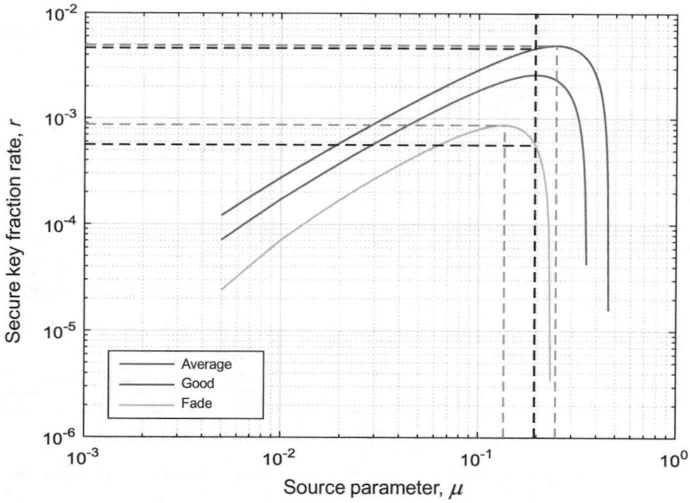

Fig. 7.5 Illustration of improvement in secure fraction rate by AR-model-based source brightness adaptation. [Modified from [21]. © IEEE 2016; reprinted with permission.]

An example of the improvement for our adaptation method is illustrated in Fig. 7.5. In calculations, we assumed that Eve located at Alice's side ($d_{AE}/d_{AB} = 0$). The blue curve in the middle corresponds to the secure fraction rate for an average channel transmittance $T = T_{\text{abs}}T_{\text{scatt}}$, in which the absorption T_{abs} and scattering T_{scatt} transmittances are set to 0.9917 and 0.365, respectively. For this mean value, the optimal source intensity $\mu = 0.195$ is chosen to maximize the secure fraction rate. When the FSO channel is in a deep fade, as illustrated by the yellow curve, we need to turn down the source brightness to ensure the security. The optimal intensity is changed to a new value $\mu = 0.135$. It can be easily seen from the figure that the secure fraction rate improves from the black dash line to the gray dash line by operating with the new optimal value. Similar to the worst case, when the channel condition becomes better, like the red curve shown at top, adapting the source intensity to the new higher optimal value $\mu = 0.25$ also increases SKR. In summary, the adaptation method allows the source to operate at the optimal source brightness (or suboptimal when the channel state predictions are not very accurate).

By employing the AR model-based adaptation method, we simulate BB84-based QKD for weak coherent source under strong turbulence regime by modeling the time-varying FSO channel as the block fading model. Here we use the same parameters introduced above. A simple first-order AR(1) model, which is trained by first 1000 samples in the channel model, is used to predict the channel states one coherence time, τ_c, and five coherence times ahead. The secure fraction rate is calculated using Eq. (7.48) for each coherence time and then averaged over total transmitted signal pulses N, and the corresponding numerical results are summarized in Fig. 7.6. The source brightness adaptation method indeed increases the secure fraction rate compared to the case when a fixed source intensity at $\mu = 0.195$ is used, which

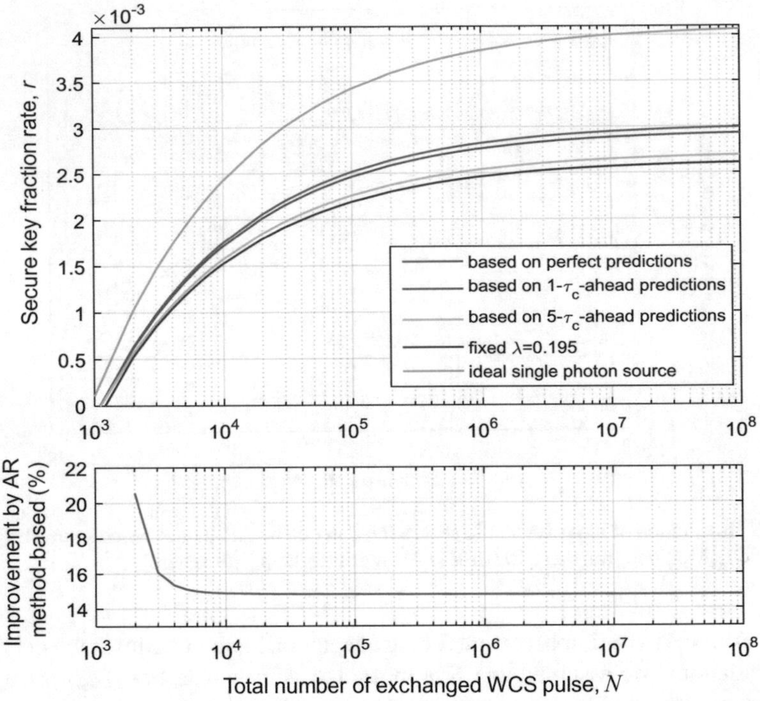

Fig. 7.6 Improvement in AR-predictor-based adaptation: (top) secure fraction rate versus the total number of exchanged WCS pulse, N; (bottom) improvement ratio for perfect compensation versus N. [Modified from [21]. © IEEE 2016; reprinted with permission.]

is optimal for the average channel conditions. For perfect predictions' curve, we determine the secure fraction rate by assuming that the channel states are perfectly predicted. Although the substantial technologies challenge of creating on-demand single-photon sources continues to limit the possibility of "pure" BB84 at high data rate, when compared to the secure fraction rate of ideal BB84 using single-photon sources, our adaptation method narrows the gap of using commercial WCS sources. The improvement ratio in secure fraction rate for the perfect compensation case is shown in Fig. 7.6. In addition, the performance of AR-method-based source brightness adaptation depends on the accuracy of channel state predictions. As the prediction accuracy degrades, when the prediction length increases, the secure fraction rate gets reduced compared to the ideal perfect prediction case. The secure fraction rate corresponding to one coherence time ahead prediction closely approaches the perfect prediction case.

7.3.2 Decoy-State Protocol Over Time-Varying FSO Channels

For optimization of the decoy-state-based protocol, we follow the protocol description given in [49]. A BB84-based decoy-state protocol with three different states, one signal state and two decoy states, is studied in this section. In [50], it has been shown that the improvement in secure fraction rate is less than one percent when employing more than two decoy states, in addition to the signal state. Here, we use $\mu = \{\mu_1, \mu_2, \mu_3\}$ to denote the three states used in the protocol, and $\lambda = \{\lambda_1, \lambda_2, \lambda_3\}$ to denote the corresponding source intensities. Each state will be chosen with probability $p_{\mu_1} > p_{\mu_2} > p_{\mu_3} = 1 - p_{\mu_1} - p_{\mu_2}$. Following the finite-key length security proof technique for decoy-state-based protocol based on the uncertainty relation for smooth entropies [28, 29, 41, 51], the secure-key fraction rate for an unbiased bases decoy-state protocol, in which two mutually bases are equally chosen, is given by

$$r \le \frac{1}{2}\left[\underline{Q}^{(0)} + \underline{Q}^{(1)} - \overline{Q}^{(1)} h\left(\overline{e}^{(1)}\right) - \frac{\text{leak}_{EC}}{N} - \Delta(N)\right], \tag{7.52}$$

where lower bound and upper bound are indicated by underbars and overbars, respectively. In (7.52), $Q^{(n)}$ denotes the detection rate at Bob's side which is contributed from signals with n photons emitted by Alice's source. With e we denoted the total QBER, while $e^{(n)}$ represents the QBER for n photon pulses. Similarly, as before, $h_2(x)$ denotes the binary entropy function. The parameter leak_{EC} corresponds to the leakage due to error correction (information reconciliation). This part of information must be subtracted when calculating secure fraction rate because Eve could determine these bits by simply tapping the classic channel. Typically, it is approximated by $\text{leak}_{EC} \approx Q f_{EC} h(e)$, where Q is the total detection rate and f_{EC} is the error correction inefficiency. The last term in (7.52), $\Delta(N)$, is related to the security under finite-size effect, which is estimated by

$$\Delta(N) = \frac{1}{N}\log_2\left(\frac{2}{\varepsilon_c}\right) + \frac{6}{N}\log_2\left(\frac{46}{\varepsilon_s}\right). \tag{7.53}$$

Similarly, as in previous section, the decoy-state-based QKD protocol is called ε-secure, or more precisely ε_c-correct and ε_s-secret, when $\varepsilon \ge \varepsilon_s + \varepsilon_c$. The lower bounds and upper bounds in Eq. (7.52) are determined by using Hoeffding's inequality [52] and constrained optimization. More specifically, measurable quantities in the protocol, such as detection counts and number of errors, are bounded from the asymptotic values by Hoeffding's inequality: $\left|x_{\text{asym}} - x\right| \le \sqrt{(x/2)\ln(1/\varepsilon)}$ with a failure probability of 2ε. Then these bounds are used as constrained limits to solve optimization problem to estimate quantities which cannot be directly measured. For example, linear programing is used to obtain the detection rate of n-photon signals $\underline{q}_{\mu_i}^{(n)}$, which originate from pulses Alice prepared in state μ_i:

$$\min \underline{q}_{\mu_i}^{(n)}, \tag{7.54}$$

subject to:

$$\begin{aligned} &\underline{Q}_{\mu_i} \le \sum_{n=0}^{\infty} \underline{q}_{\mu_i}^{(n)} e^{-\lambda_i} \frac{\lambda_i^n}{n!} \le \overline{Q}_{\mu_i}, n = \{0, 1\}, i = \{1, 2, 3\} \\ &0 \le \underline{q}_{\mu_i}^{(n)} \le 1, n = \{0, 1\}, i = \{1, 2, 3\}, \end{aligned} \tag{7.55}$$

where $Q_{\mu_i} = C_{\mu_i}/N_{\mu_i}$ is the detection rate for each state μ_i, C_{μ_i} and N_{μ_i} are the total number of signals that get detected and sent in state μ_i, respectively. Then the bounds for the detection rate of n-photon pulses are given as $\underline{Q}^{(n)} = \left\lfloor \frac{N_{\mu_i}}{N} \underline{q}_{\mu_i}^{(n)} e^{-\lambda_i} \frac{\lambda_i^n}{n!} \right\rfloor$. The optimization problem for upper bound of single-photon QBER $\overline{e}^{(1)}$ can be simplified by the following relation [53]:

$$\overline{e}^{(1)} \le \frac{e^{\lambda_i} e_{\mu_i} - \frac{1}{2} \underline{q}_{\mu_i}^{(0)}}{\lambda_i \underline{q}_{\mu_i}^{(1)}}, \tag{7.56}$$

where $e_{\mu_i} = E_{\mu_i}/N_{\mu_i}$ is the QBER from each state μ_i, E_{μ_i} is the total number of errors that get detected in state μ_i. Numerical optimization has been performed to determine the optimal parameters for the decoy-state protocol, with numerical results being summarized in Fig. 7.7 for the total number of exchanged signal pulses being $N = 10^8$. Clearly, source intensity for signal state does not need to be scaled down in linear fashion with the channel transmittance, and the optimal value is maintained in a relatively high level of $\lambda \sim 1$. As the channel loss increases, stronger decoy states are needed more often for estimating single-photon statistics since the detection rate is low. Note that, our numerical results indicate that one of the decoy states is always optimal at zero intensity, which is consistent with the analysis in [54] in which such a state is called a vacuum decoy state.

Similarly to the WCS-based BB84, FSO-based decoy-state QKD is affected by the atmospheric turbulence effects. We simulate the decoy-state protocol described above under strong turbulence with the AR-predictor-based adaptation. The same security values $\varepsilon_c = \varepsilon_s = 10^{-6}$ and background noise $p_b = 10^{-5}$ are used in simulations as in the WCS-based BB84 protocol. As shown in Fig. 7.8, secure fraction rates are improved with AR-predictor-based adaptation method and the level of secure-key fraction rates performance improvement is dependent on the accuracy of predictions. The secure fraction rates are upper bounded by the perfect prediction case. Compared to the BB84 protocols, since the secure fraction rate scales linearly with the channel transmittance for decoy-state protocol [55], the improvement of adapting source intensity is lower. For the fixed-parameter case, the optimal parameters are calculated under asymptotic assumption, and thus as expected it performs better when the total number of transmitted signal pulses N is large. The adaptation improvement is reduced from 20% at $N = 10^6$ to 5% at $N = 10^9$ in terms of the secure fraction

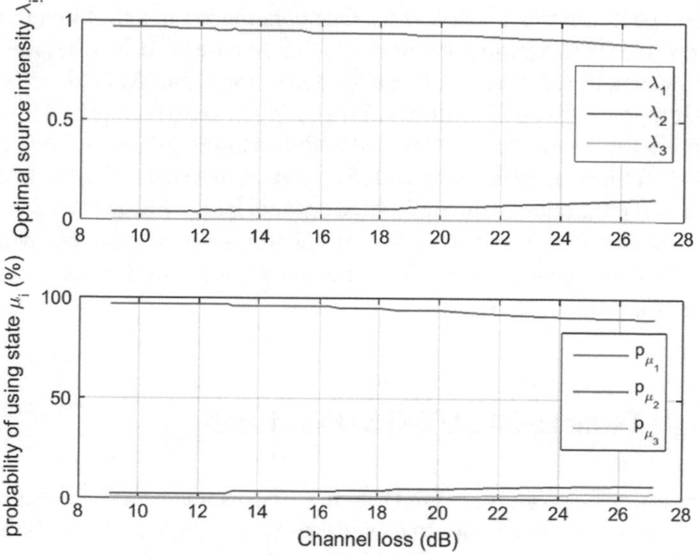

Fig. 7.7 Optimal two decoy-state protocol parameters versus instantaneous channel loss resulting from the turbulence: (top) optimal source intensity versus channel loss in dBs, (bottom) probability (%) of Alice using each state versus channel loss. [After Ref. [21]. © IEEE 2016; reprinted with permission.]

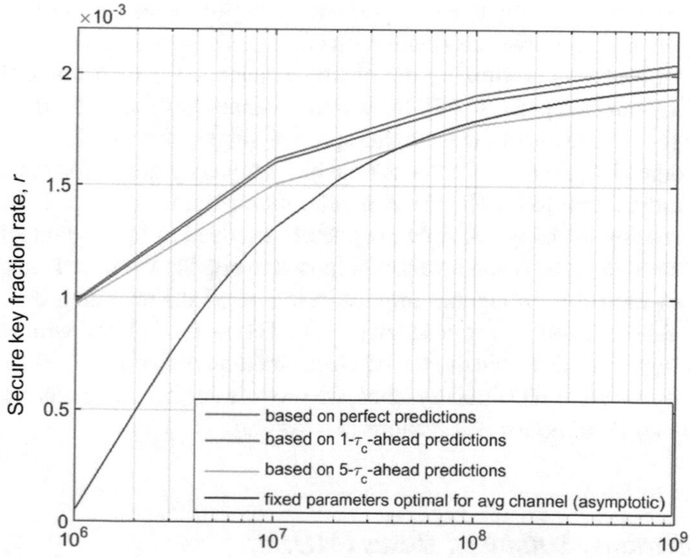

Fig. 7.8 Secure-key fraction rate versus total number of transmitted signal pulses, N, for decoy-state-based QKD with AR-method-based adaptation. [Modified from [21]. © IEEE 2016; reprinted with permission.]

rate with respect to perfect adaptations. One coherence time ahead prediction case closely approaches the perfect channel estimation case, and it is helpful for all N values. However, the accuracy of five coherence time ahead predictions is not good enough compared to BB84 for all N's. When N is large, where the fixed-parameter case performs well, the adaptation method leads to a lower secure fraction key rate due to mismatch from poor channel state predictions. In this sense, decoy-state protocol requires higher prediction accuracy than the original BB84. In other words, the decoy-state-based protocol is more robust to atmospheric turbulence effects, compared to BB84 protocol, so that the AR-predictor-based adaptation in this case does not help much for sufficiently long N.

7.4 High-Dimensional DV-QKD Protocols

Despite the appealing features of QKD, there are two fundamental and technical challenges need to be addressed prior to its widespread applications. First, the secret-key rate of QKD is fundamentally limited by channel loss, as quantified by the rate-loss trade-off [56, 57, 58]. Quantum repeaters would represent an ultimate solution to overcome the channel loss, but at this point of time, they are well beyond the reach of available technology. A near-term solution is to encode quantum information into a Hilbert space with a dimension higher than the mainstream two-dimensional encoding using, e.g., polarization states of photons. A high-dimensional (HD) encoding space ensures that multiple information bits can be delivered upon receiving each single photon, thereby optimizing the photon efficiency for high-rate QKD in the absence of quantum repeaters [59]. Several approaches have recently been pursued to increase the dimension of the encoding space of single photons. These approaches either leverage HD parameters such as time bin, frequency qubits [60, 61], position and linear momentum [62–64], orbital angular momentum (OAM) [32, 33, 65–72], or simultaneously utilizing multiple parameters encoded in hyper-entangled states [73, 74]. Moreover, it has been recently proposed to employ the fiber Bragg gratings (FBGs) with mutually orthogonal impulse response as the encoding basis for HD QKD [31]. Alternatively, as proposed in [75], FBGs can be replaced with waveguide Bragg gratings (WBGs), which can be electronically controlled. Before we continue our discussion of HD QKD systems, we provide additional details on mutually unbiased bases (MUBs) for HD systems [31, 76–79].

7.4.1 *Mutually Unbiased Bases (MUBs)*

Two orthonormal bases $\{|e_1\rangle, \ldots, |e_D\rangle\}$ and $\{|f_1\rangle, \ldots, |f_D\rangle\}$ in Hilbert space $\mathbb{C}^D$ are MUBs when the square of magnitude of the inner product between any two-basis states $\{|e_m\rangle\}$ and $\{|f_n\rangle\}$ is equal to the inverse of the dimension:

$$|\langle e_m | f_n \rangle|^2 = \frac{1}{D} \quad \forall m, n \in \{1, \cdots, D\}. \tag{7.57}$$

The key word *unbiased* means that if a system is prepared in a state belonging to one of the basis, all outcomes of the measurement with respect to the other basis will be equally likely. We say that the set of MUBs of dimensionality D is *maximal*, if there exist $D + 1$ bases that are mutually unbiased in pairwise fashion. When dimensionality is a prime number p, the construction of maximal set was provided in [76]. When the dimensionality is the prime power, the construction of maximal sets was provided in [77]. The problem of proving if there exist $D + 1$ bases for arbitrary dimensions seems to be open. In general, if D is composite number represented in prime power factorization as follows:

$$D = p_1^{n_1} p_2^{n_2} \cdots p_d^{n_d}; \; p_1^{n_1} < p_2^{n_2} < \cdots < p_d^{n_d}, \tag{7.58}$$

then the maximum number of MUBs, denoted as $\mathfrak{N}(D)$, will be bounded by

$$p_1^{n_1} + 1 \leq \mathfrak{N}(D) \leq D + 1. \tag{7.59}$$

Clearly, when $D = p^n$, then $\mathfrak{N}(D) = p^n + 1$.

As an illustration, the set of MUBs for $D = 3$ used in three-state BB84 QKD protocol is given by

$$MUB_0 = \{|0\rangle, |1\rangle\}, \quad MUB_1 = \left\{|+\rangle = \frac{|0\rangle + |1\rangle}{\sqrt{2}}, |-\rangle = \frac{|0\rangle - |1\rangle}{\sqrt{2}}\right\}, \tag{7.60}$$

$$MUB_2 = \left\{|R\rangle = \frac{|0\rangle + j|1\rangle}{\sqrt{2}}, |L\rangle = \frac{|0\rangle - j|1\rangle}{\sqrt{2}}\right\}. \tag{7.61}$$

7.4.1.1 Computational Basis and Dual Basis

For *two-bases, D-dimensional, protocols,* it is straightforward to determine two MUBs. One of the MUBs will be just the *computational basis* (CB),

$$MUB_0 = \{|0\rangle, |1\rangle, \ldots, |D - 1\rangle\}. \tag{7.62}$$

The other basis, known as the *dual basis* (DB), can be obtained by applying the discrete quantum Fourier transform to the basket in computational basis:

$$MUB_1 = \left\{\left|\hat{0}\right\rangle, \left|\hat{1}\right\rangle, \ldots, \left|\widehat{D-1}\right\rangle\right\},$$
$$\left|\hat{i}\right\rangle = \frac{1}{\sqrt{D}} \sum_{d=0}^{D-1} e^{-j\frac{2\pi}{D}di} |i\rangle = \frac{1}{\sqrt{D}} \sum_{d=0}^{D-1} W_D^{-di} |i\rangle, \; W_D = e^{j\frac{2\pi}{D}}. \tag{7.63}$$

These two MUBs have been employed in time-phase HD encoding [80]. Clearly, CB and DB are MU:

$$\left\langle \hat{i} | k \right\rangle = D^{-1/2} W_N^{ik}; \quad i, k = 0, 1, \ldots, D-1 \tag{7.64}$$

as expected. More formally, we can use the *Weyl–Schwinger* formalism as discussed in [78]. Similarly to Pauli *X*- and *Z*-operators for 2-D systems, we can define the *cyclic X*- and *Z*-operators as follows:

$$\begin{aligned} Z|k\rangle &= W_D^k |k\rangle,\ Z^D = \mathbf{1}, \\ X\left|\hat{i}\right\rangle &= W_D^i \left|\hat{i}\right\rangle,\ X^D = \mathbf{1}, \end{aligned} \tag{7.65}$$

where **1** is the identity operator. By employing the MU property (7.64), the actions of cyclic operators become

$$\begin{aligned} &X|k\rangle = |k+1\rangle;\ k = 0, 1, \ldots, D-2;\ X|D-1\rangle = |0\rangle \\ &\left\langle \hat{i} \right| Z = \left\langle \widehat{i+1} \right|;\ i = 0, 1, \ldots, D-2;\ \left\langle \widehat{D-1} \right| Z = \left\langle \hat{0} \right|. \end{aligned} \tag{7.66}$$

The cycle operators satisfy the fundamental *Weyl commutation rule*:

$$X^l Z^k = W_D^{-lk} Z^k X^l. \tag{7.67}$$

7.4.1.2 Heisenberg–Weyl Group

The products of *XZ*, together with the powers of W_N form a group, are commonly referred to as the *Heisenberg–Weyl group*. In analogy with 2-D systems, where product of *X* and *Z* Pauli operators is related to Pauli *Y* operator by $Y = \mathrm{j}XZ$, we can define the *Y*-operators in the Weyl group by

$$Y_{k,l,m} = W_D^k X^l Z^m; \quad k, l, m = 0, 1, \cdots, D-1. \tag{7.68}$$

This group of unitary operators is sometimes called the *generalized Pauli group*. The multiplication operation is, in fact, the composition:

$$\begin{aligned} Y_{k_1,l_1,m_1} Y_{k_2,l_2,m_2} &= W_D^{k_1} X^{l_1} Z^{m_1} W_D^{k_2} X^{l_2} Z^{m_2} = W_D^{k_1+k_2} X^{l_1} Z^{m_1} W_D^{-l_2 m_2} Z^{m_2} X^{l_2} \\ &= W_D^{k_1+k_2-l_2 m_2} X^{l_1} Z^{m_1+m_2} X^{l_2} = W_D^{k_1+k_2-l_2 m_2} X^{l_1} W_D^{l_2(m_1+m_2)} X^{l_2} Z^{m_1+m_2} \\ &= W_D^{k_1+k_2+m_1 l_2} X^{l_1+l_2} Z^{m_1+m_2} = Y_{k_1+k_2+m_1 l_2, l_1+l_2, m_1+m_2}, \end{aligned} \tag{7.69}$$

wherein the summation of index is per mod *D*. We can use either *Y*-operators or order products of *X*- and *Z*-operators to enumerate all group elements, D^3. For instance, for $D = 2$, eight group elements will be $\pm\mathbf{1}$, $\pm X$, $\pm Z$, $\pm XZ$. It can easily be shown that the Dth power of $Y_{k,l,m}$ is not necessarily the identity operator:

$$Y_{k,l,m}^{D} = \begin{cases} \mathbf{1}, & D \text{ odd} \\ (-1)^{lm}\mathbf{1}, & D \text{ even} \end{cases}. \tag{7.70}$$

The Heisenberg–Weyl group can also be represented as a group of unitary transformation of the arbitrary operator as follows [78]:

$$F \rightarrow YFY^{\dagger}, \tag{7.71}$$

wherein Y can be expressed in terms of X- and Z-operators as $X^{l}Z^{m}$, so that the unitary transformation becomes

$$F \rightarrow YFY^{\dagger} = X^{l}Z^{m}F(X,Z)Z^{-m}X^{-l} = F\left(W_{D}^{l}X, W_{D}^{-m}Z\right) = Z^{l}X^{m}F(X,Z)X^{-m}Z^{-l}. \tag{7.72}$$

Clearly, the phase terms can be ignored here, and the group of D^2 unitary transformation will form the abelian group. On the other hand, the group of unitary operators is non-abelian. From Eq. (7.70), we conclude that for even D, one-quarter of Y-operators will have a period of $2D$. When we observe the group of unitary transformations, this problem is not relevant. For instance, for $D = 2$, we have that $(XZ)^2 = -\mathbf{1}$, so that the following is valid:

$$(XZ)^{2}F(XZ)^{2} = F. \tag{7.73}$$

The unitary operators U that map the Heisenberg–Weyl group to itself under conjugation operation,

$$Y_{k,l,m} \rightarrow UY_{k,l,m}U^{\dagger} = Y_{k',l',m'}, \tag{7.74}$$

represent the *Clifford group* [5, 78, 81, 82]. The Clifford group contains the Weyl group as the subgroup.

7.4.1.3 MUBs for Prime Dimensionality

Let us now observe the situation where the *dimensionality is a prime number*, $D = p$. Based on Eq. (7.70), we conclude that Y-operators defined by (7.68) are periodic with a period of $D = p$ (except for identity operator). The following $D + 1$ operators are pairwise complementary, that is, their eigenvalues are nondegenerate and corresponding eigenkets satisfy the MU property [76, 78, 83]:

$$X, XZ, XZ^{2}, \ldots, XZ^{D-1}, Z. \tag{7.75}$$

The kth eigenket of the ith basis, denoted as $|i, k\rangle$, is given for odd D as follows [78]:

$$|i,k\rangle = D^{-1/2}\sum_{d=0}^{D-1}|d\rangle W_D^{-kd}W_D^{id(d-1)/2}, \qquad (7.76)$$

for the reference basis of the eigenkets of Z, which was determined from $XZ^i|i,k\rangle = W_D^k|i,k\rangle$. For $i = 0$, we obtain the eigenkets of X, that is, $|\hat{k}\rangle = |0,k\rangle$. The unitary operators X and Z can be represented, for $D = p$, by

$$X = \sum_{d=0}^{D-1}|d+1\rangle\langle d|, \quad Z = \sum_{d=0}^{D-1}W_D^d|d\rangle\langle d|. \qquad (7.77)$$

7.4.1.4 MUBs for Prime Power Dimensionality

When the *dimensionality is the prime power*, that is, $D = p^m$, we employ the finite field (Galois field), GF(p^m), to construct the maximal set of $D + 1$ MUBs [78, 83, 84]. (A reader not familiar with finite fields should refer to the Appendix.) To label the elements of the computational basis of $\mathbb{C}^D$, we use the elements from GF(p^m); in other words, the CB is denoted by $\{|a\rangle|\ a\in\mathrm{GF}(p^m)\}$. The X- and Z-operators can be defined as the generalization of (7.77) by

$$X(a) = \sum_{d\in GF(p^m)}|d+a\rangle\langle d|, Z(b) = \sum_{d\in GF(p^m)}\omega^{\mathrm{tr}(bd)}|d\rangle\langle d|, \qquad (7.78)$$

where the addition and multiplication operations are defined over GF(p^m), $\omega = \exp(\mathrm{j}2\pi/p)$ is the pth root of unity, while we use tr(x) to denote the *trace operation* from GF(p^m) to GF(p), which is defined as [5]

$$\mathrm{tr}(x) = x + x^p + x^{p^2} + \cdots + x^{p^{m-1}} = \sum_{i=0}^{m-1}x^{p^m}. \qquad (7.79)$$

Clearly, for $m = 1$ and $a = b = 1$, Eq. (7.78) reduces down to (7.77). So the actions of $X(a)$ and $Z(b)$ on $|x\rangle$ will be

$$X(a)|x\rangle = \sum_{d\in GF(p^m)}|d+a\rangle\underbrace{\langle d|x\rangle}_{\delta_{dx}} = |x+a\rangle,$$

$$Z(b)|x\rangle = Z(b) = \sum_{d\in GF(p^m)}\omega^{\mathrm{tr}(bd)}|d\rangle\underbrace{\langle d|x\rangle}_{\delta_{dx}} = \omega^{\mathrm{tr}(bx)}|x\rangle. \qquad (7.80)$$

Interestingly enough, X- and Z-operators (gates), defined above, are two basic gates, out of four required, in high-dimensional quantum error correction, which is also known as the *nonbinary quantum error correction* [5]. Let us now form $D + 1$ sets of commuting unitary operators as follows:

$$\{Z(b)|b \in GF(p^m)\}, \{X(b)Z(bc)|b \in GF(p^m)\}, \forall c \in GF(p^m). \quad (7.81)$$

The joint eigenkets of operators in one set are MU to the eigenkets in other set of operators. Clearly, there are $D + 1$ MUBs. For additional details, an interested reader is referred to [78, 83].

Let us now describe how to create the *dual basis* [78, 84]. Let further GF elements be denoted by m-tuples $(i_0, i_1, \ldots, i_{m-1})$, where $i_k \in \{0, 1, \ldots, p-1\}$. The corresponding integer representation will be $i = \sum_{l=0}^{m-1} i_l p^l$. In this notation, the states of the computational basis will be denoted by $\{|0\rangle, |1\rangle, \ldots, |D-1\rangle\}$. Let the addition and multiplication operations, in this notation, be denoted by $\oplus$ and $\odot$, respectively. The result of addition of two integers i and k, denoted as $i \oplus k$, is determined by componentwise addition per mod p, that is, $i_l + k_l$ mod p; $l = 0, 1, \ldots, p - 1$. However, the multiplication operation of two integers is more complicated, unless $D = p$, when it is mod p multiplication. The simplest way is to apply the multiplication procedure from Appendix. Namely, we need first to read the integer indices i and k, identify the power of primitive element representation, perform the multiplication in this representation, and then read out the final integer from the lookup table, which will represent the result of $i \odot k$ operation. Alternatively, the method introduced in [78] can be applied. Let us introduce the V_l^0-operator whose action on the baseket in CB is to shift the label by l,

$$V_l^0|i\rangle = |i \oplus l\rangle, \quad (7.82)$$

which is just the permutation of the kets. The dual basis can be created by applying the generalized version of the discrete quantum Fourier transform, called here as *Galois–Fourier transform* which is given as follows:

$$|\tilde{i}\rangle = D^{-1/2}\sum_{d=0}^{D-1} \omega^{\boxminus d \odot i}|d\rangle, \; \omega = e^{j\frac{2\pi}{p}}, \quad (7.83)$$

where $\boxminus d = x$ if $x \oplus d = 0$. The action of V_l^0-operator on the baseket in dual basis will be

$$V_l^0|\tilde{i}\rangle = D^{-1/2}\sum_{d=0}^{D-1} \omega^{\boxminus d \odot i}|d \oplus l\rangle = D^{-1/2}\sum_{d=0}^{D-1} \omega^{\boxminus (d' \boxminus l) \odot i}|d'\rangle = \omega^{l \odot i}|\tilde{i}\rangle, \quad (7.84)$$

indicating that the eigenvalues of V_l^0-operator are $W^{l \odot i}$. Clearly, the CB and DB are MU since

$$\langle\tilde{i}|k\rangle = \left|D^{-1/2}\omega^{i\odot k}\right|^2 = \frac{1}{D};\ \ \forall\, i, k = 0, 1, \ldots, D-1. \tag{7.85}$$

Let us now introduce the V_0^l-operator that shifts each basket in DB $\left\{|\tilde{0}\rangle, \cdots, |\tilde{i}\rangle, \ldots, \left|\widetilde{D-1}\right\rangle\right\}$ by $\boxminus l$ that is,

$$V_0^l|\tilde{i}\rangle = \left|\widetilde{i\boxminus l}\right\rangle,\ \langle\tilde{i}|V_0^l = \left\langle\widetilde{i\oplus l}\right|. \tag{7.86}$$

These permutation operators are diagonal in the CB, that is,

$$V_0^l = \sum_{d=0}^{D-1}|\tilde{d}\rangle\left\langle\widetilde{d+l}\right| = \sum_{d=0}^{D-1}\omega^{d\odot l}|d\rangle\langle d| = Z(l), \tag{7.87}$$

and therefore it corresponds to the $Z(l)$-operator, defined by Eq. (7.78), but in different notation. On the other hand, the V_l^0-operator can be represented by

$$V_l^0 = \sum_{d=0}^{D-1}|d+l\rangle\langle d| = \sum_{d=0}^{D-1}\omega^{d\odot l}|\tilde{d}\rangle\langle\tilde{d}| = X(l), \tag{7.88}$$

and therefore it corresponds to the $X(l)$-operator, defined by Eq. (7.78). By combining $X(i)$ and $Z(j)$ operators we obtain the *Heisenberg–Weyl operators*:

$$Z(k)X(i) = V_0^k V_i^0 = \sum_{d=0}^{D-1}\omega^{(d\oplus i)\odot k}|d\oplus i\rangle\langle d| = V_i^k, \tag{7.89}$$

and these operators are basic building blocks to form the *Heisenberg–Weyl group*. Similarly to the Pauli X- and Z-operators, generalized X- and Z-operators or, in other words, the shift operators do not commute since

$$V_i^0 V_0^k = \omega^{\boxminus i\odot k}V_0^k V_i^0. \tag{7.90}$$

Given that the Hilbert–Schmidt inner product is given by [78]

$$\left(V_i^j, V_k^l\right) = \mathrm{Tr}\left(\underbrace{\left(V_i^j\right)^\dagger}_{\omega^{\boxminus(i\odot j)}V_i^j} V_k^l\right) = D\delta_{ik}\delta_{jl}, \tag{7.91}$$

different Weyl operators are orthonormal, wherein we used the $\left(V_i^j\right)^\dagger = \omega^{\boxminus(i\odot j)}V_i^j$, derived from the composition property [78]:

$$V_i^j V_k^l = V_0^j V_i^0 V_0^l V_k^0 = \omega^{\boxminus i \odot l} V_0^j V_0^l V_i^0 V_k^0 = \omega^{\boxminus i \odot l} V_{i \oplus k}^{j \oplus l}. \tag{7.92}$$

The pth power of V_k^l will be [78]

$$\left(V_k^l\right)^p = \left(\omega^{\boxminus k \odot l}\right)^{1+2+\cdots+p-1} \underbrace{V_0^0}_{1} = \left(\omega^{\boxminus k \odot l}\right)^{p(p-1)/2} \mathbf{1} = \begin{cases} (-1)^{k \odot l}\mathbf{1}, & p = 2 \\ \mathbf{1}; & p = 3, 5, 7, \cdots \end{cases} \tag{7.93}$$

indicating that even prime case is different from odd prime case. The Weyl group can be decomposed into $D + 1$ abelian subgroups. The kth basis ket of the ith subgroup can be denoted by $\left|e_k^{(i)}\right\rangle$. We already determined the CB and DB, describing the abelian groups D and 0 as follows:

$$\begin{aligned} W_{0l} &= V_0^l = \sum_{d=0}^{D-1} \omega^{d \odot l} |d\rangle\langle d| = \sum_{d=0}^{D-1} \omega^{d \odot l} \left|e_d^{(D)}\right\rangle\left\langle e_d^{(D)}\right|, \\ W_{l0} &= V_l^0 = \sum_{d=0}^{D-1} \omega^{d \odot l} |\tilde{d}\rangle\langle\tilde{d}| = \sum_{d=0}^{D-1} \omega^{d \odot l} \left|e_d^{(0)}\right\rangle\left\langle e_d^{(0)}\right|. \end{aligned} \tag{7.94}$$

Other $D - 1$ commuting Weyl operators can be represented in similar fashion:

$$W_{li} = V_l^i = \sum_{d=0}^{D-1} \omega^{d \odot l} \left|e_d^{(i)}\right\rangle\left\langle e_d^{(i)}\right|; \; i = 1, 2, \cdots, D-1. \tag{7.95}$$

We now need to determine base kets to satisfy orthonormality principle within each basis, and MU property between kets from different basis, that is,

$$\left\langle e_k^{(i)} | e_l^{(j)} \right\rangle = \begin{cases} \delta_{kl}, \, i = j \\ 1/D, \, i \neq j \end{cases}. \tag{7.96}$$

It has been shown in [78] that the following base kets satisfy the conditions (7.96):

$$\left|e_k^{(i)}\right\rangle = D^{-1/2} \sum_{d=0}^{D-1} \omega^{\boxminus k \odot d} \alpha^{i}_{\boxminus d} \left|e_d^{(D)}\right\rangle, \tag{7.97}$$

where the phase constants $\alpha^{i}_{\boxminus d}$ are to be properly chosen as described in [78]. The shift operators change the kets within the same basis.

7.4.1.5 MUBs Derived from Complex Hadamard Matrices

The *complex* Hadamard *matrices* can also be used to derive the MUBs as discussed in [31, 78]. As an illustration, in four-dimensional (4-D) system, assuming that MUB_0

is given by {(1, 0, 0, 0), (0, 1, 0, 0), (0, 0, 1, 0), (0, 0, 0, 1)}, the another MUB can be derived from the following normalized complex Hadamard matrices, parametrized by parameter θ:

$$H_4^{(\mathrm{MUB}_1)}(\theta) = \frac{1}{\sqrt{4}}\begin{bmatrix} 1 & 1 & 1 & 1 \\ 1 & e^{j\theta} & -1 & -e^{j\theta} \\ 1 & -1 & 1 & -1 \\ 1 & -e^{j\theta} & -1 & e^{j\theta} \end{bmatrix}. \tag{7.98}$$

The corresponding MUB_1 can be obtained by setting $\theta = 0$ rad in (7.98), to obtain $\mathrm{MUB}_1 = \{0.5(1, 1, 1, 1), 0.5(1, 1, -1, -1), 0.5(1, -1, 1, -1), 0.5(1, -1, -1, 1)\}$. On the other hand, the MUB_2 can be obtained from complex Hadamard matrix given by equation below:

$$H_4^{(\mathrm{MUB}_2)} = \frac{1}{\sqrt{4}}\begin{bmatrix} 1 & j & j & -1 \\ 1 & j & -j & 1 \\ 1 & -j & j & 1 \\ 1 & -j & -j & 1 \end{bmatrix}. \tag{7.99}$$

Therefore, the problem of finding a set of $N + 1$ MUBs is equivalent to the problem of finding N mutually unbiased complex Hadamard matrices. Various methods to design complex Hadamard matrices are discussed in [85]. For additional MUB constructions based on complex Hadamard matrices, an interested reader is referred to [78].

7.4.2 *Generalized Bell States and High-Dimensional QKD*

It has been shown in [81, 84, 86] (see also [78]) that there exists one-to-one correspondence between the Heisenberg–Weyl group and the elements of the orthonormal basis of the generalized Bell states. The generalized Bell states have application in high-dimensional QKD [31–33], quantum teleportation, quantum swapping, and dense coding, to mention few. They can be introduced as follows [78, 81, 84, 86]. For all kets $|\psi\rangle$ and bras $\langle\phi|$, we introduce the conjugate $|\psi^*\rangle$ and bras $\langle\phi^*|$ through the dot product:

$$\langle\phi|\psi\rangle = \langle\psi|\phi\rangle^* = \langle\psi^*|\phi^*\rangle. \tag{7.100}$$

However, this introduction is not unique one; nevertheless, different versions for mapping $|\psi\rangle \to |\psi^*\rangle$ are related to each by a unitary transformation. Given that conjugate kets live in the same space as the bras, the following one-to-one correspondence is useful:

$$|\psi\rangle\langle\phi| \leftrightarrow |\phi^*\rangle|\psi\rangle = |\phi^*, \psi\rangle. \tag{7.101}$$

As a consequence, the operators $A = |\psi_1\rangle\langle\phi_1|$ and $B = |\psi_2\rangle\langle\phi_2|$ get transformed to

$$A = |\psi_1\rangle\langle\phi_1| \rightarrow |\phi_1^*\rangle|\psi_1\rangle = |\phi_1^*, \psi_1\rangle = |a\rangle, \quad B = |\psi_2\rangle\langle\phi_2| \rightarrow |\phi_2^*, \psi_2\rangle = |b\rangle. \tag{7.102}$$

Now the Hilbert–Schmidt inner product of A and B becomes

$$\begin{aligned}\mathrm{Tr}\left(A^\dagger B\right) &= \mathrm{Tr}(|\phi_1\rangle\langle\psi_1|\psi_2\rangle\langle\phi_2|) = \langle\psi_1|\psi_2\rangle \underbrace{\mathrm{Tr}(|\phi_1\rangle\langle\phi_2|)}_{\langle\phi_2|\phi_1\rangle} \\ &= \langle\phi_1^*|\phi_2^*\rangle\langle\psi_1|\psi_2\rangle = \langle a|b\rangle.\end{aligned} \tag{7.103}$$

In similar fashion, we can prove that

$$\begin{aligned}A \leftrightarrow |a\rangle \Rightarrow BA &\leftrightarrow (\mathbf{1}\otimes B)|a\rangle, \\ \Rightarrow AB^\dagger &\leftrightarrow \left(B^*\otimes\mathbf{1}\right)|a\rangle.\end{aligned} \tag{7.104}$$

The identity operator $\mathbf{1}$ gets mapped to the generalized Bell state $|B_{00}\rangle$. By applying the mapping rule given by Eq. (7.101) on completeness relation, we obtain

$$\mathbf{1} = \sum_d |d\rangle\langle d| = \sum_d \left|e_d^{(i)}\right\rangle\left\langle e_d^{(i)}\right| \leftrightarrow \sum_d |d^*\rangle|d\rangle = \sum_d \left|e_d^{(i)*}, e_d^{(i)}\right\rangle = D^{1/2}|B_{00}\rangle, \tag{7.105}$$

where $D^{-1/2}$ is the normalization factor that normalizes the generalized Bell state $|B_{00}\rangle$ to the unit length. (Namely, $\mathrm{Tr}(\mathbf{1}) = D$, so we need to normalize.) In other words, the following correspondence is valid $D^{1/2}\mathbf{1} \leftrightarrow |B_{00}\rangle$.

The generalized Bell state $|B_{00}\rangle$ is the basis invariant, so, as indicated earlier, different conjugation mappings can be used, and these are related by the unitary transformation. As an illustration, for $D = 2$, there are four alternative ways to perform the mapping $|\psi\rangle\rightarrow|\psi^*\rangle$, that is,

$$|\psi\rangle = \alpha|0\rangle + \beta|0\rangle \rightarrow |\psi^*\rangle = \begin{cases}\alpha^*|0\rangle + \beta^*|1\rangle \\ \alpha^*|0\rangle - \beta^*|1\rangle \\ \beta^*|0\rangle + \alpha^*|1\rangle \\ \beta^*|0\rangle - \alpha^*|1\rangle\end{cases}. \tag{7.106}$$

Now by applying the correspondence given by Eq. (7.101), we obtain

$$2^{-1/2}\,\mathbf{1} = 2^{-1/2}(|0\rangle\langle 0| + |1\rangle\langle 1|) \rightarrow \ |B_{00}\rangle = 2^{-1/2}\left(|0^*\rangle|0\rangle + |1^*\rangle|1\rangle\right)$$
$$= 2^{-1/2}\begin{cases} |0\rangle|0\rangle + |1\rangle|1\rangle \\ |0\rangle|0\rangle - |1\rangle|1\rangle \\ |0\rangle|1\rangle + |1\rangle|0\rangle \\ |0\rangle|1\rangle - |1\rangle|0\rangle \end{cases}, \qquad (7.107)$$

which represent the standard Bell states for 2-D systems.

Given that $V_0^0 = \mathbf{1}$ and the mapping (7.105) can be rewritten as $D^{-1/2}\ V_0^0 \leftrightarrow |B_{00}\rangle$, we can use the unitary shift operators V_m^n to obtain the *generalized Bells states* $|B_{mn}\rangle$ as follows:

$$D^{-1/2} V_m^n = D^{-1/2} \sum\nolimits_d \omega^{(d\oplus m)\odot n} |d \oplus m\rangle\langle d|$$
$$\leftrightarrow D^{-1/2} \sum\nolimits_d \omega^{(d\oplus m)\odot n} |d^*\rangle|d \oplus m\rangle = |B_{mn}\rangle. \qquad (7.108)$$

The generalized Bell states are orthonormal since

$$\langle B_{mn}|B_{m'n'}\rangle = D^{-1} Tr\left(\left(V_m^n\right)^\dagger V_m^n\right) = \delta_{mm'}\delta_{nn'}. \qquad (7.109)$$

The shift operators V_k^l permute the generalized Bell states since

$$\left(\mathbf{1} \otimes V_k^l\right)\left|B_{mn}\right\rangle = \omega^{\boxminus(k\odot n)}\left|B_{m\oplus k, n\oplus l}\right\rangle. \qquad (7.110)$$

When the *system dimensionality is the prime*, $D = p$, the Weyl operator (gate) representation (7.89) gets simplified, since addition and multiplication operations in GF become mod D operations:

$$W_{mn} = V_m^n = Z(n)X(m) = V_0^n V_m^0 = \sum_{d=0}^{D-1} \omega^{\overbrace{(d\oplus m)\odot n}^{d\oplus n}} \left|d \oplus m\right\rangle\left\langle d\right|,$$
$$= \sum_{d=0}^{D-1} \omega^{d\oplus n} \left|d \oplus m\right\rangle\left\langle d\right|, \qquad (7.111)$$

and this representation was used in [32], where different dimensions correspond to the *orbital angular momentum (OAM) modes*, which are mutually orthogonal. To impose different OAM modes on a single-photon level, we can use the spatial light modulators (SLMs) [87–89], spiral phase plates, and few-mode fibers (FMFs) [32], to mention few. The OAM modes have the azimuthal phase dependence of the form exp(j$m\phi$), where m is the azimuthal mode index, $m = 0, \pm 1, \pm 2, \ldots$ By employing the OAM modes 0, ± 1, ± 2, …, $\pm L$ the system becomes $(2L + 1)$-dimensional, that is, $D = 2L + 1$. For $D = p$, the parameter L is determined by $(D - 1)/2$.

The generalized Bell state $|B_{00}\rangle$, for D being the prime, can be simplified to

$$|B_{00}\rangle = D^{-1/2}\sum_d |d\rangle|d\rangle. \tag{7.112}$$

Finally, the generalized Bell state $|B_{mn}\rangle$ can be simplified to

$$|B_{mn}\rangle = D^{-1/2}\sum_d \omega^{d\odot n}|d\rangle|d \oplus m\rangle. \tag{7.113}$$

A reader interested to learn how to potentially generate the generalized Bell states by employing the OAM modes should refer to Ref. [32].

7.4.3 Security Analysis of Entanglement-Based High-Dimensional (HD) QKD Systems

The *entanglement-based HD QKD protocols* can be introduced as follows [32]. Alice prepares $|B_{00}\rangle$ baseket, as described above, and sends one of the qubits to Bob. Alice further performs the measurement in eigenkets of one of the W_{mn} selected at random, which were introduced by Eq. (7.95). Bob performs the measurement in eigenkets of one of the W_{mn}^* also selected at random. Only the items for which both (Alice and Bob) used the same bases are kept in sifting phase. For information reconciliation, Alice then performs (n, k) LDPC coding on positions in which both have used the same basis. Alice further sends $n - k \ll n$ parity symbols to Bob, who performs LDPC decoding. The privacy amplification is then performed to distill for a shorter key so that the correlation with Eve's string is minimized, as explained in Chap. 4 (see also Ref. [90]). The security analysis of two related classes of entanglement-based protocols, two-basis protocols and $(D + 1)$-basis protocols, is discussed in the rest of the section.

We are concerned here with the security against collective attacks, the attacks in which Eve's interaction during QKD is i.i.d. The eigenkets of W_{mn} can be used to create the MUBs. For D-dimensional systems, in total, there are D^2-1 nontrivial W_{mn}'s; however, some of them are redundant. As discussed above, there are at least two MUBs and maximum $D + 1$ MUBs for arbitrary D-dimensional system. For two-basis protocol, we can select the following set $\{W_{01}, W_{10}\}$, corresponding to CB and DB as introduced above. On the other hand, for $(D + 1)$-basis protocols, the following set $\{W_{01}; W_{10}, \ldots, W_{1,D-1}\}$ can be used. Since the commutator $\left[W_{mn} W_{mn}^*, W_{m'n'} W_{m'n'}^*\right] = 0$, it can be easily shown that the Alice–Bob density operator ρ_{AB} is diagonal in the generalized Bell basis:

$$\rho_{AB} = \sum_{m,n=0}^{D-1} \lambda_{mn}|B_{mn}\rangle\langle B_{mn}|; \quad \sum_{m,n=0}^{D-1} \lambda_{mn} = 1. \tag{7.114}$$

The parameters to be estimated in entanglement-assisted protocols are related to probabilities that Alice and Bob's outcomes, denoted as a and b, differ by $d \in \{0, 1, \ldots, D - 1\}$, and when both Alice and Bob choose to use randomly the basis of W_{mn}, observed per mod D, denoted as $q_{mn}(d)$, it can be determined as

$$q_{01}(d) = \sum_{n=0}^{D-1} \lambda_{d,n}, \; q_{1n}(d) = \sum_{n=0}^{D-1} \lambda_{m,(mn-d) \bmod D}. \tag{7.115}$$

Evidently, the probability that there is no error can be found as $q_{mn}(0) = 1 - \sum_{d=1}^{D-1} q_{mn}(d)$. The Eve's accessible information, representing the maximum of mutual information where maximization is performed over all generalized positive-operator-valued measurement (POVM) schemes, is upper bounded by Holevo information (see Chap. 5):

$$\chi(A : E|\rho_{AB}) = S(\rho_E) - \sum_{a=0}^{D-1} p(a)S(\rho_{E|a}), \; S(\rho) = -Tr(\rho \log \rho) = -\sum\nolimits_{\lambda_i} \lambda_i \log \lambda_i, \tag{7.116}$$

where $S(\rho)$, as before, denotes the von Neumann entropy and with λ_i we denoted the eigenvalues of ρ. In generalized Bell diagonal state, we have that $p(a) = 1/D$ and in order to estimate $S(\rho_{E|a})$ we need to perform the purification of ρ_{AB} as follows: $|\phi_{AB,\,E}\rangle = \sum_{m,n} \sqrt{\lambda_{mn}} |B_{mn}\rangle_{AB} |\phi_{mn}\rangle_E$, where Eve's basis $|\phi_{mn}\rangle_E$ is properly chosen so that the state $|\phi_{AB,\,E}\rangle$ is pure. Since $\rho_{E|a}$ is a diagonal in the generalized Bell basis, and given the that von Neumann entropy is equal to the Shannon entropy when quantum states are mutually orthogonal, we have that $S(\rho_{E|a}) = H(q_{01}(0), \ldots, q_{01}(D - 1))$, where $H(\cdot)$ denotes the Shannon entropy for the distribution $\{ q_{01}(0), \ldots, q_{01}(D - 1)\}$. Eve's mutual information is now given by

$$I_E = \begin{cases} \chi(A : E|\rho_{AB}), \text{ for } (D+1) - \text{basis protocols} \\ \max \chi(A : E|\rho_{AB}), \text{ for two} - \text{basis protocols} \end{cases}, \tag{7.117}$$

wherein the Holevo information between Alice and Bob, given ρ_{AB}, is determined by

$$\chi(A : E|\rho_{AB}) = H(\lambda_{mn}) - H(\underbrace{q_{01}(0), \cdots, q_{01}(D-1)}_{q_{01}}). \tag{7.118}$$

As an illustration, let us consider the *generalized depolarization channel* to model error probabilities $q_{mn}(d)$, which represent the generalization of classical D-ary symmetric channel, that is, we can write

$$q_{mn}(d) = \begin{cases} 1 - q, d = 0 \\ q/(D-1), d \neq 0 \end{cases}. \tag{7.119}$$

Notice that error probability $q_{mn}(d)$ typically decreases as d increases, indicating that this model represents the worst-case scenario. A better model was presented in [33]. In the generalized depolarization model, Eve's mutual information can be determined in closed form, which for two-basis protocols given by

$$I_E = -(1-q)\log(1-q) - (D-1)\frac{q}{D-1}\log\left(\frac{q}{D-1}\right) \doteq H(\boldsymbol{q}), \quad (7.120)$$

while for $(D+1)$-basis protocols given by

$$I_E = -\left(1-q-\frac{q}{D}\right)\log\left(1-q-\frac{q}{D}\right) - \left(q+\frac{q}{D}\right)\log\left[\frac{q}{D(D-1)}\right] - H(\boldsymbol{q}). \quad (7.121)$$

The corresponding *asymptotic secret fraction* for infinitely long keys is given by

$$r_{\text{asymptotic}} = \log_2(D) - I_E(q) - H(\boldsymbol{q}). \quad (7.122)$$

The results for infinitely long secure keys, assuming perfect information reconciliation and privacy amplification, are summarized in Fig. 7.9 for different system dimensionalities. Clearly, high-dimensional protocols can improve the secrecy fraction rates of two-dimensional-based protocols. It is also evident that the values of transition probability q for which the secrecy fraction rates become zero ($r_{\text{asymptotic}} = 0$) are higher for $(D+1)$-basis-based protocols. For instance, for $(D+1)$-basis protocol for $D = 10$, the secrecy fraction becomes zero at q of 32.64%, while for two-basis protocols, the zero rate is obtained at 26.21%.

Practical key lengths are finite, and we need to assume that various steps in entanglement-assisted protocols will fail with certain probability. To deal with such scenarios, the concept of ε-security is introduced in [28, 29], which was discussed in Sect. 7.2.1.

To calculate the secrecy fraction for finite-key lengths, we can use Eq. (7.37). The term min $S(X|E)$ is determined by $\log(D) - I_E$, but now with error probabilities $q_{mn}(d)$ subject to fluctuations:

$$\tilde{q}_{mn}(d) \in [q_{mn}(d) - \Delta q_{mn}, q_{mn}(d) + \Delta q_{mn}], \ \Delta q_{mn} = \xi(b, D)/[2(D-1)], \quad (7.123)$$

where $\xi(b, D) = [2\ln(1/\varepsilon_{PE})/b + D\ln(b+1)/b]^{1/2}$, $b = Kp_{mn}^2$, with p_{mn} being the probability of selecting base m and n. The expression for $\xi(b, D)$ follows from the law of large numbers as explained in Sect. 7.2.1. In calculations that follow, we assume the uniform distribution of $\tilde{q}_{mn}$ in (7.123), with the normalization constraint $\sum_d \Delta q_{mn}(b, d) = 0$. The results of calculations are summarized in Fig. 7.10, for fixed total error rate ε of 10^{-5}, tolerable error correction rate ε_{EC} of 10^{-10}, and fixed transition probability $q = 0.05$. The results are obtained by numerical maximization of (7.37), with respect to unknown parameters ($\varepsilon_{\text{PA}}, \varepsilon_{\text{PE}}, p_{mn}$). For sufficiently long keys, the $(D+1)$-basis protocols outperform two-basis protocols in terms of finite secret fraction. On the other hand, the two-basis protocols show earlier saturation of secure-key rates (against the length N) compared to $(D+1)$-basis protocols. Both types of HD protocols significantly outperform conventional 2-D QKD protocols.

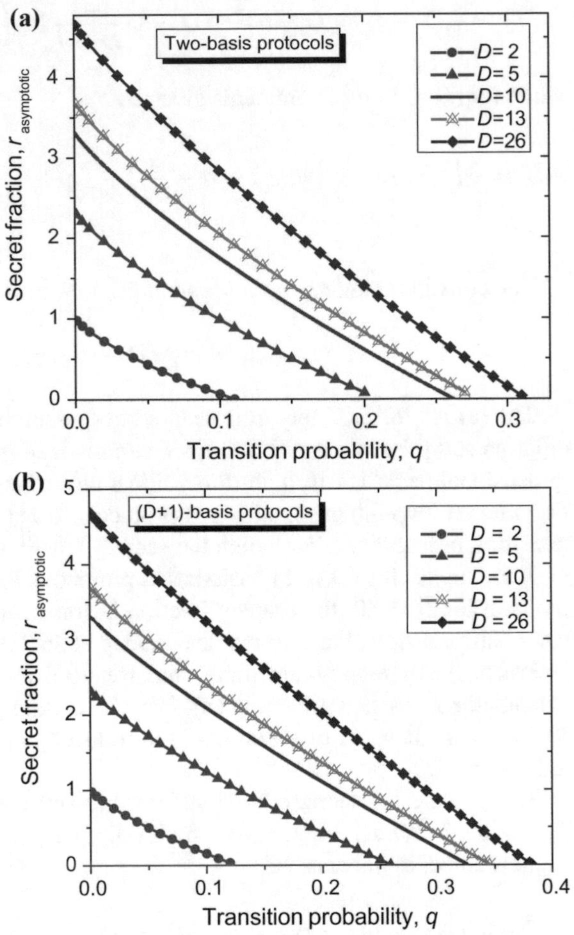

Fig. 7.9 Secure-key fraction rate versus transition probability, when system dimensionality D is used as a parameter for: **a** two-basis protocols and **b** ($D+1$)-basis protocols. [Modified from [32]. © IEEE 2013; reprinted with permission.]

Let us now describe the *K-neighbor transition probability model* [33] to determine the error probabilities $q_{mn}(d)$, which is suitable to consider the imperfect generation of OAM modes, causing the OAM cross talk. In this model, we assume that N neighbors get affected by OAM cross talk so that the probability of transition from a given state to $K-1$ of its neighbors is given by $q/(K-1)$, where q is the probability for a given symbol to be wrong. The probability of not having error will be then $1-q$. For example, the transition probability matrix Π describing this model for $D=8$ and $K=3$ is given by

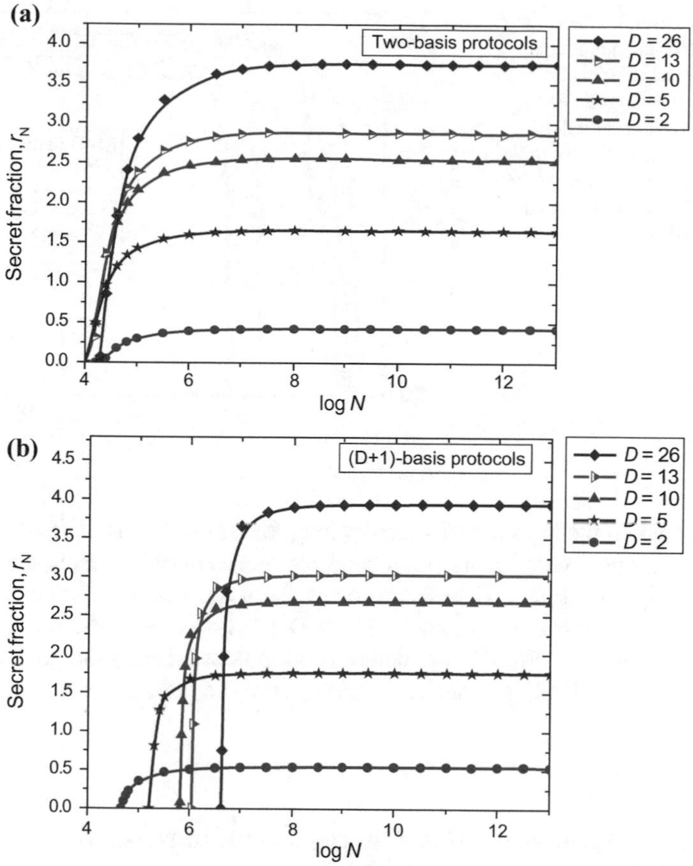

Fig. 7.10 Secret fraction rates for finite-key lengths for: **a** two-basis protocols and **b** $(D+1)$-basis protocols. The parameters are set as follows: $\varepsilon = 10^{-5}$, $\varepsilon_{EC} = 10^{-10}$, $q = 0.05$. [Modified from [32]. © IEEE 2013; reprinted with permission.]

$$\Pi = \begin{bmatrix} 1-q & q/2 & 0 & 0 & 0 & 0 & 0 & q/2 \\ q/2 & 1-q & q/2 & 0 & 0 & 0 & 0 & 0 \\ 0 & q/2 & 1-q & q/2 & 0 & 0 & 0 & 0 \\ 0 & 0 & q/2 & 1-q & q/2 & 0 & 0 & 0 \\ 0 & 0 & 0 & q/2 & 1-q & q/2 & 0 & 0 \\ 0 & 0 & 0 & 0 & q/2 & 1-q & q/2 & 0 \\ 0 & 0 & 0 & 0 & 0 & q/2 & 1-q & q/2 \\ q/2 & 0 & 0 & 0 & 0 & 0 & q/2 & 1-q \end{bmatrix}. \tag{7.124}$$

The results of calculations are summarized in Fig. 7.11, for fixed total error rate ε of 10^{-5}, tolerable error correction rate ε_{EC} of 10^{-9}, and parameter q is set to 0.05.

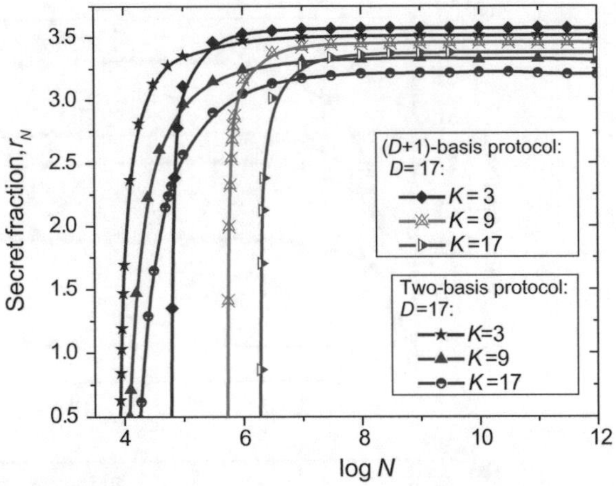

Fig. 7.11 Secret fraction rates for finite-key lengths in the presence of imperfect OAM generation, modeled by the K-neighbor model, for the system dimensionality of $D = 17$. [Modified from [33]. © IEEE 2013; reprinted with permission.]

Clearly, when the keys are sufficiently long, the $(D + 1)$-basis protocols outperform two-basis protocols. On the other hand, for shorter finite-key rates, the two-basis protocols outperform $(D + 1)$-basis protocols. Both types of proposed protocols significantly outperform conventional 2-D QKD protocols. Namely, the upper limit for secret-key rate is $\log_2 D - I_E$, where I_E is mutual information gained by Eve. Therefore, for 2-D QKD protocols, the ideal secret-key rate is always smaller than 1.

7.5 Time-Phase and Time-Energy Encoding-Based High-Dimensional (HD) QKD

In this section, we describe time-phase encoding HD QKD scheme proposed in [80, 91], followed by the time-energy HD QKD scheme introduced in [92].

7.5.1 *Time-Phase Encoding-Based HD QKD*

We have already introduced the time-phase encoding in Chap. 6, devoted to 2-D QKD schemes. Here we generalize this scheme to D-dimensional QKD system [80, 91]. The time-phasing encoding states for D-ary HD protocol are provided in Fig. 7.12. The time-basis corresponds to the computational basis, while the phase-basis to the dual basis, introduced earlier. The pulse of width Δt is localized within the time bin of duration $\tau = T/D$, that is, $\Delta t < \tau$. The time-basis is similar to D-ary pulse-position modulation (PPM). The state in which the photon is placed in the dth-time bin ($d = 0, 1, \ldots, D - 1$) is denoted by $|t_d\rangle$. The phase MUB states are defined by the quantum

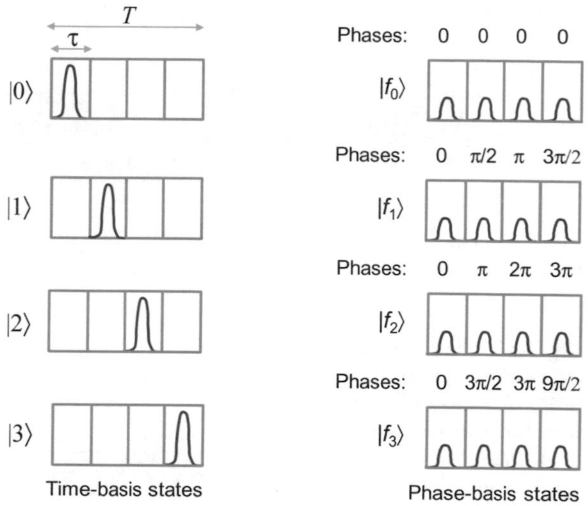

Fig. 7.12 The time-phase encoding states to be used in 4-D HD QKD protocol

discrete Fourier transform, introduced earlier, of time-basis as follows:

$$|f_d\rangle = \sum_{d=0}^{D-1} \omega^{dn}|t_n\rangle, \quad \omega = e^{-j\frac{2\pi}{D}}, \tag{7.125}$$

where now ω is the Dth root of unity. It can be shown that MU condition is satisfied.

Alice randomly selects either the time-basis or the phase-basis, followed by random selection of the basis state. The logical symbol d (wherein $d = 0, 1, \ldots, D-1$) is represented by $|t_d\rangle$, $|f_d\rangle$. Bob measures each qubit by randomly selecting the basis, the time-basis or the phase-basis. In sifting procedure, Alice and Bob announce the bases being used for each qubit and keep only instances when they used the same basis.

To implement the time-phase encoder either the electro-optical polar modulator, composed of concatenation of an amplitude modulator and phase modulator, or I/Q modulator can be used [43]. The amplitude modulator (Mach–Zehnder modulator) is used for pulse shaping and the phase modulator to introduce the desired phase shift, as illustrated in Fig. 7.13. The arbitrary waveform generator (AWG) is used to generate the corresponding RF waveforms as needed. The variable optical attenuator (VOA) is used to attenuate the modulated beam down to single-photon level. On receiver side, with the help of beam splitter, Bob randomly selects whether to measure in time-basis or phase-basis. The time-basis states can be detected with the help of single-photon counter and properly designed electronics. On the other hand, to detect the phase-basis states, the properly designed series of the time-delay interferometers can be used as described in [80, 91], see Fig. 7.14.

The basic building block is the delay interferometer, whose configuration is provided in Fig. 7.14b. The difference in the path between two arms is [80, 91] $\Delta L =$

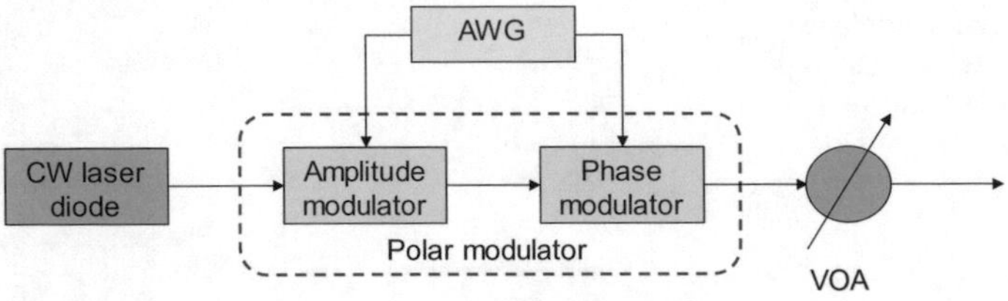

Fig. 7.13 The time-phase encoder for D-ary HD QKD protocol. AWG: arbitrary waveform generator, VOA: variable optical attenuator

Fig. 7.14 The phase-basis decoder for 4-D QKD protocol: **a** three DI-based detection scheme and **b** the configuration of DI. BS: 50:50 beam splitter, SPD: single-photon detector

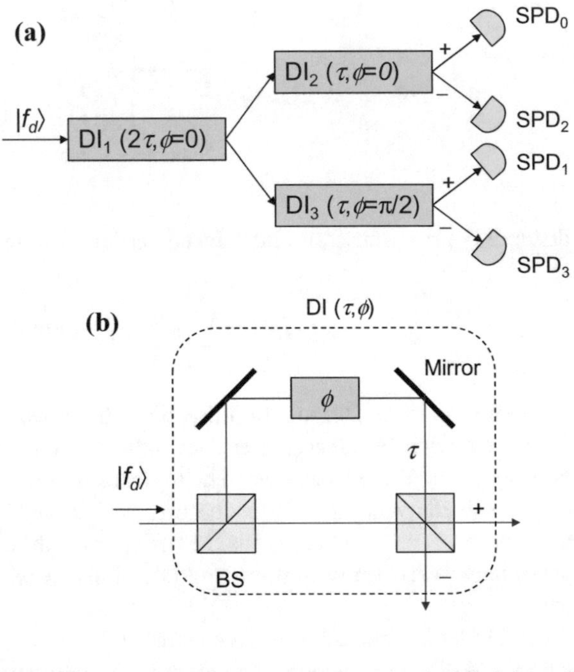

$\Delta L_0 + \delta L$, where ΔL_0 is the path difference equal to $c\tau$ (c is the speed of light), and $\delta L \ll \Delta L_0$ is small path difference used to adjust the phase since $\phi = k\delta L$ (k is the wave number). When the phase-state $|f_0\rangle$ is incident to the time-phase decoder, the outputs of second delay interferometer (DI_2) occupy seven time slots, and the + output denotes the case when interferometer outputs interfere constructively, while the – output denotes that corresponding outputs interfere destructively. For constructive interference, the middle pulse gets increased multiple times, while for destructive interference middle pulses cancel each other. Therefore, the click of SPD in the middle slot at the + output identifies the $|f_0\rangle$-state, while the corresponding click at the – output identifies the $|f_2\rangle$-state. In a similar fashion, at the outputs of DI_3, the click of SPD in the middle slot at the + output identifies the $|f_1\rangle$-state, while the corresponding click at the – output identifies the $|f_3\rangle$-state. Clearly, this scheme is suitable for implementation in integrated optics. For additional information about HD time-phase encoding and corresponding experimental demonstration, an interested reader is referred to refs. [80, 91].

In the rest of this section, we described the time-energy HD QKD scheme introduced in [92].

7.5.2 *Time-Energy Encoding-Based HD QKD*

Power-efficient modulation schemes, such as pulse-position modulation (PPM), are widely adopted in free-space optical communications, in particular, deep-space optical communications. Sufficiently large bandwidth of these links (compared to RF links) has made the low spectral efficiency of PPM less of a concern. From Shannon's theory, we know that information capacity is a *linear* function of number of dimensions, and a logarithmic function of signal-to-noise ratio. Therefore, by increasing the number of dimensions, we can dramatically improve the overall data rate. By using this approach, in Ref. [93], the D-dimensional pulse-position modulation approach has been proposed to enable ultrahigh-speed deep-space optical communication. In this scheme, D pulse-position basis functions have been used, which are defined as follows [93]:

$$\Phi_d(t) = \frac{1}{\sqrt{T_s/D}}\mathrm{rect}\left[\frac{t-(d-1)T_s/D}{T_s/D}\right]; d = 1, \ldots, D, \tag{7.126}$$

where T_s is a symbol duration, and rect(t) is defined as

$$\mathrm{rect}(t) = \begin{cases} 1, 0 \le t < 1 \\ 0, \text{otherwise} \end{cases}.$$

In Ref. [92], this concept was generalized to weak coherent state-based time-frequency HD QKD, which is illustrated in Fig. 7.15. Alice generates a sequence of 2^N-ary symbols generated in a pseudo-random fashion. As an illustration, the corresponding coordinates for $D = 3$ are selected from the set {(0, 0, 0), (0, 0, 1), (0, 1, 0), (0, 1, 1), (1, 0, 0), (1, 0, 1), (1, 1, 0), (1, 1, 1)}, and are imposed D Mach–Zehnder modulators (MZMs) as illustrated in Fig. 7.15b. The optical delay lines (DLs) are properly chosen to implement time-domain basis functions according to (7.126). Such obtained D-dimensional signal is sent to remote destination (Bob) over either free-space optical (FSO) link or single-mode fiber (SMF). With probability 0.5, the signal is first passed over all-optical inverse FFT (O-IFFT) device. In this case, the signal is imposed in frequency domain and converted to time domain by O-IFFT device. On receiver side, Bob, with probability 0.5, converts signal in electrical domain by the D-dimensional receiver whose configuration is provided in Fig. 7.15c. With probability 0.5, the signal is first converted back to frequency domain by all-optical FFT (O-FFT) device before photodetection takes place. Alice and Bob then perform sifting procedure similar to BB84 protocol, followed by conventional

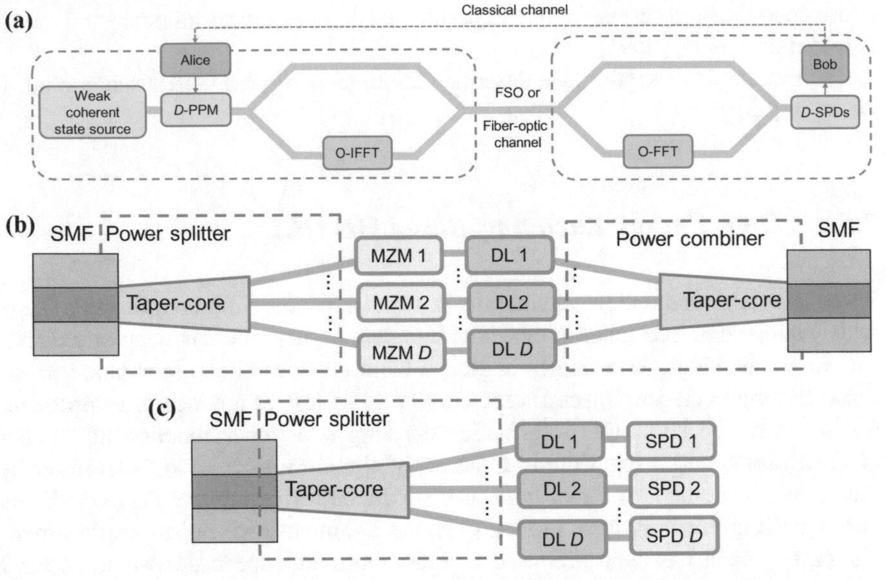

Fig. 7.15 The *D*-dimensional-PPM-based weak coherent state time-frequency QKD scheme suitable for implementation in integrated optics: **a** the operation principle of HD protocol, **b** configuration of *D*-dimensional-PPM transmitter, and **c** configuration of *D*-dimensional-PPM receiver. O-IFFT: optical inverse FFT, O-FFT: optical FFT, MZM: Mach–Zehnder modulator, DL: delay line, SPD: single-photon detector

information reconciliation and privacy amplification stages. For the serial-to-parallel and parallel-to-serial conversions required in FFT, the sequence of optical delay lines is used.

For implementation in bulk optics, instead of integrated optics, Y-junctions from Fig. 7.15 should be replaced by beam splitters. The secret-key rate of this scheme is $D/\log_2 D$-times larger than that when corresponding conventional PPM-based scheme is used. We now describe an all-optical implementation of IFFT/FFT device in integrated optics. The basic building block of all-optical implementation of FFT is Franson interferometer [94] shown in Fig. 7.16. The bulky optics version contains two beam splitters (BS) and phase shifter (PS), as described in [94]. In integrated optics version, the BS is replaced with Y-junction, while PS with phase modulator (PM). Alternatively, instead of Y-junctions, directional couplers can be used. Clearly, the Franson interferometer is similar to Mach–Zehnder delay interferometer (MZDI), routinely used in fiber optics communications [43]. From DSP, we know that the basic building block for the decimation-in-time FFT is butterfly computation block [95] as shown in Fig. 7.17a. Corresponding implementation in integrated optics is shown in Fig. 7.17b. Clearly, in addition to two Y-junctions and two Y-combiners, two phase modulators are needed to introduce phase shift $W_D = \exp(-j2\pi/D)$, where $D = T/\Delta t$ with T being symbol duration and Δt being the time-slot duration in PPM, and the second one to introduce the phase shift of π rad. By using the

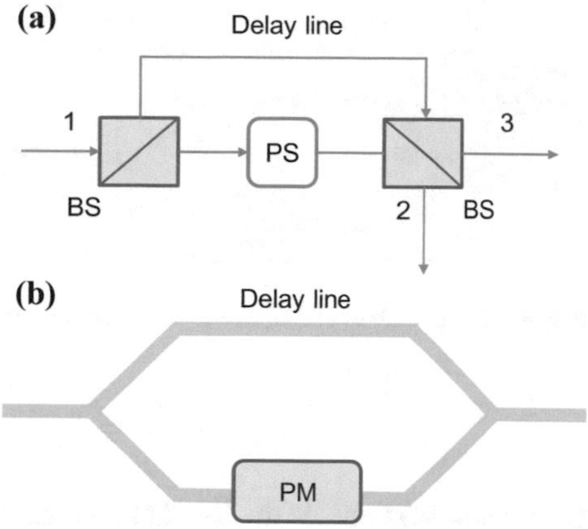

Fig. 7.16 Implementation of Franson interferometer in: **a** bulky optics and **b** integrated optics

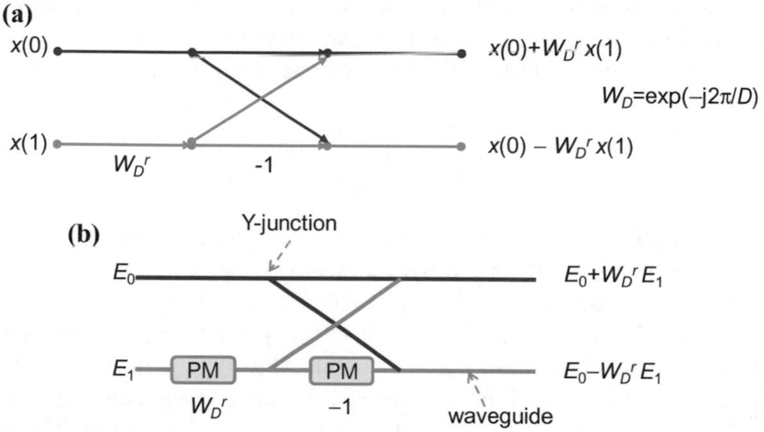

Fig. 7.17 Implementation of the butterfly computation block: **a** operation principle and **b** integrated optics implementation

butterfly computation block shown in Fig. 7.17b, 8-point decimation-in-time FFT can be implemented as shown in Fig. 8.1.6 of Ref. [95]. Some other approaches of all-optical implementation of FFT include [96, 97]. The basic building blocks described in this section are applicable to the entanglement-based time-energy HD QKD protocols, described next.

We first describe two-base entanglement-assisted time-frequency QKD protocol, which is illustrated in Fig. 7.18. The entangled photon pairs are generated by a spontaneous four-wave-mixing (SFWM) source based on highly nonlinear optical fiber (HNLF). The entangled photons are strongly correlated in both time and frequency domains. The corresponding generalized Bell baseket $|B_{00}\rangle$, introduced earlier, can be described as

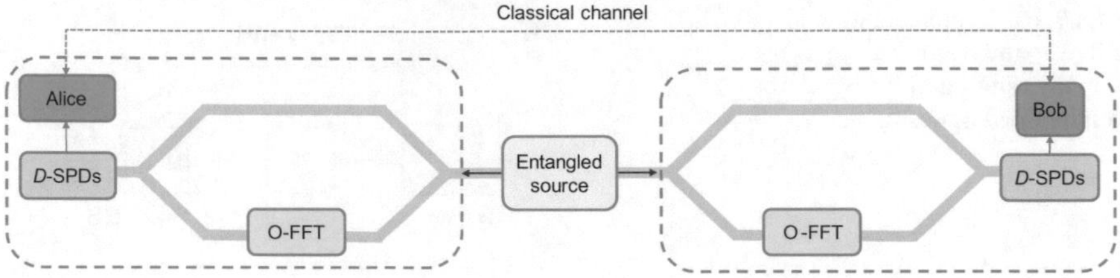

Fig. 7.18 The schematic of the two-base entanglement-based time-frequency HD QKD protocol

$$|B_{00}\rangle = D^{-1/2}\sum\nolimits_{d}|d,\ d\rangle, \tag{7.127}$$

where now $|d\rangle$ denotes the baseket corresponding to the dth position of photon in time domain. Alice and Bob measure the corresponding qubits in either time or frequency domain, with basis selected randomly and independently. This random selection of the basis has been performed by Y-junction, as shown in Fig. 7.18.

The frequency-domain basis $|\Psi_k\rangle$ (the dual basis) is related to the time-domain basis $|\psi_n\rangle$ by

$$|\Psi_d\rangle = D^{-1/2}\sum\nolimits_{n} e^{-j\frac{2\pi}{D}nd}|\psi_n\rangle. \tag{7.128}$$

With probability 0.5, Alice/Bob measures directly in time domain by using D SPDs with configuration already provided in Fig. 7.15c, and with probability 0.5 performs the measurement in frequency domain upon applying O-FFT device (also using D SPDs) as described in Fig. 7.17. In sifting procedure, Alice and Bob keep only items for which both used the same basis. Alice then performs (n, k) LDPC coding on positions in which both have used the same basis. Alice further sends $n - k \ll n$ parity symbols to Bob, who performs LDPC decoding. The privacy amplification is then performed to distill for a shorter key so that the correlation with Eve's string is minimized. The corresponding scheme for $(D+1)$-base entanglement-assisted time-frequency QKD protocol, for prime D, is shown in Fig. 7.19. The *rth frequency-domain basis* $|\Psi_k^{(r)}\rangle$ $(r = 1, 2, \ldots, D)$ is related to time-domain $|\psi_n\rangle$ basis by

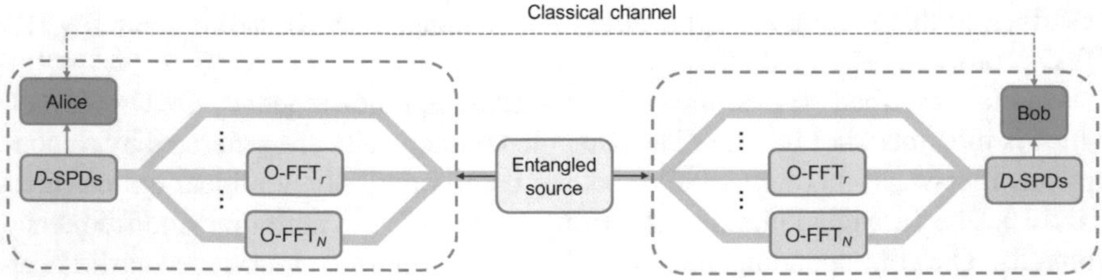

Fig. 7.19 The schematic of the $(D + 1)$-base entanglement-based time-frequency HD QKD protocol

$$\left|\Psi_d^{(r)}\right\rangle = D^{-1/2} \sum_n e^{-j\frac{2\pi}{D}n(rn+d)} |\psi_n\rangle;\ d = 0, 1, \ldots, D-1. \tag{7.129}$$

Since $rn + d \bmod D = d' \in \{0, 1, \ldots, D-1\}$, the parallel O-FFT output must be properly permuted before parallel-to-serial conversion can take place. In this protocol, Alice and Bob measure the corresponding qubits in one of $(D + 1)$ basis, according to Fig. 7.19, selected randomly and independently.

7.6 FBG/WBG-Based High-Dimensional QKD

The HD time-bin-encoded QKD, discussed in previous section, can be seamlessly incorporated into standard telecom networks, but the time-bin encoding largely sacrifices the spectral efficiency. As illustrated in Fig. 7.20a, a D-dimensional either time-bin encoding scheme or time-frequency (t.f.) encoding requires D time slots with only one time bin, on average, being occupied by a photon, but the rest time slots are left vacuum. In fact, the required optical bandwidth is proportional to D, as illustrated in Fig. 7.20a. To tackle the limitation of time-bin encoding, it has been proposed recently in [31, 75] to introduce the orthogonal Slepian sequence states, as illustrated in Fig. 7.20b. In Slepian-encoded HD quantum state, *every time slot* is encoded with a single-photon level signal. The temporal-spectral profile, described by a state in the Slepian sequence, represents the encoded HD quantum information. Note that different Slepian states are mutually orthogonal so that a Slepian sequence with D elements can be used as an encoding basis. The HD quantum communication based on Slepian states is anticipated to enjoy high spectral efficiency and low cross talk between multiplexed quantum channels. As illustrated in Fig. 7.20b, the consumed optical bandwidth ($1/T$) is independent on the system dimensionality D. However, the bandwidth required in time-frequency QKD system is proportional to $1/\tau = D/T$.

To implement this HD QKD scheme, the use of fiber Bragg gratings (FBGs) was advocated in [31], while the use of electronically controllable waveguide Bragg grat-

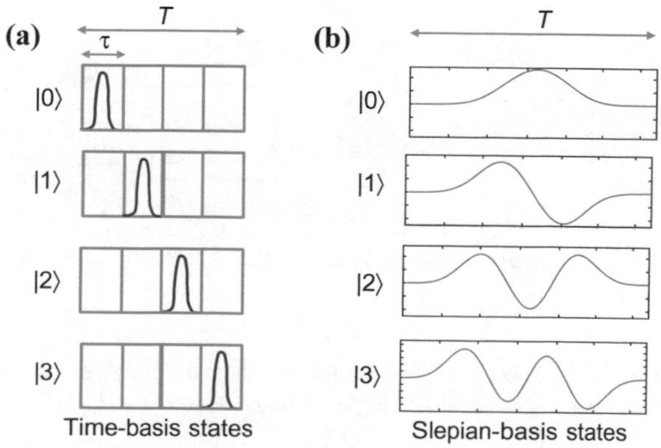

Fig. 7.20 **a** Time-basis in 4-D t.f. QKD (T: symbol duration, τ: bin duration, $\tau = T/4$). **b** Slepian states in 4-D HD QKD

ings (WBGs) was advocated in [75]. The key idea behind both versions is to employ FBGs/WBGs with orthogonal impulse responses as encoding basis for HD QKD. Both prepare-and-measure weak coherent state (WCS)-based and entanglement-based QKD protocols are possible with mutually unbiased bases (MUBs) derived from the complex Hadamard matrices.

In basic FBG-based *random base selection weak coherent state protocol*, we employ FBGs with mutually orthogonal impulse responses, denoted as $\{[0\rangle, |1\rangle, \ldots, |D-1\rangle\}$ as the encoding basis for HD QKD, as shown in Fig. 7.21. The Alice encoder is composed of 1:D optical switch, $D+1$ circulators, $D+1$ FBGs, and $(D+1)$:1 optical star coupler (power combiner). The Mach–Zehnder modulator (MZM) is optional, it is used to perform NRZ to RZ conversion. When the WCS is based on a pulse laser, the MZM can be omitted. To encode Alice, randomly select the optical switch output, say the *d*th output, and the *d*th FBG imposes the corresponding impulse response on an attenuated laser pulse, which is passed to the star coupler. With D mutually orthogonal impulse responses imposed on FBGs, Alice can transmit $\log_2 D$ bits per signaling interval. The superposition FBG is fabricated by writing the resulting impulse response, obtained by superposition of D mutual orthogonal impulse responses, on a single FBG. To probe for Eve's presence, Alice selects the $(D+1)$th optical switch output and therefore creates the superposition state $([0\rangle+|1\rangle+\ldots+|D-1\rangle)/D^{1/2}$. On receiver side, Bob employs a series of matched FBGs. The matched complex-conjugate FBG reflects the pulse back, and the SPD at port 3 of circulator detects the presence of pulse, and Bob is able to identify the transmitted symbol by employing the selected largest input logic. When Bob detects the photons on majority of SPDs, he will know that this particular symbol should be used to estimate the quantum bit error rate (QBER), and therefore to detect the presence of Eve. The mutual orthogonal impulse response for FBG-design can be derived from Slepian sequences [43] as it was proposed in [98, 99]. To implement the FBGs with mutual orthogonal impulse responses, with perfect impulse responses, the time-domain-based FBG design algorithm should be used [99, 100]. Compared to conventional discrete layer peeling algorithm (DLPA) applied in spectral domain [98], the time-domain FBG design results in FBGs with perfectly orthogonal impulse responses [99].

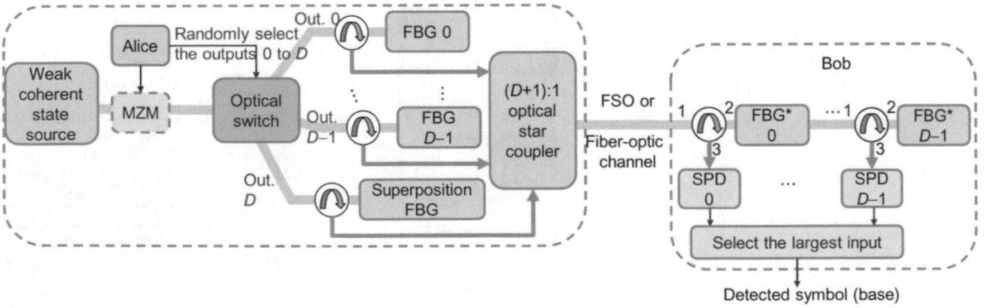

Fig. 7.21 The basic FBG-based protocol, the prepare-and-measure WCS-based protocol. FSO: free-space optical, SPD: single-photon detector

As discussed in [75], WBG-based HD QKD is suitable for implementation in integrated optics. The Alice encoder is composed of an adaptive, reconfigurable Slepian-WBG implemented in photonic integrated circuit (PIC) technology and a circulator, with the MZM being optional, as discussed above. To encode Alice randomly selects the orthogonal impulse response (IR) to be used, and a genetic algorithm (GA) is used to determine the voltages to be applied on electrodes of WBG-device shown in Fig. 7.22 to reconfigure to the desired basis function $|d\rangle$. The surface-profile diffraction grating is used as one of the substrates. By filling the grating groves with the dielectric, controlled by electrodes on another substrate, the waveguide is created as explained in [101]. The mth electrode ($m = 1, 2, \ldots, M$) together with grating waveguide below it serves as the mth segment with refractive index $n(m)$. By properly changing the control voltages, we can tune the overall impulse response to the desired Slepian sequence. In the absence of control voltages, the default grating will represent the central Slepian sequence from the set of Slepian sequences being employed. To speed up the reconfiguration process, the GA should be run in installation stage only to determine the set of voltages required for each basis function, with corresponding results being stored in a lookup table (LUT). Given that optical circulators are difficult to implement in integrated optics, better option will be to employ the *transmissive Slepian-WBGs* instead, with corresponding scheme being provided in Fig. 7.23. The transmissive FBGs have been already studied for use in picosecond optical signal processing [102], and we believe that fabrication of transmissive WBGs to generate Slepian states will be possible too.

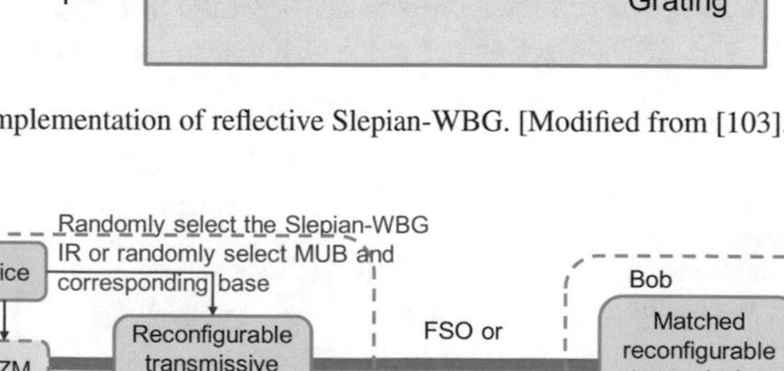

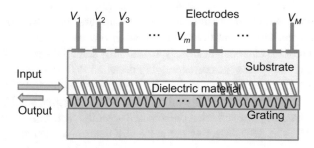

Fig. 7.22 The implementation of reflective Slepian-WBG. [Modified from [103].]

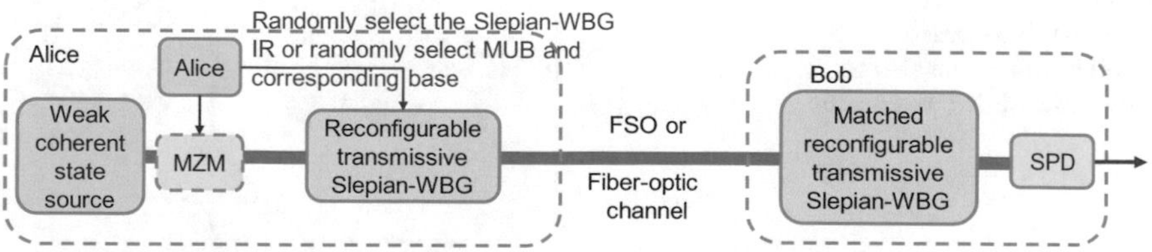

Fig. 7.23 The reconfigurable Slepian-WBG-based schemes implementing either the random base selection prepare-and-measure WCS-based protocol or MUB-based protocol for HD QKD based on transmissive Slepian-WBGs

On receiver side, Bob employs another reconfigurable matched-transmissive Slepian-WBG. The non-matched-transmissive Slepian-WBG reflects the pulse back, while the matched-transmissive-WBG passes the signal and the SPD detects the presence of pulse, and Bob is able to identify the transmitted symbol, when Alice used the same basis state $|d\rangle$. To improve the SKR, the Bob's detector shown in Fig. 7.24 should be used instead, composed of D reflective-WBGs. However, this scheme requires the employment of circulators. Bob employs a series of matched reflective-WBGs. The matched reflective-WBG reflects the pulse back, and the corresponding SPD at port 3 of circulator will be able to detect the presence of a pulse. Only one SPD will detect the presence of the pulse, unless when the signaling interval is used for QBER estimation. In sifting procedure, Alice announces the signaling intervals in which she transmitted the superposition states. These are used to check for Eve's presence/activity. If the QBER is lower than the prescribed threshold QBER value they continue with the protocol; otherwise, they abort the protocol. Other signaling intervals that are not used for QBER estimations are used for the sifted key. After that, the classical postprocessing steps are applied. The corresponding transmissive WBGs-based scheme will require the employment of the circulators as well.

The $(D + 1)$-basis protocol can also be implemented by employing WBG-based devices. Alice first randomly selects which MUB to use followed by the random selection of the base within the MUB, and after that she adjusts the voltages to impose the needed superposition state. In initialization stage, the GA has been run to determine the voltages needed to select each base within a given MUB, and the corresponding results are stored in LUT. In operation stage, selected MUB and its corresponding base are used to determine the address in LUT where the voltages' values are stored, which are further used to reconfigure the WBG to the required superposition state. Given that the reconfiguration process is electro-optical, it can be done in order of ns. On receiver side, Bob randomly selects the measurement MUB to be used and reconfigures the corresponding matched WBGs shown in Fig. 7.24 to match the states that corresponds to the MUB. The matched base WBG will reflect the pulse encoded using the corresponding base WBG on Alice's side, while other WBGs are transparent. The reflected pulse will trigger the SPD at corresponding port 3 (see Fig. 7.24). The SPD detecting the pulse will determine the base used by Alice

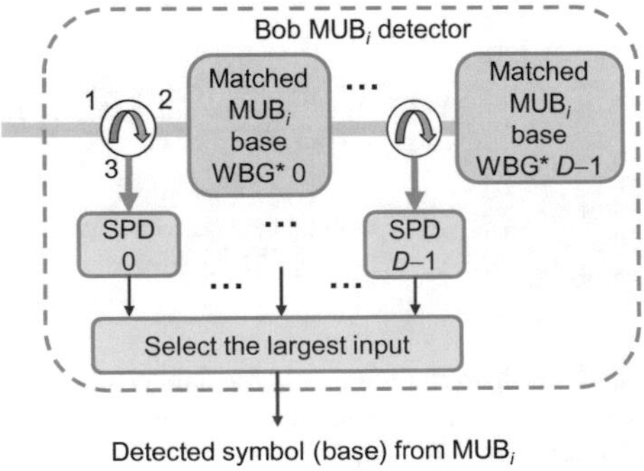

Fig. 7.24 Bob's reconfigurable reflective Slepian-WBGs-based detector, which can also be used as the ith MUB detector

for the same MUB. For additional details, in particular, related to the entanglement-based HD QKD, an interested reader is referred to refs. [31, 75].

7.7 Concluding Remarks

This chapter has been devoted to the DV-QKD protocols and represents the continuation of Chap. 6. The chapter starts with the description of BB84 and decoy-state-based protocols in Sect. 7.1, and evaluation of their secrecy fraction performance in terms of achievable distance. The next topic in Sect. 7.2 has been related to the security of DV-QKD protocols when the resources are finite. We introduce the concept of composable ε-security in Sect. 7.2.1 and describe how it can be evaluated for both collective and coherent attacks. We also discuss in Sect. 7.2.2 how the concept of correctness and secrecy can be combined to derive the tight security bounds. In Sect. 7.3, we evaluate the BB84 and decoy-state protocols for finite-key assumption over atmospheric turbulence effects. In the same section, we also describe how to deal with time-varying free-space optical channel conditions. In Sect. 7.4, the focused has been then moved to high-dimensional (HD) QKD protocols, starting with the description of mutually unbiased bases (MUBs) selection in Sect. 7.4.1, followed by the introduction of the generalized Bell states in Sect. 7.4.2. In Sect. 7.4.3, we provide the security analysis of generic HD QKD protocol.

In Sect. 7.5, we have described the time-phase encoding and time-energy encoding HD QKD protocols. In Sect. 7.6, we have described FBG-based HD QKD WBGs-based HD QKD protocols.

References

1. Bennet CH, Brassard G (1984) Quantum cryptography: public key distribution and coin tossing. In: Proceedings of the IEEE international conference on computers, systems, and signal processing, Bangalore, India, pp 175–179
2. Bennett CH (1992) Quantum cryptography: uncertainty in the service of privacy. Science 257:752–753
3. Neilsen MA, Chuang IL (2000) Quantum computation and quantum information. Cambridge: Cambridge University Press
4. Van Assche G (2006) Quantum cryptography and secrete-key distillation. Cambridge University Press, Cambridge-New York
5. Djordjevic IB (2012) Quantum information processing and quantum error correction: an engineering approach. Elsevier/Academic Press, Amsterdam-Boston
6. Bennett CH (1992) Quantum cryptography using any two nonorthogonal states. Phys Rev Lett 68(21):3121–3124
7. Devetak I, Winter A (2005) Distillation of secret key and entanglement from quantum states. Proc R Soc Lond Ser A 461(2053):207–235
8. Scarani V, Bechmann-Pasquinucci H, Cerf NJ, Dušek M, Lütkenhaus N, Peev M (2009) The security of practical quantum key distribution. Rev Mod Phys 81:1301

9. Holevo AS (1973) Bounds for the quantity of information transmitted by a quantum communication channel. Probl Inf Trans 9(3):177–183
10. Lo H-K, Ma X, Chen K (2005) Decoy state quantum key distribution. Phys Rev Lett 94:230504
11. Hwang W-Y (2003) Quantum key distribution with high loss: toward global secure communication. Phys Rev Lett 91:057901
12. Ma X, Fung C-HF, Dupuis F, Chen K, Tamaki K, Lo H-K (2006) Decoy-state quantum key distribution with two-way classical postprocessing. Phys Rev A 74:032330
13. Zhao Y, Qi B, Ma X, Lo H-K, Qian L (2006) Experimental quantum key distribution with decoy states. Phys Rev Lett 96:070502
14. Rosenberg D, Harrington JW, Rice PR, Hiskett PA, Peterson CG, Hughes RJ, Lita AE, Nam SW, Nordholt JE (2007) Long-distance decoy-state quantum key distribution in optical fiber. Phys Rev Lett 98(1):010503
15. Yuan ZL, Sharpe AW, Shields AJ (2007) Unconditionally secure one-way quantum key distribution using decoy pulses. Appl Phys Lett 90:011118
16. Hasegawa J, Hayashi M, Hiroshima T, Tanaka A, Tomita A (2007) Experimental decoy state quantum key distribution with unconditional security incorporating finite statistics. arXiv: 0705.3081
17. Tsurumaru T, Soujaeff A, Takeuchi S (2008) Exact minimum and maximum of yield with a finite number of decoy light intensities. Phys Rev A 77:022319
18. Hayashi M (2007) General theory for decoy-state quantum key distribution with an arbitrary number of intensities. New J Phys 9:284
19. Zhao Y, Qi B, Ma X, Lo H, Qian L (2006) Simulation and implementation of decoy state quantum key distribution over 60 km telecom fiber. In: Proceedings of the 2006 IEEE international symposium on information theory, Seattle, WA, pp 2094–2098
20. Wang X-B (2013) Three-intensity decoy-state method for device-independent quantum key distribution with basis-dependent errors. Phys Rev A 87(1):012320
21. Sun X, Djordjevic IB, Neifeld MA (2016) Secret key rates and optimization of BB84 and decoy state protocols over time-varying free-space optical channels. IEEE Photon J 8(3):7904713
22. Ma X (2008) Quantum cryptography: from theory to practice. PhD dissertation, University of Toronto
23. Fung C-HF, Tamaki K, Lo H-K (2006) Performance of two quantum-key-distribution protocols. Phys Rev A 73:012337
24. Gobby C, Yuan ZL, Shields AJ (2004) Quantum key distribution over 122 km of standard telecom fiber. Appl Phys Lett 84:3762
25. Ben-Or M, Horodecki M, Leung DW, Mayers D, Oppenheim J (2005) The universal composable security of quantum key distribution. In: Theory of cryptography: second theory of cryptography conference, TCC 2005. Lecture notes in computer science, vol 3378. Springer, Berlin, pp 386–406
26. König R, Renner R, Bariska A, Maurer U (2007) Small accessible quantum information does not imply security. Phys Rev Lett 98:140502
27. Renner R, König R (2005) universally composable privacy amplification against quantum adversaries. In: Theory of cryptography: second theory of cryptography conference, TCC 2005. Lecture notes in computer science, vol 3378. Springer, Berlin, pp 407–425
28. Renner R (2005) Security of quantum key distribution. PhD dissertation, Swiss Federal Institute of Technology, Zurich
29. Scarani V, Renner R (2008) Quantum cryptography with finite resources: unconditional security bound for discrete-variable protocols with one-way postprocessing. Phys Rev Lett 100(20):200501
30. Sheridan L, Le TP, Scarani V (2010) Finite-key security against coherent attacks in quantum key distribution. New J Phys 12:123019
31. Djordjevic IB (2018) FBG-based weak coherent state and entanglement assisted multidimensional QKD. IEEE Photon J 10(4):7600512
32. Djordjevic IB (2013) Multidimensional QKD based on combined orbital and spin angular momenta of photon. IEEE Photon J 5(6):7600112

33. Djordjevic IB (2016) Integrated optics modules based proposal for quantum information processing, teleportation, QKD, and quantum error correction employing photon angular momentum. IEEE Photon J 8(1):6600212
34. Ilic I, Djordjevic IB, Stankovic M (2017) On a general definition of conditional Rényi entropies. In: Proceedings of the 4th international electronic conference on entropy and its applications, 21 November–1 December 2017. Sciforum electronic conference series, vol 4. https://doi.org/10.3390/ecea-4-05030
35. Impagliazzo R, Levin LA, Luby M (1989) Pseudo-random generation from one-way functions. In: Proceedings of the 21st annual ACM symposium on theory of computing, Johnson DS (ed.), pp. 12–24, May 14–17, 1989, Seattle, Washington, USA
36. Håstad J, Impagliazzo R, Levin LA, Luby M (1999) A pseudorandom generator from any one-way function. SIAM J Comput 28(4):1364–1396
37. Tomamichel M, Schaffner C, Smith A, Renner R (2011) Leftover hashing against quantum side information. IEEE Trans Inf Theory 57(8):5524–5535
38. Cover TM, Thomas JA (1991) Elements of information theory. Wiley, New York
39. Renner R (2007) Symmetry of large physical systems implies independence of subsystems. Nat Phys 3:645–649
40. Christandl M, König R, Renner R (2009) Postselection technique for quantum channels with applications to quantum cryptography. Phys Rev Lett 102:020504
41. Tomamichel M, Lim CCW, Gisin N, Renner R (2012) Tight finite-key analysis for quantum cryptography. Nat Commun 3:634
42. Huttner B, Imoto N, Gisin N, Mor T (1995) Quantum cryptography with coherent states. Phys Rev A 51(3):1863
43. Djordjevic IB (2017) Advanced optical and wireless communications systems. Springer International Publishing, Switzerland
44. Al-Habash MA, Andrews LC, Phillips RL (2001) Mathematical model for the irradiance probability density function of a laser beam propagating through turbulent media. Opt Eng 40(8):1554–1562
45. Andrews LC, Phillips LC (2005) Laser beam propagation through random media, 2nd edn. SPIE Press, Bellingham, Washington, USA
46. Toyoshima M, Sasaki T, Takenaka H, Takayama Y (2012) Scintillation model of laser beam propagation in satellite-to-ground bidirectional atmospheric channels. Acta Astronaut 80:58–64
47. Kim K-H, Higashino T, Tsukamoto K, Komaki S (2011) Optical fading analysis considering spectrum of optical scintillation in terrestrial free-space optical channel. In: Proceedings of IEEE conference on space optical systems and applications (ICSOS) (IEEE, 2011), pp. 58–66, 11-13 May 2011, Santa Monica, CA, USA
48. Weinberg GV, Gunn L (2011) Simulation of statistical distributions using the memoryless nonlinear transform. DSTO Technical Report DSTO TR-2517 2011. https://apps.dtic.mil/dtic/tr/fulltext/u2/a541304.pdf
49. Lucamarini M, Dynes JF, Fröhlich B, Yuan ZL, Shields AJ (2015) Security bounds for efficient decoy-state quantum key distribution. IEEE J Sel Top Quantum Electron 21(2):197–204
50. Rice P, Harrington J (2009) Numerical analysis of decoy state quantum key distribution protocols. http://www.arxiv.org/abs/quant-ph/0901.0013
51. Tomamichel M, Renner R (2011) Uncertainty relation for smooth entropies. Phys Rev Lett 106(11):110506
52. Hoeffding W (1963) Probability inequalities for sums of bounded random variables. J Am Statist Assoc 58(301):13–30
53. Lucamarini M, Dynes JF, Yuan ZL, Shields AJ (2012) Practical treatment of quantum bugs. In: Proceedings of the SPIE 8542, electro-optical remote sensing, photonic technologies, and applications VI, 85421K (19 November 2012). https://doi.org/10.1117/12.977870
54. Ma X, Qi B, Zhao Y, Lo H-K (2005) Practical decoy state for quantum key distribution. Phys Rev A 72(1):012326

55. Gottesman D, Lo H-K, Lütkenhaus N, Preskill J (2004) Security of quantum key distribution with imperfect devices. In: Proceedings of international symposium on information theory (ISIT 2004), pp 325–360, 27 June–2 July 2004, Chicago, IL, USA
56. Takeoka M, Guha S, Wilde MM (2014) Fundamental rate-loss tradeoff for optical quantum key distribution. Nat Commun 5:5235
57. Pirandola S, Laurenza R, Ottaviani C, Banchi L (2017) Fundamental limits of repeaterless quantum communications Nat Commun 8:15043
58. Djordjevic IB (2019) Hybrid DV-CV QKD outperforming existing QKD protocols in terms of secret-key rate and achievable distance. In: Proceedings of the 21st international conference on transparent optical networks (ICTON 2019), Angers, France, 9–13 July 2019
59. Jiang L, Taylor JM, Nemoto K, Munro WJ, van Meter R, Lukin MD (2009) Quantum repeater with encoding. Phys Rev A 79:032325
60. de Riedmatten H, Marcikic I, Scarani V, Tittel W, Zbinden H, Gisin N (2004) Tailoring photonic entanglement in high-dimensional Hilbert spaces. Phys Rev A 69:050304
61. Thew RT, Acín A, Zbinden H, Gisin N (2004) Bell-type test of energy-time entangled qutrits. Phys Rev Lett 93:010503
62. O'Sullivan-Hale MN, Khan IA, Boyd RW, Howell JC (2005) Pixel entanglement: experimental realization of optically entangled d = 3 and d = 6 qudits. Phys Rev Lett 94:220501
63. Neves L, Lima G, Gómez JGA, Monken CH, Saavedra C, Pádua S (2005) Generation of entangled states of qudits using twin photons. Phys Rev Lett 94:100501
64. Walborn SP, Lemelle DS, Almeida MP, Ribeiro PHS (2006) Quantum key distribution with higher-order alphabets using spatially encoded qudits. Phys Rev Lett 96:090501
65. Vaziri A, Weihs G, Zeilinger A (2002) Experimental two-photon, three-dimensional entanglement for QuCom. Phys Rev Lett 89:240401
66. Langford NK, Dalton RB, Harvey MD, O'Brien JL, Pryde GJ, Gilchrist A, Bartlett SD, White AG (2004) Measuring entangled Qutrits and their use for quantum bit commitment. Phys Rev Lett 93:053601
67. Molina-Terriza G, Vaziri A, Reháček J, Hradil Z, Zeilinger A (2004) Triggered Qutrits for QuCom protocols. Phys Rev Lett 92:167903
68. Groblacher S, Jennewein T, Vaziri A, Weihs G, Zeilinger A (2006) Experimental quantum cryptography with qutrits. New J Phys 8:75
69. Bozinovic N et al (2013) Terabit-scale orbital angular momentum mode division multiplexing in fibers. Science 340(6140):1545–1548
70. Zhao Z-M et al (2013) A large-alphabet quantum key distribution protocol using orbital angular momentum entanglement. Chin Phys Lett 30(6):060305
71. Leach J, Courtial J, Skeldon K, Barnett SM, Franke-Arnold S, Padgett MJ (2004) Interferometric methods to measure orbital and spin, or the total angular momentum of a single photon. Phys Rev Lett 92(1):013601
72. Li H, Phillips D, Wang X, Ho D, Chen L, Zhou X-Q, Zhu J, Yu S, Cai X (2015) Orbital angular momentum vertical-cavity surface-emitting lasers. Optica 2:547–552
73. Barreiro JT, Langford NK, Peters NA, Kwiat PG (2005) Generation of hyperentangled photon pairs. Phys Rev Lett 95:260501
74. Barreiro JT, Wei T-C, Kwiat PG (2008) Beating the channel capacity limit for linear photonic superdense coding. Nat Phys 4:282
75. Djordjevic IB (2019) Slepian-states-based DV- and CV-QKD schemes suitable for implementation in integrated optics. In: Proceedings of the 21st European conference on integrated optics (ECIO 2019), 24–26 April, 2019, Ghent, Belgium
76. Ivanovic ID (1981) Geometrical description of quantal state determination. J Phys A 14:3241
77. Wootters WK, Fields BD (1989) Optimal state-determination by mutually unbiased measurements. Ann Phys 191(2):363–381
78. Durt D, Englert B-G, Bengtsson I, Zyczkowski K (2010) On mutually unbiased bases. Int J Quantum Inf 8(4):535–640
79. Aguilar EA, Borka JJ, Mironowicz P, Pawlowski M (2018) Connections between mutually unbiased bases and quantum random access codes. Phys Rev Lett 121:050501

80. Islam NT, Lim CCW, Cahall C, Kim J, Gauthier DJ (2017) Provably secure and high-rate quantum key distribution with time-bin qudits. Sci Adv 3(11):e1701491
81. Fivel DI (1995) Remarkable phase oscillations appearing in the lattice dynamics of Einstein-Podolsky-Rosen states. Phys Rev Lett 74:835
82. Gottesman D (1998) Theory of fault-tolerant quantum computation. Phys Rev A 57:127
83. Bandyopadhyay S, Boykin PO, Roychowdhury V, Vatan F (2002) A new proof for the existence of mutually unbiased bases. Algorithmica 34(4):512–528
84. Nagler B, Durt T (2003) Covariant cloning machines for four-level systems. Phys Rev A 68:042323
85. Tadej W, Zyczkowski K (2006) A concise guide to complex Hadamard matrices. Open Syst Inf Dyn 13:133–177
86. Durt T, Kaszlikowski D, Chen J-L, Kwek LC (2004) Security of quantum key distributions with entangled qudits. Phys Rev A 69:032313
87. Qu Z, Djordjevic IB (2016) 500 Gb/s free-space optical transmission over strong atmospheric turbulence channels. Opt Lett 41(14):3285–3288
88. Djordjevic IB, Qu Z (2016) Coded orbital-angular-momentum-based free-space optical transmission. In: Wiley encyclopedia of electrical and electronics engineering. http://onlinelibrary.wiley.com/doi/10.1002/047134608X.W8291/abstract
89. Qu Z, Djordjevic IB (2018) Orbital angular momentum multiplexed free-space optical communication systems based on coded modulation, in Novel insights into orbital angular momentum beams: from fundamentals, devices to applications. Appl Sci 8(11):2179 (Invited paper)
90. Bennett CR, Brassard G, Crepeau C, Maurer UM (1995) Generalized privacy amplification. IEEE Inform. Theory 41(6):1915–1923
91. Islam NT (2018) High-rate, high-dimensional quantum key distribution systems. PhD dissertation, Department of Physics, Duke University
92. Djordjevic IB, Zhang Y (2015) Entanglement assisted time-energy QKD employing Franson interferometers and cavity quantum electrodynamics (CQED) principles. In: Proceedings of the SPIE photonics west 2015, OPTO: advances in photonics of quantum computing, memory, and communication VIII, p 93770L, 7–12 February 2015, San Francisco, California, United States
93. Djordjevic IB (2011) Multidimensional pulse-position coded-modulation for deep-space optical communication. IEEE Photon Technol Lett 23(18):1355–1357
94. Brougham T, Barnett SM, McCusker KT, Kwiat P, Gauthier D (2013) Security of high-dimensional quantum key distribution protocols using Franson interferometers. J Phys B At Mol Opt Phys 46:104010
95. Proakis JG, Manolakis DM (2007) Digital signal processing: principles, algorithms, and applications. Fourth Edition, Pearson Prentice Hall
96. Hillerkuss D, Winter M, Teschke M, Marculescu A, Li J, Sigurdsson G, Worms K, Ben Ezra S, Narkiss N, Freude W, Leuthold J (2010) Simple all-optical FFT scheme enabling Tbit/s real-time signal processing. Opt Express 18:9324–9340
97. Liao Y, Pan W (2011) All-optical OFDM based on arrayed grating waveguides in WDM systems. In: Proceedings of the 2011 international conference on electronics, communications and control (ICECC), pp 707–710
98. Djordjevic IB, Saleh AH, Küppers F (2014) Design of DPSS based fiber Bragg gratings and their application in all-optical encryption, OCDMA, optical steganography, and orthogonal-division multiplexing. Opt Express 22(9):10882–10897
99. Djordjevic IB, Zhang S, Wang T (2016) Optically encrypted multidimensional coded modulation for multi-Pb/s optical transport. In: Proceedings of the IEEE photonics conference 2016, Paper MB3.6, pp 57–58
100. Dong L, Fortier S (2004) Formulation of time-domain algorithm for fiber Bragg grating simulation and reconstruction. IEEE J Quantum Electron 40(8):1087–1098
101. Wu C, Raymer MG (2006) Efficient picosecond pulse shaping by programmable Bragg gratings. IEEE J Quantum Electron 42(9):873–884

102. Fernández-Ruiz MR, Li M, Dastmalchi M, Carballar A, LaRochelle S, Azaña J (2013) Picosecond optical signal processing based on transmissive fiber Bragg gratings. Opt Lett 38:1247–1249
103. Djordjevic IB (2019) Proposal for slepian-states-based DV- and CV-QKD schemes suitable for implementation in integrated photonics platforms. IEEE Photonics J 11(4):7600312

Chapter 8
Continuous Variable (CV)-QKD

Abstract This chapter is devoted to the detailed description of continuous variable (CV) QKD schemes, in particular, with Gaussian modulation and discrete modulation. The chapter starts with the fundamentals of Gaussian quantum information theory, where the P-representation is introduced and applied to represent the thermal noise as well as the thermal noise plus the coherent state signal. Then, quadrature operators are introduced, followed by the phase-space representation. Further, Gaussian and squeezed states are introduced, followed by Wigner function definition as well as the definition of correlation matrices. The next subsection is devoted to the Gaussian transformation and Gaussian channels, with beam splitter operation and phase rotation operation being the representative examples. The thermal decomposition of Gaussian states is discussed next and the von Neumann entropy for thermal states is derived. The focus is then moved to the nonlinear quantum optics fundamentals, in particular, the three-wave mixing and four-wave mixing are described in detail. Further, the generation of the Gaussian states is discussed, in particular, the EPR state. The correlation matrices for two-mode Gaussian states are discussed next, and how to calculate the symplectic eigenvalues, relevant in von Neumann entropy calculation. The Gaussian state measurements and detection are discussed then, with emphasis on homodyne detection, heterodyne detection, and partial measurements. In section on CV-QKD protocols with Gaussian modulation, after the brief description of squeezed state-based protocols, the coherent state-based protocols are described in detail. We start the section with the description of both lossless and lossy transmission channels, followed by the description of how to calculate the covariance matrix under various transformations, including beam splitter, homodyne detection, and heterodyne detection. The equivalence between the prepare-and-measure (PM) and entanglement-assisted protocols with Gaussian modulation is discussed next. The focus is then moved to the secret-key rate (SKR) calculation under collective attacks. The calculation of mutual information between Alice and Bob is discussed first, followed by the calculation of Holevo information between Eve and Bob, in both cases assuming the PM protocol and reverse reconciliation. Further, entangling cloner attack is described, followed by the derivation of Eve-to-Bob Holevo information. The entanglement-assisted protocol is described next as well as the corresponding Holevo information derivation. In all these derivations, both homodyne

I. B. Djordjevic, *Physical-Layer Security and Quantum Key Distribution*,
https://doi.org/10.1007/978-3-030-27565-5_8

detection and heterodyne detection are considered. Some illustrative SKR results corresponding to the Gaussian modulation are provided as well. In section on CV-QKD with discrete modulation, after the brief introduction, we describe the four-state and eight-state CV-QKD protocols. Both the PM and entanglement-assisted protocols are discussed. The SKR calculation for discrete modulation is discussed next, with illustrative numerical results being provided. We also identify conditions under which the discrete modulation can outperform the Gaussian modulation. In section on RF-assisted CV-QKD scheme, we describe a generic RF-assisted scheme applicable to arbitrary two-dimensional modulation schemes, including M-ary PSK, and M-ary QAM. This scheme exhibits better tolerance to laser phase noise and frequency offset fluctuations compared to conventional CV-QKD schemes with discrete modulation. We then discuss how to increase the SKR through the parallelization approach. The final section in the chapter provides some relevant concluding remarks.

8.1 Gaussian Quantum Information Theory Fundamentals

To facilitate the description of continuous variable (CV)-QKD schemes, in this section, we first describe the Gaussian quantum information theory fundamentals [1–6].

8.1.1 The Field Coherent States and P-Representation

The continuous variable quantum system lives in an infinite-dimensional Hilbert space, characterized by the observables with continuous spectra [4]. The CV quantum system can be represented in terms of M quantized radiation modes of the electromagnetic field (M bosonic modes). The quantum theory of radiation treats each radiation (bosonic) mode as a harmonic oscillator [2, 3]. The bosonic modes live in a tensor product Hilbert space $\mathcal{H}^{\otimes M} = \otimes_{m=1}^{M} \mathcal{H}_m$. The *radiation (bosonic) modes* are associated with M pairs of bosonic field operators $\left\{a_m^\dagger, a_m\right\}_{m=1}^{M}$, which are known as the creation and annihilation operators, respectively. Clearly, the Hilbert space is infinitely dimensional, but separable. Namely, a single-mode Hilbert space is spanned by a countable basis of number (Fock) states $\{|n\rangle\}_{n=1}^{\infty}$, wherein the Fock state $|n\rangle$ is the eigenstate of the number operator $\hat{n} = a^\dagger a$ satisfying the eigenvalue equation $\hat{n}|n\rangle = n|n\rangle$. Now the creation and annihilation operators can be, respectively, defined as follows:

$$
\begin{aligned}
a^\dagger|n\rangle &= (n+1)^{1/2}|n+1\rangle, n \geq 0 \\
a|n\rangle &= n^{1/2}|n-1\rangle, n \geq 1, a|0\rangle = 0.
\end{aligned} \tag{8.1}
$$

Clearly, the creation operator raises the number of photons by one, while the annihilation operator reduces the number of photons by one. The right eigenvector of the annihilation operator is commonly referred to as the *coherent state*, and it is defined as follows:

$$a|\alpha\rangle = \alpha|\alpha\rangle, \alpha = \alpha_I + j\alpha_Q, \tag{8.2}$$

where $\alpha = I + \mathrm{j}Q$ is a complex number. The energy of the mth mode is given by $H_m = \hbar\omega_m|\alpha_m|^2$. As we have shown in Chap. 6, the coherent state vector can be expressed in terms of number states as follows:

$$|\alpha\rangle = e^{-|\alpha|^2/2}\sum_n \frac{\alpha^n}{\sqrt{n!}}|n\rangle, \langle\alpha|\alpha\rangle = 1. \tag{8.3}$$

The coherent states are complete since the following is valid:

$$\frac{1}{\pi}\int |\alpha\rangle\langle\alpha|d^2\alpha = \mathbf{1}, d^2\alpha = d\alpha_I d\alpha_Q, \tag{8.4}$$

where we use **1** to denote the identity operator. The coherent states $|\alpha\rangle$ and $|\beta\rangle$ are not orthogonal since their inner product is nonzero:

$$\langle\beta|\alpha\rangle = \exp\left(\alpha\beta^* - |\alpha|^2/2 - |\beta|^2/2\right). \tag{8.5}$$

The entire field is in coherent state $\{|\alpha_m\rangle\}$ when all modes $|\alpha_m\rangle$ are in coherent states, and the state vector $\{|\alpha_m\rangle\}$ is then a simultaneous right eigenket of all annihilation operators, that is, $a_m|\{\alpha_m\}\rangle = \alpha_m|\{\alpha_m\}\rangle$. The state vector can be represented as the tensor product of the individual state vectors:

$$|\{\alpha_m\}\rangle = \prod_m |\alpha_m\rangle. \tag{8.6}$$

It has been shown by Glauber [7, 8] that a large class of density operators ρ can be written in so-called *P-representation* as follows:

$$\rho = \int P(\{\alpha_m\})\prod_{m=1}^{\infty}|\alpha_m\rangle\langle\alpha_m|d^2\alpha_m, \tag{8.7}$$

where the weight function $P(\{\alpha_m\})$ has many properties of the classical probability density function, except that it is not always positive. In particular, from $\mathrm{Tr}\rho = 1$ we have that

$$\int P(\{\alpha_m\})\prod_{m=1}^{\infty} d^2\alpha_m = 1. \tag{8.8}$$

When the state of the field is described by Eq. (8.7), then the expected value of an operator A is given by

$$\langle A\rangle = \mathrm{Tr}(\rho A) = \int P(\{\alpha_m\})\langle\{\alpha_m\}|A|\{\alpha_m\}\rangle \prod_{m=1}^{\infty} d^2\alpha_m. \tag{8.9}$$

8.1.2 The Noise Representation

Let us consider the scenario in which the field is composed of *thermal radiation* being in thermal equilibrium at temperature $\mathfrak{T}$. The corresponding density operator describing the state of the mth mode in the P-representation is given by

$$\rho_m = \frac{1}{\pi\mathfrak{N}_m}\int \exp\left(-|\alpha|^2/\mathfrak{N}_m\right)|\alpha\rangle\langle\alpha|d^2\alpha,\ \mathfrak{N}_m = Tr\left(\rho a_m^\dagger a_m\right) = \frac{1}{e^{\hbar\omega/k_B\mathfrak{T}}-1}, \tag{8.10}$$

where $\mathfrak{N}_m$ denotes the average number of photons in the mth mode with frequency ω_m and k_B is the Boltzmann constant. In the classical limit, when $k_B\mathfrak{T} \gg \hbar\omega_m$, the weight function $P(\alpha_m)$ becomes the classical probability density function, and we can write

$$P(\alpha_m) = \frac{1}{\pi\mathfrak{N}_m}\exp\left(-|\alpha_m|^2/\mathfrak{N}_m\right) = \frac{1}{\pi\mathfrak{N}_m}\exp\left[-(\alpha_{mI}^2+\alpha_{mQ}^2)/\mathfrak{N}_m\right]. \tag{8.11}$$

The average number of photons in classical limit is $k_B\mathfrak{T}/\hbar\omega_m$ and the energy of the mth mode is $k_B\mathfrak{T}$, which is independent of the frequency, and corresponds to the additive white Gaussian noise (AWGN). By substituting Eq. (8.3) into (8.10), we can express the density operator ρ_m in terms of number states $|n_m\rangle$ as follows:

$$\rho_m = \sum_{n_m=0}^{\infty}(1-v_m)v_m^{n_m}|n_m\rangle\langle n_m|,\ v_m = \frac{\mathfrak{N}_m}{\mathfrak{N}_m+1} = e^{-\hbar\omega/k_B\mathfrak{T}}. \tag{8.12}$$

The density operator ρ of the whole field, composed of thermal radiation in equilibrium of M modes, can be represented as the direct product of modes:

$$\rho = \pi^{-M}\prod_m \int e^{-\frac{|\alpha_m|^2}{\mathfrak{N}_m}}|\alpha_m\rangle\langle\alpha_m|d^2\alpha_m/\mathfrak{N}_m. \tag{8.13}$$

When the received signal contains both the *thermal noise* and the *coherent signal* represented by the coherent state vector $|\mu_m\rangle$, the center of Gaussian weight function in the P-representation is shifted from the origin by a phasor μ_m, that is, $P_m =$

$(\pi \mathcal{N}_m)^{-1} \exp\left(-|\alpha - \mu_m|^2/\mathcal{N}_m\right)$. The corresponding density operator ρ_m in the P-representation is given by

$$\rho_m = \frac{1}{\pi \mathcal{N}_m} \int e^{-\frac{|\alpha-\mu_m|^2}{N_m}} |\alpha\rangle\langle\alpha| d^2\alpha. \tag{8.14}$$

Based on Eqs. (8.3) and (8.13), we can express the elements of the density operator ρ_m in terms of number states $|n_m\rangle$ as follows:

$$\langle l|\rho_m|k\rangle = \begin{cases} (1-v_m)v_m^k\left(\mu_m^*/\mathcal{N}_m\right)^{k-l}(l!/k!)e^{-(1-v_m)|\mu_m|^2}L_l^{k-l}\left[-(1-v_m)^2|\mu_m|^2/v_m\right], & l \geq k \\ \langle k|\rho_m|l\rangle^*, & \end{cases} \tag{8.15}$$

where $v_m = \mathcal{N}_m/(\mathcal{N}_m + 1)$, as before. The density operator of the entire received field composed of the coherent signal and Gaussian thermal noise can be written in P-representation as follows:

$$\begin{aligned} \rho &= \int \cdots \int P(\{\alpha_m\}) \prod_{m=1}^{M} |\alpha_m\rangle\langle\alpha_m| d^2\alpha_m, \\ P(\{\alpha_m\}) &= \frac{1}{\pi^M |\det \boldsymbol{\Sigma}|} e^{-\sum_k \sum_l (\alpha_k^* - \mu_k^*)(\boldsymbol{\Sigma}^{-1})(\alpha_l^* - \mu_l^*)}, \end{aligned} \tag{8.16}$$

where μ_k is the complex amplitude of the kth coherent signal and $\boldsymbol{\Sigma}$ represents the *mode correlation matrix* whose elements are determined by

$$\Sigma_{lk} = \mathrm{Tr}\left(\rho a_k^\dagger a_l\right) - \mathrm{Tr}\left(\rho a_k^\dagger\right)\mathrm{Tr}(\rho a_l). \tag{8.17}$$

When the modes are statistically independent, the mode correlation matrix is diagonal, with elements being

$$\Sigma_{lk} = \mathcal{N}_k \delta_{lk}, \tag{8.18}$$

where $\mathcal{N}_k$ is the average number of thermal photons in mode k. The P-representation of the density operator ρ is not the unique one. Let $\{|\beta_m\rangle\}$ be coherent states spanning the same vector space as spanned by the vectors $\{|\alpha_m\rangle\}$. The P-representation in terms of coherent states $\{|\beta_m\rangle\}$ will be

$$\rho = \int \cdots \int P(\{\beta_m\}) \prod_m |\beta_m\rangle\langle\beta_m| d^2\beta_m, \tag{8.19}$$

where the weight function $P(\{\beta_m\})$ is related to the previous weight function $P(\{\alpha_m\})$ by the following substitution relation:

$$\alpha_m = \sum_{j=1}^{\infty} \beta_j U_{jm}^{\dagger}, \tag{8.20}$$

where U_{jm} are elements of unitary matrix $\boldsymbol{U}$. The unitary matrix $\boldsymbol{U}$ should be chosen in such a way that the matrix $\boldsymbol{U}\Sigma^{-1}\boldsymbol{U}^{\dagger}$ is diagonal such that the density operator ρ can be simplified to

$$\rho = \pi^{-v} \int e^{-\sum_m \frac{|\beta_m - \mu'_m|^2}{\mathfrak{N}'_m}} |\{\beta_m\}\rangle\langle\{\beta_m\}| \prod_m d^2\beta_m/\mathfrak{N}'_m, \tag{8.21}$$

where $\mathfrak{N}'_m$ is the mth diagonal element of $\boldsymbol{U}\Sigma^{-1}\boldsymbol{U}^{\dagger}$, while

$$\mu'_m = \sum_{i=1}^{\infty} U_{mi}\mu_i. \tag{8.22}$$

Clearly, for any coherent signal in Gaussian thermal noise, the receiver can be properly chosen such that the individual modes are uncorrelated.

8.1.3 *Quadrature Operators and Phase-Space Representation, Gaussian States, Squeezed States*

The creation and annihilation operators can also be expressed in terms of *quadrature operators* $\hat{I}$ and $\hat{Q}$, defined as follows:

$$a^{\dagger} = \frac{\hat{I} - j\hat{Q}}{2\sqrt{N_0}}, \quad a = \frac{\hat{I} + j\hat{Q}}{2\sqrt{N_0}}, \tag{8.23}$$

where N_0 represents the variance of vacuum fluctuation (shot noise). Very often the quadrature operators are expressed in terms of the *shot-noise unit* (SNU), where N_0 is normalized to 1, that is, $a^{\dagger} = \left(\hat{I} - j\hat{Q}\right)/2, \ a = \left(\hat{I} + j\hat{Q}\right)/2$. In other words, the quadrature operators, related to the in-phase and quadrature components of the electromagnetic field, are related to the creation and annihilation operators (in SNU) by $\hat{I} = a^{\dagger} + a, \ \hat{Q} = j\left(a^{\dagger} - a\right)$. The quadrature operators satisfy the commutation relation:

$$\begin{aligned}
\left[\hat{I}, \hat{Q}\right] &= \left[a^{\dagger} + a, j\left(a^{\dagger} - a\right)\right] = \left[a^{\dagger}, j\left(a^{\dagger} - a\right)\right] + \left[a, j\left(a^{\dagger} - a\right)\right] \\
&= j\underbrace{\left[a^{\dagger}, a^{\dagger}\right]}_{=0} - j\underbrace{\left[a^{\dagger}, a\right]}_{-\mathbf{1}} + j\underbrace{\left[a, a^{\dagger}\right]}_{\mathbf{1}} - j\underbrace{\left[a, a\right]}_{0} \\
&= 2j\mathbf{1}
\end{aligned} \tag{8.24}$$

where in the first line we employed the following property of the commutators [9] $[A + B, C] = [A, C] + [B, C]$, while in the second line the following two commutation relations of the creation/annihilation operators of the harmonic oscillator [10] $\left[a_j, a_k^\dagger\right] = \delta_{jk}\mathbf{1}$, $\left[a_j^\dagger, a_k^\dagger\right] = \left[a_j, a_k\right] = 0$.

As discussed in Chap. 6, for the coherent state, both quadratures have the same uncertainty, that is, the same variance:

$$V\left(\hat{I}\right) = \mathrm{Var}\left(\hat{I}\right) = V\left(\hat{Q}\right) = \mathrm{Var}\left(\hat{Q}\right). \tag{8.25}$$

The expected value of the in-phase operator, with respect to coherent state $|\alpha\rangle$, is given by

$$\left\langle \hat{I} \right\rangle_\alpha = \langle\alpha|\hat{I}|\alpha\rangle = \langle\alpha|a + a^\dagger|\alpha\rangle = \langle\alpha|\alpha + \alpha^*|\alpha\rangle = 2\mathrm{Re}\{\alpha\} = 2\alpha_I. \tag{8.26}$$

On the other hand, the expected value of the square of in-phase operator is given by

$$\begin{aligned}\left\langle \hat{I}^2 \right\rangle_\alpha &= \langle\alpha|\hat{I}^2|\alpha\rangle = \langle\alpha|\left(a + a^\dagger\right)\left(a + a^\dagger\right)|\alpha\rangle = \langle\alpha|a^2 + \underbrace{aa^\dagger}_{a^\dagger a + \mathbf{1}} + a^\dagger a + \left(a^\dagger\right)^2|\alpha\rangle \\ &= \langle\alpha|\alpha^2 + 2\alpha^*\alpha + \alpha^{*2} + 1|\alpha\rangle = \left(\alpha + \alpha^*\right)^2 + 1 = 4\mathrm{Re}^2\{\alpha\} + 1 = 4\alpha_I^2 + 1,\end{aligned} \tag{8.27}$$

where we used the following property of annihilation and creation operators [10] $aa^\dagger = a^\dagger a + \mathbf{1}$. Now the variance of in-phase operator will be

$$V\left(\hat{I}\right) = \mathrm{Var}\left(\hat{I}\right) = \left\langle \hat{I}^2 \right\rangle_\alpha - \left\langle \hat{I} \right\rangle_\alpha^2 = 4\alpha_I^2 + 1 - 4\alpha_I^2 = 1, \tag{8.28}$$

which represents the minimum uncertainty expressed in SNU, commonly referred to as the shot noise. Therefore, the *uncertainty relation in SNU* can be represented as

$$V\left(\hat{I}\right)V\left(\hat{Q}\right) \geq 1. \tag{8.29}$$

Given that the number state $|n\rangle$ can be represented in terms of ground state by [9, 10], $|n\rangle = \left(a^\dagger\right)^n|0\rangle/\sqrt{n!}$, another convenient form of the coherent state will be

$$\begin{aligned}|\alpha\rangle &= e^{-|\alpha|^2/2}\sum_n \frac{\alpha^n}{\sqrt{n!}}\left(a^\dagger\right)^n|0\rangle/\sqrt{n!} = e^{-|\alpha|^2/2}\underbrace{\sum_n \frac{\left(\alpha a^\dagger\right)^n}{n!}}_{\exp\left(\alpha a^\dagger\right)}|0\rangle \\ &= \underbrace{\exp\left(\alpha a^\dagger - \alpha^* a\right)}_{D(\alpha)}|0\rangle = D(\alpha)|0\rangle, \quad D(\alpha) = \exp\left(\alpha a^\dagger - \alpha^* a\right),\end{aligned} \tag{8.30}$$

where $D(\alpha)$ is the *displacement operator*. The displacement operator is used to describe one coherent state as the displacement of another coherent state in the I-Q diagram.

The CV system composed of M bosonic modes can also be described by the quadrature operators $\left\{\hat{I}_m, \hat{Q}_m\right\}_{m=1}^{M}$, which can be arranged in the vector form:

$$\hat{\boldsymbol{x}} \doteq \left[\hat{I}_1\, \hat{Q}_1 \,\cdots\, \hat{I}_M\, \hat{Q}_M\right]^T. \tag{8.31}$$

As a generalization of Eq. (8.24), these quadrature operators satisfy the commutation relation:

$$\left[\hat{x}_m, \hat{x}_n\right] = 2j\Omega_{mn}, \tag{8.32}$$

where Ω_{mn} is the mth row, nth column element of $2M \times 2M$ matrix

$$\boldsymbol{\Omega} \doteq \bigoplus_{m=1}^{M} \boldsymbol{\omega} = \begin{bmatrix} \boldsymbol{\omega} & & \\ & \ddots & \\ & & \boldsymbol{\omega} \end{bmatrix}, \boldsymbol{\omega} = \begin{bmatrix} 0 & 1 \\ -1 & 0 \end{bmatrix}, \tag{8.33}$$

commonly referred to as the *symplectic form*.

For the quadrature operators, the first moment of $\hat{x}$ represents the displacement and can be determined as

$$\bar{\boldsymbol{x}} = \langle\hat{\boldsymbol{x}}\rangle = \mathrm{Tr}(\hat{\boldsymbol{x}}\hat{\rho}), \tag{8.34}$$

where the density operator $\hat{\rho}$ is the trace-one positive operator, which maps $\mathcal{H}^{\otimes M}$ to $\mathcal{H}^{\otimes M}$. The space of all density operators is known as the *state space*. When the density operator is the projector operator, that is, $\hat{\rho}^2 = \hat{\rho}$, we say that the density operator is *pure* and can be represented as $\hat{\rho} = |\phi\rangle\langle\phi|,\ |\phi\rangle \in \mathcal{H}^{\otimes M}$. On the other hand, we can determine the elements of *covariance matrix* $\boldsymbol{\Sigma} = (\Sigma_{mn})_{2N\times 2N}$ as follows [4]:

$$\Sigma_{mn} = \frac{1}{2}\langle\{\Delta\hat{x}_m, \Delta\hat{x}_n\}\rangle = \frac{1}{2}\left(\langle\hat{x}_m, \hat{x}_n\rangle + \langle\hat{x}_n, \hat{x}_m\rangle\right) - \langle\hat{x}_m\rangle\langle\hat{x}_n\rangle,\ \ \Delta\hat{x}_m = \hat{x}_m - \langle\hat{x}_m\rangle, \tag{8.35}$$

where $\{\cdot, \cdot\}$ denotes the anticommutator. The diagonal elements of the covariance matrix are clearly the variances of the quadratures:

$$V(\hat{x}_i) = \Sigma_{ii} = \left\langle(\Delta\hat{x}_i)^2\right\rangle = \left\langle(\hat{x}_i)^2\right\rangle - \langle\hat{x}_i\rangle^2. \tag{8.36}$$

The covariance matrix is real, symmetric, and positive-definite ($\boldsymbol{\Sigma} > 0$) matrix that satisfies the uncertainty principle, expressed as $\boldsymbol{\Sigma} + \mathrm{j}\boldsymbol{\Omega} \geq 0$. When we observe only diagonal terms, we arrive at usual uncertainty principle of quadratures, that is, $V\left(\hat{I}_m\right)V\left(\hat{Q}_m\right) \geq 1$.

When the first two moments are enough to completely characterize the density operator, that is, $\hat{\rho} = \hat{\rho}(\bar{\boldsymbol{x}}, \boldsymbol{\Sigma})$, we say that they represent the *Gaussian states*. The Gaussian states are bosonic states for which the quasi-probability distribution, commonly referred to as the *Wigner function*, is multivariate Gaussian distribution:

$$W(\boldsymbol{x}) = \frac{1}{(2\pi)^M (\det \boldsymbol{\Sigma})^{1/2}} \exp\left[-(\boldsymbol{x} - \bar{\boldsymbol{x}})^T \boldsymbol{\Sigma}^{-1}(\boldsymbol{x} - \bar{\boldsymbol{x}})/2\right], \tag{8.37}$$

where $\boldsymbol{x} \in \mathfrak{R}^{2M}$ ($\mathfrak{R}$-th set of real numbers) are eigenvalues of the quadrature operators $\hat{\boldsymbol{x}}$. The Wigner function of the thermal state, given by Eq. (8.12), is zero-mean Gaussian with covariance matrix $\boldsymbol{\Sigma} = (2\mathfrak{N} + 1)\mathbf{1}$, where $\mathfrak{N}$ is the average number of photons. The Gaussian CVs span a real symplectic space $(\mathfrak{R}^{2M}, \boldsymbol{\Omega})$ commonly referred to as the *phase space*. The corresponding *characteristic function* is given by

$$\chi(\boldsymbol{\zeta}) = \exp\left[-j(\boldsymbol{\Omega}\,\bar{\boldsymbol{x}})^T\boldsymbol{\zeta} - \frac{1}{2}\boldsymbol{\zeta}^{\mathrm{T}}(\boldsymbol{\Omega}\boldsymbol{\Sigma}\boldsymbol{\Omega}^T)\boldsymbol{\zeta}\right],\ \boldsymbol{\zeta} \in \mathfrak{R}^{2M}. \tag{8.38}$$

Not surprisingly, the Wigner function and the characteristic function are related by the Fourier transform:

$$W(\boldsymbol{x}) = \frac{1}{(2\pi)^{2M}} \int\limits_{\mathfrak{R}^{2M}} \chi(\boldsymbol{\zeta}) e^{-j\boldsymbol{x}^T\boldsymbol{\Omega}\boldsymbol{\zeta}} d^{2M}\boldsymbol{\zeta}. \tag{8.39}$$

Let us introduce the *Weyl operator* as follows:

$$D(\boldsymbol{\zeta}) = e^{j\hat{\boldsymbol{x}}^T\boldsymbol{\Omega}\boldsymbol{\zeta}}. \tag{8.40}$$

Clearly, the displacement operator $D(\alpha)$ is just the complex version of the Weyl operator. The characteristic function is related to the Weyl operator by

$$\chi(\boldsymbol{\zeta}) = \mathrm{Tr}\left[\hat{\rho}D(\boldsymbol{\zeta})\right], \tag{8.41}$$

and represents the expected value of the Weyl operator. It can be applied to both Gaussian and non-Gaussian states. The Wigner function can be determined as the Fourier transform of the characteristic function, given by Eq. (8.39). Therefore, the Wigner function is an equivalent representation of an arbitrary quantum state, defined over $2M$-dimensional phase space. When the Wigner function is nonnegative, the pure state will be Gaussian.

A *squeezed state* is a Gaussian state which has unequal fluctuations in quadratures:

$$V\left(\hat{I}\right) = s, \quad V\left(\hat{Q}\right) = s^{-1}, \tag{8.42}$$

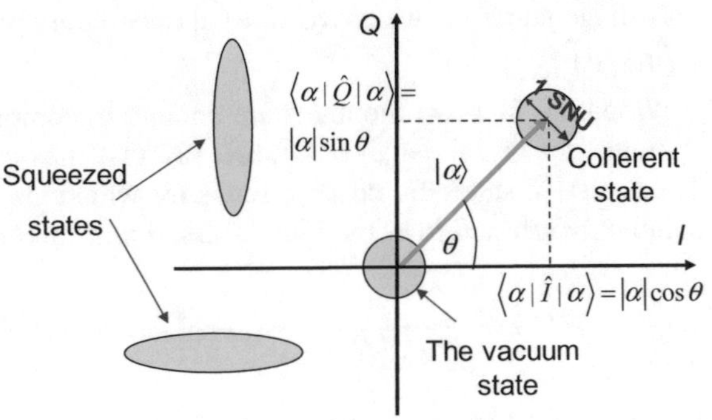

Fig. 8.1 The illustration of coherent and squeezed states

where s is the squeezing parameter. (The fluctuations above are expressed in SNU.)

For a single mode, the probability density function (PDF) of in-phase component, denoted as $f(I)$, can be determined by averaging the Wigner function over the quadrature component, that is, $f(I) = \int_Q W(I, Q)dQ$. On the other hand, if someone measures the quadrature component, the result will follow the following PDF: $f(Q) = \int_I W(I, Q)dI$. Therefore, we can interpret the Wigner function of the coherent state $|\alpha\rangle$ as the 2-D Gaussian distribution centered at (Re$\{\alpha\}$, Im$\{\alpha\}$) $= (\alpha_I, \alpha_Q)$ with variance being 1 in SNU (and of zero covariance). Therefore, in the (I, Q)-plane, the coherent state is represented by the circle, with radius related to the standard deviation (uncertainty), which is illustrated in Fig. 8.1. For the squeezed states, the I-coordinate is scaled by factor $s^{1/2}$, while the Q-coordinate by factor $s^{-1/2}$.

The squeezed coherent state $|\alpha, s\rangle$ can also be represented in terms of the ground state $|0\rangle$ as follows:

$$|\alpha, s\rangle = D(\alpha)S(s)|0\rangle, \;\; S(s) = e^{-\frac{1}{2}\left(s a^{\dagger 2} - s^* a^2\right)}, \tag{8.43}$$

where $S(s)$ is the *squeezing operator*. The squeezing parameter s in quantum optics is typically a real number, so that squeezing operator gets simplified to $S(s) = \exp\left[-s\left(a^{\dagger 2} - a^2\right)/2\right]$. By applying the squeezing operator on the vacuum state, we obtain the *squeezed vacuum state* [4, 11]:

$$|0, s\rangle = D(0)S(s)|0\rangle = (\cosh s)^{-1/2} \sum_{n=0}^{\infty} \frac{\sqrt{(2n)!}}{2^n n!} \tanh s^n |2n\rangle. \tag{8.44}$$

8.1.4 Gaussian Transformations and Gaussian Channels

As discussed in Chap. 5, a given quantum state undergoes the transformation described by the quantum operation (superoperator) [9]. Let the composite system C be composed of quantum register Q and environment E. This kind of system can be modeled as a closed quantum system. When the composite system is closed, its dynamic is unitary, and final state is specified by a unitary operator U as follows: $U(\hat{\rho} \otimes \varepsilon_0)U^{\dagger}$, where ε_0 is the initial density operator of the environment E. The reduced density operator of Q upon interaction ρ_f can be obtained by tracing out the environment:

$$\hat{\rho}_f = \mathrm{Tr}_E\left[U(\hat{\rho} \otimes \varepsilon_0)U^{\dagger}\right] \doteq \xi(\hat{\rho}). \tag{8.45}$$

The *superoperator* (*quantum operation*) is the mapping of initial density operator ρ to the final density operator ρ_f, denoted as $\xi : \hat{\rho} \to \hat{\rho}_f$. We say that the *superoperator is Gaussian* when it maps Gaussian states into Gaussian states as well. The *Gaussian channels* are such channel that preserves the Gaussian nature of the quantum state. The unitary transformations for Gaussian channels are like the time-evolution operator, that is, $U = \exp(-j\hat{H})$ with $\hat{H}$ being the Hamiltonian. The Hamiltonian is the second-order polynomials of the creation and annihilation operators' vectors, defined as $\boldsymbol{a}^{\dagger} = \left[a_1^{\dagger} \cdots a_M^{\dagger}\right]$, $\boldsymbol{a} = \left[a_1 \cdots a_M\right]$, and can be represented as follows [4]:

$$\hat{H} = j\left(\boldsymbol{a}^{\dagger}\boldsymbol{\alpha} + \boldsymbol{a}^{\dagger}\boldsymbol{F}_1\boldsymbol{a} + \boldsymbol{a}^{\dagger}\boldsymbol{F}_2\boldsymbol{a}\right) + H.c., \tag{8.46}$$

where $\alpha \in \mathbb{C}^M$ ($\mathbb{C}$—the set of complex numbers), while $\boldsymbol{F}_1$ and $\boldsymbol{F}_2$ are $M \times M$ complex matrices (and H.c. stands for the Hermitian conjugate). In the Heisenberg picture, these kinds of transformations correspond to a unitary *Bogoliubov transformation* [4, 12, 13]:

$$\boldsymbol{a} \to \boldsymbol{a}_f = U^{\dagger}\boldsymbol{a}U = \boldsymbol{A}\boldsymbol{a} + \boldsymbol{B}\boldsymbol{a}^{\dagger} + \boldsymbol{\alpha}, \tag{8.47}$$

where $\boldsymbol{A}$ and $\boldsymbol{B}$ are $M \times M$ complex matrices satisfying $\boldsymbol{A}\boldsymbol{B}^{\boldsymbol{T}} = \boldsymbol{B}\boldsymbol{A}^{\boldsymbol{T}}, \boldsymbol{A}\boldsymbol{A}^{\dagger} = \boldsymbol{B}\boldsymbol{B}^{\dagger} + \mathbf{1}$. Bogoliubov transformation induces an autoequivalence on the respective representations. As an illustration, let us consider the creation and annihilation operators, which satisfy the commutation relation $\left[a, a^{\dagger}\right] = \mathbf{1}$. Let us define new pairs of operators $\hat{b} = ua + va^{\dagger}$, $\hat{b}^{\dagger} = u^*a^{\dagger} + v^*a$, where u and v are complex constants. Given that Bogoliubov transformation preserves the commutativity relations, the new commutator $\left[\hat{b}, \hat{b}^{\dagger}\right]$ must be 1, and consequently $|u|^2 - |v|^2$ must be 1. Since the hyperbolic identity $\cosh^2 x - \sinh^2 x = 1$ has the same form, we can parametrize u and v by $u = e^{j\theta_u}\cosh s$, $u = e^{j\theta_v}\sinh s$. This transformation is, therefore, a linear symplectic transformation of the phase space. Two angles correspond to the orthogonal symplectic transformations (such as rotations) and the squeezing factor s corresponds to the diagonal transformation. For the quadrature operators, Gaussian transformation is described by an affine map $U_{\boldsymbol{S},\boldsymbol{c}}$:

$$U_{\boldsymbol{S},\boldsymbol{c}} : \boldsymbol{x} \to \boldsymbol{x}_f = \hat{\mathbf{S}}\boldsymbol{x} + \boldsymbol{c}, \tag{8.48}$$

where $\boldsymbol{S}$ is $2M \times 2M$ real matrix and $\boldsymbol{c}$ is a real vector of length $2M$. This transformation preserves the commutation relations, given by Eq. (8.32), when the matrix $\boldsymbol{S}$ is a symplectic matrix, that is, when it satisfies the following property:

$$\boldsymbol{S}\boldsymbol{\Omega}\boldsymbol{S}^T = \boldsymbol{\Omega}. \tag{8.49}$$

The affine mapping can be decomposed as

$$U_{S,\boldsymbol{c}} = D(\boldsymbol{c})U_S,\ D(\boldsymbol{c}) : \boldsymbol{x} \to \boldsymbol{x} + \boldsymbol{c},\ U_S : \boldsymbol{x} \to \boldsymbol{S}\boldsymbol{x}, \tag{8.50}$$

where $D(\boldsymbol{c})$ is the Weyl operator representing the translation in the phase space, while U_S is a linear symplectic mapping. The action of Gaussian transformation on the first two moments of the quadrature field operators will be simply

$$\bar{\boldsymbol{x}} \to \boldsymbol{S}\bar{\boldsymbol{x}} + c,\ \boldsymbol{\Sigma} \to \boldsymbol{S}\boldsymbol{\Sigma}\boldsymbol{S}^T. \tag{8.51}$$

Clearly, the action on the Gaussian state is completely characterized by Eq. (8.51).

As an illustration, let us consider the *beam splitter (BS) transformation*. The *beam splitter* is a partially silvered piece of glass with reflection coefficient parameterized by $\cos^2\theta$. The transmissivity of the beam splitter is given by $T = \cos^2\theta \in [0, 1]$. The beam splitter acts on two modes, which can be described by the creation (annihilation) operators a ($a^\dagger$) and b ($b^\dagger$). In the Heisenberg picture, the annihilation operators of the modes get transformed by the following Bogoliubov transformation:

$$\begin{bmatrix} a \\ b \end{bmatrix} \to \begin{bmatrix} a_f \\ b_f \end{bmatrix} = \begin{bmatrix} \sqrt{T} & \sqrt{1-T} \\ -\sqrt{1-T} & \sqrt{T} \end{bmatrix} \begin{bmatrix} a \\ b \end{bmatrix} = \begin{bmatrix} \cos\theta & \sin\theta \\ -\sin\theta & \cos\theta \end{bmatrix} \begin{bmatrix} a \\ b \end{bmatrix}. \tag{8.52}$$

The action of beam splitter can be described by the following transformation:

$$BS(\theta) = e^{-j\hat{H}} = e^{\theta(a^\dagger b - ab^\dagger)} \tag{8.53}$$

The quadrature operators of the modes $\hat{\boldsymbol{x}} = \left(\hat{x}_1\, \hat{x}_2\, \hat{x}_3\, \hat{x}_4\right)^T = \left(\hat{I}_a\, \hat{Q}_a\, \hat{I}_b\, \hat{Q}_b\right)^T$ get transformed by the following symplectic transformation:

$$\hat{\boldsymbol{x}} \to \hat{\boldsymbol{x}}_f = \begin{bmatrix} \sqrt{T}\mathbf{1} & \sqrt{1-T}\mathbf{1} \\ -\sqrt{1-T}\mathbf{1} & \sqrt{T}\mathbf{1} \end{bmatrix} \hat{\boldsymbol{x}}, \tag{8.54}$$

where now $\mathbf{1}$ is 2×2 identity matrix.

As another illustrative example, let us consider the *phase rotation* operator. The phase rotation can be described by the free-propagation Hamiltonian $\hat{H} = \theta a^\dagger a$ as follows $\hat{R}(\theta) = e^{-j\hat{H}} = e^{-j\theta a^\dagger a}$. For the annihilation operator, the corresponding transformation is simply $a \to a_f = ae^{-j\theta}$. For the quadrature operators, the corresponding transformation is symplectic mapping:

$$\hat{\boldsymbol{x}} \to \hat{\boldsymbol{x}}_f = \hat{R}(\theta)\hat{\boldsymbol{x}}, \hat{R}(\theta) = \begin{bmatrix} \cos\theta & \sin\theta \\ -\sin\theta & \cos\theta \end{bmatrix}. \quad (8.55)$$

8.1.5 *Thermal Decomposition of Gaussian States and von Neumann Entropy*

The *thermal state* is a bosonic state that maximizes the von Neumann entropy:

$$S(\hat{\rho}) = -\mathrm{Tr}(\hat{\rho}\log\hat{\rho}) = -\sum_i \eta_i \log \eta_i, \quad (8.56)$$

where η_i are eigenvalues of the density operator. The thermal state, as shown by Eq. (8.12), can represented in terms of the number states as follows:

$$\hat{\rho}_{\text{thermal}} = \sum_{n=0}^{\infty}(1-v)v^n|n\rangle\langle n| = \frac{1}{\mathcal{N}+1}\sum_{n=0}^{\infty}\left(\frac{\mathcal{N}}{\mathcal{N}+1}\right)^n|n\rangle\langle n|, \; v = \frac{\mathcal{N}}{\mathcal{N}+1}. \quad (8.57)$$

So, the entropy of the thermal state is given by [14]

$$\begin{aligned} S(\hat{\rho}_{\text{thermal}}) &= -\sum_i \eta_i \log \eta_i = \mathcal{N}\log\left(\frac{\mathcal{N}+1}{\mathcal{N}}\right) - \log(\mathcal{N}+1) \\ &= g(\mathcal{N}), \; g(x) = (x+1)\log(x+1) - x\log x, \; x > 0, \end{aligned} \quad (8.58)$$

where the g-function is a monotonically increasing concave function. The density operator of the whole field, composed of M thermal states, can be represented as the direct product of modes, as given by Eq. (8.13), which can also be written as

$$\hat{\rho}_{\text{total}} = \bigotimes_{m=1}^{M} \hat{\rho}_{\text{thermal}}(\mathcal{N}_m). \quad (8.59)$$

Given that thermal states are independent, the von Neumann entropy is additive, and we can write

$$S(\hat{\rho}_{\text{total}}) = S\left[\bigotimes_{m=1}^{M}\hat{\rho}_{\text{thermal}}(\mathcal{N}_m)\right] = \sum_{m=1}^{M} S(\hat{\rho}_{\text{thermal}}(\mathcal{N}_m)) = \sum_{m=1}^{M} g(\mathcal{N}_m), \quad (8.60)$$

where $\mathcal{N}_m$ is the average number of photons in the mth thermal state.

By applying the Williamson's theorem [15] on the M-mode covariance matrix $\boldsymbol{\Sigma}$, claiming that every positive-definite real matrix can be placed in a diagonal form by a suitable symplectic transformation, we can represent the covariance matrix $\boldsymbol{\Sigma}$, with the help of a symplectic matrix $\boldsymbol{S}$, as follows [4, 14]:

$$\boldsymbol{\Sigma} = \boldsymbol{S}\boldsymbol{\Sigma}^{\oplus}\boldsymbol{S}^{T}, \boldsymbol{\Sigma}^{\oplus} = \bigoplus_{m=1}^{M} \nu_m \mathbf{1}, \tag{8.61}$$

where $\boldsymbol{\Sigma}^{\oplus}$ is the diagonal matrix known as the Williamson's form, and ν_m ($m = 1, \ldots, M$) are symplectic eigenvalues of the covariance matrix, determined as the modulus of ordinary eigenvalues of the matrix $\mathrm{j}\boldsymbol{\Omega}\boldsymbol{\Sigma}$. Given that only modulus of the eigenvalues is relevant, the $2M \times 2M$ covariance matrix $\boldsymbol{\Sigma}$ will have exactly M symplectic eigenvalues. The representation (8.61) is in fact the thermal decomposition of a zero-mean Gaussian state, described by the covariance matrix $\boldsymbol{\Sigma}$. The uncertainty principle can now be expressed as $\boldsymbol{\Sigma}^{\oplus} \geq 1$, $\boldsymbol{\Sigma} > 0$. Given that the covariance matrix is positive definite, its symplectic eigenvalues are larger or equal to 1, and the entropy can be calculated by using Eq. (8.60) as follows:

$$S(\hat{\rho}(\mathbf{0}, \boldsymbol{\Sigma}^{\oplus})) = S\left[\bigotimes_{m=1}^{M} \hat{\rho}_{\text{thermal}}\left(\frac{\nu_m - 1}{2}\right)\right] = \sum_{m=1}^{M} S\left(\hat{\rho}_{\text{thermal}}\left(\frac{\nu_m - 1}{2}\right)\right) = \sum_{m=1}^{M} g\left(\frac{\nu_m - 1}{2}\right). \tag{8.62}$$

The *thermal decomposition* of an arbitrary Gaussian state can be represented as

$$\hat{\rho}(\bar{\boldsymbol{x}}, \boldsymbol{\Sigma}^{\oplus}) = D(\bar{\boldsymbol{x}})U_S[\hat{\rho}(\mathbf{0}, \boldsymbol{\Sigma}^{\oplus})]U_S^{\dagger}D(\bar{\boldsymbol{x}})^{\dagger}, \tag{8.63}$$

where $D(\bar{\boldsymbol{x}})$ represents the phase-space translation, while U_S corresponds to the symplectic mapping described by Eq. (8.61).

8.1.6 Nonlinear Quantum Optics Fundamentals and Generation of Quantum States

Most of the optical media are dielectric [16–18], so that the magnetic flux density $\boldsymbol{B}$ and the magnetic field $\boldsymbol{H}$ are related by $\boldsymbol{B} = \mu\boldsymbol{H}$ (with μ being the permittivity of the optical medium), while the electric flux density $\boldsymbol{D}$, the electric field $\boldsymbol{E}$, and induced electric density (polarization) $\boldsymbol{P}$ are related by the constitutive relation $\boldsymbol{D} = \varepsilon\boldsymbol{E} + \boldsymbol{P}$ (with ε being the permeability). For the nonlinear optical medium, the electric polarization is related to the electric field by

$$\boldsymbol{P} = \vec{\chi}^{(1)} : \boldsymbol{E} + \underbrace{\vec{\chi}^{(2)} : \boldsymbol{E}\boldsymbol{E} + \vec{\chi}^{(3)} : \boldsymbol{E}\boldsymbol{E}\boldsymbol{E} + \cdots}_{\boldsymbol{P}_{NL}} \tag{8.64}$$

where $\vec{\chi}^{(i)}$ ($i = 1, 2, 3, \ldots$) are tensors and the operation (:) denotes the tensor product. The first term is the linear term and it is applicable in linear optics. The second and higher terms are responsible for different phenomena in the nonlinear (NL) optics, as illustrated in Fig. 8.2. The second-order term dominates in bulk optic materials such as various crystals, while the third-order term dominates in optical fibers and gaseous atomic/molecular media and due to symmetry of randomly oriented particles (atoms and molecules) the second term becomes zero. The nonlinear devices introduce interaction among field modes, with interaction Hamiltonian being represented as

$$\hat{H}_{\text{interaction}} = \frac{1}{2}\int \hat{\boldsymbol{P}}_{NL}(\boldsymbol{r},t)\cdot \hat{\boldsymbol{E}}(r,t), \hat{\boldsymbol{P}}_{NL} = \vec{\chi}^{(2)} : \hat{\boldsymbol{E}}\hat{\boldsymbol{E}} + \vec{\chi}^{(3)} : \hat{\boldsymbol{E}}\hat{\boldsymbol{E}}\hat{\boldsymbol{E}} + \cdots. \tag{8.65}$$

The *evolution of the quantum state*, in interaction picture, is described by the following unitary operator:

$$\hat{U} = \exp\left(-j\int \hat{H}_{\text{interaction}}dt\right), \tag{8.66}$$

which gives rise to the momentum conservation. The total electromagnetic energy localized in the medium of volume V is given by [19]

$$W = \int_V [\boldsymbol{E}(\boldsymbol{r},t)\cdot \boldsymbol{D}(\boldsymbol{r},t) + \boldsymbol{H}(\boldsymbol{r},t)\cdot \boldsymbol{B}(\boldsymbol{r},t)]d^3r, \tag{8.67}$$

and gives rise to the energy conservation.

In the second-order dominating devices, as illustrated in Fig. 8.2a, the strong pump photon will generate the signal and idler photons. This interaction is commonly referred to as the *three-wave mixing (TWM)*, and can be described by the following interaction Hamiltonian:

$$\hat{H}_{TWM} = j\eta a_s^\dagger a_i^\dagger a_p + H.c., \tag{8.68}$$

where we use a_p to denote the pump photon annihilation operators, while $a_s^\dagger$ and $a_i^\dagger$ to denote the signal and idle photons creation operators. The parameter η is related to the conversion efficiency. On the other hand, in the $\chi^{(3)}$-device, see Fig. 8.2b, two bright pump photons interact and generate the signal and idler photons. This

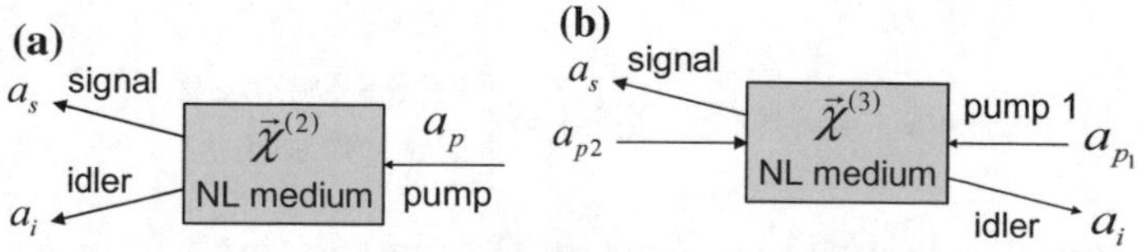

Fig. 8.2 Illustration of NL optics phenomena: **a** three-wave mixing and **b** four-wave mixing

interaction is known as the *four-wave mixing (FWM)*, and can be described by the following interaction Hamiltonian:

$$\hat{H}_{FWM} = j\eta a_s^\dagger a_i^\dagger a_{p_1} a_{p_2} + H.c., \tag{8.69}$$

The second- and third-order nonlinear effects are typically weak, and we need to use the strong pump beams to generate any observable quantum effect. When the pump beam is very strong, they become classical and can be represented by the corresponding complex variables, for instance, we can use the complex variable A_p instead of a_p. The corresponding two-photon processes are known as the *parametric process*. By substituting ηA_p in TWM Hamiltonian with ζ, and $\eta A_{p1} A_{p2}$ in FWM Hamiltonian with as well, the parametric processes Hamiltonian can be written as follows:

$$\hat{H}_P = j\zeta a_s^\dagger a_i^\dagger + H.c.. \tag{8.70}$$

The corresponding evolution operator becomes

$$\hat{U}_P = \exp\left(-j\int \hat{H}_P dt\right) = \exp\left(\underbrace{\zeta t}_{-\xi^*} a_s^\dagger a_i^\dagger - \zeta^* t a_s a_i\right) = \exp\left(-\xi^* a_s^\dagger a_i^\dagger + \xi a_s a_i\right), \xi = -\zeta^* t. \tag{8.71}$$

When both signal and idler photons are in vacuum states, the state at the output of the NL device is the two-mode squeeze (TMS) state:

$$|\Psi\rangle = \exp\left(-\xi^* a_s^\dagger a_i^\dagger + \xi a_s a_i\right)|0\rangle|0\rangle. \tag{8.72}$$

In this case, the parameter ξ is real number and can be denoted by s to specify the level of squeezing. The corresponding Hamiltonian in this case is commonly referred to as the *two-mode squeezing operator*:

$$S_2(s) = \exp\left(-s\left(a_s^\dagger a_i^\dagger - a_s a_i\right)\right). \tag{8.73}$$

In the Heisenberg picture, the quadrature operators $\hat{\boldsymbol{x}} = \left(\hat{x}_1\hat{x}_2\hat{x}_3\hat{x}_4\right)^T = \left(\hat{I}_a\hat{Q}_a\hat{I}_b\hat{Q}_b\right)^T$ get transformed to [4]

$$\hat{\boldsymbol{x}} \to \hat{\boldsymbol{x}}_f = \hat{\mathbf{S}}_2(s)\hat{\boldsymbol{x}},\ \hat{\mathbf{S}}_2(s) = \begin{bmatrix} \cosh s\,\mathbf{1} & \sinh s\,\mathbf{Z} \\ \sinh s\,\mathbf{Z} & \cosh s\,\mathbf{1} \end{bmatrix}, \tag{8.74}$$

where $\mathbf{Z} = \text{diag}(1, -1)$ is the Pauli Z-matrix. Now, by applying $S_2(s)$ on two vacuum sates, as described by Eq. (8.72), we can represent the TMS state, also known as the

Einstein–Podolsky–Rosen (EPR) state, as follows [4]:

$$|\Psi\rangle = \sqrt{1-\lambda^2}\sum_{n=0}^{\infty}\lambda^n|n\rangle_a|n\rangle_b = \sqrt{\frac{2}{v+1}}\sum_{n=0}^{\infty}\left(\frac{v-1}{v+1}\right)^{n/2}|n\rangle_a|n\rangle_b,\ \lambda^2 = \frac{v-1}{v+1}, \tag{8.75}$$

with $\lambda = \tanh(s)$ and $v = \cosh(2s)$ representing the variances of the quadratures. The corresponding density operator is given by $\rho_{\text{EPR}} = |\Psi\rangle\langle\Psi|$. The EPR state represents the Gaussian state with zero-mean and covariance matrix given by

$$\mathbf{\Sigma}_{\text{EPR}} = \begin{bmatrix} v\mathbf{1} & \sqrt{v^2-1}\,\mathbf{Z} \\ \sqrt{v^2-1}\,\mathbf{Z} & v\mathbf{1} \end{bmatrix},\quad v = \cosh 2s, \tag{8.76}$$

wherein the elements of the covariance matrix are determined by applying the definition, Eqs. (8.35) and (8.36), to the EPR state. The EPR state (8.75) is directly applicable to the CV-QKD systems.

When the signal and idler quantum fields are identical, that is, $a_s = a_i = a$, the interaction Hamiltonian (8.70) simplifies to $\hat{H}_{\text{p}} = j(\zeta a^{\dagger 2} - \zeta^* a^2)$, and the corresponding evolution operator becomes $\hat{U}_P = \exp\left(-j\hat{H}_P t\right) = \exp(-\xi^* a^{\dagger 2} + \xi a^2)$, $\xi = -\zeta^* t$. The squeezing parameter ξ in quantum optics is typically a real number, say s, so that the *squeezing operator* becomes $S(s) = \exp\left[-s\left(a^{\dagger 2} - a^2\right)/2\right]$. By applying the squeezing operator on the vacuum state, we obtain the *squeezed vacuum state* given already by Eq. (8.44). In the Heisenberg picture, the annihilation operator gets transformed by the linear unitary Bogoliubov transformation as follows $a \to a_f = (\cosh s)a - (\sinh s)a^\dagger$. On the other hand, the quadrature operators $\hat{\mathbf{x}} = (\hat{x}_1\hat{x}_2)^T = \left(\hat{I}\hat{Q}\right)^T$ get transformed to [4]

$$\hat{\mathbf{x}} \to \hat{\mathbf{x}}_f = \mathbf{S}(s)\hat{\mathbf{x}},\ \mathbf{S}(s) = \begin{bmatrix} e^{-s} & 0 \\ 0 & e^{-s} \end{bmatrix}. \tag{8.77}$$

The covariance matrix for the squeezed vacuum state is given by $\mathbf{\Sigma} = \mathbf{S}(s)[\mathbf{S}(s)]^T = \mathbf{S}(2s)$ and has different variances for the quadratures, with one of the variances being squeezed below the quantum shot noise and the other one being antisqueezed above the quantum shot noise.

For $|\xi| \ll 1$, by Taylor expansion in the evaluation operator of Eq. (8.72), we obtain the following approximation for TMS state:

$$|\Psi\rangle \cong \left(\mathbf{1} - \xi^* a_s^\dagger a_i^\dagger + \xi^{*2} a_s^{\dagger 2} a_i^{\dagger 2} + \cdots\right)|0\rangle|0\rangle = |0\rangle_s|0\rangle_i - \xi^*|1\rangle_s|1\rangle_i + \xi^{*2}|2\rangle_s|2\rangle_i + \cdots \tag{8.78}$$

Clearly, for sufficiently small ξ, the third and higher terms are negligible, the two-photon states term dominates, and these are widely used in two-photon interference studies. On the other hand, in the four-photon coincidence measurements, the first two terms are irrelevant. Now, by employing the beam splitter on two-photon and four-photon states, various entangled states can be generated.

Let us now consider the case in which a_s field in TWM and a_s and a_{p2} fields in FWM are strong enough to be considered classical, then the interaction Hamiltonian can be represented as follows:

$$\hat{H}_{FC} = j\zeta a_i^\dagger a_p - j\zeta^* a_i a_p^\dagger, \zeta = \begin{cases} \eta A_s^*, \text{TWM} \\ \eta A_s^* A_{p2}, \text{FWM} \end{cases} \tag{8.79}$$

and describes *one-photon process* in which one photon is annihilated or created at a given time instance. This process annihilates the photon in mode a_p and creates one photon in mode a_i. When these two photons have different frequencies, the *photon conversion process* takes place. The corresponding evolution operator will be

$$\hat{U}_{FC} = \exp\left(\xi a_i^\dagger a_p - \xi^* a_i a_p^\dagger\right), \xi = \zeta t, \tag{8.80}$$

which is for real ξ equivalent to the beam splitter evolution operator.

8.1.7 Correlation Matrices of Two-Mode Gaussian States

Two-mode Gaussian states $\hat{\rho} = \hat{\rho}(\bar{\boldsymbol{x}}, \boldsymbol{V})$ have been intensively studied [4, 20–24], and simple analytical formulas have been derived. The correlation matrix for the two-mode Gaussian states can be represented as [4, 22, 24] follows:

$$\boldsymbol{\Sigma} = \begin{bmatrix} \boldsymbol{A} & \boldsymbol{C} \\ \boldsymbol{C}^T & \boldsymbol{B} \end{bmatrix}, \tag{8.81}$$

where $\boldsymbol{A}$, $\boldsymbol{B}$, and $\boldsymbol{C}$ are 2×2 real matrices, while $\boldsymbol{A}$ and $\boldsymbol{B}$ are also orthogonal matrices. The symplectic eigenvalues' spectrum $\{\lambda_1, \lambda_2\} = \{\lambda_-, \lambda_+\}$ can be computed by [4, 22, 24]

$$\lambda_{1,2} = \sqrt{\frac{\Delta(\boldsymbol{\Sigma}) \mp \sqrt{\Delta^2(\boldsymbol{\Sigma}) - 4\det\boldsymbol{\Sigma}}}{2}},$$
$$\det\boldsymbol{\Sigma} = \lambda_1^2\lambda_2^2, \Delta(\boldsymbol{\Sigma}) = \det\boldsymbol{A} + \det\boldsymbol{B} + 2\det\boldsymbol{C} = \lambda_1^2 + \lambda_2^2. \tag{8.82}$$

The uncertainty principle is equivalent to the following conditions [24]:

$$\boldsymbol{\Sigma} > 0, \det\boldsymbol{\Sigma} \geq 1, \Delta(\boldsymbol{\Sigma}) \leq 1 + \det\boldsymbol{\Sigma}. \tag{8.83}$$

Important classes of two-mode Gaussian states are those whose covariance matrix can be represented in the *standard form*, such as the EPR state, as follows [20, 21]:

$$\boldsymbol{\Sigma} = \begin{bmatrix} a\mathbf{1} & \boldsymbol{C} \\ \boldsymbol{C} & b\mathbf{1} \end{bmatrix}, \quad \boldsymbol{C} = \begin{bmatrix} c_1 & 0 \\ 0 & c_2 \end{bmatrix}; a, b, c \in \Re. \tag{8.84}$$

In particular, when $c_1 = -c_2 = c \geq 0$, the symplectic eigenvalues get simplified to

$$\lambda_{1,2} = \left[\sqrt{(a+b)^2 - 4c^2} \mp (b-a)\right]/2. \tag{8.85}$$

The corresponding symplectic matrix $\boldsymbol{S}$ performing the symplectic decomposition $\boldsymbol{\Sigma} = \boldsymbol{S}\boldsymbol{\Sigma}^{\oplus}\boldsymbol{S}^T$ is given by [4]

$$\boldsymbol{S} = \begin{bmatrix} d_+\mathbf{1} & d_-\boldsymbol{Z} \\ d_-\boldsymbol{Z} & d_+\mathbf{1} \end{bmatrix}, \quad d_\pm = \sqrt{\frac{a+b \pm \sqrt{(a+b)^2 - 4c^2}}{2\sqrt{(a+b)^2 - 4c^2}}}. \tag{8.86}$$

8.1.8 Gaussian State Measurement and Detection

A quantum measurement is specified by the set of measurement operators given by $\{M_m\}$, where index m stands for possible measurement result, satisfying the completeness relation $\sum_m M_m^\dagger M_m = \mathbf{1}$. The probability of finding the measurement result m, given the state $|\psi\rangle$, is given by [9, 25, 26]

$$p_m = \Pr(m) = \mathrm{Tr}\left(\hat{\rho} M_m^\dagger M_m\right). \tag{8.87}$$

After the measurement, the system will be left in following state:

$$\hat{\rho}_f = M_m \hat{\rho} M_m^\dagger / p_m. \tag{8.88}$$

If we are only interested in the outcome of the measurement, we can define a set of positive operators $\left\{\Pi_m = M_m^\dagger M_m\right\}$ and describe the measurement as the *positive-operator-valued measure* (POVM). The probability of getting the mth measurement result will be then [25, 26]

$$p_m = \Pr(m) = \mathrm{Tr}\left(\hat{\rho}\Pi_m\right). \tag{8.89}$$

In addition to being positive, the Π_m-operators satisfy completeness relation $\sum_m \Pi_m = \mathbf{1}$.

In the case of CV systems, the measurement results are real, that is, $m \in R$, so that the probability of outcome becomes the probability density function. When measuring the Gaussian states, the measurement distribution will be Gaussian. If we perform the Gaussian measurements on M modes out of $M + N$ available, the distribution of measurements will be Gaussian, while remaining unmeasured N modes will be in a Gaussian state.

8.1.8.1 Homodyne Detection

In homodyne detection, the most common measurement in CV systems, we measure one of the two quadratures, since we cannot measure both quadratures simultaneously, according to the uncertainty principle. The measurement operators are projectors over the quadrature basis $|I\rangle\langle I|$ or $|Q\rangle\langle Q|$, and corresponding outcomes have PDF $f(I)$ or $f(Q)$, obtained by the marginal integration of the Wigner function:

$$f(I) = \int_Q W(I, Q)dQ, \; f(Q) = \int_I W(I, Q)dI. \tag{8.90}$$

When more than one bosonic mode are in the optical beam, we have to apply partial homodyning and also perform integration over both quadratures of all other modes.

In Gaussian-modulation-based CV-QKD with homodyne detection [1, 27], shown in Fig. 8.3, only one quadrature component is measured at a time. This protocol is commonly referred to as the *GG02 CV-QKD protocol*, named after the authors Grosshans and Grangier who proposed it in 2002 [28]. Alice sufficiently attenuates the laser beam, which serves as a weak coherent state (WCS) source, and employs the balanced beam splitter (BBS) to split the signal into two parts. The upper output of BBS is used as an input to the I/Q modulator to place the Gaussian state to the desired location in I-Q plane to get $|\alpha\rangle = |I + jQ\rangle$ as well as to control the Alice's variance. The second output of BBS is used to send the local oscillator (LO) reference signal for Bob. Since the action of beam splitter is described by $BBS = \frac{1}{\sqrt{2}}\begin{bmatrix} 1 & 1 \\ -1 & 1 \end{bmatrix}$, the annihilation operators at the out of Bobs' BBS will be

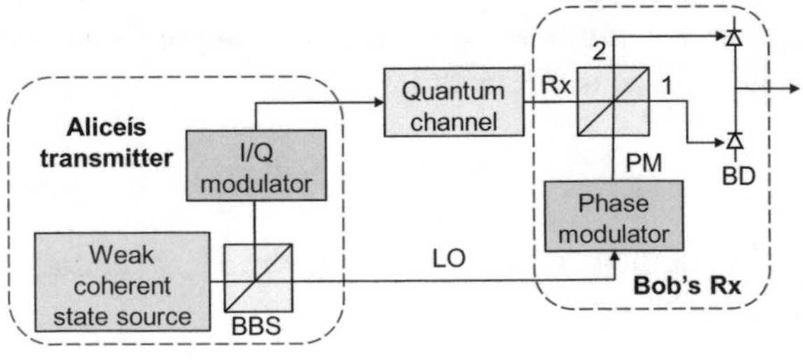

Fig. 8.3 Illustrating the Gaussian modulation-based CV-QKD protocols with homodyne detection. BBS: balanced beam splitter, BD: balanced detector

$$\begin{bmatrix} a_1 \\ a_2 \end{bmatrix} = \frac{1}{\sqrt{2}} \begin{bmatrix} 1 & 1 \\ -1 & 1 \end{bmatrix} \begin{bmatrix} a_{Rx} \\ a_{PM} \end{bmatrix} = \frac{1}{\sqrt{2}} \begin{bmatrix} a_{Rx} + a_{PM} \\ -a_{Rx} + a_{PM} \end{bmatrix}, \tag{8.91}$$

where the creation (annihilation) operators at receive BBS inputs Rx and PM are denoted by $a_{Rx}^{\dagger}$ (a_{Rx}) and $a_{PM}^{\dagger}(a_{PM})$, respectively. The number operators at receive BBS outputs 1 and 2 can be expressed now as follows:

$$\hat{n}_1 = a_1^{\dagger} a_1 = \frac{1}{2}\left(a_{PM}^{\dagger} + a_{Rx}^{\dagger}\right)(a_{PM} + a_{Rx}), \hat{n}_2 = a_2^{\dagger} a_2 = \frac{1}{2}\left(a_{PM}^{\dagger} - a_{Rx}^{\dagger}\right)(a_{PM} - a_{Rx}). \tag{8.92}$$

The output of balanced detector (BD), assuming the photodiode responsivity of $P_{\mathrm{ph}} = 1$A/W, is given by (in SNU):

$$\hat{i}_{BD} = 4(\hat{n}_1 - \hat{n}_2) = 4\left(a_{PM}^{\dagger} a_{Rx} + a_{Rx}^{\dagger} a_{PM}\right). \tag{8.93}$$

The phase modulation output can be represented in SNU by $a_{PM} = |\alpha| \exp(j\theta)\mathbf{1}/2$, and after substitution into (8.93) we obtain

$$\hat{i}_{BD} = 4\left(\frac{|\alpha| e^{-j\theta}}{2} a_{Rx} + a_{Rx}^{\dagger} \frac{|\alpha| e^{j\theta}}{2}\right) = 2|\alpha|\left(e^{j\theta} a_{Rx}^{\dagger} + e^{-j\theta} a_{Rx}\right). \tag{8.94}$$

In the absence of Eve, the receive signal constellation will be $\boldsymbol{a}_{\mathrm{Rx}} = (I, Q)/2$, which after substitution into (8.93) yields

$$\begin{aligned} \hat{i}_{BD} &= |\alpha|\left[(\cos\theta, \sin\theta)\mathbf{1} \cdot \left(I\hat{I}, -Q\hat{Q}\right) + (\cos\theta, \sin\theta)\mathbf{1} \cdot \left(I\hat{I}, Q\hat{Q}\right)\right] \\ &= |\alpha|\left(\cos\theta \cdot I\hat{I} + \sin\theta \cdot Q\hat{Q}\right). \end{aligned} \tag{8.95}$$

Bob measures in-phase component $\hat{I}$ by selecting $\theta = 0$ to get $|\alpha|I$ and the quadrature component $\hat{Q}$ by setting $\theta = \pi/2$ to get $|\alpha|Q$.

8.1.8.2 Heterodyne Detection

The theory of optical heterodyne detection has been developed in [29]. The heterodyne detection corresponds to the projection onto coherent states, that is, $M(\alpha) \sim |\alpha\rangle\langle\alpha|$. This theory has been generalized to the POVM based on projections over pure Gaussian states [4, 30]. For the heterodyne detection-based Gaussian modulation CV-QKD, the transmitter side is identical to coherent detection version. In heterodyne detection receiver, shown in Fig. 8.4, the quantum signal is split using the BBS, one output is used to measure the in-phase component (upper branch), while the other output (lower branch) is used as input to the second BD to detect the quadrature component. The LO classical reference signal is also split by BBS,

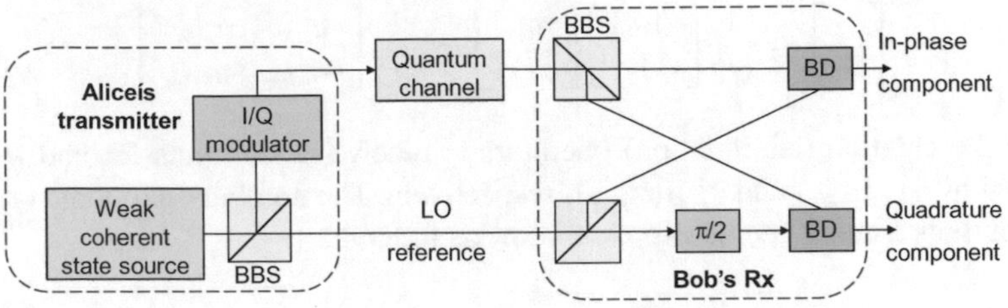

Fig. 8.4 Illustrating the Gaussian modulation-based CV-QKD protocols with heterodyne detection. BBS: balanced beam splitter, BD: balanced detector, shown in Fig. 8.3

with upper output being used as input to the in-phase BD, and the other output, after the phase modulation to introduce the π/2-phase shift, is used as the LO input to the quadrature BD to detect the quadrature component. On such a way, both quadratures can be measured simultaneously and mutual information can be doubled. On the other hand, the BBS introduced 3 dB attenuation of the quantum signal.

8.1.8.3 Partial Measurements

As indicated earlier, the measurement on one mode will affect the remaining modes of Gaussian state, depending on corresponding correlations [4, 30]. Let B be the mode on which measurement is performed and A denotes remaining $M - 1$ modes of the Gaussian state described by $2M \times 2M$ covariance matrix $\boldsymbol{\Sigma}$. Similar to Eq. (8.81), we can decompose the covariance matrix as follows [4, 30]:

$$\boldsymbol{\Sigma} = \begin{bmatrix} \boldsymbol{A} & \boldsymbol{C} \\ \boldsymbol{C}^T & \boldsymbol{B} \end{bmatrix}, \tag{8.96}$$

where $\boldsymbol{B}$ is the real matrix of size 2×2, $\boldsymbol{A}$ is the real matrix of size $2(M - 1) \times 2(M - 1)$, and $\boldsymbol{C}$ is the real matrix of size $2(M - 1) \times 2$. For homodyne detection of in-phase I or quadrature Q, the partial measurement of B transforms the covariance matrix of A as follows [4, 30]:

$$\boldsymbol{\Sigma}_{\boldsymbol{A}|\boldsymbol{B}}^{\left(\hat{I},\hat{Q}\right)} = \boldsymbol{A} - \boldsymbol{C}\left(\boldsymbol{\Pi}_{I,Q}\boldsymbol{B}\boldsymbol{\Pi}_{I,Q}\right)^{-1}\boldsymbol{C}^T,\ \boldsymbol{\Pi}_I = \mathrm{diag}(1,0),\ \boldsymbol{\Pi}_Q = \mathrm{diag}(0,1), \tag{8.97}$$

where we use $(\cdot)^{-1}$ to denote the pseudo-inverse operation. Given that $\boldsymbol{\Pi}_I\boldsymbol{B}\boldsymbol{\Pi}_I = \begin{bmatrix} B_{11} & 0 \\ 0 & 0 \end{bmatrix} = \begin{bmatrix} V\left(\hat{I}\right) & 0 \\ 0 & 0 \end{bmatrix}$ and $\boldsymbol{\Pi}_Q\boldsymbol{B}\boldsymbol{\Pi}_Q = \begin{bmatrix} 0 & 0 \\ 0 & B_{22} \end{bmatrix} = \begin{bmatrix} 0 & 0 \\ 0 & V\left(\hat{Q}\right) \end{bmatrix}$ matrices are diagonal, the previous equation simplifies to

$$\boldsymbol{\Sigma}_{A|B}^{(\hat{I},\hat{Q})} = \boldsymbol{A} - \frac{1}{V\left(\hat{I},\hat{Q}\right)}\boldsymbol{C}\boldsymbol{\Pi}_{I,Q}\boldsymbol{C}^T. \tag{8.98}$$

In heterodyne detection, we measure both quadratures, and given that additional balanced beam splitter is used, with one of inputs being in vacuum state, we need to account for fluctuations due to vacuum state and modify the partial correlation matrix as follows [4, 30]:

$$\boldsymbol{\Sigma}_{A|B} = \boldsymbol{A} - \boldsymbol{C}(\boldsymbol{B}+\boldsymbol{1})^{-1}\boldsymbol{C}^T. \tag{8.99}$$

8.2 CV-QKD Protocols with Gaussian Modulation

As indicated above, the CV-QKD can be implemented by employing either homodyne detection, where only one quadrature component is measured at a time (because of the uncertainty principle), or with heterodyne detection (HD), where one beam splitter (BS) and two balanced photodetectors are used to measure both quadrature components simultaneously. HD can double the mutual information between Alice and Bob compared to the homodyne detection scheme at the expense of additional 3 dB loss of the beam splitter (BS). In order to reduce the phase noise, the quantum signals are typically co-propagated together with the time-domain-multiplexed high-power pilot tone (PT) to align Alice's and Bob's measurement bases. CV-QKD studies are getting momentum, by judging by increasing number of papers related to CV-QKD [1, 27, 28, 31–51], thanks to their compatibility with the state-of-the-art optical communication technologies. The CV-QKD system experiences the 3 dB loss limitation in transmittance when the direct reconciliation is used. To avoid this problem either reverse reconciliation [32] or the postelection [39] methods are used. Different eavesdropping strategies discussed in Sect. 6.4.1 are applicable to CV-QKD systems as well. It has been shown that for Gaussian modulation, Gaussian attack is optimum attack for both individual attacks [45] and collective attacks [46, 47]. In this section of the chapter, we focus our attention to the CV-QKD with Gaussian modulation. To implement CV-QKD, both squeezed states and coherent states can be employed.

Squeezed state-based protocols employ the Heisenberg uncertainty principle, claiming that it is impossible to measure both quadratures with complete precision. To impose the information, Alice randomly selects to use either in-phase or quadrature degree of freedom (DOF), as illustrated in Fig. 8.5a. In Alice encoding rule I (when in-phase DOF is used), the squeezed state is imposed on the in-phase component (with squeezed parameter $s_I < 1$) by (in SNU)

$$X_{A,I} \to \left|X_{A,I}, s_I\right\rangle, s_I < 1, X_{A,I} \sim \mathfrak{N}\left(0, \sqrt{v_{A,I}}\right) \tag{8.100}$$

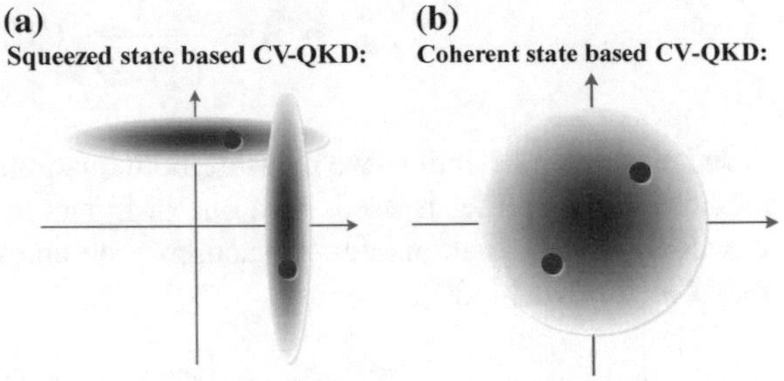

Fig. 8.5 Illustration of operation principles of squeezed (**a**) and coherent (**b**) state-based protocols

The amplitude is selected from a zero-mean Gaussian source of variance $v_{A,I}N_0$. On the other hand, in Alice encoding rule Q (when the quadrature is used), the squeezed state is imposed on the quadrature (with squeezed parameter $s_Q > 1$) and we can write (in SNU)

$$X_{A,Q} \to \left|jX_{A,Q}, s_Q\right\rangle, s_Q > 1, X_{A,Q} \sim \mathfrak{N}\left(0, \sqrt{v_{A,Q}}\right) \tag{8.101}$$

The variances of squeezed states I and Q will be $V\left(\hat{I}\right) = \sigma_I^2 = s_I$ and $V\left(\hat{Q}\right) = \sigma_Q^2 = 1/s_Q$, respectively (as described above). On the receiver side, Bob randomly selects whether to measure either $\hat{I}$ or $\hat{Q}$. Alice and Bob exchange the encoding rules being used by them to measure the quadrature for every squeezed state and keep only instances when they measured the same quadrature in the sifting procedure. Therefore, this protocol is very similar to the BB84 protocol. After that, the information reconciliation takes place, followed by the privacy amplification.

From Eve's point of view, the corresponding mixed states when Alice employed either I or Q DOF will be [1]

$$\begin{aligned}\rho_I &= \int f_N\left(0, \sqrt{v_{A,I}}\right)|x, s_I\rangle\langle x, s_I|dx, \\ \rho_Q &= \int f_N\left(0, \sqrt{v_{A,Q}}\right)\left|jQ, s_Q\right\rangle\!\left\langle jQ, s_Q\right|dx\end{aligned} \tag{8.102}$$

where $f_N(0, \sigma)$ is zero-mean Gaussian distribution with variance σ^2. When the condition $\rho_I = \rho_Q$ is satisfied, Eve cannot distinguish whether Alice was imposing the information on the squeezed state I or Q. Given that squeezed states are much more difficult to generate than the coherent states, we focus our attention to the coherent state-based CV-QKD protocols.

8.2.1 Coherent State-Based CV-QKD Protocols

In coherent state-based protocols, there is only one encoding rule for Alice:

$$(X_{A,I}, X_{A,Q}) \to |X_{A,I} + jX_{A,Q}\rangle; \; X_{A,I} \sim \mathcal{N}(0, \sqrt{v_A}), \; X_{A,Q} \sim \mathcal{N}(0, \sqrt{v_A}). \tag{8.103}$$

In other words, Alice randomly selects a point in 2-D (I, Q) space from a zero-mean circular symmetric Gaussian distribution, with the help of apparatus provided in Fig. 8.4. Clearly, both quadratures have the same uncertainty. Here, we again employ the Heisenberg uncertainty, claiming that it is impossible to measure both quadratures with complete precision. For homodyne receiver, Bob performs the random measurement on either $\hat{I}$ or $\hat{Q}$, and let the result of his measurement be denoted by Y_B. When Bob measures $\hat{I}$, his measurement Y_B is correlated $X_{A,I}$. On the other hand, when Bob measures $\hat{Q}$, his measurement result Y_B is correlated $X_{A,Q}$. Clearly, Bob is able to measure a single coordinate of the signal constellation point sent by Alice using Gaussian coherent state. Bob then announces which quadrature he measured in each signaling interval and Alice selects the coordinate that agrees with Bob's measurement quadrature. The rest of the protocol is the same as for the squeezed state-based protocol.

From Eve's point of view, the mixed state will be

$$\rho = \int f_N(0, \sqrt{v_A}) f_N(0, \sqrt{v_A}) |I + jQ\rangle\langle I + jQ| dI dQ, \tag{8.104}$$

where v_A is the Alice modulation variance in SNU. For lossless transmission, the total variance of Bob's random variable in SNU will be

$$V(Y_B) = \mathrm{Var}(Y_B) = v_A + 1, \tag{8.105}$$

which is composed of Alice's modulation term v_A and the variance of intrinsic coherent state fluctuations is 1.

For lossy transmission, the channel can be modeled as a beam splitter with attenuation $0 \leq T \leq 1$, as illustrated in Fig. 8.6. Based on Fig. 8.6, we conclude that the output state can be represented in terms of input state $\boldsymbol{x}_A$ and ground state $\boldsymbol{x}_0$ as follows:

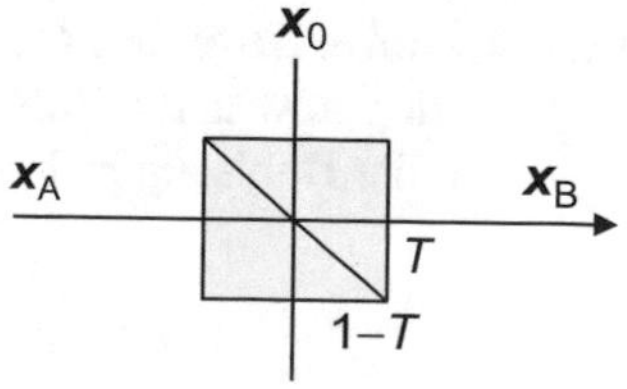

Fig. 8.6 The illustration of lossy transmission channel

$$\boldsymbol{x}_B = \sqrt{T}\boldsymbol{x}_A + \sqrt{1-T}\boldsymbol{x}_0 = \sqrt{T}\left(\boldsymbol{x}_A + \sqrt{\chi_{\text{line}}}\boldsymbol{x}_0\right),\ \chi_{\text{line}} = (1-T)/T = 1/T - 1 \tag{8.106}$$

For lossy transmission, when the Gaussian modulation of coherent states is employed instead of vacuum states, the previous expression needs to be modified as follows:

$$\boldsymbol{x}_B = \sqrt{T}\left(\boldsymbol{x}_A + \sqrt{\chi_{\text{line}}}\boldsymbol{x}_0\right),\ \chi_{\text{line}} = (1-T)/T + \varepsilon = 1/T - 1 + \varepsilon,\ \varepsilon \geq 0. \tag{8.107}$$

The first term in χ_{line} is contributed to the channel attenuation, while the second term ε, commonly referred to as the excess noise, is due to other noisy factors such as the laser phase noise, imperfect orthogonality of quadratures, and so on. The total variance of Bob's random variable will be now composed of three terms:

$$V(Y_B) = \text{Var}(Y_B) = T\left(\underbrace{v_A + 1}_{v} + \chi_{\text{line}}\right) = T(v + \chi_{\text{line}}), \tag{8.108}$$

where $v = v_{\text{A}} + 1$.

Mutual information between Alice and Bob, based on Shannon's formula, can be calculated by

$$\begin{aligned} I(A;B) &\doteq I(X_A;Y_B) = \frac{1}{2}\log\left(1 + \frac{Tv_A}{T(1+\chi_{\text{line}})}\right) \\ &= \frac{1}{2}\log\left(1 + \frac{v_A}{1+\chi_{\text{line}}}\right). \end{aligned} \tag{8.109}$$

The Eve's information, in CV-QKD schemes, is determined by reconciliation strategy.

Assuming that all sources of excess noise are statistically independent and additive, the total variance (in SNU) can be represented as a summation of individual contributions:

$$\varepsilon = \varepsilon_{\text{modulator}} + \varepsilon_{PN} + \varepsilon_{RIN} + \cdots \tag{8.110}$$

where the first term denotes the modulation imperfections, the second term corresponds to the phase noise (PN), while the third term corresponds to the relative intensity noise (RIN) of LO reference signal.

Referring now to the Bob's input (or equivalently to the *channel output*), we can represent the Bob's equivalent detector noise variance by

$$\chi_{\text{det}} = \frac{d(1+v_{\text{el}}) - \eta}{\eta}, \tag{8.111}$$

where $d = 1$ for homodyne detection (either in-phase or quadrature can be measured at given signaling interval in homodyne detection scheme) and $d = 2$ for heterodyne detection (both quadratures can be measured simultaneously). The parameter η denotes the detection efficiency, while v_{el} denotes the variance of equivalent electrical noise. The variance of the total noise can be determined by summing up the variance of the channel noise and detection noise, which can be expressed by referring to the *channel input* as follows:

$$\chi_{\mathrm{total}} = \chi_{\mathrm{line}} + \frac{\chi_{\mathrm{det}}}{T} = \underbrace{\frac{1-T}{T} + \varepsilon}_{\chi_{\mathrm{line}}} + \frac{d(1+v_{el}) - \eta}{T\eta}. \tag{8.112}$$

When referring to the channel output, variance of the total noise at Bob's side can be determined by

$$\begin{aligned}\chi_{\mathrm{total,\,Bob}} &= T\eta\left(\chi_{\mathrm{line}} + \frac{\chi_{\mathrm{det}}}{T}\right) = T\eta\frac{1-T}{T} + T\eta\varepsilon + T\eta\frac{d(1+v_{el}) - \eta}{T\eta} \\ &= \eta(1-T) + T\eta\varepsilon + d(1+v_{el}) - \eta.\end{aligned} \tag{8.113}$$

8.2.1.1 The Covariance Matrices

For the bipartite system composed of two separable subsystems A and B, each described by corresponding density matrices ρ_{A} and ρ_{B}, the density operator of the bipartite system can be represented by $\rho = \rho_{\mathrm{A}} \otimes \rho_{\mathrm{B}}$. For Gaussian bipartite system, composed of two separable subsystems A and B, with A subsystem being a Gaussian state with M modes represented by quadrature operators $\hat{\boldsymbol{x}}_A$, while B subsystem being a Gaussian state with N modes represented by quadrature operators $\hat{\boldsymbol{x}}_B$, the Gaussian state of bipartite system will contain $M + N$ modes and can be described by the quadrature operators $\hat{\boldsymbol{x}} = \left[\hat{\boldsymbol{x}}_A \; \hat{\boldsymbol{x}}_B\right]^T$, with corresponding covariance matrix of size $2(M+N) \times 2(M+N)$ being represented as

$$\boldsymbol{\Sigma} = \boldsymbol{\Sigma}_{\mathbf{A}} \oplus \boldsymbol{\Sigma}_{\mathbf{B}} = \begin{bmatrix} \boldsymbol{\Sigma}_A & \mathbf{0} \\ \mathbf{0} & \boldsymbol{\Sigma}_B \end{bmatrix}, \tag{8.114}$$

where $\mathbf{0}$ denotes the all-zeros matrix.

Let us assume that the EPR state, defined by Eq. (8.75), is used in CV-QKD system, with covariance matrix defined by (8.76). The identity matrix can be used to describe the thermal state at the input of beam splitter, see Fig. 8.6. Based on previous equation, corresponding bipartite system will have the following correlation matrix:

$$\boldsymbol{\Sigma} = \begin{bmatrix} \boldsymbol{\Sigma}_{\text{EPR}} & \mathbf{0} \\ \mathbf{0} & \mathbf{1} \end{bmatrix} = \begin{bmatrix} v\mathbf{1} & \sqrt{v^2-1}\mathbf{Z} & \mathbf{0} \\ \sqrt{v^2-1}\mathbf{Z} & v\mathbf{1} & \mathbf{0} \\ \mathbf{0} & \mathbf{0} & \mathbf{1} \end{bmatrix}. \tag{8.115}$$

The beam splitter acts on Bob's qubit and thermal state, while leaves Alice qubit unaffected; therefore, its action can be described as

$$BS_{\text{combined}}(T) = \mathbf{1} \oplus \begin{bmatrix} \sqrt{T}\mathbf{1} & \sqrt{1-T}\mathbf{1} \\ -\sqrt{1-T}\mathbf{1} & \sqrt{T}\mathbf{1} \end{bmatrix} = \begin{bmatrix} \mathbf{1} & \mathbf{0} & \mathbf{0} \\ \mathbf{0} & \sqrt{T}\mathbf{1} & \sqrt{1-T}\mathbf{1} \\ \mathbf{0} & -\sqrt{1-T}\mathbf{1} & \sqrt{T}\mathbf{1} \end{bmatrix}. \tag{8.116}$$

To determine the covariance matrix after the beam splitter, we apply the symplectic operation by BS_{combined} on input covariance matrix to obtain

$$\begin{aligned} \boldsymbol{\Sigma}' &= BS_{\text{combined}}(T)\boldsymbol{\Sigma}[BS_{\text{combined}}(T)]^T \\ &= \begin{bmatrix} \mathbf{1} & \mathbf{0} & \mathbf{0} \\ \mathbf{0} & \sqrt{T}\mathbf{1} & \sqrt{1-T}\mathbf{1} \\ \mathbf{0} & -\sqrt{1-T}\mathbf{1} & \sqrt{T}\mathbf{1} \end{bmatrix} \begin{bmatrix} v\mathbf{1} & \sqrt{v^2-1}\mathbf{Z} & \mathbf{0} \\ \sqrt{v^2-1}\mathbf{Z} & v\mathbf{1} & \mathbf{0} \\ \mathbf{0} & \mathbf{0} & \mathbf{1} \end{bmatrix} \begin{bmatrix} \mathbf{1} & \mathbf{0} & \mathbf{0} \\ \mathbf{0} & \sqrt{T}\mathbf{1} & \sqrt{1-T}\mathbf{1} \\ \mathbf{0} & -\sqrt{1-T}\mathbf{1} & \sqrt{T}\mathbf{1} \end{bmatrix}^T. \end{aligned} \tag{8.117}$$

Now by keeping Alice and Bob submatrices, we obtain

$$\boldsymbol{\Sigma}'_{\text{AB}} = \begin{bmatrix} v\mathbf{1} & \sqrt{T(v^2-1)}\mathbf{Z} \\ \sqrt{T(v^2-1)}\mathbf{Z} & (vT+1-T)\mathbf{1} \end{bmatrix}. \tag{8.118}$$

In the presence of excess noise, we need to replace the $(\Sigma'_{\text{AB}})_{22}$-element by $vT + 1 + \varepsilon' - T$ to get the following *covariance matrix for homodyne detection*:

$$\boldsymbol{\Sigma}_{\text{AB}}^{(\text{homodyne})} = \begin{bmatrix} v\mathbf{1} & \sqrt{T(v^2-1)}\mathbf{Z} \\ \sqrt{T(v^2-1)}\mathbf{Z} & (vT+1+\varepsilon'-T)\mathbf{1} \end{bmatrix}. \tag{8.119}$$

In *heterodyne detection*, Bob employs the balanced beam splitter (BBS) to split the quantum signal, while the second input to the BBS is the vacuum state, and the operation of the BBS can be represented by Eq. (8.116) by setting $T = 1/2$. The corresponding covariance matrix at the input of BBS will be

$$\boldsymbol{\Sigma}_{\text{in}}^{(\text{BBS})} = \begin{bmatrix} \boldsymbol{\Sigma}_{\text{AB}}^{(\text{homodyne})} & \mathbf{0} \\ \mathbf{0} & \mathbf{1} \end{bmatrix} = \begin{bmatrix} v\mathbf{1} & \sqrt{T(v^2-1)}\mathbf{Z} & \mathbf{0} \\ \sqrt{T(v^2-1)}\mathbf{Z} & (vT+1+\varepsilon'-T)\mathbf{1} & \mathbf{0} \\ \mathbf{0} & \mathbf{0} & \mathbf{1} \end{bmatrix}. \tag{8.120}$$

The covariance matrix and the BBS output can be determined by applying the symplectic operation by $BS_{\text{combined}}(1/2)$ on input covariance matrix to obtain

$$\begin{aligned}
\boldsymbol{\Sigma}_{AB}^{(\text{het})} &= BS_{\text{combined}}(1/2)\boldsymbol{\Sigma}_{\text{in}}^{(\text{BBS})}[BS_{\text{combined}}(1/2)]^{T} \\
&= \begin{bmatrix} \mathbf{1} & \mathbf{0} & \mathbf{0} \\ \mathbf{0} & \frac{1}{\sqrt{2}}\mathbf{1} & \frac{1}{\sqrt{2}}\mathbf{1} \\ \mathbf{0} & -\frac{1}{\sqrt{2}}\mathbf{1} & \frac{1}{\sqrt{2}}\mathbf{1} \end{bmatrix} \begin{bmatrix} v\mathbf{1} & \sqrt{T(v^2-1)}\mathbf{Z} & \mathbf{0} \\ \sqrt{T(v^2-1)}\mathbf{Z} & (vT+1+\varepsilon'-T)\mathbf{1} & \mathbf{0} \\ \mathbf{0} & \mathbf{0} & \mathbf{1} \end{bmatrix} \begin{bmatrix} \mathbf{1} & \mathbf{0} & \mathbf{0} \\ \mathbf{0} & \frac{1}{\sqrt{2}}\mathbf{1} & \frac{1}{\sqrt{2}}\mathbf{1} \\ \mathbf{0} & -\frac{1}{\sqrt{2}}\mathbf{1} & \frac{1}{\sqrt{2}}\mathbf{1} \end{bmatrix}^{T} \\
&= \begin{bmatrix} v\mathbf{1} & \sqrt{\frac{T}{2}(v^2-1)}\mathbf{Z} & -\sqrt{\frac{T}{2}(v^2-1)}\mathbf{Z} \\ \sqrt{\frac{T}{2}(v^2-1)}\mathbf{Z} & \frac{vT+2+\varepsilon'-T}{2}\mathbf{1} & -\frac{vT+\varepsilon'-T}{2}\mathbf{1} \\ -\sqrt{\frac{T}{2}(v^2-1)}\mathbf{Z} & -\frac{vT+\varepsilon'-T}{2}\mathbf{1} & \frac{vT+2+\varepsilon'-T}{2}\mathbf{1} \end{bmatrix}.
\end{aligned} \tag{8.121}$$

8.2.1.2 On the Equivalence of Prepare-and-Measure (PM) and Entanglement-Assisted (EA) Protocols

As discussed in previous sections, instead of employing EPR state, Alice employs the prepare-and-measure (PM) scheme, in which she employs a WCS source, and with the help of an I/Q modulator generates the Gaussian state with variance v_A. Given that the minimum uncertainty of the coherent source is 1 SNU, Bob's variance will be $v_A + 1$, and we can represent Bob's in-phase operator by

$$\hat{I}_B \sim \hat{I}_A + \mathcal{N}(0,1), \tag{8.122}$$

where $\mathcal{N}(0,1)$ is a zero-mean Gaussian source of variance 1 SNR. Clearly, the variance of Bob's in-phase operator is $V\left(\hat{I}_B\right) = V\left(\hat{I}_A\right) + 1 = v_A + 1$, as expected. On the other hand, the covariance of Alice and Bob's in-phase operators will be

$$\begin{aligned}
\text{Cov}\left(\hat{I}_A, \hat{I}_B\right) &= \left\langle \left(\hat{I}_A - \underbrace{\left\langle \hat{I}_A \right\rangle}_{=0}\right)\left(\hat{I}_B - \underbrace{\left\langle \hat{I}_B \right\rangle}_{0}\right)\right\rangle = \left\langle \hat{I}_A \underbrace{\hat{I}_B}_{\hat{I}_A + N(0,1)} \right\rangle = \left\langle \hat{I}_A^2 \right\rangle + \underbrace{\left\langle \hat{I}_A N(0,1) \right\rangle}_{=0} \\
&= v_A = \text{Cov}\left(\hat{I}_B, \hat{I}_A\right).
\end{aligned} \tag{8.123}$$

Therefore, the *covariance matrix for the PM protocol* is given as follows:

$$\boldsymbol{\Sigma}_{AB}^{(\text{PM protocol})} = \begin{bmatrix} \overbrace{V\left(\hat{I}_A\right)}^{v_A}\mathbf{1} & \overbrace{\text{Cov}\left(\hat{I}_A, \hat{I}_B\right)}^{v_A}\mathbf{1} \\ \underbrace{\text{Cov}\left(\hat{I}_B, \hat{I}_A\right)\mathbf{1}}_{v_A} & \underbrace{V\left(\hat{I}_B\right)\mathbf{1}}_{v_B = v_A + 1} \end{bmatrix} = \begin{bmatrix} v_A\mathbf{1} & v_A\mathbf{1} \\ v_A\mathbf{1} & \underbrace{(v_A+1)}_{v}\mathbf{1} \end{bmatrix} = \begin{bmatrix} (v-1)\mathbf{1} & (v-1)\mathbf{1} \\ (v-1)\mathbf{1} & v\mathbf{1} \end{bmatrix}. \tag{8.124}$$

In the *entanglement-assisted (EA) protocol*, Alice performs heterodyne measurement on her qubit, by employing the BBS, in similar fashion as for derivation of Eq. (8.121), so that the corresponding correlation matrix becomes

$$\begin{aligned}\boldsymbol{\Sigma}_{AB}^{(EA)} &= BS_{\text{combined}}^{(\text{Alice})}(1/2)\left(\mathbf{1}\oplus\boldsymbol{\Sigma}_{\text{EPR}}\right)\left[BS_{\text{combined}}^{(\text{Alice})}(1/2)\right]^T\\ &= \begin{bmatrix}\frac{1}{\sqrt{2}}\mathbf{1} & \frac{1}{\sqrt{2}}\mathbf{1} & \mathbf{0}\\ -\frac{1}{\sqrt{2}}\mathbf{1} & \frac{1}{\sqrt{2}}\mathbf{1} & \mathbf{0}\\ \mathbf{0} & \mathbf{0} & \mathbf{1}\end{bmatrix}\begin{bmatrix}\mathbf{1} & \mathbf{0} & \mathbf{0}\\ \mathbf{0} & v\mathbf{1} & \sqrt{v^2-1}\mathbf{Z}\\ \mathbf{0} & \sqrt{v^2-1}\mathbf{Z} & v\mathbf{1}\end{bmatrix}\begin{bmatrix}\frac{1}{\sqrt{2}}\mathbf{1} & \frac{1}{\sqrt{2}}\mathbf{1} & \mathbf{0}\\ -\frac{1}{\sqrt{2}}\mathbf{1} & \frac{1}{\sqrt{2}}\mathbf{1} & \mathbf{0}\\ \mathbf{0} & \mathbf{0} & \mathbf{1}\end{bmatrix}^T\\ &= \begin{bmatrix}\frac{v+1}{2}\mathbf{1} & \frac{v-1}{2}\mathbf{1} & \sqrt{\frac{1}{2}(v^2-1)}\mathbf{Z}\\ \frac{v-1}{2}\mathbf{1} & \frac{v+1}{2}\mathbf{1} & \sqrt{\frac{1}{2}(v^2-1)}\mathbf{Z}\\ \sqrt{\frac{1}{2}(v^2-1)}\mathbf{Z} & \sqrt{\frac{1}{2}(v^2-1)}\mathbf{Z} & v\mathbf{1}\end{bmatrix}.\end{aligned} \tag{8.125}$$

Given that Alice' both quadrature modes are equivalent, we can observe the corner elements as the Alice–Bob reduced covariance matrix for EA protocol:

$$\boldsymbol{\Sigma}_{AB,\text{ reduced}}^{(EA)} = \begin{bmatrix}\frac{v+1}{2}\mathbf{1} & \sqrt{\frac{1}{2}(v^2-1)}\mathbf{Z}\\ \sqrt{\frac{1}{2}(v^2-1)}\mathbf{Z} & v\mathbf{1}\end{bmatrix}. \tag{8.126}$$

When Alice scales her measurement operators by $\pm\sqrt{2(v-1)/(v+1)}$, the scaled covariance matrix becomes

$$\begin{aligned}\boldsymbol{\Sigma}_{AB,\text{reduced,scaled}}^{\prime(EA)} &= \begin{bmatrix}\overbrace{2\frac{v-1}{v+1}\frac{v+1}{2}}^{v-1}\mathbf{1} & \overbrace{\sqrt{2\frac{v-1}{v+1}\frac{1}{2}(v^2-1)}}^{v-1}\mathbf{Z}\\ \underbrace{\sqrt{2\frac{v-1}{v+1}\frac{1}{2}(v^2-1)}}_{v-1}\mathbf{Z} & v\mathbf{1}\end{bmatrix} = \begin{bmatrix}(v-1)\mathbf{1} & (v-1)\mathbf{Z}\\ (v-1)\mathbf{Z} & v\mathbf{1}\end{bmatrix}\\ &= \boldsymbol{\Sigma}_{AB}^{(\text{PM protocol})}.\end{aligned} \tag{8.127}$$

Clearly, by proper scaling of the quadrature components in EA protocol, Alice is able to obtain the same correlation matrix as in PM protocol. Conversely, when in PM protocol, Alice scales the quadrature components by $\pm\sqrt{2(v+1)/(v-1)}$, and Alice is able to obtain the same correlation matrix as in EA protocol, that is,

$$\Sigma_{AB}^{\prime\,(\text{PM protocol})} = \begin{bmatrix} \overbrace{\frac{v+1}{2(v-1)}(v-1)\,\mathbf{1}}^{(v+1)/2} & \overbrace{\sqrt{\frac{v+1}{2(v-1)}(v-1)^2}\,\mathbf{Z}}^{\sqrt{(v^2-1)/2}} \\ \underbrace{\sqrt{\frac{v+1}{2(v-1)}(v-1)^2}\,\mathbf{Z}}_{\sqrt{(v^2-1)/2}} & v\mathbf{1} \end{bmatrix} = \begin{bmatrix} \frac{v+1}{2}\mathbf{1} & \sqrt{\frac{v^2-1}{2}}\mathbf{Z} \\ \sqrt{\frac{v^2-1}{2}}\mathbf{Z} & v\mathbf{1} \end{bmatrix}$$
$$= \Sigma_{\text{AB, reduced, scaled}}^{(\text{EA})}. \tag{8.128}$$

8.2.2 *Secret-Key Rate of CV-QKD with Gaussian Modulation Under Collective Attacks*

In this subsection, we describe how to calculate the secret fraction for coherent state homodyne- detection-based CV-QKD scheme when Eve employs the Gaussian collective attack [43–47], which is known to be optimum for Gaussian modulation. This attack is completely characterized by estimating the covariance matrix by Alice and Bob Σ_{AB}, which is based on Eq. (8.119) for homodyne detection and Eq. (8.121) for heterodyne detection. Let us rewrite the covariance matrix for homodyne detection as follows:

$$\Sigma_{AB}^{(\text{homodyne})} = \begin{bmatrix} v\mathbf{1} & \sqrt{T(v^2-1)}\mathbf{Z} \\ \sqrt{T(v^2-1)}\mathbf{Z} & T\left(v + 1/\text{T} - 1 + \underbrace{\varepsilon'/T}_{\varepsilon}\right)\mathbf{1} \end{bmatrix}$$
$$= \begin{bmatrix} v\mathbf{1} & \sqrt{T(v^2-1)}\mathbf{Z} \\ \sqrt{T(v^2-1)}\mathbf{Z} & T\left(v + \underbrace{1/\text{T} - 1 + \varepsilon}_{\chi_{\text{line}}}\right)\mathbf{1} \end{bmatrix} = \begin{bmatrix} v\mathbf{1} & \sqrt{T(v^2-1)}\mathbf{Z} \\ \sqrt{T(v^2-1)}\mathbf{Z} & T(v+\chi_{\text{line}})\mathbf{1} \end{bmatrix}. \tag{8.129}$$

where $\varepsilon = \varepsilon'/T$ is the excess noise and $1/T - 1$ is channel loss term, both observed from Alice side (at the channel input). The total variance of both terms is denoted by $\chi_{\text{line}} = 1/T - \varepsilon$. Now, by adding the detector noise term, the covariance matrix for homodyne detection becomes

$$\Sigma_{AB}^{(\text{hom., complete})} = \begin{bmatrix} v\mathbf{1} & \sqrt{T\eta(v^2-1)}\mathbf{Z} \\ \sqrt{T\eta(v^2-1)}\mathbf{Z} & T\eta(v+\chi_{\text{total}})\mathbf{1} \end{bmatrix}, v = v_A + 1, \tag{8.130}$$

where η is detection efficiency and χ_{total} is defined by Eq. (8.112), that is, $\chi_{\text{total}} = \chi_{\text{line}} + \chi_{\text{det}}/T = \chi_{\text{line}} + [d(1 + v_{el}) - \eta]/(T\eta)$, where v_{el} is the variance of the equivalent electrical noise.

As discussed in Chap. 6, in CV-QKD for direct reconciliation with nonzero SKR, the channel loss is lower limited by $T \geq 1/2$, and for higher channel losses we need to use the reverse reconciliation. In the reverse reconciliation, the *secret fraction* is given by

$$r = \beta I(A; B) - \chi(B; E), \tag{8.131}$$

where β is the reconciliation efficiency and $I(A;B)$ is the mutual information between Alice and Bob that is identical for both individual and collective attacks [41–45]:

$$I(A; B) = \frac{d}{2} \log_2\left(\frac{v + \chi_{\text{total}}}{1 + \chi_{\text{total}}}\right), d = \begin{cases} 1, \text{homodyne detection} \\ 2, \text{heterodyne detection} \end{cases}. \tag{8.132}$$

which is nothing else by the definition formula for channel capacity of Gaussian channel, as described in Chap. 2 (see also [52, 53]). What remains to determine is the Holevo information between Bob and Eve, denoted as $\chi(B;E)$, which is much more challenging to derive, and it is a subject of subsequent sections. After that, the SKR can be determined as the product of the secret fraction and signaling rate. Given that homodyne/heterodyne detection is not limited by the dead time, typical SKRs for CV-QKD schemes are significantly higher than that of DV-QKD schemes.

8.2.2.1 Holevo Information Between Eve and Bob

Assuming the reverse reconciliation, based on Chap. 5, the Holevo information between Eve and Bob can be calculated by

$$\chi(E; B) = S_E - S_{E|B}, \tag{8.133}$$

where S_E is the von Neumann entropy of Eve's density operator $\hat{\rho}_E$, defined by Eq. (8.56), that is, $S(\hat{\rho}_E) = -\text{Tr}(\hat{\rho}_E \log \hat{\rho}_E)$, while $S_{E|B}$ is the von Neumann entropy of Eve's state when Bob performs the projective measurement, be homodyne or heterodyne. In principle, the symplectic eigenvalues of corresponding correlation matrices can be determined first to calculate the von Neumann entropies, as described in Sect. 8.1.7. To simplify the analysis, typically we assume that Eve possesses the purification [9, 54] of Alice and Bob's joint state $\hat{\rho}_{AB}$, so that the resulting state is a pure state $|\psi\rangle$ and corresponding density operator will be

$$\hat{\rho}_{ABE} = |\psi\rangle\langle\psi|. \tag{8.134}$$

By tracing out Eve's subspace, we obtain Alice–Bob mixed state, that is, $\hat{\rho}_{AB} = \mathrm{Tr}_E \hat{\rho}_{ABE}$. In similar fashion, by tracing out Alice–Bob subspace, we get Eve's mixed state $\hat{\rho}_E = \mathrm{Tr}_{AB} \hat{\rho}_{ABE}$. By applying the *Schmidt decomposition theorem* [9, 54], we can represent the bipartite state $|\psi\rangle \in \mathcal{H}_{\mathrm{AB}} \otimes \mathcal{H}_{\mathrm{E}}$ in terms of *Schmidt bases* for subsystem AB, denoted as $\{|i_{\mathrm{AB}}\rangle\}$, and subsystem E, denoted as $\{|i_{\mathrm{E}}\rangle\}$, as follows:

$$|\psi\rangle = \sum_i \sqrt{p_i} |i_{AB}\rangle |i_E\rangle, \tag{8.135}$$

where $\sqrt{p_i}$ are Schmidt coefficients satisfying relationship $\sum_i p_i = 1$. By tracing out Eve's subsystem, we obtain the following mixed density operator for Alice–Bob subsystem:

$$\hat{\rho}_{AB} = \mathrm{Tr}_E(|\psi\rangle\langle\psi|) = \sum_i p_i |i_{AB}\rangle\langle i_{AB}|. \tag{8.136}$$

On the other hand, by tracing out Alice–Bob subsystem, we obtained the following mixed state for Eve:

$$\hat{\rho}_E = \mathrm{Tr}_{AB}(|\psi\rangle\langle\psi|) = \sum_i p_i |i_E\rangle\langle i_E|. \tag{8.137}$$

Clearly, both Alice–Bob and Eve's density operators have the same spectra (of eigenvalues), and therefore the same von Neumann entropy:

$$S_{AB} = S_E = -\sum_i p_i \log p_i. \tag{8.138}$$

The calculation of Holevo information is now greatly simplified since the following is valid:

$$\chi(E; B) = S_E - S_{E|B} = S_{AB} - S_{A|B}. \tag{8.139}$$

To determine the von Neumann entropy for Alice–Bob subsystem, which is independent on Bob's measurement, we use the following covariance matrix from Sect. 8.2.1.1, observed at the channel input:

$$\mathbf{\Sigma}_{\mathrm{AB}} = \begin{bmatrix} v\mathbf{1} & \sqrt{T(v^2-1)}\mathbf{Z} \\ \sqrt{T(v^2-1)}\mathbf{Z} & T(v+\chi_{\mathrm{line}})\mathbf{1} \end{bmatrix}. \tag{8.140}$$

To determine the symplectic eigenvalues, we employ the theory from Sect. 8.1.6. Clearly, the covariance matrix has the same form as Eq. (8.84), with $a = v$, $b = T(v + \chi_{\mathrm{line}})$, and $c = \sqrt{T(v^2-1)}$. We can directly apply Eq. (8.85) to determine the symplectic eigenvalues. Alternatively, given that symplectic eigenvalues satisfy the following:

$$\lambda_1^2 + \lambda_2^2 = a^2 + b^2 - 2c^2,$$
$$\lambda_1^2 \lambda_2^2 = \left(ab - c^2\right)^2, \tag{8.141}$$

By introducing

$$A = a^2 + b^2 - 2c^2 = v^2(1 - 2T) + 2T + T^2(v + \chi_{\text{line}}),$$
$$B = \left(ab - c^2\right)^2 = T^2(1 + v\chi_{\text{line}}), \tag{8.142}$$

the corresponding symplectic eigenvalues for Alice–Bob covariance matrix can be expressed as

$$\lambda_{1,2} = \sqrt{\frac{1}{2}\left(A \pm \sqrt{A^2 - 4B}\right)}, \tag{8.143}$$

which represents a form very often found in open literature [41–43, 55, 56]. The von Neumann entropy, based on thermal state decomposition theorem and Eq. (8.60), is given by

$$S_{AB} = S(\hat{\rho}_{AB}) = g\left(\frac{\lambda_1 - 1}{2}\right) + g\left(\frac{\lambda_2 - 1}{2}\right). \tag{8.144}$$

The calculation of $S_{A|B}$ is dependent on Bob's measurement strategy, homodyne or heterodyne, and we can apply the theory of partial measurements from Sect. 8.1.8.3. For homodyne detection, based on Eq. (8.140), we conclude that $\boldsymbol{A} = v\mathbf{1}$, $\boldsymbol{B} = T(v + \chi_{\text{line}})\mathbf{1}$, and $\boldsymbol{C} = \sqrt{T\left(v^2 - 1\right)}\boldsymbol{Z}$ so that the Bob's homodyne measurement transforms the covariance matrix above to

$$\boldsymbol{\Sigma}_{A|B}^{(\hat{\imath})} = \boldsymbol{A} - \frac{1}{V\left(\hat{I}\right)}\boldsymbol{C}\boldsymbol{\Pi}_I\boldsymbol{C}^T = \begin{bmatrix} v & 0 \\ 0 & v \end{bmatrix} - \frac{1}{T(v + \chi_{\text{line}})}T\left(v^2 - 1\right) - \underbrace{\begin{bmatrix} 1 & 0 \\ 0 & -1 \end{bmatrix}\begin{bmatrix} 1 & 0 \\ 0 & 0 \end{bmatrix}\begin{bmatrix} 1 & 0 \\ 0 & -1 \end{bmatrix}}_{\begin{bmatrix} 1 & 0 \\ 0 & 0 \end{bmatrix}}$$
$$= \begin{bmatrix} v - \frac{v^2 - 1}{v + \chi_{\text{line}}} & 0 \\ 0 & v \end{bmatrix}. \tag{8.145}$$

The symplectic eigenvalue is determined now by solving

$$\det\left(j\Omega\Sigma_{\mathrm{A|B}}-\lambda\mathbf{1}\right)=\det\left(j\overbrace{\begin{bmatrix}0 & 1\\ -1 & 0\end{bmatrix}\begin{bmatrix}v-\frac{v^2-1}{v+\chi_{\mathrm{line}}} & 0\\ 0 & v\end{bmatrix}}^{\begin{bmatrix}0 & v\\ \frac{v^2-1}{v+\chi_{\mathrm{line}}}-v & 0\end{bmatrix}}-\lambda\begin{bmatrix}1 & 0\\ 0 & 1\end{bmatrix}\right)$$

$$=\det\left(\begin{bmatrix}-\lambda & jv\\ j\left(\frac{v^2-1}{v+\chi_{\mathrm{line}}}-v\right) & -\lambda\end{bmatrix}\right)=\lambda^2+v\left(\frac{v^2-1}{v+\chi_{\mathrm{line}}}-v\right)=0, \quad (8.146)$$

and by solving λ, we obtain

$$\lambda_3=\sqrt{v\left(v-\frac{v^2-1}{v+\chi_{\mathrm{line}}}\right)}, \quad (8.147)$$

so that $S_{\mathrm{A|B}}$ becomes $g(\lambda_3)$. This expression is consistent with that provided in [55], except that a slightly different correlation matrix was used. Finally, for homodyne detection, the Holevo information is given by

$$\chi(E;B)=S_{AB}-S_{A|B}=g\left(\frac{\lambda_1-1}{2}\right)+g\left(\frac{\lambda_2-1}{2}\right)-g\left(\frac{\lambda_3-1}{2}\right). \quad (8.148)$$

In heterodyne detection, Bob measures both quadratures, and given that additional balanced beam splitter is used, with one of inputs being in vacuum state, we need to account for fluctuations due to vacuum state and modify the partial correlation matrix as follows [29]:

$$\boldsymbol{\Sigma}_{A|B}=\boldsymbol{A}-\boldsymbol{C}(\boldsymbol{B}+\mathbf{1})^{-1}\boldsymbol{C}^T$$

$$=\begin{bmatrix}v & 0\\ 0 & v\end{bmatrix}-T\left(v^2-1\right)-\begin{bmatrix}1 & 0\\ 0 & -1\end{bmatrix}\overbrace{\begin{bmatrix}T(v+\chi_{\mathrm{line}})+1 & 0\\ 0 & T(v+\chi_{\mathrm{line}})+1\end{bmatrix}^{-1}}^{\frac{1}{T(v+\chi_{\mathrm{line}})}\mathbf{1}}\begin{bmatrix}1 & 0\\ 0 & -1\end{bmatrix}$$

$$=\underbrace{\left[v-\frac{T\left(v^2-1\right)}{T(v+\chi_{\mathrm{line}})+1}\right]}_{K}\mathbf{1}=K\mathbf{1},\ K=v-\frac{T\left(v^2-1\right)}{T(v+\chi_{\mathrm{line}})+1}. \quad (8.149)$$

The symplectic eigenvalue can be determined from the following characteristic equation:

$$\det\left(j\boldsymbol{\Omega}\boldsymbol{\Sigma}_{\mathrm{A|B}}^{(\mathrm{heterodyne})}-\lambda\mathbf{1}\right)=\det\left(jK\overbrace{\begin{bmatrix}0&1\\-1&0\end{bmatrix}\begin{bmatrix}1&0\\0&1\end{bmatrix}}^{\begin{bmatrix}0&1\\-1&0\end{bmatrix}}-\lambda\begin{bmatrix}1&0\\0&1\end{bmatrix}\right)$$

$$=\det\left(\begin{bmatrix}-\lambda & jK\\ jK & -\lambda\end{bmatrix}\right)=\lambda^2-K^2=0, \tag{8.150}$$

to get

$$\lambda_3^{(\mathrm{heterodyne})}=K=v-\frac{T\left(v^2-1\right)}{T(v+\chi_{\mathrm{line}})+1}. \tag{8.151}$$

Finally, for heterodyne detection, the Holevo information is given by

$$\chi(E;B)^{(\mathrm{heterodyne})}=g\left(\frac{\lambda_1-1}{2}\right)+g\left(\frac{\lambda_2-1}{2}\right)-g\left(\frac{\lambda_3^{(\mathrm{heterodyne})}-1}{2}\right). \tag{8.152}$$

8.2.2.2 Entangling Cloner Attack and Corresponding Holevo Information

The optimum Gaussian eavesdropping strategy is based on *entangling cloning machine*, which is illustrated in Fig. 8.7, based on Refs. [1, 4, 46]. In Eve's apparatus, the EPR source is used with one qubit being kept by Eve in the quantum

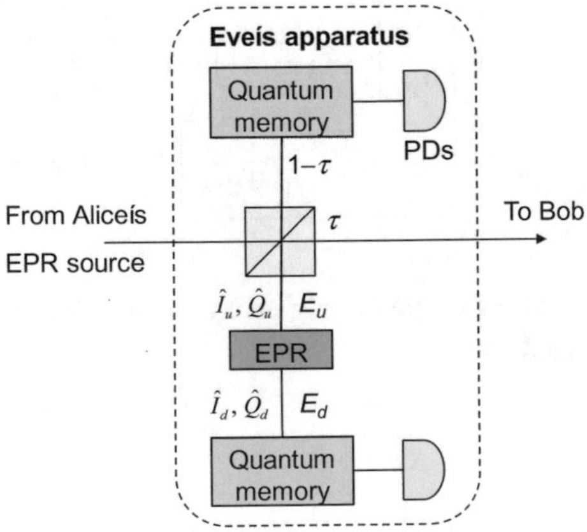

Fig. 8.7 The optimum Gaussian eavesdropping strategy based on entangling cloning machine. PDs: photodetectors

memory (QM), and the second qubit (mode) of variance u being injected into Alice-to-Bob channel by a beam splitter of transmissivity τ. The beam splitter also extracts the portion of the coherent state sent by Alice and Eve stores into another QM, until Bob announces which quadrature he has measured at every signaling interval. Let the down quadratures (kept by Eve in QM) be denoted by $\hat{I}_d$, $\hat{Q}_d$ with corresponding output denoted as E_d, while the quadratures injected into Alice-to-Bob channel by $\hat{I}_u$, $\hat{Q}_u$ with corresponding output being E_u. Clearly, Eve can simultaneously correlate the quadratures sent to Bob with quadratures kept in her QM, since $\left[\hat{I}_u - \hat{I}_d, \hat{Q}_u + \hat{Q}_d\right] = 0$. Assumption is that Alice generates the two-mode squeezed vacuum state of variance v, with corresponding covariance matrix given by Eq. (8.76), that is,

$$\mathbf{\Sigma}_{\mathrm{AB}} = \begin{bmatrix} v\mathbf{1} & \sqrt{v^2-1}\mathbf{Z} \\ \sqrt{v^2-1}\mathbf{Z} & v\mathbf{1} \end{bmatrix}. \tag{8.153}$$

Eve's EPR state, shown in Fig. 8.7, is described by a similar covariance matrix:

$$\mathbf{\Sigma}_{E_u E_d} = \begin{bmatrix} u\,\mathbf{1} & \sqrt{u^2-1}\,\mathbf{Z} \\ \sqrt{u^2-1}\,\mathbf{Z} & u\,\mathbf{1} \end{bmatrix}. \tag{8.154}$$

The combined covariance matrix of a bipartite system can be represented as

$$\begin{aligned} \mathbf{\Sigma}_{ABE_uE_d} &= \mathbf{\Sigma}_{E_uE_d} \oplus \mathbf{\Sigma}_{E_uE_d} = \begin{bmatrix} v\,\mathbf{1} & \sqrt{v^2-1}\,\mathbf{Z} \\ \sqrt{v^2-1}\,\mathbf{Z} & v\,\mathbf{1} \end{bmatrix} \oplus \begin{bmatrix} u\,\mathbf{1} & \sqrt{u^2-1}\,\mathbf{Z} \\ \sqrt{u^2-1}\,\mathbf{Z} & u\,\mathbf{1} \end{bmatrix} \\ &= \begin{bmatrix} v\,\mathbf{1} & \sqrt{v^2-1}\,\mathbf{Z} & \mathbf{0} & \mathbf{0} \\ \sqrt{v^2-1}\,\mathbf{Z} & v\,\mathbf{1} & \mathbf{0} & \mathbf{0} \\ \mathbf{0} & \mathbf{0} & u\,\mathbf{1} & \sqrt{u^2-1}\,\mathbf{Z} \\ \mathbf{0} & \mathbf{0} & \sqrt{u^2-1}\,\mathbf{Z} & u\,\mathbf{1} \end{bmatrix}. \end{aligned} \tag{8.155}$$

Eve's beam splitter (BS) of transmissivity τ combines Bob's mode with one of Eve's modes E_u, and its operation can be described by

$$BS_{BE_u} = \begin{bmatrix} \mathbf{1} & \mathbf{0} & \mathbf{0} & \mathbf{0} \\ \mathbf{0} & \sqrt{\tau}\mathbf{1} & \sqrt{1-\tau}\mathbf{1} & \mathbf{0} \\ \mathbf{0} & -\sqrt{1-\tau}\mathbf{1} & \sqrt{\tau}\mathbf{1} & \mathbf{0} \\ \mathbf{0} & \mathbf{0} & \mathbf{0} & \mathbf{1} \end{bmatrix}. \tag{8.156}$$

The entangling cloning operation is a symplectic operation, which is described by

$$\mathbf{\Sigma}'_{ABE_uE_d} = BS_{BE_u}\mathbf{\Sigma}_{ABE_uE_d}\left[BS_{BE_u}\right]^T = \begin{bmatrix} \mathbf{1} & \mathbf{0} & \mathbf{0} & \mathbf{0} \\ \mathbf{0} & \sqrt{\tau}\mathbf{1} & \sqrt{1-\tau}\mathbf{1} & \mathbf{0} \\ \mathbf{0} & -\sqrt{1-\tau}\mathbf{1} & \sqrt{\tau}\mathbf{1} & \mathbf{0} \\ \mathbf{0} & \mathbf{0} & \mathbf{0} & \mathbf{1} \end{bmatrix}$$

$$\cdot\begin{bmatrix} v\mathbf{1} & \sqrt{v^2-1}\,\mathbf{Z} & \mathbf{0} & \mathbf{0} \\ \sqrt{v^2-1}\,\mathbf{Z} & v\mathbf{1} & \mathbf{0} & \mathbf{0} \\ \mathbf{0} & \mathbf{0} & u\mathbf{1} & \sqrt{u^2-1}\,\mathbf{Z} \\ \mathbf{0} & \mathbf{0} & \sqrt{u^2-1}\,\mathbf{Z} & u\mathbf{1} \end{bmatrix} \begin{bmatrix} \mathbf{1} & \mathbf{0} & \mathbf{0} & \mathbf{0} \\ \mathbf{0} & \sqrt{\tau}\mathbf{1} & \sqrt{1-\tau}\mathbf{1} & \mathbf{0} \\ \mathbf{0} & -\sqrt{1-\tau}\mathbf{1} & \sqrt{\tau}\mathbf{1} & \mathbf{0} \\ \mathbf{0} & \mathbf{0} & \mathbf{0} & \mathbf{1} \end{bmatrix}^T$$
$$= \begin{bmatrix} v\mathbf{1} & \sqrt{\tau(v^2-1)}\mathbf{Z} & -\sqrt{(1-\tau)(v^2-1)}\mathbf{Z} & \mathbf{0} \\ \sqrt{\tau(v^2-1)}\mathbf{Z} & [u(1-\tau)+v\tau]\mathbf{1} & (u-v)\sqrt{\tau(1-\tau)}\mathbf{1} & \sqrt{(1-\tau)(u^2-1)}\mathbf{Z} \\ -\sqrt{(1-\tau)(v^2-1)}\mathbf{Z} & (u-v)\sqrt{\tau(1-\tau)}\mathbf{1} & [v(1-\tau)+u\tau]\mathbf{1} & \sqrt{\tau(u^2-1)}\,\mathbf{Z} \\ \mathbf{0} & \sqrt{(1-\tau)(u^2-1)}\mathbf{Z} & \sqrt{\tau(u^2-1)}\,\mathbf{Z} & u\mathbf{1} \end{bmatrix}. \tag{8.157}$$

The covariance matrix corresponding to Alice–Bob state is represented by top-left 4×4 submatrix, that is,

$$\boldsymbol{\Sigma}'_{\text{AB}} = \begin{bmatrix} v\mathbf{1} & \sqrt{\tau(v^2-1)}\,\mathbf{Z} \\ \sqrt{\tau(v^2-1)}\,\mathbf{Z} & [\tau v + (1-\tau)u]\mathbf{1} \end{bmatrix}. \tag{8.158}$$

When Eve sets the variance of her EPR state to be $u = \varepsilon'/(1-\tau) + 1$, the covariance matrix above becomes

$$\boldsymbol{\Sigma}'_{\text{AB}} = \begin{bmatrix} v\mathbf{1} & \sqrt{\tau(v^2-1)}\,\mathbf{Z} \\ \sqrt{\tau(v^2-1)}\,\mathbf{Z} & \tau[v + 1/\tau - 1 + \varepsilon\prime/\tau]\mathbf{1} \end{bmatrix}. \tag{8.159}$$

which is the same as the covariance matrix given by Eq. (8.140) for $T = \tau$, indicating that Alice and Bob will only see between them the lossy channel of transmissivity τ, and will not detect the presence of Eve at all. Therefore, the entangling cloner attack is the optimum one from Eve's point of view.

The covariance matrix corresponding to Eve's state is represented by bottom-right 4×4 submatrix in Eq. (8.157), that is,

$$\Sigma'_{E_u E_d} = \begin{bmatrix} [\tau u + (1-\tau)v]\mathbf{1} & \sqrt{\tau(u^2-1)}\mathbf{Z} \\ \sqrt{\tau(u^2-1)}\mathbf{Z} & u\mathbf{1} \end{bmatrix}. \tag{8.160}$$

Given that this covariance matrix has the same form as Eq. (8.84), with $a = \tau u + (1-\tau)v$, $b = u$, and $c = \sqrt{\tau(u^2-1)}$, the symplectic eigenvalues can be determined by employing Eq. (8.85). Alternatively, we can use Eq. (8.143), that is,

$$\lambda'_{1,2} = \sqrt{\frac{1}{2}\left(A' \pm \sqrt{A'^2 - 4B'}\right)}, \tag{8.161}$$

wherein the parameters A' and B' are defined as

$$A' = a^2 + b^2 - 2c^2 = \left(u^2 + v^2\right)(1-\tau)^2 + [u\tau + v(1-\tau)]^2 - u^2\tau^2 + 2\tau,$$
$$B' = \left(ab - c^2\right)^2 = [(1-\tau)uv + v\tau]^2. \tag{8.162}$$

The von Neumann entropy for S_E will be then

$$S_E = S(\hat{\rho}_E) = g\left(\frac{\lambda_1' - 1}{2}\right) + g\left(\frac{\lambda_2' - 1}{2}\right). \tag{8.163}$$

To determine the conditional von Neumann entropy of $S_{\mathrm{E|B}}$, we need to calculate the symplectic eigenvalues of Eve's covariance matrix upon Bob's measurement. The starting point will be Eve-to-Bob's covariance matrix, which can be obtained from Eq. (8.157), by removing Alice's quadratures to get

$$\boldsymbol{\Sigma}'_{BE_uE_d} = \begin{bmatrix} [u(1-\tau) + v\tau]\mathbf{1} & (u-v)\sqrt{\tau(1-\tau)}\,\mathbf{1} & \sqrt{(1-\tau)\left(u^2-1\right)}\mathbf{Z} \\ (u-v)\sqrt{\tau(1-\tau)}\,\mathbf{1} & [v(1-\tau) + u\tau]\mathbf{1} & \sqrt{\tau\left(u^2-1\right)}\,\mathbf{Z} \\ \sqrt{(1-\tau)\left(u^2-1\right)}\mathbf{Z} & \sqrt{\tau\left(u^2-1\right)}\,\mathbf{Z} & u\mathbf{1} \end{bmatrix}. \tag{8.164}$$

To do so, we must follow a very similar procedure to Sect. (8.2.2.1), which will be omitted here, followed by calculation of $S_{\mathrm{E|B}}$. Finally, the Holevo information can be determined by $\chi(\mathrm{E;B}) = S_{\mathrm{E}} - S_{\mathrm{E|B}}$.

8.2.2.3 Entanglement-Assisted CV-QKD Protocol and Corresponding Holevo Information

The entanglement-assisted Gaussian coherent state-based CV-QKD scheme is illustrated in Fig. 8.8, based on [55–58]. Alice employs the EPR source of variance v and performs heterodyne measurement on the first qubit A, while sending the second qubit B_0 toward Bob, over the quantum channel. After heterodyne detection, Alice gets the in-phase component I_A and quadrature component Q_A. Quantum channel is modeled by transmissivity T and noise variance χ_{line} composed of loss term $1/T - 1$ and excess noise term ε.

Eve might employ entanglement cloner described in previous section, and Alice and Bob will not be able to notice that. Bob's detector inefficiency is modeled by the beam splitter with transmission η. To model electronic noise source of variance v_{el}, we use an EPR source of variance $w = \eta\chi_{\mathrm{det}}$, with χ_{det} defined by Eq. (8.111), with modification as follows:

$$w = \begin{cases} \eta\chi_{\mathrm{hom}}/(1-\eta) = 1 + v_{el}/(1-\eta), & \text{for homodyne detection} \\ (\eta\chi_{\mathrm{het}} - 1)/(1-\eta) = 1 + 2v_{el}/(1-\eta), & \text{for heterodyne detection} \end{cases} \tag{8.165}$$

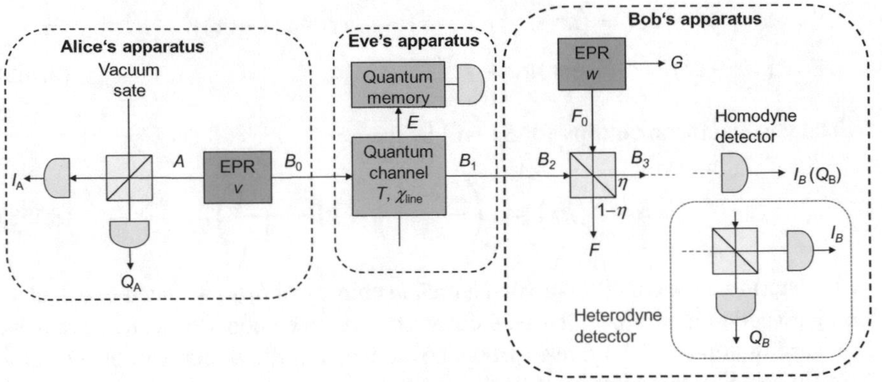

Fig. 8.8 The illustration of entanglement-assisted Gaussian coherent state-based CV-QKD scheme

where we removed 1 SNU for heterodyne detection, since it was already accounted for in Eq. (8.111) (for heterodyne detection).

Similar to Sect. 8.2.2.1, we assume that Eve possesses the purification of Alice and Bob's joint state so that $S_E = S_{AB}$, which has been already determined by Eq. (8.144). What remains to determine is the $S_{A|B}$, which based on Fig. 8.8 can be determined as $S(\hat{\rho}_{AFG|B})$.

For *homodyne detection*, the corresponding partial covariance matrix, based on Eq. (8.97), can be determined as

$$\boldsymbol{\Sigma}_{AFG|B} = \boldsymbol{\Sigma}_{AFG} - \boldsymbol{C}_{AFGB_3}(\boldsymbol{\Pi}_{\rm I}\boldsymbol{\Sigma}_{B_3}\boldsymbol{\Pi}_{\rm I})^{-1}\boldsymbol{C}^T_{AFGB_3}, \ \boldsymbol{\Pi}_{\rm I} = \mathrm{diag}(1, 0), \quad (8.166)$$

where corresponding covariance submatrices can be determined from the following correlation matrix:

$$\boldsymbol{\Sigma}_{AFGB_3} = \begin{bmatrix} \boldsymbol{\Sigma}_{AFG} & \boldsymbol{C}_{AFGB_3} \\ \boldsymbol{C}^T_{AFGB_3} & \boldsymbol{\Sigma}_{B_3} \end{bmatrix}. \quad (8.167)$$

From Fig. 8.8, we can determine the covariance matrix $\boldsymbol{\Sigma}_{AB_3FG}$ by the following symplectic transformation:

$$\boldsymbol{\Sigma}_{AB_3FG} = BS[\boldsymbol{\Sigma}_{AB_1} \oplus \boldsymbol{\Sigma}_{F_0G}]BS^T, \ BS = \mathbf{1}_A \oplus BS_{B_2F_0} \oplus \mathbf{1}_B, \quad (8.168)$$

where $\boldsymbol{\Sigma}_{AB_1}$ is given by Eq. (8.140), that is, $\boldsymbol{\Sigma}_{\rm AB_1} = \begin{bmatrix} v\mathbf{1} & \sqrt{T(v^2-1)}\boldsymbol{Z} \\ \sqrt{T(v^2-1)}\boldsymbol{Z} & T(v+\chi_{\rm line})\mathbf{1} \end{bmatrix}$, while $\boldsymbol{\Sigma}_{F_0G}$ is given by

$$\boldsymbol{\Sigma}_{F_0G} = \begin{bmatrix} w\,\mathbf{1} & \sqrt{w^2-1}\,\boldsymbol{Z} \\ \sqrt{w^2-1}\,\boldsymbol{Z} & w\,\mathbf{1} \end{bmatrix}. \quad (8.169)$$

The action of beam splitter $BS_{B_2F_0}$, describing the imperfections in detection, is simply

$$BS_{B_2F_0} = \begin{bmatrix} \sqrt{\eta}\mathbf{1} & \sqrt{1-\eta}\,\mathbf{1} \\ -\sqrt{1-\eta}\,\mathbf{1} & \sqrt{\eta}\,\mathbf{1} \end{bmatrix}. \tag{8.170}$$

After substituting (8.169) and (8.170) into (8.168), we determine the covariance matrix $\mathbf{\Sigma}_{AB_3FG}$:

$$\begin{aligned}\mathbf{\Sigma}_{AB_3FG} &= BS(\mathbf{\Sigma}_{AB_1} \oplus \mathbf{\Sigma}_{F_0G})[BS]^T = (\mathbf{1}_A \oplus BS_{B_2F_0} \oplus \mathbf{1}_B)(\mathbf{\Sigma}_{AB_1} \oplus \mathbf{\Sigma}_{F_0G})(\mathbf{1}_A \oplus BS_{B_2F_0} \oplus \mathbf{1}_B)^T \\
&= \begin{bmatrix} \mathbf{1} & \mathbf{0} & \mathbf{0} & \mathbf{0} \\ \mathbf{0} & \sqrt{\eta}\mathbf{1} & \sqrt{1-\eta}\mathbf{1} & \mathbf{0} \\ \mathbf{0} & -\sqrt{1-\eta}\mathbf{1} & \sqrt{\eta}\mathbf{1} & \mathbf{0} \\ \mathbf{0} & \mathbf{0} & \mathbf{0} & \mathbf{1} \end{bmatrix} \\
&\cdot \begin{bmatrix} v\mathbf{1} & \sqrt{T(v^2-1)}\,\mathbf{Z} & \mathbf{0} & \mathbf{0} \\ \sqrt{T(v^2-1)}\,\mathbf{Z} & T(v+\chi_l)\mathbf{1} & \mathbf{0} & \mathbf{0} \\ \mathbf{0} & \mathbf{0} & w\mathbf{1} & \sqrt{w^2-1}\,\mathbf{Z} \\ \mathbf{0} & \mathbf{0} & \sqrt{w^2-1}\,\mathbf{Z} & w\mathbf{1} \end{bmatrix} \begin{bmatrix} \mathbf{1} & \mathbf{0} & \mathbf{0} & \mathbf{0} \\ \mathbf{0} & \sqrt{\eta}\mathbf{1} & \sqrt{1-\eta}\mathbf{1} & \mathbf{0} \\ \mathbf{0} & -\sqrt{1-\eta}\mathbf{1} & \sqrt{\eta}\mathbf{1} & \mathbf{0} \\ \mathbf{0} & \mathbf{0} & \mathbf{0} & \mathbf{1} \end{bmatrix}^T \\
&= \begin{bmatrix} v\mathbf{1} & \sqrt{T\eta(v^2-1)}\mathbf{Z} & -\sqrt{T(1-\eta)(v^2-1)}\mathbf{Z} & \mathbf{0} \\ \sqrt{T\eta(v^2-1)}\mathbf{Z} & [w(1-\eta)+T\eta(v+\chi_l)]\mathbf{1} & (w-T(v+\chi_l))\sqrt{\eta(1-\eta)}\,\mathbf{1} & \sqrt{(1-\eta)(w^2-1)}\mathbf{Z} \\ -\sqrt{T(1-\eta)(v^2-1)}\mathbf{Z} & (w-T(v+\chi_l))\sqrt{\eta(1-\eta)}\,\mathbf{1} & [T(v+\chi_l)(1-\eta)+w\eta]\mathbf{1} & \sqrt{\eta(w^2-1)}\,\mathbf{Z} \\ \mathbf{0} & \sqrt{(1-\eta)(w^2-1)}\mathbf{Z} & \sqrt{\eta(w^2-1)}\,\mathbf{Z} & w\mathbf{1} \end{bmatrix}.\end{aligned} \tag{8.171}$$

where we used χ_l instead of χ_{line}, to save the space. After the rearrangement of the covariance matrix $\mathbf{\Sigma}_{AB_3FG}$, we obtain the desired covariance matrix $\mathbf{\Sigma}_{AFGB_3}$ as follows:

$$\begin{aligned}\mathbf{\Sigma}_{AFGB_3} &= \begin{bmatrix} \mathbf{\Sigma}_{AFG} & \mathbf{C}_{AFGB_3} \\ \mathbf{C}^T_{AFGB_3} & \mathbf{\Sigma}_{B_3} \end{bmatrix} \\
&= \begin{bmatrix} v\mathbf{1} & -\sqrt{T(1-\eta)(v^2-1)}\mathbf{Z} & \mathbf{0} & \sqrt{T\eta(v^2-1)}\mathbf{Z} \\ -\sqrt{T(1-\eta)(v^2-1)}\mathbf{Z} & [T(v+\chi_l)(1-\eta)+w\eta]\mathbf{1} & \sqrt{\eta(w^2-1)}\,\mathbf{Z} & (w-T(v+\chi_l))\sqrt{\eta(1-\eta)}\,\mathbf{1} \\ \mathbf{0} & \sqrt{\eta(w^2-1)}\,\mathbf{Z} & w\mathbf{1} & \sqrt{(1-\eta)(w^2-1)}\mathbf{Z} \\ \sqrt{T\eta(v^2-1)}\mathbf{Z} & (w-T(v+\chi_l))\sqrt{\eta(1-\eta)}\,\mathbf{1} & \sqrt{(1-\eta)(w^2-1)}\mathbf{Z} & [w(1-\eta)+T\eta(v+\chi_l)]\mathbf{1} \end{bmatrix}.\end{aligned} \tag{8.172}$$

Clearly, the covariance submatrices can be determined as

$$\mathbf{\Sigma}_{AFG} = \begin{bmatrix} v\mathbf{1} & -\sqrt{T(1-\eta)(v^2-1)}\mathbf{Z} \\ -\sqrt{T(1-\eta)(v^2-1)}\mathbf{Z} & [T(v+\chi_l)(1-\eta)+w\eta]\mathbf{1} \end{bmatrix},$$

$$\mathbf{\Sigma}_{B_3} = \begin{bmatrix} w\mathbf{1} & \sqrt{(1-\eta)(w^2-1)}\mathbf{Z} \\ \sqrt{(1-\eta)(w^2-1)}\mathbf{Z} & [w(1-\eta)+T\eta(v+\chi_l)]\mathbf{1} \end{bmatrix},$$

$$\boldsymbol{C}_{AFGB_3} = \begin{bmatrix} \mathbf{0} & \sqrt{T\eta\left(v^2-1\right)}\,\mathbf{Z} \\ \sqrt{\eta\left(w^2-1\right)}\,\mathbf{Z} & (w - T(v+\chi_l))\sqrt{\eta(1-\eta)}\,\mathbf{1} \end{bmatrix}. \tag{8.173}$$

The corresponding partial covariance matrix for homodyne detection, based on (8.166), becomes

$$\boldsymbol{\Sigma}_{AFG|B} = \boldsymbol{\Sigma}_{AFG} - \frac{1}{V\left(\hat{I}_{B_3}\right)}\boldsymbol{C}_{AFGB_3}\boldsymbol{C}^T_{AFGB_3} = \begin{bmatrix} v\mathbf{1} & -\sqrt{T(1-\eta)\left(v^2-1\right)}\mathbf{Z} \\ -\sqrt{T(1-\eta)\left(v^2-1\right)}\mathbf{Z} & [T(v+\chi_l)(1-\eta)+w\eta]\mathbf{1} \end{bmatrix}$$
$$-\frac{1}{w}\begin{bmatrix} \mathbf{0} & \sqrt{T\eta\left(v^2-1\right)}\,\mathbf{Z} \\ \sqrt{\eta\left(w^2-1\right)}\,\mathbf{Z} & (w-T(v+\chi_l))\sqrt{\eta(1-\eta)}\,\mathbf{1} \end{bmatrix}\begin{bmatrix} \mathbf{0} & \sqrt{T\eta\left(v^2-1\right)}\,\mathbf{Z} \\ \sqrt{\eta\left(w^2-1\right)}\,\mathbf{Z} & (w-T(v+\chi_l))\sqrt{\eta(1-\eta)}\,\mathbf{1} \end{bmatrix}^T. \tag{8.174}$$

Now, we can calculate symplectic eigenvalues of $\boldsymbol{\Sigma}_{AFG|B}$ as follows:

$$\lambda_{3,4} = \sqrt{\frac{1}{2}\left(C \pm \sqrt{C^2-4D}\right)}, \tag{8.175}$$

where C and D parameters are defined as [55, 56, 58]

$$C = \frac{A\chi_{\text{homodyne}} + v\sqrt{B} + T(v+\chi_{\text{line}})}{T(v+\chi_{\text{total}})},\quad D = \sqrt{B}\frac{v+\chi_{\text{homodyne}}\sqrt{B}}{T(v+\chi_{\text{total}})},$$
$$B = T^2(1+v\chi_{\text{line}})^2,\quad \chi_{\text{homodyne}} = (1-\eta+v_{el})/\eta. \tag{8.176}$$

The Holevo information between Bob and Eve can be now determined by

$$\chi(B;E) = g\left(\frac{\lambda_1-1}{2}\right) + g\left(\frac{\lambda_2-1}{2}\right) - g\left(\frac{\lambda_3-1}{2}\right) - g\left(\frac{\lambda_4-1}{2}\right). \tag{8.177}$$

where λ_1 and λ_2 were determined in Sect. 8.2.2.1, see Eq. (8.143).

For *heterodyne detection*, the corresponding partial covariance matrix, based on Eq. (8.99), can be determined as

$$\boldsymbol{\Sigma}_{AFG|B} = \boldsymbol{\Sigma}_{AFG} - \boldsymbol{C}_{AFGB_3}\left(\boldsymbol{\Sigma}_{B_3}+\mathbf{1}\right)^{-1}\boldsymbol{C}^T_{AFGB_3}. \tag{8.178}$$

where corresponding covariance submatrices are provided in Eq. (8.172). We then calculate symplectic eigenvalues by Eq. (8.175), wherein the parameters C and D for heterodyne detection are defined as [55, 56, 58]

$$C = \frac{A\chi^2_{\text{het}} + B + 1 + 2\chi_{\text{het}}\left[v\sqrt{B} + T(v+\chi_{\text{line}})\right] + 2T\left(v^2-1\right)}{T^2(v+\chi_{\text{total}})^2},\quad D = \frac{\left(v+\chi_{\text{het}}\sqrt{B}\right)^2}{T^2(v+\chi_{\text{total}})^2},$$
$$B = T^2(1+v\chi_{\text{line}})^2,\quad \chi_{\text{het}} = (1+(1-\eta)+2v_{el})/\eta. \tag{8.179}$$

Finally, the Holevo information between Eve and Bob is calculated by employing Eq. (8.177).

8.2.3 *Illustrative Reverse Reconciliation SKR Results for CV-QKD with Gaussian Modulation (GM)*

In this section, we provide illustrative secrecy fractions (normalized SKRs) for different scenarios, by employing the expression for Holevo information between Eve and Bob based on Sect. 8.2.2.3. We consider reverse reconciliation case only. In Fig. 8.9, we provide normalized SKR results for pure loss channel assuming that reconciliation efficiency is set to $\beta = 0.98$. We also study the influence of quantization effects. Given that in heterodyne detection both quadratures are measured simultaneously, the mutual information can be double compared to homodyne detection. However, we need to use the balanced beam splitter and then measure both quadratures, which introduces about 3 dB loss. For low channel loss, the heterodyne detection outperforms the homodyne one in terms of secret fractions. For medium and high channel losses, these two schemes perform comparably. When fixed point representation is used for five dynamic bits and three precision bits, there is no difference compared to the case with double precision.

In Fig. 8.10, we provide normalized SKRs versus channel loss when electrical noise variance in SNU is used as a parameter. The excess noise variance is set to $\varepsilon = 10^{-3}$, detector efficiency is $\eta = 0.85$, and reconciliation efficiency is set to $\beta = 0.85$. Interestingly enough, when electrical noise variance $v_{el} \leq 0.1$, there is no much degradation in SKRs, which indicates that commercially available p-i-n photodetectors can be used in CV-QKD.

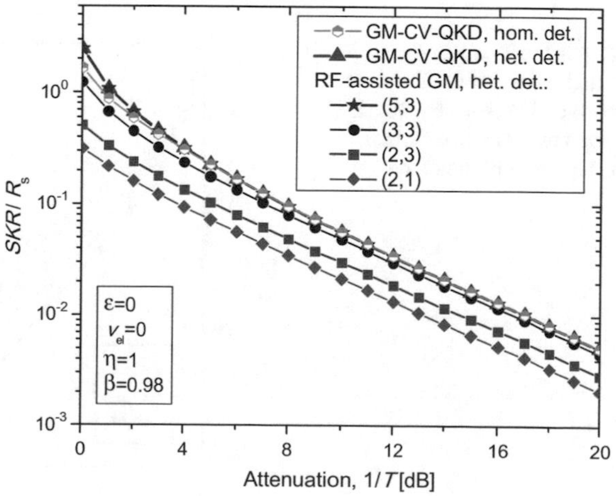

Fig. 8.9 The secrecy fraction for CV-QKD scheme with Gaussian modulation (GM) assuming pure channel loss

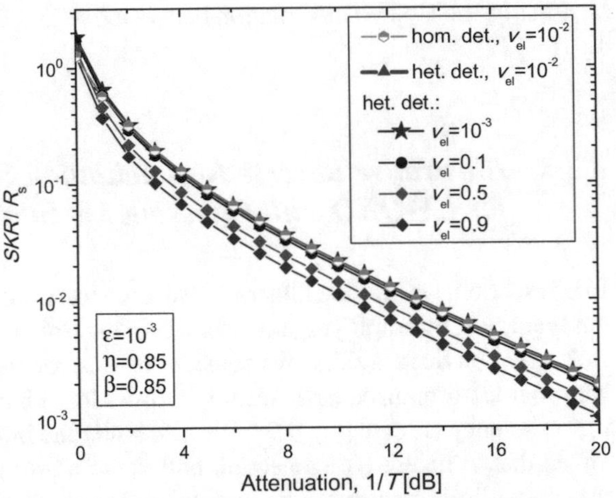

Fig. 8.10 The secrecy fraction for CV-QKD scheme with GM versus channel loss when electrical noise variance is used as a parameter

In Fig. 8.11, we show normalized SKRs versus channel loss when excess noise variance (expressed in SNU) is used as a parameter. The electrical noise variance is set to $v_{\rm el}=10^{-2}$, detector efficiency is $\eta=0.85$, and reconciliation efficiency is set to $\beta=0.85$. Evidently, the excess noise variance has high impact on SKR performance.

In Fig. 8.12, we provide normalized SKRs versus channel loss when detector efficiency η is used as a parameter. The excess noise variance is set to $\varepsilon=10^{-3}$, the electrical noise variance is set to $v_{\rm el}=10^{-2}$, and reconciliation efficiency is set to $\beta=0.85$. Clearly, the detection efficiency does not have high impact on SKRs.

In Fig. 8.13, we show normalized SKRs versus channel loss when reconciliation efficiency β is used as a parameter. The excess noise variance is set to $\varepsilon=10^{-3}$, the electrical noise variance is set to $v_{\rm el}=10^{-2}$, and detector efficiency is set to $\eta=0.85$. Evidently, the reconciliation efficiency does have a relevant impact on SKRs.

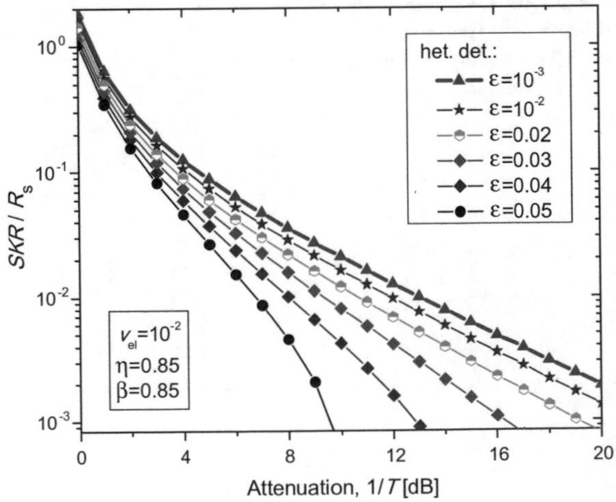

Fig. 8.11 The secrecy fraction for CV-QKD scheme with GM versus channel loss when excess noise variance is used as a parameter

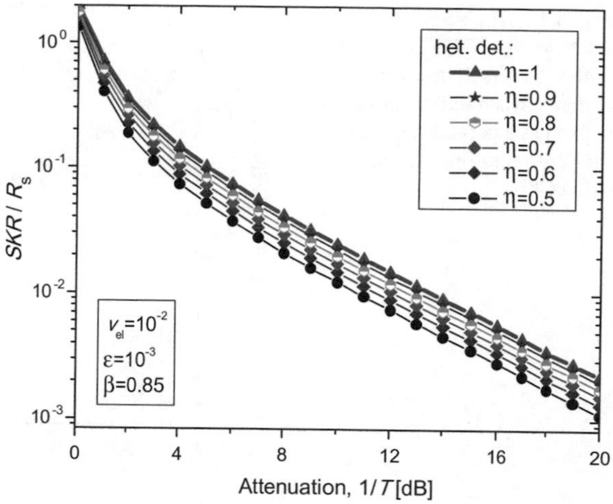

Fig. 8.12 The secrecy fraction for CV-QKD scheme with GM versus channel loss when detector efficiency is used as a parameter

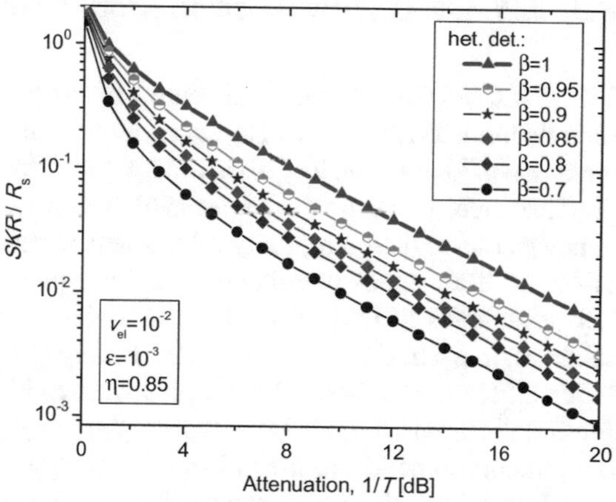

Fig. 8.13 The secrecy fraction for CV-QKD scheme with GM versus channel loss when reconciliation efficiency is used as a parameter

Finally, in Fig. 8.14, we provide SKR versus transmission distance when excess noise variance (expressed in SNU) is used as a parameter. The electrical noise variance is set to $v_{\mathrm{el}} = 0.1$, detector efficiency is $\eta = 0.9$, and reconciliation efficiency is set to $\beta = 0.9$. For transmission medium, the ultralow-loss fiber with attenuation coefficient $\alpha = 0.1419$ dB/km, described in [59], is assumed in calculations. Evidently, for the excess noise variance of 10^{-3}, the SKR of 1 Mb/s is achievable for distance of 240 km.

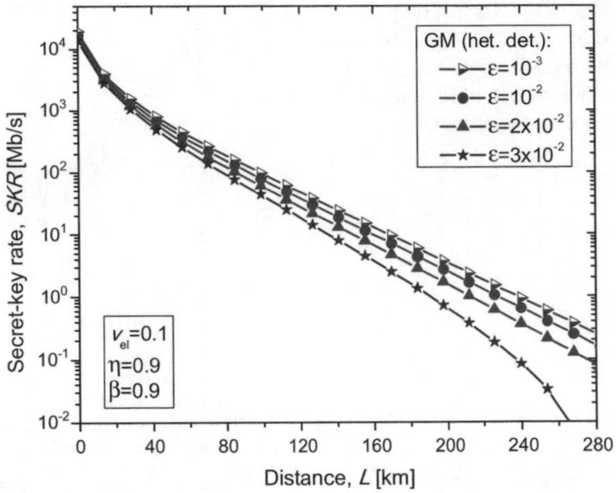

Fig. 8.14 The secrecy fraction for CV-QKD scheme with GM versus transmission distance when excess noise is used as a parameter. The raw transmission rate was set to 10 Gb/s

8.3 CV-QKD with Discrete Modulation

The first CV-QKD protocols appeared in early 2000s and were based on discrete modulation (DM) [27, 60]. However, with occurrence of GM, the interest for DM went down. One of the key disadvantages of GM is low reconciliation efficiency of practical error correction schemes [56]. Recently, CV-QKD DM experiences resurgency thanks to their simplicity for implementation and compatibility with the state-of-the-art fiber optics communications' equipment [33–44, 51, 61, 62]. One of the key advantages of DM over GM is in availability of high reconciliation efficiency schemes, in particular, those on LDPC coding [17, 18, 63, 64]. Namely, the reconciliation efficiency for DM schemes β_{DM} is much higher than reconciliation efficiency for GM protocols β_{GM}; in other words, $\beta_{DM} \gg \beta_{\mathrm{GM}}$ [4, 34, 65]. Another interesting observation comes from the information theory, where we learned that in very low signal-to-noise ratio (SNR) regime, the Shannon's channel capacity can be achieved with small signal constellation sizes [17, 18, 63, 64]. This idea was exploited in [34, 65] to show that in low SNR regime DM protocols can outperform corresponding GM protocols thanks to much better reconciliation efficiency. Moreover, by employing the RF-assisted CV-QKD proposed in [33], which is insensitive to the laser phase noise and frequency offset, the DM protocols with much lower excess noise can be employed, thus further outperforming GM protocols of bad reconciliation efficiency.

8.3.1 Four-State and Eight-State CV-QKD Protocols

The four-state and eight-state prepare-and-measure (PM) protocols can be formulated as follows [33–35, 38, 41, 65–67]:

(i) For *four-state protocol*, Alice sends at random one of four coherent states $|\alpha_k\rangle = |\alpha \exp[\mathrm{j}(2k+1)\pi/4]\rangle$ ($k = 0, 1, 2, 3$) to Bob over the quantum channel. For *eight-state protocol*, Alice sends at random one of eight coherent states $|\alpha_k\rangle = |\alpha \exp(\mathrm{j}k\pi/4)\rangle$ ($k = 0, 1, \ldots, 7$) to Bob over the quantum channel. The channel is characterized by transmissivity T and excess noise ε so that the total channel added noise, referred at the channel input, can be expressed in SNU as $\chi_{\text{line}} = 1/T - 1 + \varepsilon$.

(ii) On receiver side, once the coherent state is received, Bob can perform either homodyne or heterodyne detection, with a detector being characterized by the detector efficiency η and electric noise variance v_{el}. Let the detection added noise variance referred to Bob's input (channel output) be denoted as χ_{h}. For *homodyne detection*, we have that $\chi_{\text{h}} = [(1-\eta) + v_{\text{el}}]/\eta$. On the other hand, for *heterodyne detection*, we have that $\chi_{\text{h}} = [1 + (1-\eta) + 2v_{\text{el}}]/\eta$. Now, the total noise variance, referred at the channel input, can be expressed as $\chi_{\text{total}} = \chi_{\text{line}} + \chi_{\text{h}}/T$.

Given that in the PM DM protocol Alice sends the corresponding coherent states randomly, with the same probability, Bob will see the mixture states. The *mixture state* for the *four-state protocol* will be

$$\hat{\rho}_4 = \frac{1}{4}\sum_{k=0}^{3} |\alpha_k\rangle\langle\alpha_k|, \tag{8.180}$$

which can be expressed in terms of the following states:

$$|\phi_k\rangle = \frac{\exp\left[-\alpha^2/2\right]}{\sqrt{\xi_k}} \sum_{n=0}^{\infty} (-1)^k \frac{\alpha^{4n+k}}{\sqrt{(4n+k)!}} |4n+k\rangle, \tag{8.181}$$

by

$$\hat{\rho}_4 = \sum_{k=0}^{3} \xi_k |\phi_k\rangle\langle\phi_k|, \tag{8.182}$$

wherein

$$\begin{aligned} \zeta_{0,2} &= 0.5\exp\left(-\alpha^2\right)\left[\cosh\left(\alpha^2\right) \pm \cos\left(\alpha^2\right)\right], \\ \zeta_{1,3} &= 0.5\exp\left(-\alpha^2\right)\left[\sinh\left(\alpha^2\right) \pm \sin\left(\alpha^2\right)\right]. \end{aligned} \tag{8.183}$$

By applying the Alice mode annihilation operator $\hat{a}$ on $|\phi_k\rangle$, we obtain [65]

$$\hat{a}|\phi_k\rangle = \sqrt{\frac{\xi_{k-1}}{\xi_k}}\alpha|\phi_{k-1}\rangle; \; k = 1, 2, 3$$

$$\hat{a}|\phi_0\rangle = -\sqrt{\frac{\xi_3}{\xi_0}}\alpha|\phi_3\rangle. \tag{8.184}$$

By performing the purification of $\hat{\rho}_4$, we obtain [65]

$$|\Phi_4\rangle = \sum_{k=0}^{3}\sqrt{\xi_k}|\phi_k\rangle|\phi_k\rangle, \tag{8.185}$$

and this state can also be represented into the following form:

$$|\Phi_4\rangle = \frac{1}{2}(|\psi_0\rangle|\alpha_0\rangle + |\psi_1\rangle|\alpha_1\rangle + |\psi_2\rangle|\alpha_2\rangle + |\psi_3\rangle|\alpha_3\rangle), \tag{8.186}$$

where the states $|\psi_k\rangle$ are non-Gaussian but orthogonal states:

$$|\psi_k\rangle = \frac{1}{2}\sum_{l=0}^{3}\exp[j(2k+1)l\pi/4]|\phi_l\rangle. \tag{8.187}$$

The scheme for *entanglement-assisted (EA) four-state CV-QKD* is illustrated in Fig. 8.15, which is almost the same as corresponding scheme for GM except for entanglement source at Alice side, which generates the entangled state $|\Phi_4\rangle$ of variance v, defined by Eq. (8.186). She performs the projective measurements $\{|\psi_k\rangle\langle\psi_k|\}$ on her qubit, thus preparing the corresponding coherent state $|\alpha_k\rangle$ when her measurement gives the result k, as sending it to Bob over the quantum channel. Bob's detector is modeled by the beam splitter of transmissivity η, while the electronic noise is modeled as an EPR state of variance w. For homodyne detection, $w = 1 + v_{el}/(1-\eta)$, while for heterodyne detection $w = 1 + 2v_{el}/(1-\eta)$.

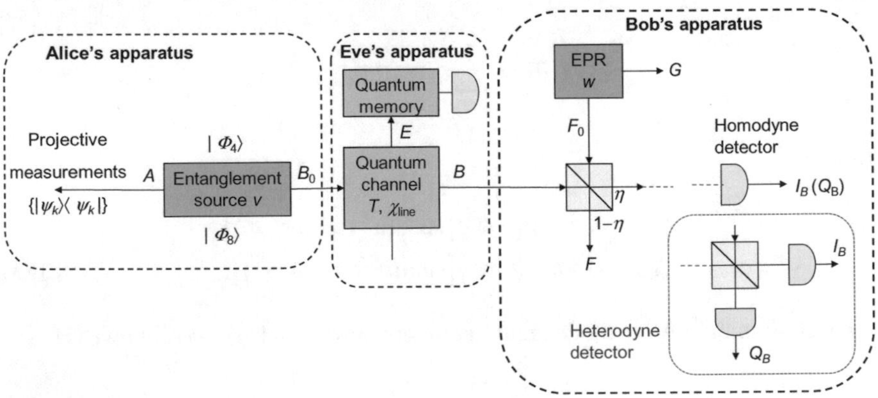

Fig. 8.15 The illustration of entanglement-assisted four-state and eight-state CV-QKD schemes

Regarding the *EA eight-state CV-QKD* the corresponding scheme is identical to the previous one, except for Alice entanglement source that now generates the entanglement state $|\Phi_8\rangle$, defined as

$$|\Phi_8\rangle = \frac{1}{4}\sum_{k=0}^{7}|\omega_k\rangle|\alpha_k\rangle, \tag{8.188}$$

where the states $|\omega_k\rangle$ are non-Gaussian but orthogonal states:

$$|\omega_k\rangle = \frac{1}{2}\sum_{l=0}^{7}\exp[j(4k+1)l\pi/4]|\varphi_l\rangle, \tag{8.189}$$

with $|\varphi_l\rangle$-states being defined as

$$|\varphi_l\rangle = \frac{\exp[-\alpha^2/2]}{\sqrt{\zeta_l}}\sum_{n=0}^{\infty}(-1)^l\frac{\alpha^{8n+l}}{\sqrt{(8n+l)!}}|8n+l\rangle, \tag{8.190}$$

The entangled state can also be expressed in terms of $|\varphi_l\rangle$-states as follows:

$$|\Phi_8\rangle = \sum_{k=0}^{7}\sqrt{\zeta_k}|\varphi_k\rangle|\varphi_k\rangle, \tag{8.191}$$

wherein the ζ-parameters are determined by [35]

$$\begin{aligned}
\zeta_{0,4} &= \frac{1}{4}\exp(-\alpha^2)\left[\cosh(\alpha^2)+\cos(\alpha^2)\pm 2\cos\left(\frac{\alpha^2}{\sqrt{2}}\right)\cosh\left(\frac{\alpha^2}{\sqrt{2}}\right)\right],\\
\zeta_{2,6} &= \frac{1}{4}\exp(-\alpha^2)\left[\cosh(\alpha^2)-\cos(\alpha^2)\pm 2\sin\left(\frac{\alpha^2}{\sqrt{2}}\right)\sinh\left(\frac{\alpha^2}{\sqrt{2}}\right)\right],\\
\zeta_{1,5} &= \frac{1}{4}\exp(-\alpha^2)\left[\sinh(\alpha^2)+\sin(\alpha^2)\pm\sqrt{2}\cos\left(\frac{\alpha^2}{\sqrt{2}}\right)\sinh\left(\frac{\alpha^2}{\sqrt{2}}\right)\pm\sqrt{2}\sin\left(\frac{\alpha^2}{\sqrt{2}}\right)\cosh\left(\frac{\alpha^2}{\sqrt{2}}\right)\right],\\
\zeta_{3,7} &= \frac{1}{4}\exp(-\alpha^2)\left[\sinh(\alpha^2)-\sin(\alpha^2)\mp\sqrt{2}\cos\left(\frac{\alpha^2}{\sqrt{2}}\right)\sinh\left(\frac{\alpha^2}{\sqrt{2}}\right)\pm\sqrt{2}\sin\left(\frac{\alpha^2}{\sqrt{2}}\right)\cosh\left(\frac{\alpha^2}{\sqrt{2}}\right)\right].
\end{aligned} \tag{8.192}$$

Similar to the four-state protocol, $|\Phi_8\rangle$-state is the purification of the mixed state:

$$\hat{\rho}_8 = \frac{1}{8}\sum_{k=0}^{7}|\alpha_k\rangle\langle\alpha_k|. \tag{8.193}$$

The mixed density state can also be expressed in terms of $|\varphi_l\rangle$-states as follows:

$$\hat{\rho}_8 = \sum_{k=0}^{7} \zeta_k |\varphi_k\rangle\langle\varphi_k|. \tag{8.194}$$

8.3.2 Secret-Key Rates for Four-State and Eight-State Protocols

The starting point is to determine the covariance matrices for the bipartite states $|\Phi_4\rangle$ and $|\Phi_8\rangle$. From signal constellation design [17, 18, 64, 68, 69], we know that optimum approximation with four points of Gaussian source is quadriphase-shift keying (PSK), so that we intuitively expect the correlation matrix for the bipartite state to be similar to that with GM:

$$\mathbf{\Sigma}_4 = \begin{bmatrix} a\mathbf{1} & Z_4\mathbf{Z} \\ Z_4\mathbf{Z} & b\mathbf{1} \end{bmatrix}, \tag{8.195}$$

wherein the following is valid [65]:

$$\begin{aligned} a = b &= \langle\Phi_4|\mathbf{1} + 2\hat{a}^\dagger\hat{a}|\Phi_4\rangle = \langle\Phi_4|\mathbf{1} + 2\hat{b}^\dagger\hat{b}|\Phi_4\rangle \\ &= \mathrm{Tr}\big(\mathbf{1} + 2\hat{a}^\dagger\hat{a}\hat{\rho}_4\big) = \mathrm{Tr}\left(\mathbf{1} + 2\sum_{k=0}^{3}\xi_k\hat{a}^\dagger\hat{a}|\phi_k\rangle\langle\phi_k|\right) \\ &= 1 + 2\sum_{k=0}^{3}\xi_k \underbrace{\langle\phi_k|\hat{a}^\dagger\hat{a}|\phi_k\rangle}_{\alpha^2\xi_{k-1}/\xi_k} = 1 + 2\alpha^2 = v, \end{aligned} \tag{8.196}$$

where $\hat{b}(\hat{b}^\dagger)$ denotes the Bob's annihilation (creation) operator. In similar fashion, the correlation term Z_4 can be determined as [65]

$$\begin{aligned} Z_4 &= \langle\Phi_4|\hat{a}\hat{b} + \hat{a}^\dagger\hat{b}^\dagger|\Phi_4\rangle = 2\mathrm{Re}\big(\langle\Phi_4|\hat{a}\hat{b}|\Phi_4\rangle\big) = 2\mathrm{Re}\left(\langle\Phi_4|\hat{a}\hat{b}\sum_{k=0}^{3}\sqrt{\xi_k}|\phi_k\rangle|\phi_k\rangle\right) \\ &= 2\mathrm{Re}\left(\langle\Phi_4|\sum_{k=0}^{3}\sqrt{\xi_k}\underbrace{\hat{a}|\phi_k\rangle}_{\sqrt{\frac{\xi_{k-1}}{\xi_k}}\alpha|\phi_{k-1}\rangle}\underbrace{\hat{b}|\phi_k\rangle}_{\sqrt{\frac{\xi_{k-1}}{\xi_k}}\alpha|\phi_{k-1}\rangle}\right) = \underbrace{2\alpha^2}_{v-1}\mathrm{Re}\left(\langle\Phi_4|\sum_{k=0}^{3}\frac{\xi_{k-1}}{\sqrt{\xi_k}}|\phi_{k-1}\rangle|\phi_{k-1}\rangle\right) \\ &= (v-1)\sum_{k=0}^{3}\xi_{k-1}^{3/2}\xi_k^{-1/2},\ \xi_{-1} = \xi_3. \end{aligned} \tag{8.197}$$

Therefore, the correlation matrix for $|\Phi_4\rangle$-state becomes

$$\boldsymbol{\Sigma}_4 = \begin{bmatrix} v\mathbf{1} & Z_4\mathbf{Z} \\ Z_4\mathbf{Z} & v\mathbf{1} \end{bmatrix}, \tag{8.198}$$

Following the similar procedure for $|\Phi_8\rangle$-state, see Ref. [35], we can determine the correlation matrix for $|\Phi_8\rangle$-state as follows:

$$\boldsymbol{\Sigma}_8 = \begin{bmatrix} v\mathbf{1} & Z_8\mathbf{Z} \\ Z_8\mathbf{Z} & v\mathbf{1} \end{bmatrix}. \tag{8.199}$$

where the correlation term Z_8 is determined by

$$Z_8 = \langle \Phi_8 | \hat{a}\hat{b} + \hat{a}^\dagger \hat{b}^\dagger | \Phi_8 \rangle = (v-1) \sum_{k=0}^{7} \zeta_{k-1}^{3/2} \zeta_k^{-1/2}, \quad \zeta_{-1} = \zeta_7. \tag{8.200}$$

As an illustration in Fig. 8.16, we show the behavior of correlation terms Z_4 and Z_8 against that for EPR state (denoted as Z_{EPR}) as a function of Alice's modulation variance $v_{\mathrm{A}} = v - 1$. Evidently, for modulation variance $v_{\mathrm{A}} \leq 1$, the Z_4 correlation term shows nice agreement with Z_{EPR}. On the other hand, for modulation variance $v_{\mathrm{A}} < 5$, the Z_8 correlation term shows reasonably good agreement with Z_{EPR}.

Following the similar approach to GM, the covariance matrix corresponding to joint Alice–Bob density state ρ_{AB} can be written as

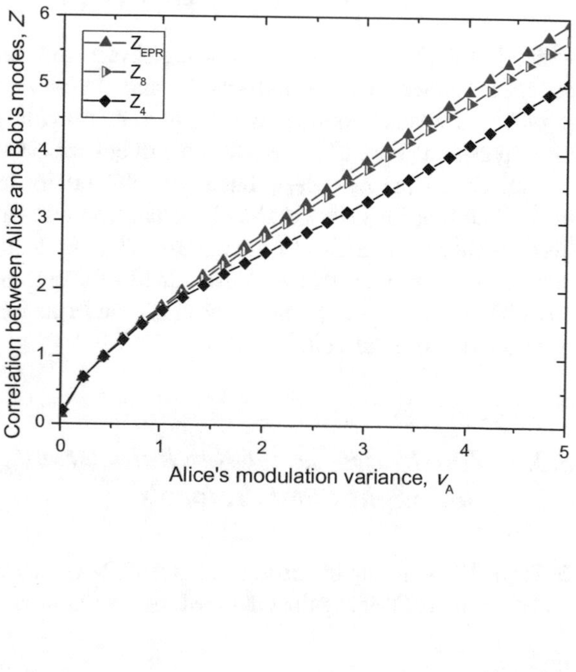

Fig. 8.16 The correlation between Alice and Bob's modes for four-state (Z_4) and eight-state (Z_8) DM against that for GM (Z_{EPR}) versus Alice's modulation variance V_A

$$\mathbf{\Sigma}_{AB,4,8} = \begin{bmatrix} v\mathbf{1} & \sqrt{T}Z_{4,8}\mathbf{Z} \\ \sqrt{T}Z_{4,8}\mathbf{Z} & T(v+\chi_{\text{line}})\mathbf{1} \end{bmatrix}. \tag{8.201}$$

Another alternative form for Alice–Bob covariance matrix is obtained by replacing the expression for $\chi_{\text{line}} = 1/T - 1 + \varepsilon$ to get

$$\mathbf{\Sigma}_{AB,4,8} = \begin{bmatrix} v\mathbf{1} & \sqrt{T}Z_{4,8}\mathbf{Z} \\ \sqrt{T}Z_{4,8}\mathbf{Z} & T(v+\chi_{\text{line}})\mathbf{1} \end{bmatrix} = \begin{bmatrix} v\mathbf{1} & \sqrt{T}Z_{4,8}\mathbf{Z} \\ \sqrt{T}Z_{4,8}\mathbf{Z} & T\left(\underbrace{v}_{v_A+1} + \frac{1}{T} - 1 + \varepsilon\right)\mathbf{1} \end{bmatrix}$$
$$= \begin{bmatrix} v\mathbf{1} & \sqrt{T}Z_{4,8}\mathbf{Z} \\ \sqrt{T}Z_{4,8}\mathbf{Z} & T\left(v_A+\varepsilon+\frac{1}{T}\right)\mathbf{1} \end{bmatrix} = \begin{bmatrix} v\mathbf{1} & \sqrt{T}Z_{4,8}\mathbf{Z} \\ \sqrt{T}Z_{4,8}\mathbf{Z} & [T(v_A+\varepsilon)+1]\mathbf{1} \end{bmatrix}. \tag{8.202}$$

Similar to Ref. [66], given the good agreement of Z_8 with Z_{EPR}, for $v_A < 5$, and since always $Z_8 < Z_{\text{EPR}}$, we could normalize the transmissivity T and redefine excess noise variance ε as follows:

$$T = \left(\frac{Z_{EPR}}{Z_8}\right)^2 T', \quad \varepsilon = \left(\frac{Z_8}{Z_{EPR}}\right)^2 (v_A + \varepsilon') - v_A \tag{8.203}$$

so that the corresponding correlation matrix between Alice and Bob becomes

$$\mathbf{\Sigma}_{AB_8} = \begin{bmatrix} (v_A+1)\mathbf{1} & \sqrt{T'}Z_{EPR}\mathbf{Z} \\ \sqrt{T'}Z_{EPR}\mathbf{Z} & [T'(v_A+\varepsilon')+1]\mathbf{1} \end{bmatrix}, \tag{8.204}$$

which is identical to the corresponding covariance matrix for Gaussian modulation, except for deferent transmissivity T' and excess noise variance ε' values. These transformations are applicable to Z_4 as well, providing that $v_A \le 1$. On such a way, the expressions for SKRs we derived earlier for GM are directly applicable for DM as well. Given that $Z_8 < Z_{\text{EPR}}$, but approaching it, the transmissivity T' will be a little lower than that for GM. On the other hand, new ε' will be higher than ε, and given high sensitivity of SKRs to the excess noise, so that the SKR values for DM will always be lower than that for GM. Finally, these transformations are applicable to any higher order discrete modulation of size M as long as the correlation term Z_M has been determined first.

8.3.3 Illustrative Secret-Key Rates Results for Four-State and Eight-State Protocols

In Fig. 8.17, we compare the normalized SKRs versus channel loss, for both four-state and eight-state DMs, against the GM. In calculations, the electrical noise variance is

Fig. 8.17 DM against GM CV-QKD protocols in terms of secrecy fraction versus the channel loss

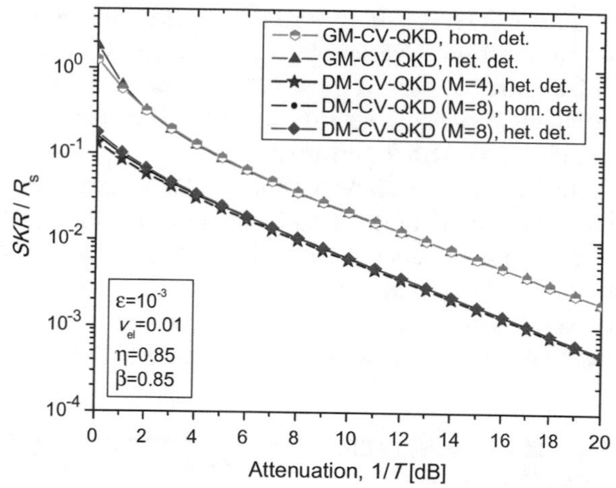

set to $v_{el} = 10^{-2}$, the excess noise variance is set to $\varepsilon = 10^{-3}$, detector efficiency is set to $\eta = 0.85$, and reconciliation efficiency is set to $\beta = 0.85$. Clearly, eight-state protocol slightly outperforms the four-state protocol. For the same set of parameters, the GM significantly outperforms the DM protocols. However, as mentioned earlier, such high reconciliation efficiency is quite difficult to achieve with GM protocols.

In Fig. 8.18, we compare the DM against the GM protocols when typical reconciliation efficiencies are employed. In calculations, the electrical noise variance is set to $v_{el} = 10^{-2}$, the excess noise variance is set to $\varepsilon = 10^{-3}$, and detector efficiency is set to $\eta = 0.85$. Evidently, when GM with reconciliation efficiency 0.5 is used initially, for very low attenuation, it outperforms the DM protocol with β ranging from 0.85 to 0.95. However, for medium and high channel losses, the DM protocols significantly outperform the GM protocols with typical reconciliation efficiencies.

Fig. 8.18 DM CV-QKD outperforming GM CV-QKD in terms of the secrecy fraction versus the channel loss, for typical reconciliation efficiencies

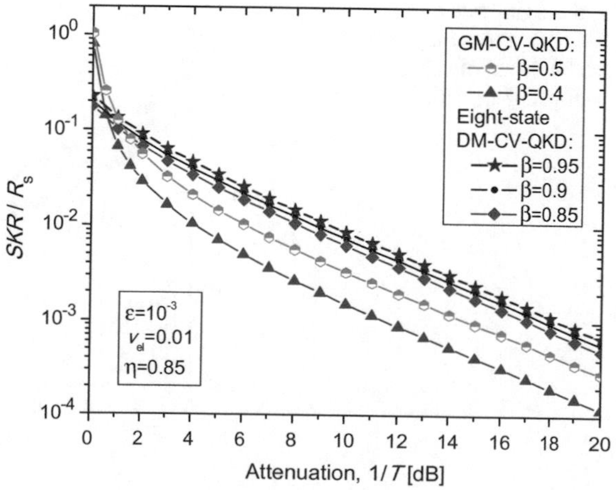

In Fig. 8.19, we show normalized SKRs versus channel loss, for eight-state DM, when excess noise variance (expressed in SNU) is used as a parameter. The electrical noise variance is set to $v_{\mathrm{el}} = 10^{-2}$, detector efficiency is set to $\eta = 0.85$, and reconciliation efficiency is set to $\beta = 0.9$. Similar to the GM, the excess noise variance has high impact on SKR performance.

In Fig. 8.20, we show normalized SKRs versus channel loss, for eight-state DM, when the electrical noise variance (expressed in SNU) is used as a parameter. The excess noise variance is set to $\varepsilon = 10^{-3}$, detector efficiency is set to $\eta = 0.85$, and reconciliation efficiency is set to $\beta = 0.9$. Evidently, the electrical noise variance does not have a high impact on SKR performance. Compared to the GM protocols, the DM protocols exhibit better tolerance to the electrical noise variance.

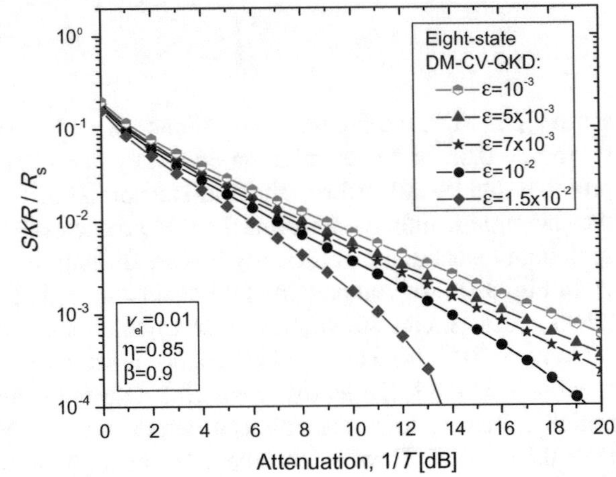

Fig. 8.19 Normalized SKRs for the eight-state DM CV-QKD versus the channel loss for different values of excess noise variance

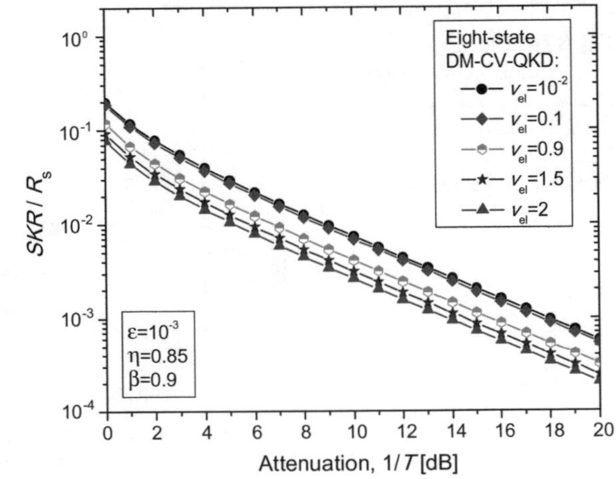

Fig. 8.20 Normalized SKRs for the eight-state DM CV-QKD versus the channel loss for different values of electrical noise variance

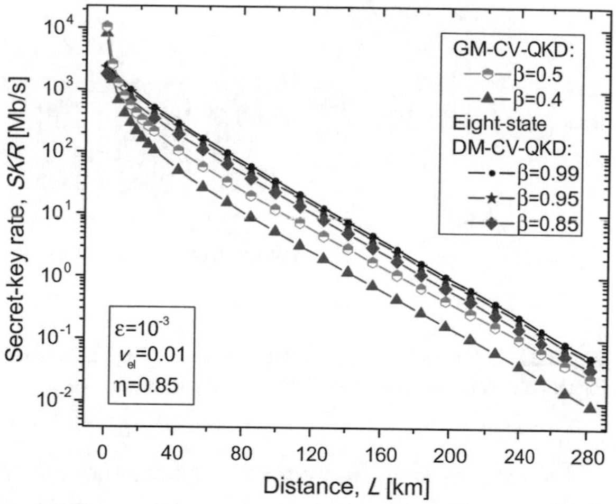

Fig. 8.21 Eight-state DM CV-QKD and GM SKR versus the transmission distance for typical reconciliation efficiencies. The raw transmission rate was set to 10 Gb/s

Finally, in Fig. 8.21, we provide SKR versus transmission distance for both DM and GM protocols assuming typical reconciliation efficiencies. The electrical noise variance is set to $v_{\mathrm{el}} = 0.01$, detector efficiency is set to $\eta = 0.85$, and excess noise variance is set to $\varepsilon = 10^{-3}$. For transmission medium, the ultralow-loss fiber with attenuation coefficient $\alpha = 0.1419$ dB/km, described in [59], is assumed in calculations. Evidently, the DM protocols outperform the GM protocols for typical reconciliation efficiency, in terms of SKR versus distance dependence. With DM protocol, the SKR of 1 Mb/s is achievable for distance of 197 km.

8.4 RF-Subcarrier-Assisted CV-QKD Schemes

In this section, we describe the RF-assisted CV-QKD scheme introduced in [33], in context of four-state protocols, and later on applied to eight-state protocols as well [38, 41–43]. Here, we describe a slightly different implementation proposed in [33], which was introduced in [70]. In particular, different I/Q modulator and optical hybrid types are used.

8.4.1 Description of Generic RF-Assisted CV-QKD Scheme

In this scheme, illustrated in Fig. 8.22, on Alice's side, the discrete modulated signals, such as M-ry PSK and quadrature-amplitude modulation (QAM) signals, are first imposed on the RF subcarrier and are then converted to optical domain with the help of an optical I/Q modulator and sent to Bob over either fiber optics or free-space optical (FSO) link.

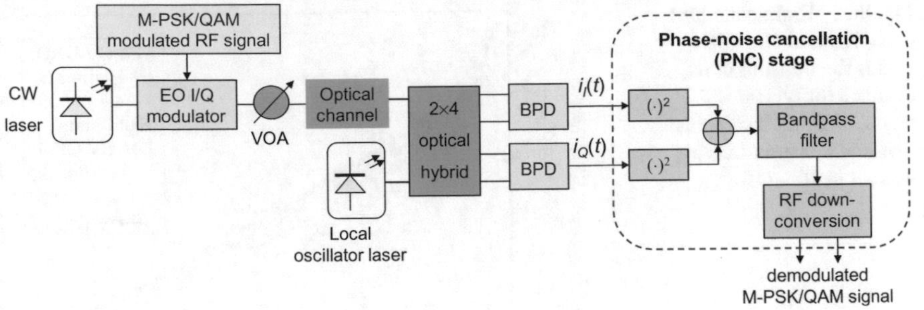

Fig. 8.22 Eight-state/four-state RF-assisted CV-QKD scheme. VOA: variable optical attenuator, BPD: balanced photodetector, BPF: band-pass filter

On receiver side, Bob employs the heterodyne coherent detection together with a phase noise cancelation (PNC) stage to control the level of excess noise. The PNC stage first squares the reconstructed in-phase and quadrature signals and after that either adds or subtracts them depending on the optical hybrid type. The PNC stage further performs band-pass filtering to remove DC component and double-frequency term, followed by the down-conversion, implemented with the help of multipliers and low-pass filters (LPFs). Given that PNC stage cancels the phase noise and frequency offset fluctuations, it exhibits better tolerance to the excess noise compared to traditional schemes discussed in previous section.

The in-phase and quadrature components of RF-assisted eight-state/four-state signal are represented as

$$\begin{aligned} s_I(t) &= A\mathrm{Re}\left\{(I(t)+jQ(t))e^{j\omega_{RF}t}\right\} = A[I(t)\cos(\omega_{RF}t) - Q(t)\sin(\omega_{RF}t)], \\ s_Q(t) &= A\mathrm{Im}\left\{(I(t)+jQ(t))e^{j\omega_{RF}t}\right\} = A[Q(t)\cos(\omega_{RF}t) + I(t)\sin(\omega_{RF}t)], \end{aligned} \tag{8.205}$$

where ω_{RF} is the RF radial frequency [rad/s], while $I(t)$ and $Q(t)$ represent the in-phase and quadrature components of the RF signal. The modulation A is used to control the modulation variance of the signal v_{A}, typically expressed in SNU. For instance, for QPSK, we have that $(I, Q) \in \{(-1, -1), (-1, 1), (1, -1), (1, 1)\}$. On the other hand, for 8PSK, we have that $(I, Q) \in \{(1, 0), (1/\sqrt{2}, 1/\sqrt{2}), (0, 1), (-1/\sqrt{2}, 1/\sqrt{2}), (-1, 0), (-1/\sqrt{2}, -1/\sqrt{2}), (0, -1), (1/\sqrt{2}, -1/\sqrt{2})\}$. The RF signal can be generated with the help of an arbitrary waveform generator (AWG). The optical I/Q modulator performs electro-optical conversion, wherein output electric field and input electric field are related by [17]

$$\frac{E_{\mathrm{out}}(t)}{E_{\mathrm{in}}(t)} = \frac{1}{2}\cos\left[\frac{\pi}{2}\frac{V_I(t)+V_{\mathrm{DC}}}{V_\pi}\right] + j\frac{1}{2}\cos\left[\frac{\pi}{2}\frac{V_Q(t)+V_{\mathrm{DC}}}{V_\pi}\right], \tag{8.206}$$

where V_π is the voltage required to introduce the phase shift of π rad, V_{DC} is the DC bias voltage, and input electric field is the CW laser unmodulated signal:

$$E_{\text{in}}(t) = \sqrt{P_s} e^{j(\omega_T t + \phi_T)}, \tag{8.207}$$

with P_s being the laser output power, ω_T is the transmit laser radial frequency, and ϕ_T represents the transmit laser phase noise. By biasing both in-phase and quadrature branches at π/4-point (that is, $V_{\text{DC}} = V_\pi/4$), the in-phase RF input of I/Q modulator can be represented by $V_I(t) = (2/\pi)V_\pi s_I(t)$, while the quadrature RF input by $V_Q(t) = (2/\pi)V_\pi s_Q(t)$, so that we can rewrite Eq. (8.206) as follows:

$$\frac{E_{\text{out}}(t)}{E_{\text{in}}(t)} = \frac{1}{2}\cos[s_I(t) + \pi/4] + j\frac{1}{2}\cos\left[s_Q(t) + \pi/4\right], \tag{8.208}$$

After substituting (8.205) into (8.206), applying the Jacobi–Anger expansion [71] and assuming small signal analysis, relevant in our case, we obtain the complex representation of I/Q modulator output–input function:

$$\frac{E_{\text{out}}(t)}{E_{\text{in}}(t)} \cong \frac{\sqrt{2}}{2} e^{j\pi/4} - A[I(t) + jQ(t)]e^{j\omega_{RF}t}. \tag{8.209}$$

By substituting Eq. (8.207) into (8.209), we obtain the following I/Q modulator output signal:

$$E_{\text{out}}(t) \cong \frac{1}{2}\sqrt{2P_s} e^{j(\omega_{Tx}t + \phi_{Tx} + \pi/4)} - A[I(t) + jQ(t)]\sqrt{P_s} e^{j[(\omega_{Tx} + \omega_{RF})t + \phi_{Tx}]}. \tag{8.210}$$

The local oscillator (LO) laser signal, at Bob's side, can be represented as

$$E_{\text{LO}}(t) = \sqrt{P_{LO}} e^{j(\omega_{LO}t + \phi_{LO})}, \tag{8.211}$$

where P_{LO}, ω_{LO}, and ϕ_{LO} represent the power, angular frequency, and the laser phase noise of the LO laser, respectively. As shown in Ref. [17] (see Chap. 6, Sect. 6.8.1), when 2 × 4 optical hybrid based on two Y-junctions and two 2 × 2 optical hybrids with properly selected phase trimers is used (see Ref. [17], Fig. 6.29), the balanced photodetector (BPD) outputs in Fig. 8.22 can be represented as

$$\begin{aligned} i_I(t) &= \frac{1}{2}R\left(E_{out}E_{LO}^* + E_{out}^*E_{LO}\right) \\ &= R\sqrt{\frac{P_s P_{LO}}{2}}\cos\left[(\omega_T - \omega_{LO})t + \phi_{PN} + \frac{\pi}{4}\right] \\ &\quad - RAI(t)\sqrt{P_s P_{LO}}\cos[(\omega_T - \omega_{LO} + \omega_{RF})t + \phi_{PN}] \\ &\quad + RAQ(t)\sqrt{P_s P_{LO}}\sin[(\omega_T - \omega_{LO} + \omega_{RF})t + \phi_{PN}] + n_I(t), \end{aligned} \tag{8.212}$$

$$i_Q(t) = \frac{1}{2}R\left(jE_{out}E_{LO}^* - jE_{out}^*E_{LO}\right)$$

$$= -R\sqrt{\frac{P_s P_{LO}}{2}}\sin\left[(\omega_T - \omega_{LO})t + \phi_{PN} + \frac{\pi}{4}\right]$$
$$+ RAI(t)\sqrt{P_s P_{LO}}\sin[(\omega_T - \omega_{LO} + \omega_{RF})t + \phi_{PN}]$$
$$+ RAQ(t)\sqrt{P_s P_{LO}}\cos[(\omega_T - \omega_{LO} + \omega_{RF})t + \phi_{PN}] + n_Q(t), \quad (8.213)$$

where R is the photodiode responsivity and $\phi_{\mathrm{PN}} = \phi_{\mathrm{T}} - \phi_{\mathrm{LO}}$ is the total phase noise (PN). With n_I and n_Q we represented the additive noise components of quadratures.

By squaring and subtracting the in-phase and quadrature photocurrents, followed by band-pass filtering (BPF) to remove the DC component and double-frequency terms, we obtain

$$r(t) = \frac{1}{R^2 P_s P_{LO}\sqrt{2}} BPF\left[i_I^2(t) - i_Q^2(t)\right]$$
$$= A[I(t)\cos(\omega_{RF}t - \pi/4) - Q(t)\sin(\omega_{RF}t - \pi/4)] + \underbrace{n_{NB,I}\cos\omega_{RF}t - n_{NB,Q}\sin\omega_{RF}t}_{n_{NB}(t)}, \quad (8.214)$$

where $n_{\mathrm{NB}}(t)$ denotes the equivalent narrowband noise at RF subcarrier. Now, we perform the down-conversion process as follows:

$$r_I(t) \cong LPF[r(t)2\cos(\omega_{RF}t - \pi/4)] \cong AI(t) + n'_I,$$
$$r_Q(t) \cong LPF[r(t)2\sin(\omega_{RF}t - \pi/4)] \cong -AQ(t) + n'_Q, \quad (8.215)$$

where n'_{I} and n'_{Q} are equivalent in-phase and quadrature low-pass additive noise processes.

Even though this scheme is described in context of 2D modulation schemes, such as M-ary PSK and M-ary QAM, this scheme is also applicable to higher dimensional signaling schemes. The experimental demonstration of RF-assisted four-state CV-QKD scheme for single channel was described in [33]. To improve the SKRs, we can employ the multidimensional QKD schemes, such as those proposed in [72–79]. However, as shown in [75–77] in multidimensional QKD protocols, the SKR is a logarithmic function of the system dimensionality. To improve further the SKR, the parallel QKD protocols, in which several QKD schemes are operated independently and in parallel to achieve the linear increase in SKRs, were advocated in [38, 41, 51, 80]. As an illustration, a high-SKR-RF-assisted four-state CV-QKD system, which is enabled by three-dimensional (3D) multiplexing, employing the polarization, wavelength, and OAM degrees of freedom, has been proposed in [38] and evaluated in the presence of atmospheric turbulence effects. The atmospheric turbulence channel was emulated with the help of two spatial light modulators (SLMs) on which two randomly generated azimuthal phase patterns yielding Andrews' spectrum [81] were recorded. The validity of this polarization-insensitive atmospheric turbulence emulator was verified in terms of on-axis intensity probability density function (PDF) and intensity correlation function (ICF). Further, the CV-QKD free-space optical (FSO) system enabled by 24 multiplexed/demultiplexed QPSK channels was studied, employing two polarization states, six wavelengths, and four OAM modes. The post-processing noise was effectively reduced by the PNC stage introduced above. Upon the system calibration, transmittance fluctuation monitoring, and residual excess noise estimation, the total maximum SKR of 1.68 Gbit/s was achieved in a back-to-back configuration. In the presence of turbulence effects, it was found that the minimum channel transmittances of 0.21 and 0.29 are required for OAM states of 2 (or −2) and 6 (or −6), respectively, to guarantee the secure transmission, with a total SKR of 120 Mbit/s being reported for the mean transmittances, by employing commercial photodetectors for state-of-the-art fiber optics communications. Additional details can be found in [38, 51]. In the rest of the section, we describe the four-dimensionally (4-D) multiplexed eight-state CV-QKD scheme proposed in [41].

8.4.2 4-D Multiplexed Eight-State CV-QKD Scheme

The 4-D multiplexed eight-state CV-QKD scheme [41] is depicted in Fig. 8.23. At Alice's side, six CW sources with linewidth of 10 kHz are generated on a 50-GHz grid, with details specified in [41], multiplexed together by a WDM multiplexer, and modulated by an optical I/Q modulator. A series of 2.5 GBaud 8PSK symbols were generated and upconverted to the 5 GHz RF domain, which was implemented with the help of an arbitrary waveform generator AWG from Tektronix. The resulting RF signals were used as RF inputs to the optical I/Q modulator, which is properly biased. A wavelength interleaver (IL) is used to separate the odd and even wavelength channels, each of which is decorrelated and recombined by a 3 dB coupler.

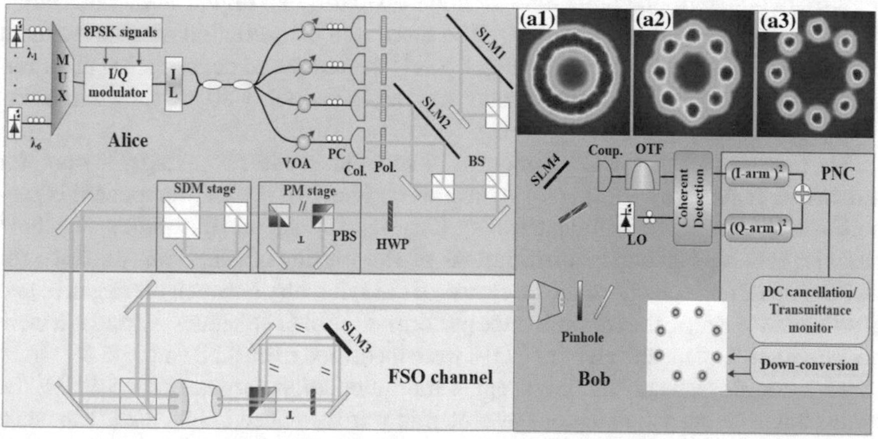

Fig. 8.23 The 4-D multiplexed eight-state CV-QKD scheme. PNC: phase noise cancelation stage. Insets (**a1**–**a3**) represent the constellation diagrams without PNC stage, and with PNC stage for modulation variances $V_A = 40$ and $v_A = 80$, respectively (After Ref. [41]; © IEEE 2017; reprinted with permission)

The decorrelated signals were further split into four branches, where the modulation variance v_A of the signals in each branch was adjusted with the help of a variable optical attenuator (VOA). The four beams were then collimated and imposed on OAM states ±2 and ±6 by two reflective phase-only SLMs form Holoeye, where the left and right halves of the SLM screen were continuously upgraded with specific random phase patterns. The four OAM modes were combined with the help of three beam splitters (BSs) to form the centrally aligned OAM multiplexed beams. A half-waveplate (HWP) was then used to adjust the polarization state. These beams were split, decorrelated, and recombined by employing two polarization beam splitters (PBSs). After the polarization multiplexing stage, the beams were further split, decorrelated, and recombined again with a spatial offset of 10 mm to shape the 10 mm-spaced spatial polarization multiplexed beams. The 4-D multiplexed beams then pass the polarization-insensitive turbulent emulator [38] comprised of a BS, a beam expander (BE), a PBS, a HWP, two mirrors, and a polarization-sensitive SLM. The BE is used in the emulator to adjust the beam size to fit the SLM screen, where a series of phase patterns following Andrews' spectrum [81] have been continuously displayed at a frame rate of 50 Hz.

After the FSO channel transmission, the distorted beams are collected by a compressing telescope, which is followed by a HWP to select one polarization for OAM demultiplexing. A spiral phase pattern is recorded on the SLM4 to back-convert the incoming OAM modes to a Gaussian-like beam. The desirable Gaussian mode can be efficiently coupled to an AR-coated SMF patch after precious calibration. One wavelength channel is selected by the OTF with 3 dB bandwidth of 25 GHz, and mixed with the LO at the coherent receiver for offline signal processing. The in-phase and quadrature photocurrents are digitized by a real-time oscilloscope with 100 GSa/s sampling rate, and then passed through the digital PNC stage configured by two square operators, one addition operator, one digital D.C. cancelation block, and a digital down-converter. The recovered constellation diagram without PNC stage is illustrated in inset of Fig. 8.23a1; the recovered constellation diagrams with PNC stage for modulation variances $v_A = 40$ and $v_A = 80$ are shown in insets of Fig. 8.23a2, a3, respectively.

We first employ the DC component, a module in the PNC stage, to monitor turbulence-induced transmittance fluctuations, given that the DC component is proportional T^2, where T is transmissivity [38, 41, 51]. As an illustration, Fig. 8.24 provides recorded statistical distributions of the varying channel transmittance for OAM states 2 and 6, respectively. Because of the symmetry between the negative and positive OAM states, the transmittance performances of OAM states −2 and −6 were not provided. The mean values of PDFs were measured to be 0.58 and 0.45 for OAM states 2 and 6, respectively. By keeping the modulation variance at 0.5 SNU, the excess noise variance fluctuations over 50 min were provided in Fig. 8.25. The mean excess noise variances were measured to be 0.03 and 0.031 for OAM states 2 and 6, respectively. The excess noises can be contributed to the imperfect azimuthal phase patterns recorded on Alice's SLM screens and the inter-mode cross talk induced by the atmospheric turbulence effects. Inter-mode cross talk has a greater impact on OAM state 2, while OAM state 6 is more sensitive to the imperfect phase patterns.

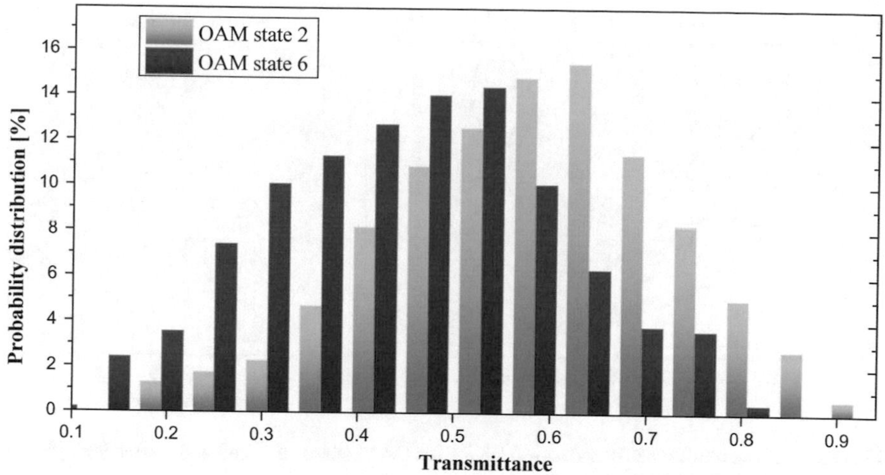

Fig. 8.24 The measured PDFs of fluctuating channel transmittances for OAM channels 2 and 6, respectively (After Ref. [41]; © IEEE 2017; reprinted with permission)

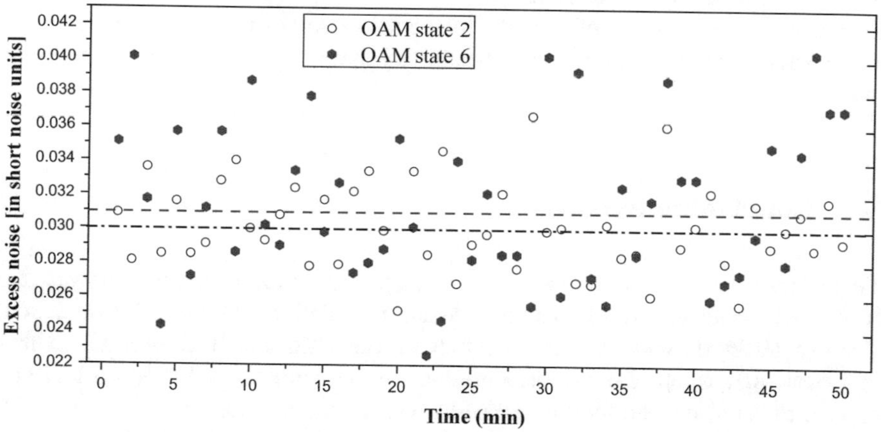

Fig. 8.25 The measured excess noise variances of OAM states 2 and 6 for a period over 50 min (After Ref. [41]; © IEEE 2017; reprinted with permission)

In addition, the quantization noise and the bias dithering of the I/Q modulator are also relevant contributors to the excess noise.

The experimentally measured SKRs for the reconciliation efficiency of $\beta = 0.9$ were summarized in Fig. 8.26. The SKRs of >50 Mb/s and 28 Mb/s can be reached for OAM states 2 and 6, respectively, for transmissivity close to 1. In the presence of atmospheric turbulence effects, the corresponding SKRs of 16 Mb/s and 4 Mb/s can be obtained at the mean channel transmittances of $T = 0.58$ and 0.45 for OAM states 2 and 6, respectively. The total SKRs can be summed up from six wavelength

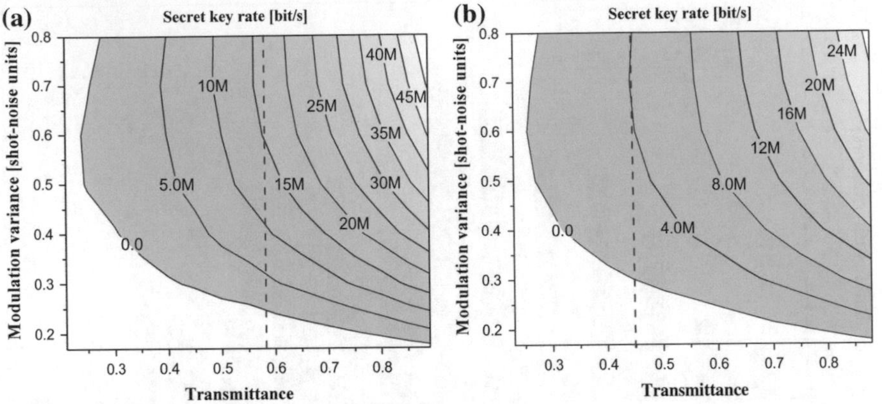

Fig. 8.26 The experimentally measured SKRs for OAM states: **a** 2 and **b** 6 (After Ref. [41]; © IEEE 2017; reprinted with permission)

channels, four OAM modes, two polarization states, and two spatially positioned modes to achieve a maximum SKR of 3.744 Gb/s when transmissivity tends to 1. In the presence of turbulence effects, the total SKR of 960 Mb/s SKR can be achieved at the mean channel transmittances. Additional details of this scheme can be found in [41].

8.5 Concluding Remarks

This chapter has been devoted to the detailed description of continuous variable (CV) QKD schemes, with focus on Gaussian modulation and discrete modulation. The chapter starts with the fundamentals of Gaussian quantum information theory, Sect. 8.1, where the *P*-representation has been introduced (Sect. 8.1.1) and applied to represent the thermal noise as well as the thermal noise together with the coherent state signal (Sect. 8.1.2). Then, quadrature operators have been introduced (Sect. 8.1.3) followed by the phase-space representation. Further, Gaussian and squeezed states have been introduced followed by Wigner function definition as well as the definition of correlation matrices. Section 8.1.4 has been devoted to the Gaussian transformation and Gaussian channels, with beam splitter operation and phase rotation operation being the representative examples. The thermal decomposition of Gaussian states has been discussed in Sect. 8.1.5 and the von Neumann entropy for thermal states has been derived. In Sect. 8.1.6, the focus has been moved to the nonlinear quantum optics fundamentals, in particular, the three-wave mixing and four-wave mixing are described in detail. Further, the generation of the Gaussian states has been discussed in the same section, in particular, the EPR state. The correlation matrices for two-mode Gaussian states have been then discussed in Sect. 8.1.7, and how to calculate the symplectic eigenvalues, relevant in von Neumann entropy calculation. The Gaussian state measurements and detection have been discussed in Sect. 8.1.8, with emphasis being on homodyne detection, heterodyne detection, and partial measurements.

In section on CV-QKD protocols with Gaussian modulation (Sect. 8.2), after the brief description of squeezed state-based protocols, the coherent state-based protocols have been described in detail in Sect. 8.2.1. This section starts with the description of both lossless and lossy transmission channels followed by the description of how to calculate the covariance matrix under various transformations, including beam splitter, homodyne detection, and heterodyne detection. The equivalence between the prepare-and-measure and entanglement-assisted protocols with Gaussian modulation has been established in this section. In Sect. 8.2.2, the focus has been moved to the SKR calculation under collective attacks. The calculation of mutual information between Alice and Bob has been discussed first followed by the calculation of Holevo information between Eve and Bob, in both cases assuming the PM protocol and reverse reconciliation. Further, entangling cloner attack has been described followed by the derivation of Eve-to-Bob Holevo information. The entanglement-assisted protocol has been described next as well as the corresponding Holevo information derivation. In all these derivations, both homodyne detection and heterodyne detection have been studied. Some illustrative SKR results, corresponding to the Gaussian modulation, have been provided in Sect. 8.2.3.

In section on CV-QKD with discrete modulation (Sect. 8.3), after the brief introduction, in Sect. 8.3.1, we have described both four-state and eight-state CV-QKD protocols. Both the PM and entanglement-assisted protocols have been studied. The SKR calculation for discrete modulation has been subject of investigation in Sect. 8.3.2. Some illustrative numerical results related to CV-QKD with discrete modulation have been provided in Sect. 8.3.3. In the same section, the conditions under which the discrete modulation can outperform the Gaussian modulation have been identified.

In section on RF-assisted CV-QKD scheme (Sect. 8.4), a generic RF-assisted scheme applicable to arbitrary two-dimensional modulation schemes, including M-ary PSK and M-ary QAM, has been introduced (see Sect. 8.4.1). This scheme exhibits better tolerance to laser phase noise and frequency offset fluctuations compared to conventional CV-QKD schemes with discrete modulation. Section 8.4.2 has been devoted to increase the SKR through the parallelization approach. In particular, 4-D multiplexing scheme, wherein multiplexing degrees of freedom include polarization, OAM, wavelength, and spatial position, has been introduced and experimental verification has been described.

References

1. Van Assche G (2006) Quantum cryptography and secrete-key distillation. Cambridge University Press, Cambridge, New York
2. Helstrom CW (1976) Quantum detection and estimation theory. Academic Press, New York

3. Helstrom CW, Liu JWS, Gordon JP (1970) Quantum-mechanical communication theory. Proc IEEE 58(10):1578–1598
4. Weedbrook C, Pirandola S, García-Patrón R, Cerf NJ, Ralph TC, Shapiro JH, Seth Lloyd L (2012) Gaussian quantum information. Rev Mod Phys 84:621
5. Mandel L, Wolf E (1995) Optical coherence and quantum optics. Cambridge University Press, Cambridge, New York, Melbourne
6. Scully MO, Zubairy MS (1997) Quantum optics. Cambridge University Press, Cambridge, New York, Melbourne
7. Glauber RJ (1963) Coherent and incoherent states of the radiation field. Phys Rev 131:2766
8. Glauber RJ (2007) Quantum theory of optical coherence. Selected papers and lectures, Wiley-VCH, Weinheim
9. Djordjevic IB (2012) Quantum information processing and quantum error correction: an engineering approach. Elsevier/Academic Press, Amsterdam, Boston
10. Sakurai JJ (1994) Modern quantum mechanics. Addison-Wesley
11. Yuen HP (1976) Two-photon coherent states of the radiation field. Phys Rev A 13:2226
12. Bogoliubov NN (1947) On the theory of superfluidity. J Phys (USSR) 11:23 (Izv. Akad. Nauk Ser. Fiz. 11:77 (1947))
13. Bogoljubov NN (1958) On a new method in the theory of superconductivity. Il Nuovo Cimento 7(6):794–805
14. Holevo AS, Sohma M, Hirota O (1999) Capacity of quantum Gaussian channels. Phys Rev A 59(3):1820–1828
15. Williamson J (1936) On the algebraic problem concerning the normal forms of linear dynamical systems. Am J Math 58(1):141–163
16. Ou Z-YJ (2017) Quantum optics for experimentalists. World Scientific, Hackensack, London
17. Djordjevic IB (2017) Advanced optical and wireless communications systems. Springer International Publishing, Switzerland
18. Cvijetic M, Djordjevic IB (2013) Advanced optical communications and networks. Artech House
19. Velickovic DM (1999) Electromagnetics, vol. I, 2nd edn. Faculty of Electronic Engineering, University of Nis, Serbia
20. Simon R (2000) Peres-Horodecki separability criterion for continuous variable systems. Phys Rev Lett 84:2726
21. Duan L-M, Giedke G, Cirac JI, Zoller P (2000) Inseparability criterion for continuous variable systems. Phys Rev Lett 84:2722
22. Serafini A, Illuminati F, De Siena S (2004) Symplectic invariants, entropic measures and correlations of Gaussian states. J Phys B 37:L21–L28
23. Serafini A (2006) Multimode uncertainty relations and separability of continuous variable states. Phys Rev Lett 96:110402
24. Pirandola S, Serafini A, Lloyd S (2009) Correlation matrices of two-mode bosonic systems. Phys Rev A 79:052327
25. Djordjevic IB (2016) Quantum biological information theory. Springer International Publishing Switzerland, Cham, Heidelberg, New York
26. McMahon D (2008) Quantum computing explained. Wiley, Hoboken, NJ
27. Ralph TC (1999) Continuous variable quantum cryptography. Phys Rev A 61:010303(R)
28. Grosshans F, Grangier P (2002) Continuous variable quantum cryptography using coherent states. Phys Rev Lett 88:057902
29. Yuen HP, Shapiro JH (1980) Optical communication with two-photon coherent states–part III: quantum measurements realizable with photoemissive detectors. IEEE Trans Inf Theory 26(1):78–92
30. Giedke G, Cirac JL (2002) Characterization of Gaussian operations and distillation of Gaussian states. Phys Rev A 66:032316
31. Grosshans F (2005) Collective attacks and unconditional security in continuous variable quantum key distribution. Phys Rev Lett 94:020504

32. Grosshans F, Grangier P (2002) Reverse reconciliation protocols for quantum cryptography with continuous variables. arXiv:quant-ph/0204127
33. Qu Z, Djordjevic IB, Neifeld MA (2016) RF-subcarrier-assisted four-state continuous-variable QKD based on coherent detection. Opt Lett 41(23):5507–5510
34. Leverrier A, Grangier P (2009) Unconditional security proof of long-distance continuous-variable quantum key distribution with discrete modulation. Phys Rev Lett 102:180504
35. Becir A, El-Orany FAA, Wahiddin MRB (2012) Continuous-variable quantum key distribution protocols with eight-state discrete modulation. Int J Quantum Inform 10:1250004
36. Xuan Q, Zhang Z, Voss PL (2009) A 24 km fiber-based discretely signaled continuous variable quantum key distribution system. Opt Express 17(26):24244–24249
37. Huang D, Huang P, Lin D, Zeng G (2016) Long-distance continuous-variable quantum key distribution by controlling excess noise. Sci Rep 6:19201
38. Qu Z, Djordjevic IB (2017) High-speed free-space optical continuous-variable quantum key distribution enabled by three-dimensional multiplexing. Opt Express 25(7):7919–7928
39. Silberhorn C, Ralph TC, Lütkenhaus N, Leuchs G (2002) Continuous variable quantum cryptography: beating the 3 dB Loss Limit. Phys Rev A 89:167901
40. Patel KA, Dynes JF, Choi I, Sharpe AW, Dixon AR, Yuan ZL, Penty RV, Shields J (2012) Coexistence of high-bit-rate quantum key distribution and data on optical fiber. Phys Rev X 2:041010
41. Qu Z, Djordjevic IB (2017) Four-dimensionally multiplexed eight-state continuous-variable quantum key distribution over turbulent channels. IEEE Photonics J 9(6):7600408
42. Qu Z, Djordjevic IB (2017) Approaching Gb/s secret key rates in a free-space optical CV-QKD system affected by atmospheric turbulence. In: Proceedings of ECOC 2017, P2.SC6.32, Gothenburg, Sweden
43. Qu Z, Djordjevic IB (2018) High-speed free-space optical continuous variable-quantum key distribution based on Kramers-Kronig scheme. IEEE Photonics J 10(6):7600807
44. Heid M, Lütkenhaus N (2006) Efficiency of coherent state quantum cryptography in the presence of loss: influence of realistic error correction. Phys Rev A 73:052316
45. Grosshans F, Cerf NJ (2004) Continuous-variable quantum cryptography is secure against non-Gaussian attacks. Phys Rev Lett 92:047905
46. García-Patrón R, Cerf NJ (2006) Unconditional optimality of Gaussian attacks against continuous-variable quantum key distribution. Phys Rev Lett 97:190503
47. Navascués M, Grosshans F, Acín A (2006) Optimality of Gaussian attacks in continuous-variable quantum cryptography. Phys Rev Lett 97:190502
48. Grosshans F, Van Assche G, Wenger J, Tualle-Brouri R, Cerf NJ, Grangier P (2003) Quantum key distribution using Gaussian-modulated coherent states. Nature 421:238
49. Grosshans F (2002) Communication et Cryptographie Quantiques avec des Variables Continues. PhD Dissertation, Université Paris XI
50. Wegner J (2004) Dispositifs Imulsionnels pour la Communication Quantique à Variables Contines. PhD Dissertation, Université Paris XI
51. Qu Z (2018) Secure high-speed optical communication systems. PhD Dissertation, University of Arizona
52. Shannon CE (1948) A mathematical theory of communication. Bell Syst Tech J 27:379–423 and 623–656
53. Cover TM, Thomas JA (1991) Elements of information theory. Wiley, New York
54. Neilsen MA, Chuang IL (2000) Quantum computation and quantum information. Cambridge University Press, Cambridge
55. Fossier S, Diamanti E, Debuisschert T, Tualle-Brouri R, Grangier P (2009) Improvement of continuous-variable quantum key distribution systems by using optical preamplifiers. J Phys B 42:114014
56. Garcia-Patron R (2007) Quantum information with optical continuous variables: from Bell tests to key distribution. PhD thesis, Université Libre de Bruxelles

57. Scarani V, Bechmann-Pasquinucci H, Cerf NJ, Dušek M, Lütkenhaus N, Peev M (2009) The security of practical quantum key distribution. Rev Mod Phys 81:1301
58. Lodewyck J, Bloch M, García-Patrón R, Fossier S, Karpov E, Diamanti E, Debuisschert T, Cerf NJ, Tualle-Brouri R, McLaughlin SW, Grangier P (2007) Quantum key distribution over 25 km with an all-fiber continuous-variable system. Phys Rev A 76:042305
59. Tamura, Y., et al.: The first 0.14-dB/km loss optical fiber and its impact on submarine transmission. J Lightw Technol 36:44–49 (2018)
60. Namiki R, Hirano T (2003) Security of quantum cryptography using balanced homodyne detection. Phys Rev A 67:022308
61. Zhao Y-B, Heid M, Rigas J, Lütkenhaus N (2009) Asymptotic security of binary modulated continuous-variable quantum key distribution under collective attacks. Phys Rev A 79:012307
62. Sych D, Leuchs G (2010) Coherent state quantum key distribution with multi letter phase-shift keying. New J Phys 12:053019
63. Ryan WE, Lin S (2009) Channel codes: classical and modern. Cambridge University Press, Cambridge, New York
64. Djordjevic IB (2016) On advanced FEC and coded modulation for ultra-high-speed optical transmission. IEEE Commun Surv Tutor 18(3):1920–1951
65. Leverrier A, Grangier P (2010) Continuous-variable quantum key distribution protocols with a discrete modulation, arXiv:1002.4083
66. Shen Y, Zou H, Tian L, Chen P, Yuan J (2010) Experimental study on discretely modulated continuous-variable quantum key distribution. Phys Rev A 82:022317
67. Zhang H, Fang J, He G (2012) Improving the performance of the four-state continuous-variable quantum key distribution by using optical amplifiers. Phys Rev A 86:022338
68. Batshon HG, Djordjevic IB, Xu L, Wang T (2010) Iterative polar quantization based modulation to achieve channel capacity in ultra-high-speed optical communication systems. IEEE Photon J 2(4):593–599
69. Liu T, Djordjevic IB (2012) On the optimum signal constellation design for high-speed optical transport networks. Opt Express 20(18):20396–20406
70. Djordjevic IB (2019) On the discretized Gaussian modulation (DGM)-based continuous variable-QKD. IEEE Access 7:65342–65346
71. Cuyt A, Petersen VB, Verdonk B, Waadeland H, Jones WB (2008) Handbook of continued fractions for special functions. Springer Science+Business Media B.V
72. De Riedmatten H, Marcikic I, Scarani V, Tittel W, Zbinden H, Gisin N (2004) Tailoring photonic entanglement in high-dimensional Hilbert spaces. Phys Rev A 69:050304
73. O'Sullivan-Hale MN, Khan IA, Boyd RW, Howell JC (2005) Pixel entanglement: experimental realization of optically entangled d = 3 and d = 6 qudits. Phys Rev Lett 94:220501
74. Barreiro JT, Langford NK, Peters NA, Kwiat PG (2005) Generation of hyperentangled photon pairs. Phys Rev Lett 95:260501
75. Djordjevic IB (2013) Multidimensional QKD based on combined orbital and spin angular momenta of photon. IEEE Photonics J 5(6):7600112
76. Djordjevic IB (2016) Integrated optics modules based proposal for quantum information processing, teleportation, QKD, and quantum error correction employing photon angular momentum. IEEE Photonics J 8(1):6600212
77. Djordjevic IB (2018) FBG-based weak coherent state and entanglement assisted multidimensional QKD. IEEE Photonics J 10(4):7600512
78. Islam NT, Lim CCW, Cahall C, Kim J, Gauthier DJ (2017) Provably secure and high-rate quantum key distribution with time-bin qudits. Sci Adv 3(11):e1701491
79. Islam NT (2018) High-rate, high-dimensional quantum key distribution systems. PhD Dissertation, Duke University

80. Djordjevic IB (2019) Slepian-states-based DV- and CV-QKD schemes suitable for implementation in integrated optics. In: Proceedings of 21st European Conference on Integrated Optics (ECIO 2019), 24–26 April, 2019, Ghent, Belgium
81. Andrews LC, Phillips RL, Hopen CY (2001) Laser beam scintillation with applications. SPIE Press, Bellingham, Washington

Chapter 9
Recent Quantum-Key Distribution Schemes

Abstract This chapter is devoted to the recently proposed both discrete variable (DV) and continuous variable (CV)-QKD schemes. The chapter starts with the description of Hong–Ou–Mandel effect and photonic Bell state measurements (BSMs). Both polarization state-based and time-bin-state-based BSMs are introduced. After that, the BB84 and decoy-state protocols are briefly revisited. The next topic in the chapter is devoted to the measurement-device-independent (MDI)-QKD protocols. Both polarization state-based and time-phase-state-based MDI-QKD protocols are described. Further, the twin-field (TF)-QKD protocols are described, capable of beating the Pirandola–Laurenza–Ottaviani–Banchi (PLOB) bound on a linear key rate. Floodlight (FL) CV-QKD protocol is then described, which is capable of achieving record secret-key rates (SKRs). Finally, Kramers–Kronig (KK)-receiver-based CV-QKD scheme is introduced, representing high-SKR scheme of low complexity.

9.1 Hong–Ou–Mandel Effect and Photonic Bell State Measurements

9.1.1 Hong–Ou–Mandel (HOM) Effect

The Hong–Ou–Mandel (HOM) effect [1, 2] is one of the most known two-photon interference phenomena. It has found the applications in linear optical quantum computing [3] and quantum communications [4], among others. In Fig. 9.1, we illustrate the beam splitter action on a single photon. Assuming that amplitude reflection and transmission coefficients are denoted by A_r and A_t, respectively, the beam splitter (BS) operator can be represented by

$$U_{BS} = A_t|d_2\rangle\langle l| + A_r|d_1\rangle\langle l| + A_r|d_2\rangle\langle r| + A_t|d_1\rangle\langle r|,$$
$$|A_r|^2 + |A_t|^2 = 1, \ \ A_t A_r^* + A_r A_t^* = 0, \tag{9.1}$$

I. B. Djordjevic, *Physical-Layer Security and Quantum Key Distribution*,
https://doi.org/10.1007/978-3-030-27565-5_9

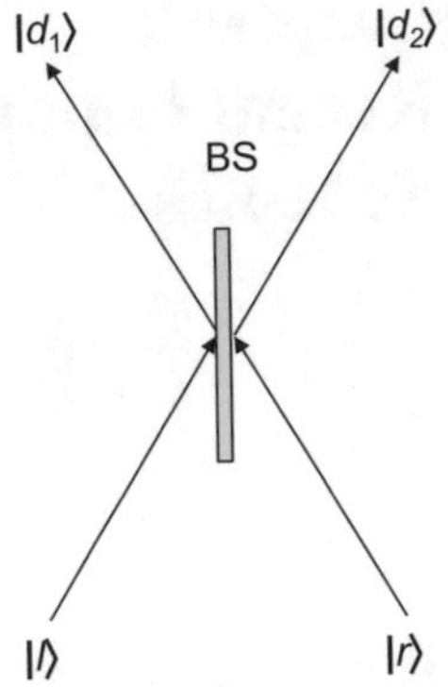

Fig. 9.1 Illustration of the beam splitter (BS) operation

where we used $|l\rangle$-state ($|r\rangle$-state) to denote that the photon is coming from the left (the right) side of the BS while $|d_i\rangle$ ($i = 1, 2$) to denote the corresponding outputs of the BS . The reflection and transmission amplitude coefficients can be parameterized by $A_r = \cos\theta$, $A_t = \text{j} \sin\theta$, and the BS operator action can also be represented as

$$U_{BS} = \cos\theta(|d_1\rangle\langle l| + |d_2\rangle\langle r|) + j\sin\theta(|d_2\rangle\langle l| + |d_1\rangle\langle r|). \tag{9.2}$$

In similar fashion, the polarization beam splitter (PBS) action on either horizontal photon $|\text{H}\rangle$ or vertical photon $|\text{V}\rangle$, coming from the left or right side of PBS, can be described by

$$U_{PBS} = A_r\left|H_l^{(\text{out})}\right\rangle\!\left\langle H_l^{(\text{in})}\right| + A_r\left|H_r^{(\text{out})}\right\rangle\!\left\langle H_r^{(\text{in})}\right| + A_t\left|V_l^{(\text{out})}\right\rangle\!\left\langle V_r^{(\text{in})}\right| + A_t\left|V_r^{(\text{out})}\right\rangle\!\left\langle V_l^{(\text{in})}\right|. \tag{9.3}$$

In other words, the horizontal photon gets reflected by the PBS while the vertical photon goes through the PBS. Finally, the action of the half-wave plate at 45° can be described as

$$U_{\lambda/2,\,45^\circ} = \left|V^{(\text{out})}\right\rangle\!\left\langle H^{(\text{in})}\right| + \left|H^{(\text{out})}\right\rangle\!\left\langle V^{(\text{in})}\right|. \tag{9.4}$$

In other words, the horizontal photon gets converted to the vertical photon and vice versa.

Let us now observe two photons at BS inputs, as shown in Fig. 9.1. By using the formalism given by Eq. (9.1), we can represent joint output state of the BS as follows:

$$\begin{aligned}|\psi\rangle_{\text{out}} &= (U_{BS}|l\rangle) \otimes (U_{BS}|r\rangle) = (A_r|d_1\rangle + A_t|d_2\rangle)(A_r|d_2\rangle + A_t|d_1\rangle)\\ &= A_r^2|d_1\rangle|d_2\rangle + A_rA_t|d_1\rangle|d_1\rangle + A_tA_r|d_2\rangle|d_2\rangle + A_t^2|d_2\rangle|d_1\rangle.\end{aligned} \tag{9.5}$$

Now by replacing $A_r = \cos\theta$, $A_t = \text{j}\sin\theta$, we obtain

$$|\psi\rangle_{\text{out}} = \cos^2\theta|d_1\rangle|d_2\rangle + j\sin\theta\cos\theta|d_1\rangle|d_1\rangle + j\sin\theta\cos\theta|d_2\rangle|d_2\rangle - \sin^2\theta|d_2\rangle|d_1\rangle. \tag{9.6}$$

For 50:50 BS, we have that $\cos\theta = \sin\theta = 1/\sqrt{2}$, so that the first and last terms cancel each other, which is commonly referred to as the *two-photon destructive interference*, and the output state becomes

$$|\psi\rangle_{\text{out}} = \frac{j}{2}|d_1\rangle|d_1\rangle + \frac{j}{2}|d_2\rangle|d_2\rangle, \tag{9.7}$$

indicating that both photons arrive together at the same output port. This effect is commonly referred to as the HOM effect.

Let us now explain the HOM effect by employing the formalism we introduced in Sect. 8.1.4 of Chap. 8 (related to Gaussian transformations). The corresponding annihilation operators of photons at the input of the BS, see Fig. 9.2, are denoted by $\hat{a}_1$ and $\hat{a}_2$, respectively. Let the photons at the output of the BS be represented by the annihilation operators denoted by $\hat{b}_1$ and $\hat{b}_2$, respectively. In the Heisenberg picture, as shown in Sect. 8.1.4, the annihilation operators of input modes get transformed by the BS according to the following Bogoliubov transformation:

$$\begin{bmatrix}\hat{b}_1\\ \hat{b}_2\end{bmatrix} = \begin{bmatrix}\hat{U}a_1\hat{U}^\dagger\\ \hat{U}a_2\hat{U}^\dagger\end{bmatrix} = \begin{bmatrix}\sqrt{T} & \sqrt{1-T}\\ -\sqrt{1-T} & \sqrt{T}\end{bmatrix}\begin{bmatrix}\hat{a}_1\\ \hat{a}_2\end{bmatrix} = \begin{bmatrix}t & r\\ -r & t\end{bmatrix}\begin{bmatrix}\hat{a}_1\\ \hat{a}_2\end{bmatrix}. \tag{9.8}$$

So, the output state of the BS splitter will be

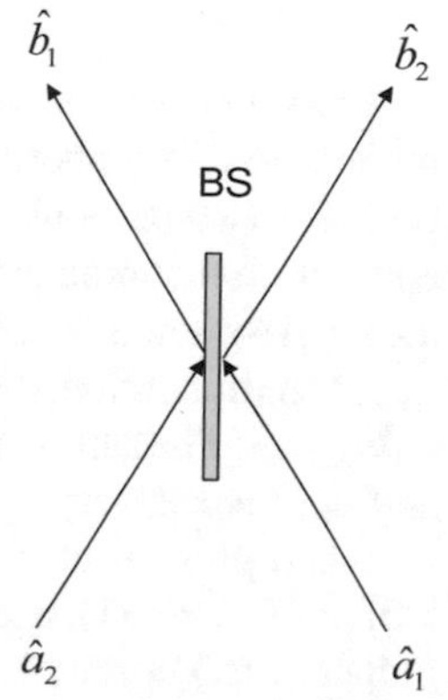

Fig. 9.2 Illustration of the Hong–Ou–Mandel (HOM) effect for two photons

$$
\begin{aligned}
|\psi_{\text{out}}\rangle &= \hat{U}|1\rangle_1|1\rangle_2 = \hat{U}\hat{a}_1^\dagger|0\rangle_1\hat{a}_2^\dagger|0\rangle_2 = \underbrace{\hat{U}\hat{a}_1^\dagger\hat{U}^\dagger}_{\hat{b}_1^\dagger}\underbrace{\hat{U}\hat{a}_2^\dagger\hat{U}^\dagger}_{\hat{b}_1^\dagger}\underbrace{\hat{U}|0\rangle_1|0\rangle_2}_{|0\rangle_1|0\rangle_2} \\
&= \left(t\hat{a}_1^\dagger + r\hat{a}_2^\dagger\right)\left(-r\hat{a}_1^\dagger + t\hat{a}_2^\dagger\right)|0\rangle_1|0\rangle_2 \\
&= \left[-tr\left(\hat{a}_1^\dagger\right)^2 + t^2\hat{a}_1^\dagger\hat{a}_2^\dagger - r^2\hat{a}_2^\dagger\hat{a}_1^\dagger + rt\left(\hat{a}_2^\dagger\right)^2\right]|0\rangle_1|0\rangle_2 \\
&= -tr\sqrt{2}|2\rangle_1|0\rangle_2 + (t^2 - r^2)|1\rangle_1|1\rangle_2 + rt\sqrt{2}|0\rangle_1|2\rangle_2,
\end{aligned}
\tag{9.9}
$$

where we used a different type of BS (consistent with Chap. 8), and we use $|n\rangle$ ($n = 0, 1, 2$) to denote the number state. For 50:50 beam splitter, we have that $t = r = 1/\sqrt{2}$, so that two photons at different outputs cancel, and the output state becomes

$$
|\psi_{\text{out}}\rangle = -\frac{1}{\sqrt{2}}|2\rangle_1|0\rangle_2 + \frac{1}{\sqrt{2}}|0\rangle_1|2\rangle_2, \tag{9.10}
$$

which represents the two-photon NOON state. In other words, two photons arrive together at the same output port with the same probability.

9.1.2 Photonic Bell State Measurements (BSMs)

The Bell state measurements (BSMs) play a key role in photonic quantum computation and communication including quantum teleportation [5], superdense coding [6], quantum swapping [7], quantum repeaters [8], and QKD [9–12], to mention few. A *complete BSM* represents a projection of any two-photon state to maximally entangled Bell states, which are defined by

$$
|\psi^\pm\rangle = \frac{1}{\sqrt{2}}(|01\rangle \pm |10\rangle), \quad |\phi^\pm\rangle = \frac{1}{\sqrt{2}}(|00\rangle \pm |11\rangle). \tag{9.11}
$$

Unfortunately, the complete BSM is impossible to achieve using only linear optics without auxiliary photons. It has been shown in [13] that the probability of successful BSM for two photons being in complete mixed input states is limited to 50%. The conventional approach to the Bell state analysis is to employ 50:50 BS, followed by the single-photon detectors that allow to discriminate between orthogonal states.

As an illustration, the setup to perform the *BSM on polarization states* is provided in Fig. 9.3. The states emitted by Alice and Bob are denoted by density operators ρ_A and ρ_B, respectively.

When photons at BS inputs 1 and 2 are of the same polarization, that is, $|HH\rangle = |00\rangle$ or $|VV\rangle = |11\rangle$, according to Eq. (9.7), they will appear simultaneously at either output port 3 or port 4, and the Bell states $|\phi^\pm\rangle$ get transformed to

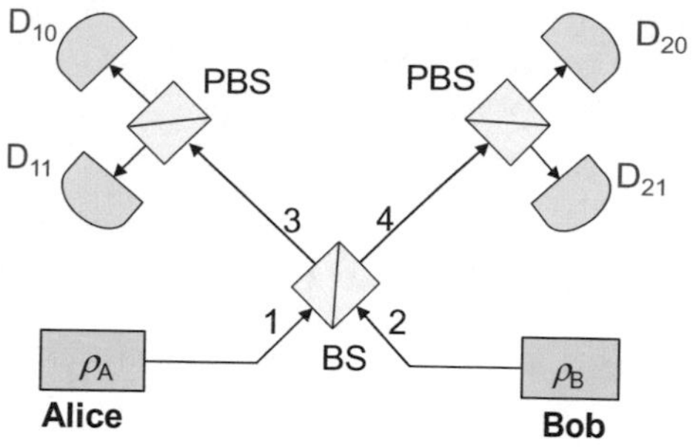

Fig. 9.3 Illustration of the setup to perform the BSMs on polarization states

$$|\phi^{\pm}\rangle = \frac{1}{\sqrt{2}}(|00\rangle_{12} \pm |11\rangle_{12}) \overset{BS}{\rightarrow} |\phi^{\pm}_{\text{out}}\rangle = \frac{j}{2\sqrt{2}}(|00\rangle_{33} + |00\rangle_{44} \pm |11\rangle_{33} \pm |11\rangle_{44}), \tag{9.12}$$

and clearly cannot be distinguished by corresponding PBSs.

On the other hand, when photons at the BS input ports 1 and 2 are of different polarizations, they can exit either the same output port or different ports. When both photons exit the output port 3, after the PBS the single-photon detectors (SPDs) D_{10} and D_{11} will simultaneously click. When both photons exit the output port 4, after the PBS the SPDs D_{20} and D_{21} will simultaneously click. When input photons of different polarizations exit different ports there exist two options. When horizontal photon exits port 3 and vertical photon port 4, the SPDs D_{11} and D_{20} will simultaneously click. Finally, when horizontal photon exits port 4 and vertical photon port 3, the SPDs D_{10} and D_{21} will simultaneously click. Given that the Bell states $|\psi^{\pm}\rangle$ get transformed to (by the BS)

$$\begin{aligned} |\psi^{+}\rangle &= \frac{1}{\sqrt{2}}(|01\rangle_{12} + |10\rangle_{12}) \overset{BS}{\rightarrow} |\psi^{+}_{\text{out}}\rangle = j(|01\rangle_{33} + |10\rangle_{44}), \\ |\psi^{-}\rangle &= \frac{1}{\sqrt{2}}(|01\rangle_{12} - |10\rangle_{12}) \overset{BS}{\rightarrow} |\psi^{-}_{\text{out}}\rangle = |01\rangle_{34} - |10\rangle_{34}, \end{aligned} \tag{9.13}$$

they can be distinguished by PBSs and SPDs. Clearly, based on the above discussion we conclude that the projection on $|\psi^{+}\rangle$ occurs when both SPDs after any of two PBSs simultaneously click. In other words, when either D_{10} and D_{11} or D_{20} and D_{21} simultaneously click, the projection on $|\psi^{+}\rangle$ is identified. Similarly, when either D_{10} and D_{21} or D_{20} and D_{21} simultaneously click, the projection on $|\psi^{-}\rangle$-state is identified. In conclusion, when ideal SPDs are used, the BSM efficiency is ½. If the SPDs have nonideal detection efficiency η, the BSM efficiency will be $0.5\eta^2$.

Another widely used degree of freedom (DOF) to encode the quantum states is *time-bin DOF*. For 2-D quantum communications, the photon can be encoded as the superposition of early state, denoted as $|0\rangle$, and late state, denoted as $|1\rangle$, as follows $|\xi\rangle = \alpha|0\rangle + \beta|1\rangle$, $|\alpha|^2 + |\beta|^2 = 1$. To perform the BSM, we can use the setup provided in Fig. 9.4, which requires only two SPDs in addition to the BS. Similar to

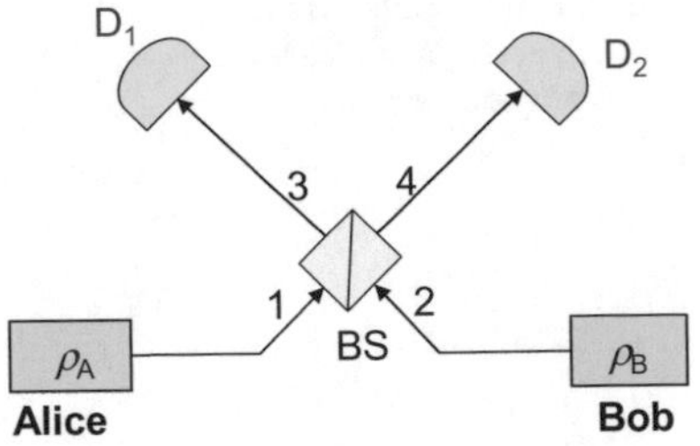

Fig. 9.4 Illustration of the setup to perform the BSMs on time-bin states

polarization, the Bell states are defined by Eq. (9.11), but with different basis kets. Again, when two photons arrive at the same time at BS input ports, they will exit the same output port according to the HOM effect, and we cannot distinguish the $|\phi^{\pm}\rangle$ states.

On the other hand, the projection on $|\psi^{+}\rangle$ is identified when either detector registers the click in early and late time bins. Similarly, the projection on $|\psi^{-}\rangle$ occurs when either detectors D_1 and D_2 or D_2 and D_1 click in consecutive (early–late) time bins.

9.2 BB84 and Decoy-State Protocols Revisited

9.2.1 The BB84 Protocol Revisited

In BB84 protocol [14–18], Alice randomly selects the computational basis (Z-basis) $\{|0\rangle\ |1\rangle\}$ or diagonal basis (X-basis) $\{|+\rangle, |-\rangle\}$, followed by random selection of the basis state. The logical 0 is represented by $|0\rangle, |+\rangle$ states, while logical one by $|1\rangle, |-\rangle$ states. Bob measures each qubit by randomly selecting the basis, computational or diagonal. In sifting procedure, Alice and Bob announce the bases being used for each qubit and keep only instances when they used the same basis. From remaining bits, Alice selects a subset of bits to be used against Eve's interference and channel errors and informs Bob which ones. Alice and Bob announce and compare the values of these bits used for quantum bit error (QBER) rate estimation. If more than acceptable number of bits disagree, dictated by the error correction capability of the code, they abort protocol. Otherwise, Alice and Bob perform information reconciliation and privacy amplification on the remaining bits to obtain shared key bits. The *secret fraction* that can be achieved with BB84 protocol is lower bounded by

$$r = q^{(Z)}\left[1 - h_2\left(e^{(X)}\right)\right] - q^{(Z)} f_e h_2\left(e^{(Z)}\right), \tag{9.14}$$

where $q^{(Z)}$ denotes the probability of declaring a successful result ("the gain") when Alice sent a single photon and Bob detected it in the Z-basis, f_e denotes the error correction inefficiency ($f_e \geq 1$), $e^{(X)}$ [$e^{(Z)}$] denotes the QBER in the X-basis (Z-basis), and $h_2(x)$ is the binary entropy function defined as

$$h_2(x) = -x\log_2(x) - (1-x)\log_2(1-x). \tag{9.15}$$

The second term $q^{(Z)}h_2[e^{(X)}]$ denotes the amount of information Eve was able to learn during the raw key transmission, and this information is typically removed from the final key during the privacy amplification stage. The last term $q^{(Z)}f_e h_2[e^{(Z)}]$ denotes the amount of information revealed during the information reconciliation (error correction) stage, typically related to the parity bits exchanged over an authenticated noiseless public channel (when systematic error correction is used).

9.2.2 *The Decoy-State Protocols Revisited*

When weak coherent state (WCS) source is used instead of single-photon source, this scheme is sensitive to the photon number splitting (PNS) attack, as discussed in Chap. 6. Given that the probability that multiple photons are transmitted from WCS is now nonzero, Eve can exploit this fact to gain information from the channel. To overcome the PNS attack, the use of *decoy-state* quantum-key distribution systems can be used [19–30]. In decoy-state-based QKD, the average number of photons transmitted is increased during random timeslots, allowing Alice and Bob to detect if Eve is stealing photons when multiple photons are transmitted. There exist different versions of decoy-state-based protocols. In one-decoy-state protocol, in addition to signal state with mean photon number μ, the decoy state with mean photon number $\nu < \mu$ is employed. Alice first decides on signal and decoy mean photon number levels and can then determine the optimal probabilities for these two levels to be used based on corresponding SKR expression. In weak plus vacuum decoy-state protocol, in addition to μ and ν levels, the vacuum state is also used as the decoy state. Both protocols have been evaluated in [28], both theoretically and experimentally. It was concluded that the use of one signal state and two decoy states is enough, which agrees with findings in [29, 30].

The probability that Alice can successfully detect the photon when Alice employs the WCS with mean photon number μ, denoted as q_μ, can be determined by

$$q_\mu = \sum_{n=0}^{\infty} y_n e^{-\mu}\frac{\mu^n}{n!}, \tag{9.16}$$

where y_n is the probability of Bob's successful detection when Alice has sent n photons. The similar expression holds for decoy states with mean photon number σ_i, that is,

$$q_{\sigma_i} = \sum_{n=0}^{\infty} y_n e^{-\sigma_i}\frac{\sigma_i^n}{n!}. \tag{9.17}$$

For two decoy-state-based protocol, we can measure the QBERs corresponding to σ_1 and σ_2 and we can lower bound y_1-probability by the following expression:

$$y_1 \geq \frac{\mu}{\mu\sigma_1 - \mu\sigma_2 - \sigma_1^2 + \sigma_2^2}\left[q_{\sigma_1}e^{\sigma_1} - q_{\sigma_2}e^{\sigma_2} - \frac{\sigma_1^2 - \sigma_2^2}{\mu^2}\left[q_\mu e^\mu - \max\left(\frac{\sigma_1 q_{\sigma_2}e^{\sigma_2} - \sigma_2 q_{\sigma_1}e^{\sigma_1}}{\sigma_1 - \sigma_2}, 0\right)\right]\right]. \tag{9.18}$$

The secrecy fraction can be lower bounded by modifying the corresponding expression for BB84 protocols as follows:

$$r = q_1^{(Z)}\left[1 - h_2\left(e^{(X)}\right)\right] - q_\mu^{(Z)} f_e h_2\left(e_\mu^{(Z)}\right), \tag{9.19}$$

where we used the subscript 1 to denote the single-photon pulses and μ to denote the pulse with the mean photon number μ.

9.3 Measurement-Device-Independent (MDI)-QKD Protocols

Any discrepancy in parameters of devices from assumptions made in QKD protocols can be exploited by adversaries through side-channel and quantum hacking attacks [31–35] to compromise security of the corresponding protocols. One way to solve this problem is to invent different ways to overcome the known side-channel/quantum hacking attacks. Unfortunately, the eavesdropper can always come up with a more sophisticated attack. Another, more effective strategy is to invent new protocols in which minimal assumptions have been made on devices being employed in the protocol. One such protocol is known as the device-independent QKD (DI-QKD) protocol [10, 36–38] in which the statistics of detection has been determined to secure the key without any assumption being made on the operation of devices.

9.3.1 Description of MDI-QKD Protocol

In measurement-device-independent (MDI)-QKD protocols [11, 39–42], Alice and Bob are connected to a third party, Charlie, through a quantum channel, such as a free-space optical (FSO) or a fiber optics link. The single-photon detectors are located at Charlie's side, as illustrated in Fig. 9.5, for polarization-based MDI-QKD protocol.

Alice and Bob can employ either single-photon sources or WCS sources, and randomly select one of four states $|0\rangle$, $|1\rangle$, $|+\rangle$, $|-\rangle$ to be sent to Charlie, with the help of polarization modulator (PolM), which is very similar to that used in BB84 protocol. The intensity modulator (IM) is employed to impose different decoy states. Charlie can essentially be Eve as well. Charlie performs the partial BSM with the

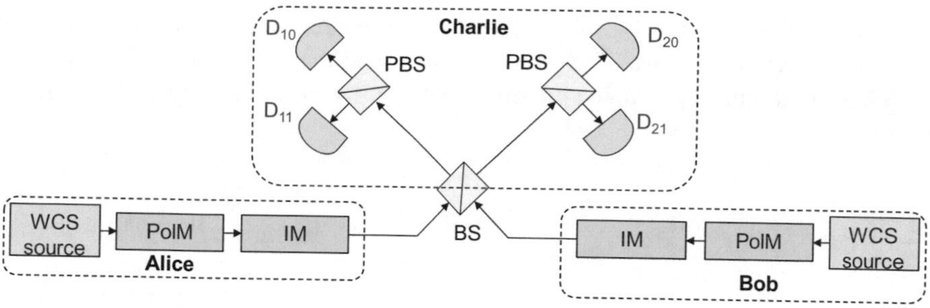

Fig. 9.5 The polarization state-based MDI-QKD system configuration

help of one BS, two PBSs, and four SPDs as described in Sect. 9.1, and announces the events for each measurement resulted in either $|\psi^+\rangle = (|01\rangle + |10\rangle)/\sqrt{2}$ state or $|\psi^-\rangle = (|01\rangle - |10\rangle)/\sqrt{2}$ state. Charlie also discloses the result of the Bell state measurement. Given that Charlies' BSMs are used only to check the parity of Alice and Bob's bits, these measurements will not disclose any information related to the individual bits themselves [39].

A successful BSM corresponds to the observation when precisely two detectors (with orthogonal polarizations) are triggered as follows:

- Detectors' D_{10} and D_{21} click identify the Bell state $|\psi^-\rangle$ and
- On the other hand, detectors' D_{11} and D_{20} click identify the Bell state $|\psi^+\rangle$.

Other detection patterns are related to $|\phi^\pm\rangle$-states, which cannot be distinguished with SPDs, and therefore cannot be considered as successful and as such are discarded from consideration. Clearly, based on description above, we conclude that the MDI-QKD protocol represents generalization of both the EPR-based QKD protocol converted into prepare-and-measure scheme and the decoy-state-based protocol. By comparing the portion of their sequences, Alice and Bob can verify Charlie's honesty.

Let us now describe one possible implementation of the PolM, that is, decoy-state-based BB84 encoder configuration, which is provided in Fig. 9.6. The laser beam signal is split by 1:4 coupler into four branches, each containing the polarization controller to impose the following polarization states $|0\rangle$, $|1\rangle$, $|+\rangle$, $|-\rangle$. With the help of 4:1 optical space switch, we randomly select the polarization state to be sent to Charlie. For this purpose, the use of switches with electro-optic switching is desirable,

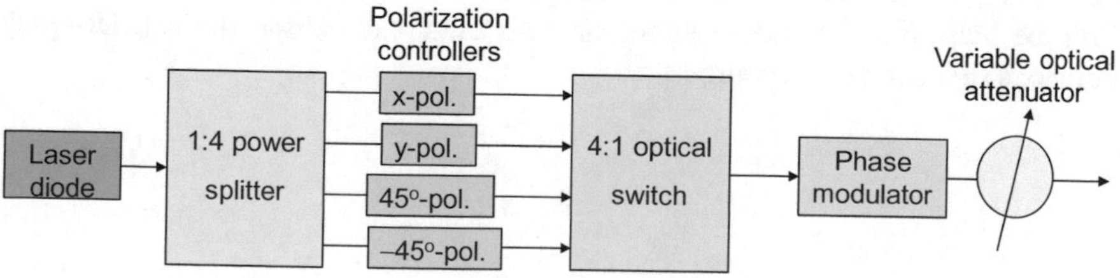

Fig. 9.6 The decoy-state-based BB84 encoder (PolM) configuration

such as those reported in [43, 44], since in these switches, the switching state can be changed within few ns. The phase modulator at the output is used to perform the phase randomization of the coherent state $|\alpha\rangle = \exp(-|\alpha|^2/2)\sum_n \alpha^n |n\rangle/\sqrt{n!}$ as follows:

$$
\begin{aligned}
\frac{1}{2\pi}\int_0^{2\pi} \left| |\alpha| e^{j\vartheta}\right\rangle\left\langle |\alpha| e^{-j\vartheta}\right| d\vartheta &= \frac{1}{2\pi}\sum_{n=0}^{\infty}\sum_{m=0}^{\infty}\frac{|\alpha|^{m+n}}{\sqrt{m!n!}} e^{-|\alpha|^2}|n\rangle\langle m| \underbrace{\int_0^{2\pi} e^{j(n-m)\vartheta}\, d\vartheta}_{\delta_{nm}} \\
&= \sum_{n=0}^{\infty}\frac{\left(|\alpha|^2\right)^n}{n!} e^{-|\alpha|^2}|n\rangle\langle n| = \sum_{n=0}^{\infty}\frac{\mu^n}{n!}e^{-\mu}|n\rangle\langle n|, \quad \mu = |\alpha|^2,
\end{aligned}
\tag{9.20}
$$

where $\mu = |\alpha|^2$ is mean photon number per pulse. Variable optical attenuator is then used to reduce the mean photon number down to the quantum level. Other PolM configurations can be used, such as one described in [39], employing four laser diodes.

Alice and Bob then exchange the information over an authenticated classical, noiseless, channel about the bases being used: Z-basis spanned by $|0\rangle$ and $|1\rangle$ [computational basis (CB)] states or X-basis spanned by $|+\rangle$ and $|-\rangle$ (diagonal basis) states. Alice and Bob then keep only events in which they used the same basis, while at the same time Charlie announced the BSM states $|\psi^-\rangle$ or $|\psi^+\rangle$. One of them (say Bob) flips the sent bits except when both used diagonal basis and Charlie announces the state $|\psi^+\rangle$. The remaining events are discarded.

Further, the X-key is formed out of those bits when Alice and Bob prepared their photons in X-basis (diagonal basis). The error rate on these bits is used to bound the information obtained by Eve. Next, Alice and Bob form the Z-key out of those bits for which they both used the Z-basis (computational basis). Finally, they perform information reconciliation and privacy amplification on the Z-key to get the secret key.

In addition to being insensitive to the SPDs used by Charlie, the MDI-QKD protocol allows to connect multiple participants in a star-like QKD network, in which Charlie is located in the center (serving as a hub) [45, 46]. In this QKD network, illustrated in Fig. 9.7, the most expensive equipment (SPDs) is shared among multiple users, and the QKD network can easily be upgraded without disruption to other users. With the help of N:2 optical switch, any two users can participate in MDI-QKD protocol to exchange the secure keys.

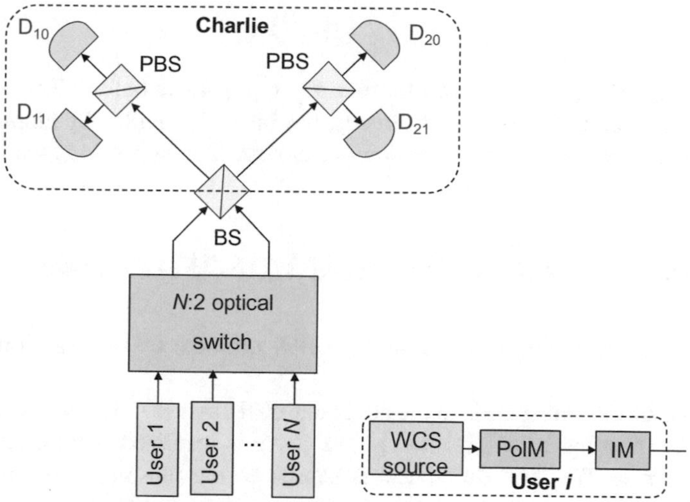

Fig. 9.7 The MDI-QKD-based network

9.3.2 The Secrecy Fraction of MDI-QKD Protocols

The secret fraction of the MDI-QKD protocol, when both Alice and Bob employ single-photon sources, is similar to that for BB84 protocol:

$$r = q^{(Z)}\left[1 - h_2\left(e^{(X)}\right)\right] - q^{(Z)} f_e h_2\left(e^{(Z)}\right), \tag{9.21}$$

where $q^{(Z)}$ now denotes the probability of successful BSM result ("the gain"). On the other hand, when both Alice and Bob employ the WCS sources, the decoy-state method must be employed, so that the secret fraction can be estimated by [40]

$$r = q_{11}^{(Z)}\left[1 - h_2\left(e_{11}^{(X)}\right)\right] - q_{\mu\sigma}^{(Z)} f_e h_2\left(e_{\mu\sigma}^{(Z)}\right), \tag{9.22}$$

where we introduced the following notation:

- $q_{11}^{(Z)}$ denotes the probability that Charlie declares a successful result ("the gain") when both Alice and Bob sent single photon each in the Z-basis,
- $e_{11}^{(X)}$ denotes the phase-error rate of the single-photon signals, both sent in the Z-basis, and
- $q_{\mu\sigma}^{(Z)}$ denotes the gain in the Z-basis when Alice sent the weak coherent state (WCS) of intensity μ to Charlie, while Bob at the same time sent WCS of intensity σ.

In the equation above, the information removed from the final key in privacy amplification step is described by the following term $q_{11}^{(Z)} h_2\left(e_{11}^{(X)}\right)$. The information revealed by Alice in the information reconciliation's step is described by the following term:

$$q_{\mu\sigma}^{(Z)} f_e h_2\left(e_{\mu\sigma}^{(Z)}\right),$$

with f_e being the error correction inefficiency ($f_e \geq 1$) as before. The error rates $q_{11}^{(Z)}$ and $e_{11}^{(X)}$ cannot be measured directly, but instead bounded by employing the decoy-state approach, in similar fashion to that described in previous section.

9.3.3 Time-Phase-Encoding-Based MDI-QKD Protocol

The time-phase encoding basis states for BB84 protocol ($N = 2$) are provided in Fig. 9.8.

The time-basis corresponds to the computational basis $\{|0\rangle, |1\rangle\}$ while the phase-basis to the diagonal basis $\{|+\rangle, |-\rangle\}$. The pulse is localized within the time bin of duration $\tau = T/2$. The time-basis is similar to the pulse-position modulation (PPM). The state in which the photon is placed in the first time bin (early state) is denoted by $|0\rangle = |e\rangle$, while the state in which the photon is placed in the second time bin (late state) is denoted by $|1\rangle = |l\rangle$. The time-phase states are defined by $|+\rangle = (|0\rangle + |1\rangle)/\sqrt{2}, |-\rangle = (|0\rangle - |1\rangle)/\sqrt{2}$. Alice randomly selects either the time-basis or the phase-basis, followed by random selection of the basis state. The logical 0 is represented by $|0\rangle$, $|+\rangle$ while logical one by $|1\rangle$, $|-\rangle$. Alice and Bob randomly select one of four states $|0\rangle$, $|1\rangle$, $|+\rangle$, $|-\rangle$ to be sent to Charlie. Charlie performs the partial BSM with the help of one 50:50 BS, and two SPDs as shown in Fig. 9.4, and announces the events for each measurement resulted into either $\left|\psi^+\right\rangle = (|01\rangle + |10\rangle)/\sqrt{2}$ state or $\left|\psi^-\right\rangle = (|01\rangle - |10\rangle)/\sqrt{2}$ state. Charlie also discloses the result of the Bell state measurement. A successful BSM corresponds to the observation when the SPDs are triggered as follows:

- Two consecutive (both early and late time bins) clicks on one of SPDs identify the Bell state $|\psi^-\rangle$.
- On the other hand, when one SPD detects the presence of photon in early time bin and the second one in the late time bin, we identify the Bell state $|\psi^+\rangle$.
- Other detection patterns are related to $|\phi^{\pm}\rangle$-states and are discarded from consideration.

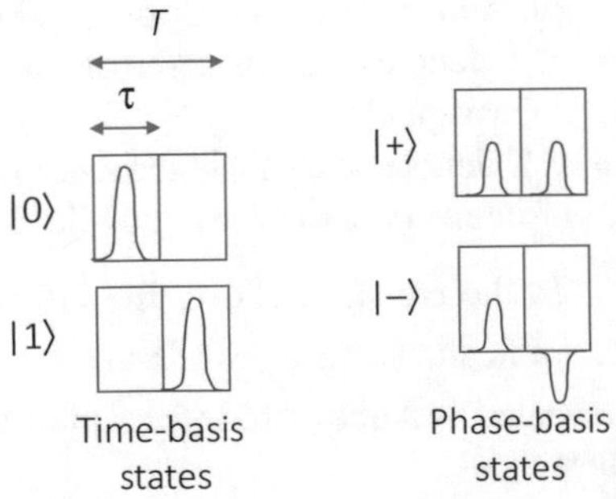

Fig. 9.8 The time-basis states and phase-basis states used in time-phase encoding for BB84 protocol

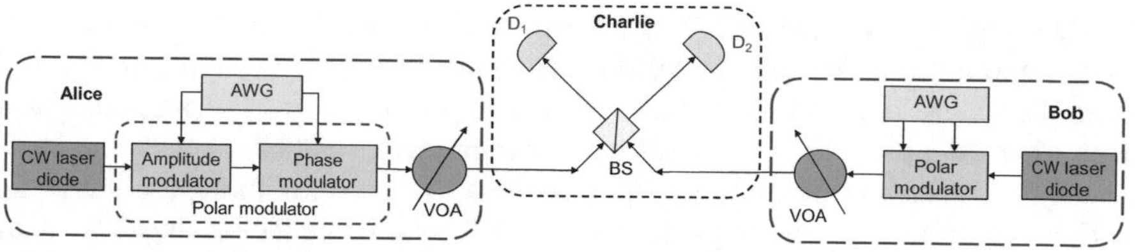

Fig. 9.9 The time-phase encoding-based MDI-QKD system configuration

The rest of the protocol is similar to the polarization state-based MDI-QKD protocol. To implement the time-phase encoder, Alice and Bob can employ either the electro-optical polar modulator, composed of concatenation of an amplitude modulator and phase modulator, which is illustrated in Fig. 9.9, or alternatively an electro-optical I/Q modulator can be used [47]. The amplitude modulator (Mach–Zehnder modulator) is used for pulse shaping and the phase modulator is used to introduce the desired phase shift according to phase-basis states (see Fig. 9.8). The arbitrary waveform generators (AWGs) are used to generate the corresponding RF waveforms as needed. The variable optical attenuators (VOAs) are used to attenuate the modulated beam down to single-photon level. For proper operation, the AWGs must be synchronized. Further, CW laser diodes, used by Alice and Bob, should be frequency locked. Finally, incoming photons (from Alice and Bob) to the BSM must be of the same polarization.

9.4 Twin-Field (TF) QKD Protocols

The quantum channel can be characterized by the transmissivity T, which can also be interpreted as the probability of a photon being successfully transmitted by the channel. The secret-key capacity of the quantum channel can be used as an upper bound on maximum possible secret-key rate. Very often, the Pirandola–Laurenza–Ottaviani–Banchi (PLOB) bound on a linear key rate is used given by [48] $r_{\mathrm{PLOB}} = -\log_2(1 - T)$.

To overcome the rate-distance limit of DV-QKD protocols, two approaches have been pursued until recently: (i) development of quantum relays [49] and (ii) the employment of the trusted relays [50]. Unfortunately, the quantum relays require the use of long-time quantum memories and high-fidelity entanglement distillation, which are still out of reach with current technology. On the other hand, the trusted-relay methodology assumes that the relay between two users can be trusted, and this assumption is difficult to verify in practice. The MDI-QKD approach, described in previous section, was able to close the detection loopholes and extend the transmission distance; however, its SKR is still bounded by $O(T)$-dependence of the upper limit.

Recently, the twin-field (TF) QKD has been proposed to overcome the rate-distance limit [51]. The authors in [51] have shown that TF-QKD upper limit scales with the square root of transmittance, that is, $r \sim O(\sqrt{T})$. In TF-QKD, Alice and Bob generate a pair of optical fields with random phases and phase encoding in X- and Y-basis at remote destinations and send them to untrusted Charlie (or Eve), who is then performing the single-photon detection. This scheme retains the advantages of MDI-QKD scheme such as immunity to detector attacks, multiplexing of SPDs, and star network architecture, but at the same time overcomes the PLOB bound. The unconditional security proof was missing in [51], and in series of papers it was proved that this scheme is unconditionally secure [52–55]. The TF-QKD scheme can be interpreted as the generalization of both (i) the time-phase-encoding-based MDI-QKD scheme shown in Fig. 9.9 and (ii) the phase-encoding-based BB84 scheme described in Chap. 6 (see Sect. 6.4.1.4 and Fig. 6.6).

Authors have shown in [55] that TF-QKD scheme can be interpreted as a particular MDI-QKD scheme, implemented in a two-dimensional subspace of vacuum state $|0\rangle$ and one-photon state $|1\rangle$, as illustrated in Fig. 9.10, representing an equivalent model for TF-QKD with single-photon sources and laser diodes. Given that the vacuum state is insensitive to the channel loss, this state can always be detected, namely, no click means successful detection of the vacuum state. This improvement of detection probability of the vacuum state actually yields to the higher SKR compared to the traditional MDI-QKD and beating up the linear key rate bound. The computational basis or Z-basis is defined by $\{|0\rangle, |1\rangle\}$ (with $|0\rangle$ and $|1\rangle$ representing the vacuum and one-photon state, respectively, as mentioned above). The X-basis (diagonal basis) is defined by $\{|+\rangle = (|0\rangle + |1\rangle)/\sqrt{2}, |-\rangle = (|0\rangle - |1\rangle)/\sqrt{2}\}$, while the Y-basis is defined by $\{|+j\rangle = (|0\rangle + j|1\rangle)/\sqrt{2}, |-j\rangle = (|0\rangle - j|1\rangle)/\sqrt{2}\}$. The Bell states are defined in similar fashion as before, that is, $|\psi^{\pm}\rangle = (|01\rangle \pm |10\rangle)/\sqrt{2}$.

Alice (Bob) with the help of second optical switch randomly selects whether to use Z-basis or X/Y-basis. When Z-basis is selected, Alice (Bob) employs the first optical switch to randomly select either the vacuum state $|0\rangle$ or single-photon state, and selected state is sent over an optical channel (SMF or FSO link) toward Charlie. When X/Y-basis branch is selected, CW laser generates the coherent state $|\alpha\rangle$, which gets attenuated by VOA, so that superposition state of vacuum and single-photon state is obtained, that is, $|+\rangle = (|0\rangle + |1\rangle)/\sqrt{2}$. Alice (Bob) with the help of a phase modulator randomly selects one of four phases from the set $\{0, \pi, \pi/2, 3\pi/2\}$. On such a way, both states for X-basis and Y-basis are selected at random. (Alternatively, instead of two 2:1 optical switches, one 4:1 optical can be used.) Charlie performs

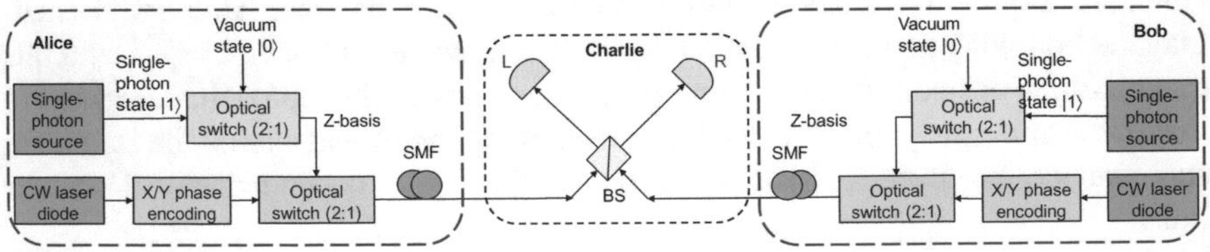

Fig. 9.10 The conceptual twin-field (TF)-QKD system with single-photon and CW laser sources

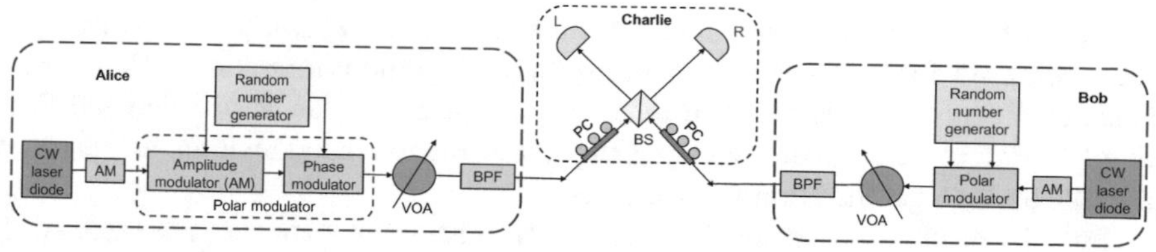

Fig. 9.11 The practical TF-QKD system configuration. BPF: optical band-pass filter, PC: polarization controller

the BSM measurements. The coincidence detection with SPD R click and no click on SPD L indicates the Bell state $|\psi^-\rangle$. On the other hand, the coincidence detection with SPD L click and no click on SPD R indicates the Bell state $|\psi^+\rangle$. Alice and Bob then exchange the basis being used through an authenticated public channel. Bob will flip his sequence when Z-basis was used and Charlie identified the Bell states $|\psi^\pm\rangle$. Bob will also flip his sequence when either X- or Y-basis was used and Charlie identified the Bell state $|\psi^-\rangle$. After the sifting procedure is completed, Alice and Bob will use the sequences corresponding to Z-basis for the key, while the sequences corresponding to X-basis/Y-basis for quantum bit error rate estimation. Finally, Alice and Bob will perform classical postprocessing steps, information reconciliation, and privacy amplification to get the common secure key.

We move our attention now to more realistic, *practical TF-QKD* scheme, shown in Fig. 9.11, which does not require the use of single-photon sources. To explain this scheme, we use interpretation similar to Ref. [55]. Both Alice and Bob employ stabilized CW lasers of low linewidth to generate the global phase-stabilized optical pulses with the help of amplitude modulator. With the help of phase modulators, Alice and Bob choose the random phases $\phi_A \in [0, 2\pi]$ and $\phi_B \in [0, 2\pi]$, respectively. The random phase difference between Alice and Bob is discretized so that

$$\phi_{A,B} \in \Delta\phi_{k_{A,B}} = \left[\frac{2\pi}{M}k_{A,B}, \frac{2\pi}{M}(k_{A,B}+1)\right), \tag{9.23}$$

where the phase-bin $\Delta\phi_{k_{A,B}}$ is discretized by $k_{A,B} \in \{0, 1, \ldots, M-1\}$. Alice (Bob) then randomly selects whether to use Z-basis or X/Y-basis. When Z-basis is selected, the phase-randomized coherent state is sent with intensity either μ or 0, representing logic bits 1 and 0, with probability t and $1-t$, respectively. Given that for lossy channel, the TF state $|1\rangle_A|1\rangle_B$ can create Bell state detection with errors, logic 0 and 1 are not sent with the same probability but with optimized probabilities instead. When X(Y)-basis encoding is selected, Alice and Bob employ corresponding phase and amplitude modulators to randomly select 0 ($\pi/2$) and π ($3\pi/2$) representing logic bits 0 and 1, and such phase-encoded pulses are sent with randomly selected intensities $\{v/2, w/2, 0\}$. The corresponding quantum states, generated by Alice and Bob, are sent toward Charlie over the optical (SMF or FSO) link. The polarization controllers (PCs) ensure that Alice and Bob's pulses have the same polarization. Charlie performs the BSM and announces the results, in the same fashion as described above. Alice and Bob then disclose their phase information, that is, $k_{A,B}$ and intensities, and these are used for parameter estimation. They keep the information related to Z-basis confidential to Charlie, and these data are used for raw key. Alice and Bob then perform classical postprocessing steps, information reconciliation, and privacy amplification to get the common secure key.

For TF-based BB84 protocol, Alice and Bob keep raw data for phase encoding only if $|k_A - k_B| = 0$ or $M/2$, when both used the X-basis, and other raw data results are discarded. For $|k_A - k_B| = 0$, Bob flips his key bit when Charlie detects the Bell state $|\psi^-\rangle$. In similar fashion, for $|k_A - k_B| = M/2$ Bob flips his key bit when Charlie detects the Bell state $|\psi^+\rangle$. The secret fraction for TF-based BB84 protocol can be estimated by [55]

$$r_{\text{TF - BB84}} = 2t(1-t)\mu e^{-\mu} y_{1ZZ}^{(TF)}\left[1 - h_2\left(e_{XX}^{(b_1)}\right)\right] - Q_{ZZ} f\, h_2(E_{ZZ}), \tag{9.24}$$

where Q_{ZZ} is the gain when both Alice and Bob used Z-basis and can be determined experimentally. The parameter f denotes error reconciliation inefficiency ($f \geq 1$). [$h_2(\cdot)$ denotes the binary entropy function defined above.] The gain Q_{ZZ} and the QBER for Z-basis can be determined by [55]

$$\begin{aligned}
Q_{ZZ} &= 2p_d(1-p_d)(1-t)^2 + 4(1-p_d)e^{-\frac{\mu}{2}T_{tot}}\left[1-(1-p_d)e^{-\frac{\mu}{2}T_{tot}}\right]t(1-t) \\
&\quad + 2(1-p_d)e^{-\mu T_{tot}}\left[I_0(\mu T_{tot}) - (1-p_d)e^{-\mu T_{tot}}\right]t^2, \\
E_{ZZ} &= \frac{2p_d(1-p_d)(1-t)^2 + 2(1-p_d)e^{-\mu T_{tot}}\left[I_0(\mu T_{tot}) - (1-p_d)e^{-\mu T_{tot}}\right]t^2}{Q_{ZZ}},
\end{aligned} \tag{9.25}$$

where $T_{\text{tot}} = T\eta_d$, with T being transmissivity and η_d being detector efficiency. Parameter p_d denotes the dark current rate of the threshold detector. Based on Eq. (9.18), by setting $\mu = v/2$, $\sigma_1 = w/2$, and $\sigma_2 = 0$, the yield $y_{1ZZ}^{(TF)}$ becomes lower bounded by

$$y_{1ZZ}^{(TF)} \geq \frac{2v}{vw - w^2}\left[q_{w/2}e^{w/2} - \frac{w^2}{v^2}q_{v/2}e^{v/2} - q_0\left(1 - \frac{w^2}{v^2}\right)\right], \tag{9.26}$$

where $q_0 = 2p_d(1-p_d)$. By employing the decoy-state method [19–30], [55], the yield $y_{1XX}^{(TF)}$ and QBER $e_{XX}^{(b_1)}$ are bounded as follows:

$$\begin{aligned}
y_{1XX}^{(TF)} &\geq \frac{v}{vw-w^2}\left[q_w^{(XX)}e^w - \frac{w^2}{v^2}q_v^{(XX)}e^v - q_0\left(1-\frac{w^2}{v^2}\right)\right] = y_{1XX}^{(TF,\,LB)}, \\
e_{XX}^{(b_1)} &\leq \frac{q_w^{(XX)}e^w Q_w^{(XX)} - q_0 e^{b_0}}{w y_{1XX}^{(TF,\,LB)}},
\end{aligned} \tag{9.27}$$

where for TF vacuum state $e^{b_0} = 1/2$ and $Q_w^{(XX)}$ should be determined experimentally.

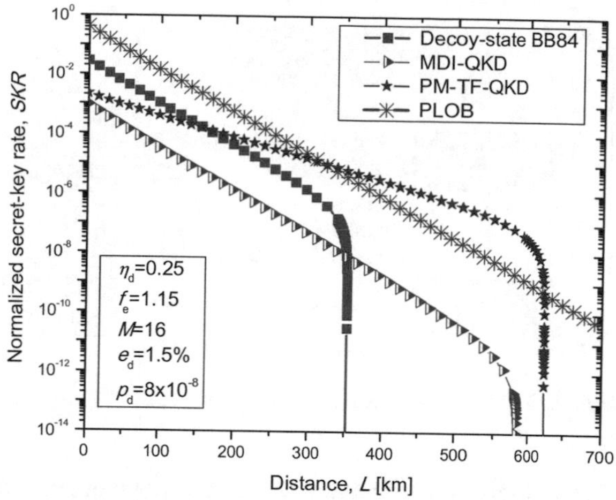

Fig. 9.12 The PM TF-QKD protocol against MDI-QKD and decoy-state BB84 protocols in terms of normalized SKR versus the transmission distance

As an illustration, in Fig. 9.12, we provide comparisons of phase-matching (PM) TF-QKD protocol introduced in [52] against the MDI-QKD protocol [9] and decoy-state-based BB84 protocol [19]. The system parameters are selected as follows: the detector efficiency $\eta_d = 0.25$, reconciliation inefficiency $f_e = 1.15$, the dark count rate $p_d = 8\times10^{-8}$, misalignment error $e_d = 1.5\%$, and the number of phase slices for PM TF-QKD is set to $M = 16$. Regarding the transmission medium, it is assumed that recently reported ultralow-loss fiber of attenuation 0.1419 dB/km (at 1560 nm) is used [56]. In the same figure, the PLOB bound on a linear key rate is provided as well. Clearly, the PM TF-QKD scheme outperforms the decoy-state BB84 protocol for distances larger than 162 km. It outperforms MDI-QKD protocol for all distances, and exceeds the PLOB bound at distance 322 km. Finally, the PM TF-QKD protocol can achieve the distance of even 623 km. Unfortunately, the normalized SKR value is rather low at this distance. For calculations, the expressions provided in Appendix B of Ref. [52] are used with few obvious typos being corrected.

9.5 Floodlight (FL)-QKD

Floodlight (FL)-QKD, as depicted in Fig. 9.13, represents a two-way CV-QKD protocol that can provide orders-of-magnitude higher SKRs compared to conventional QKD protocols [57, 58]. The key idea behind FL-QKD is to employ the multimode encoding. In other words, each message is carried by many optical modes so that the SKR per encoding is substantially increased, even though the SKR per optical mode remains restricted by the PLOB bound $-\log_2(1 - T)$ (with T being the transmissivity of the channel) bits-per-mode limit. A 55 Mbit/s experimental SKR over a 10-dB-loss channel has been reported in [58], and more recently, a 1.3 Gbit/s SKR in the presence of a 10-dB attenuation has been demonstrated [59]. In FL-QKD,

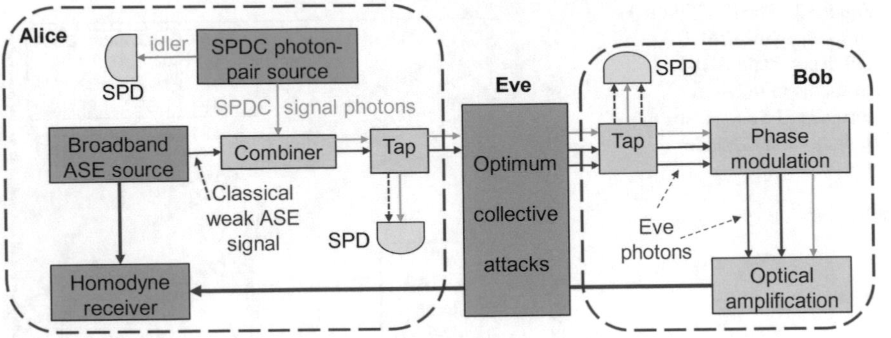

Fig. 9.13 The concept of FL-QKD system. SPD: single-photon detector

Alice employs a broadband amplified spontaneous emission (ASE) source to generate correlated reference and signal light beams. The weak-signal light beam is sent to Bob for phase modulation followed by the optical amplification while strong-ASE portion of the beam is retained at Alice side as the local oscillator (LO) reference beam. Bob performs the phase modulation on weak-signal light beam, followed by amplification by an EDFA. The EDFA not only compensates for the channel loss but also adds bright noise to mask his phase-modulated signal into noise and such prevents Eve's passive eavesdropping, given that Eve does not have the right reference signal. Additionally, Alice does not need a shot-noise limited homodyne detector but commercial homodyne detector instead to detect the Bob's sequence. In addition to broadband ASE classical signal, Alice employs a spontaneous parametric down-conversion (SPDC) photon-pair source to generate the time-correlated signal and idler photons, with idler photon being detected by the SPD, and signal photon being sent toward Bob together with the weak classical ASE signal. Alice ensures that her single photons have the same polarization and spectra as the ASE weak signal. After combining the SPDC's signal photons with the weak ASE-noise signal, Alice taps the portion of combined signal and applies the second SPD. These two SPDs comprise Alice's channel monitoring circuit. Bob taps the portion of received signal and passes it to the SPD. Alice and Bob then perform the coincidence measurements to estimate the extent of Eve's activity. Authors claim that this protocol is secure against the passive attacks, in which Eve makes use of the lost light. However, it is sensitive to *active attacks*, when Eve injects her own light to Bob's terminal and decodes Bob's beam stream using her own stored reference. To overcome this problem, authors claim that Eve's photons are uncorrelated with the idler photons and do contribute to the coincidence measurement but introduce the noise photons. This increase in noise photons can be used to quantify the level of Eve's intrusion as follows [57–59]:

$$f_E = 1 - \frac{C_{IB} - \tilde{C}_{IB}}{C_{IA} - \tilde{C}_{IA}} \frac{S_A}{S_B}, \tag{9.28}$$

where C_{IB} (C_{IA}) denotes time-aligned coincidence rate of Bob's (Alice's) tap, while $\tilde{C}_{IB}$ ($\tilde{C}_{IA}$) denotes the corresponding time-shifted coincidence rate. We use S_A (S_B) to denote the singles rate of Bob's (Alice's) tap. The SKR of FL-QKD scheme can be lower bounded by

$$SKR \geq \left[\beta I_{AB}(P_e) - \chi_{BE}^{(UB)}(f_E)\right]R, \tag{9.29}$$

where β is the reconciliation efficiency, I_{AB} is Alice–Bob mutual information determined by

$$I_{AB}(P_e) = 1 + P_e \log_2 P_e + (1 - P_e)\log_2(1 - P_e), \tag{9.30}$$

with P_e being Alice's bit error probability of phase-modulated channel. We use R to denote the signal rate and $\chi_{BE}^{(UB)}(f_E)$ to denote the upper bound of Bob–Eve Holevo's information, as described in [57].

9.6 CV-QKD Based on Kramers–Kronig (KK) Receiver

A high-speed low-cost CV-QKD scheme based on Kramers–Kronig (KK) receiver has been proposed in [60, 61]. Let us first describe the KK coherent optical receiver.

9.6.1 KK Coherent Optical Receiver

In a KK receiver, the phase information can be extracted from its intensity providing that the minimum phase condition is satisfied [62]. Namely, from linear modulation theory [63, 64] we know that single-sideband (SSB) signal can be represented by

$$s_{SSB}(t) = s(t)\cos(2\pi f_c t) \mp \hat{s}(t)\sin(2\pi f_c t), \tag{9.31}$$

where $s(t)$ is the message signal of bandwidth $B/2$, f_c is the carrier frequency, and we use $\hat{s}(t)$ to denote the Hilbert transform of $s(t)$, HT[$s(t)$], that is,

$$\hat{s}(t) = \frac{1}{\pi}\int_{-\infty}^{\infty} \frac{s(\tau)}{t - \tau} d\tau. \tag{9.32}$$

In other words, the Hilbert transformer is a filter of impulse response $1/\pi\tau$ and transfer function $-j\text{sgn}(\omega)$, where sgn(·) is the sign function. The minus sign in (9.31) corresponds to the case when upper sideband (USB) is transmitted, while the plus sign when the lower sideband (LSB) is transmitted. The corresponding *complex envelope* representation of SSB-signal will be

$$\tilde{s}_{SSB}(t) = s(t) \pm j\hat{s}(t), \tag{9.33}$$

and from (9.32) we conclude that imaginary part of complex envelope is related to the real part by the HT, and vise versa. These two relations are commonly referred to as the KK relations [65, 66]. Let us now define $A + \tilde{s}_{SSB}(t) = |A + \tilde{s}_{SSB}(t)|e^{j\phi(t)}$ and take the complex logarithm to get

$$\log\left[A + \tilde{s}_{SSB}(t)\right] = \log\left[|A + \tilde{s}_{SSB}(t)|e^{j\phi(t)}\right] = \log|A + \tilde{s}_{SSB}(t)| + j\phi(t). \tag{9.34}$$

If the condition $|A| > |\tilde{s}_{SSB}(t)|$ is satisfied, the spectrum of $A + \tilde{s}_{SSB}(t)$ will be the SSB. Therefore, the imaginary part is related to the real part by the HT:

$$\phi(t) = \frac{1}{\pi}\int_{-\infty}^{\infty} \frac{\log|A + \tilde{s}_{SSB}(t)|}{t - \tau} d\tau. \tag{9.35}$$

More efficient method to determine the phase is to employ the Fourier transform (FT) as follows:

$$\phi(t) = FT^{-1}\{-j\,\mathrm{sgn}(\omega) FT(\log|A + \tilde{s}_{SSB}(t)|)\}. \tag{9.36}$$

Let us now describe the KK receiver with coherent optical detection, proposed in [62], employing a single photodetector, followed by analog-to-digital converter (ADC) of sampling rate $f_s = 2B$, as illustrated in Fig. 9.14.

The pilot tone (PT) is added to the information-carrying signal as illustrated in Fig. 9.14. This pilot tone will serve the purpose of the LO signal on receiver side. The transmitted electric field can be represented by $E(t) = E_s(t) + E_0 e^{jBt}$, where E_s denotes the information-carrying signal and E_0 is a positive constant. The photodiode photocurrent can be represented by $I(t) = R|E(t)|^2$, where R is the photodiode responsivity. For simplicity, we will set it to $R = 1$ A/W. Since $E(t) =$

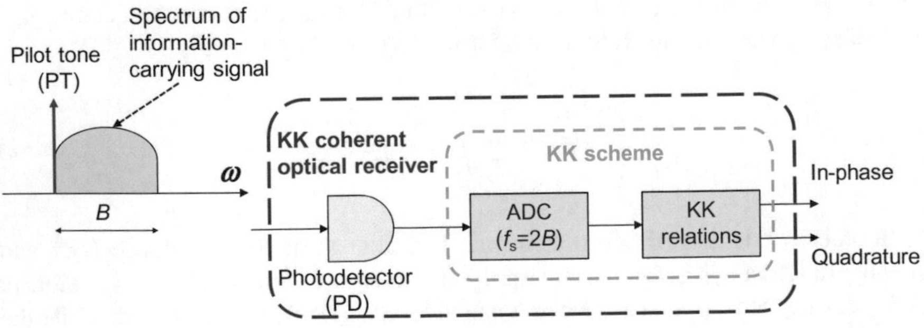

Fig. 9.14 The conceptual diagram of KK coherent optical receiver

$|E(t)|e^{j\phi(t)}$, the information-carrying signal can be determined by

$$E_s(t) = E(t) - E_0 e^{jBt} = \sqrt{I(t)}e^{j\phi(t)} - E_0 e^{jBt}, \tag{9.37}$$

where

$$\phi(t) = FT^{-1}\left\{-j\,\mathrm{sgn}(\omega) FT\left(\log\sqrt{I(t)}\right)\right\}. \tag{9.38}$$

9.6.2 KK-Receiver-Based CV-QKD

The corresponding conceptual diagram of KK-receiver-based CV-QKD scheme is provided in Fig. 9.15. Compared to the heterodyne detection scheme, described in Chap. 8, which requires 3 dB coupler before the measurements of in-phase and quadrature components, the KK-receiver-based CV-QKD does not require the 3 dB coupler at all. In this scheme, we multiplex the PT together with the information-bearing signal by sufficiently large frequency guard band, and set up a high power of PT to meet the minimum phase condition, as well as to minimize the influence of the electrical noise variance v_{el}. The quantum channel is characterized by transmissivity T and excess noise variance ε. The detection efficiency is denoted by η.

Based on Chap. 8, when the reverse reconciliation is used, the secret fraction for KK-receiver-based CV-QKD scheme is given by

$$r = \beta I(A;B) - \chi(B;E), \tag{9.39}$$

where β is the reconciliation efficiency and $I(A;B)$ is the mutual information between Alice and Bob, determined in the same fashion as for individual attacks:

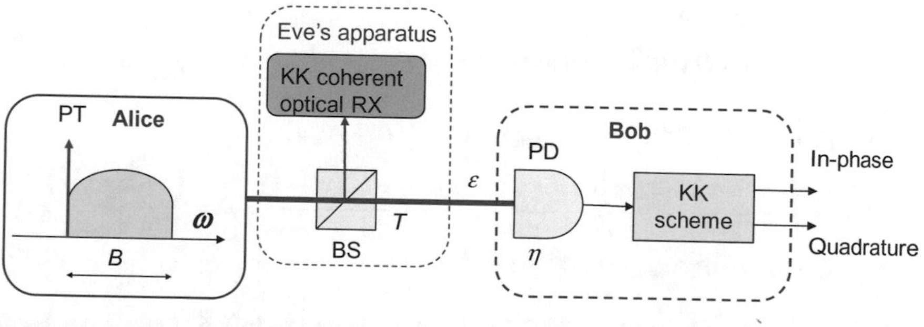

Fig. 9.15 The conceptual diagram of KK receiver-based CV-QKD scheme. T: transmissivity, ε: excess noise, η: PD efficiency

$$I(A;B) = \log_2\left(\frac{v + \chi_{\text{total}}}{1 + \chi_{\text{total}}}\right), \tag{9.40}$$

where $v = v_A + 1$, with v_A being the Alice's source variance expressed in shot-noise units. We use χ_{total} to denote the total variance observed at the channel input defined below. From Chap. 8, we learned that the Holevo information between Bob and Eve, denoted as $\chi(B;E)$, is determined by

$$\chi(B;E) = g\left(\frac{\lambda_1 - 1}{2}\right) + g\left(\frac{\lambda_2 - 1}{2}\right) - g\left(\frac{\lambda_3 - 1}{2}\right) - g\left(\frac{\lambda_4 - 1}{2}\right), \tag{9.41}$$

where $g(x) = (x + 1)\log_2(x + 1) - x\log_2 x$ is the entropy of a thermal state with the mean number of photons being x, and the λ-parameters are defined below. Similar to Chap. 8 and Ref. [67], for discrete modulation (DM) schemes with correlation coefficient between in-phase and quadrature components denoted as Z_{DM}, we normalize the transmissivity T and redefine excess noise variance ε as follows:

$$T = \left(\frac{Z_{EPR}}{Z_{DM}}\right)^2 T', \quad \varepsilon = \left(\frac{Z_{DM}}{Z_{EPR}}\right)^2 (v_A + \varepsilon') - v_A, \quad Z_{EPR} = \sqrt{v^2 - 1}, \tag{9.42}$$

so that the corresponding correlation matrix between Alice and Bob becomes

$$\boldsymbol{\Sigma}_{AB_8} = \begin{bmatrix} (v_A + 1)\mathbf{1} & \sqrt{T'}Z_{EPR}\mathbf{Z} \\ \sqrt{T'}Z_{EPR}\mathbf{Z} & [T'(v_A + \varepsilon') + 1]\mathbf{1} \end{bmatrix}, \tag{9.43}$$

which is identical to the corresponding covariance matrix for Gaussian modulation, except for deferent transmissivity T' and excess noise variance ε' values. The symplectic eigenvalues for KK-receiver-based CV-QKD can be determined by

$$\lambda_{1,2} = \sqrt{\frac{1}{2}\left(A \pm \sqrt{A^2 - 4B}\right)}, \quad \lambda_{3,4} = \sqrt{\frac{1}{2}\left(C \pm \sqrt{C^2 - 4D}\right)}, \tag{9.44}$$

where A, B, C, and D parameters are defined by

$$\begin{aligned} &A = v^2(1 - 2T') + 2T' + T'^2(v + \chi_{\text{line}})^2, \quad B = T'^2(1 + v\chi_{\text{line}})^2, \\ &C = \frac{A\chi_{\text{KK}}^2 + B + 1 + 2\chi_{\text{KK}}\left[v\sqrt{B} + T(v + \chi_{\text{line}})\right] + 2T\left(v^2 - 1\right)}{T^2(v + \chi_{\text{total}})^2}, \quad D = \frac{\left(v + \chi_{\text{KK}}\sqrt{B}\right)^2}{T^2(v + \chi_{\text{total}})^2}, \\ &\chi_{\text{KK}} = (1 + v_{el} - \eta)/\eta, \quad \chi_{\text{total}} = 1/T' - 1 + \varepsilon'. \end{aligned} \tag{9.45}$$

The KK-receiver-based CV-QKD has been evaluated in [60, 61] and it has been shown that this scheme can outperform the eight-state heterodyne CV-QKD scheme. The KK-receiver-based CV-QKD has been further studied in the presence of atmospheric turbulence effects in [61]. By employing 10 wavelength channels with raw data rate of 6 GBd each, it has been shown that total SKR of 2.1 Gb/s can be achieved in a weak turbulence regime.

9.7 Concluding Remarks

This chapter has been devoted to the recently proposed both DV-QKD and CV-QKD schemes. The chapter starts with the description of Hong–Ou–Mandel effect and photonic Bell state measurements in Sect. 9.1. Both polarization state-based and time-bin-state-based BSMs have been introduced. After that, the BB84 and decoy-state protocols have been briefly revisited in Sect. 9.2. The next topic in the chapter has been devoted to the MDI-QKD protocols introduced in Sect. 9.3. Both polarization-based and time-phase-encoding-based MDI-QKD protocols have been described. Further, the twin-field (TF)-QKD protocols have been described in Sect. 9.4, capable of beating the Pirandola–Laurenza–Ottaviani–Banchi bound on a linear key rate. In Sect. 9.5, the floodlight CV-QKD protocol has been described, capable of achieving record secret-key rates. In Sect. 9.6, Kramers–Kronig (KK)-receiver-based CV-QKD scheme have been introduced, representing high-SKR scheme of low complexity.

References

1. Hong CK, Ou ZY, Mandel L (1987) Measurement of subpicosecond time intervals between two photons by interference. Phys Rev Lett 59:2044
2. Ou Z-YJ (2017) Quantum optics for experimentalists. World Scientific, New Jersey, London, Singapore
3. Knill E, Laflamme R, Milburn GJ (2001) A scheme for efficient quantum computation with linear optics. Nature 409:46–52
4. Zeilinger A (1999) Experiment and the foundations of quantum physics. In: Bederson B (ed) More things in heaven and earth. Springer, New York, NY, pp 482–498
5. Bennett CH, Brassard G, Crépeau C, Jozsa R, Peres A, Wootters WK (1993) Teleporting an unknown quantum state via dual classical and Einstein-Podolsky-Rosen channels. Phys Rev Lett 70(13):1895
6. Mattle K, Weinfurter H, Kwiat PG, Zeilinger A (1996) Dense coding in experimental quantum communication. Phys Rev Lett 76(25):4656
7. Żukowski M, Zeilinger A, Horne MA, Ekert AK (1993) "Event-ready-detectors" Bell experiment via entanglement swapping. Phys Rev Lett 71(26):4287
8. Sangouard N, Simon C, De Riedmatten H, Gisin N (2011) Quantum repeaters based on atomic ensembles and linear optics. Rev Mod Phys 83(1):33
9. Lo H-K, Curty M, Qi B (2012) Measurement-device-independent quantum key distribution. Physical Rev Lett 108(13):130503
10. Lim CCW, Portmann C, Tomamichel M, Renner R, Gisin N (2013) Device-independent quantum key distribution with local Bell test. Physical Rev X 3(3):031006
11. Yin H-L et al (2016) Measurement-device-independent quantum key distribution over a 404 km optical fiber. Phys Rev Lett 117:190501
12. Valivarthi R, Lucio-Martinez I, Rubenok A, Chan P, Marsili F, Verma VB, Shaw MD, Stern JA, Slater JA, Oblak D, Nam SW, Tittel W (2014) Efficient Bell state analyzer for time-bin qubits with fast-recovery WSi superconducting single photon detectors. Opt Express 22:24497–24506

13. Lütkenhaus N, Calsamiglia J, Suominen K-A (1999) Bell measurements for teleportation. Phys Rev A 59:3295–3300
14. Bennet CH, Brassard G (1984) Quantum cryptography: public key distribution and coin tossing. In: Proceedings IEEE international conference on computers, systems, and signal processing. Bangalore, India, pp 175–179
15. Bennett CH (1992) Quantum cryptography: uncertainty in the service of privacy. Science 257:752–753
16. Neilsen MA, Chuang IL (2010) Quantum computation and quantum information. Cambridge University Press, Cambridge
17. Van Assche G (2006) Quantum cryptography and secrete-key distillation. Cambridge University Press, Cambridge, New York
18. Djordjevic IB (2012) Quantum information processing and quantum error correction: an engineering approach. Elsevier/Academic Press, Amsterdam, Boston
19. Lo H-K, Ma X, Chen K (2005) Decoy state quantum key distribution. Phys Rev Lett 94:230504
20. Hwang W-Y (2003) Quantum key distribution with high loss: toward global secure communication. Phys Rev Lett 91:057901
21. Ma X, Fung C-HF, Dupuis F, Chen K, Tamaki K, Lo H-K (2006) Decoy-state quantum key distribution with two-way classical postprocessing. Phys Rev A 74:032330
22. Zhao Y, Qi B, Ma X, Lo H-K, Qian L (2006) Experimental quantum key distribution with decoy states. Phys Rev Lett 96:070502
23. Rosenberg D, Harrington JW, Rice PR, Hiskett PA, Peterson CG, Hughes RJ, Lita AE, Nam SW, Nordholt JE (2007) Long-distance decoy-state quantum key distribution in optical fiber. Phys Rev Lett 98(1):010503
24. Yuan ZL, Sharpe AW, Shields AJ (2007) Unconditionally secure one-way quantum key distribution using decoy pulses. Appl Phys Lett 90:011118
25. Hasegawa J, Hayashi M, Hiroshima T, Tanaka A, Tomita A (2007) Experimental decoy state quantum key distribution with unconditional security incorporating finite statistics. arXiv:0705.3081
26. Tsurumaru T, Soujaeff A, Takeuchi S (2008) Exact minimum and maximum of yield with a finite number of decoy light intensities. Phys Rev A 77:022319
27. Hayashi M (2007) General theory for decoy-state quantum key distribution with an arbitrary number of intensities. New J Phys 9:284
28. Zhao Y, Qi B, Ma X, Lo H, Qian L (2006) Simulation and implementation of decoy state quantum key distribution over 60 km TELECOM FIBER. In: Proceedings of 2006 IEEE international symposium on information theory. Seattle, WA, pp 2094–2098
29. Wang X-B (2013) Three-intensity decoy-state method for device-independent quantum key distribution with basis-dependent errors. Phys Rev A 87(1):012320
30. Sun X, Djordjevic IB, Neifeld MA (2016) Secret key rates and optimization of BB84 and decoy state protocols over time-varying free-space optical channels. IEEE Photonics J 8(3):7904713
31. Lucamarini M, Choi I, Ward MB, Dynes JF, Yuan ZL, Shields AJ (2015) Practical security bounds against the Trojan-Horse attack in quantum key distribution. Phys Rev X 5:031030
32. Kurtsiefer C, Zarda P, Mayer S, Weinfurter H (2001) The breakdown flash of silicon avalanche photodiodes—back door for eavesdropper attacks? J Mod Opt 48(13):2039–2047
33. Qi B, Fung C-HF, Lo H-K, Ma X (2006) Time-shift attack in practical quantum cryptosystems. Quantum Inf Comput 7(1):73–82. https://arxiv.org/abs/quant-ph/0512080
34. Lydersen L, Wiechers C, Wittmann S, Elser D, Skaar J, Makarov V (2010) Hacking commercial quantum cryptography systems by tailored bright illumination. Nat Photonics 4(10):686
35. Lütkenhaus N, Jahma M (2002) Quantum key distribution with realistic states: photon-number statistics in the photon-number splitting attack. New J Phys 4:44
36. Masanes L, Pironio S, Acín A (2011) Secure device-independent quantum key distribution with causally independent measurement devices. Nat Commun 2:238
37. Vazirani U, Vidick T (2014) Fully device-independent quantum key distribution. Phys Rev Lett 113(14):140501

38. Pironio S, Acin A, Brunner N, Nicolas Gisin N, Massar S, Scarani V (2009) Device-independent quantum key distribution secure against collective attacks. New J Phys 11(4):045021
39. Xu F, Curty M, Qi B, Lo H-K (2015) Measurement-device-independent quantum cryptography. IEEE J Sel Top Quantum Electron 21(3):148–158
40. Chan P, Slater JA, Lucio-Martinez I, Rubenok A, Tittel W (2014) Modeling a measurement-device-independent quantum key distribution system. Opt Express 22(11):12716–12736
41. Curty M, Xu F, Cui W, Lim CCW, Tamaki K, Lo H-K (2014) Finite-key analysis for measurement-device-independent quantum key distribution. Nat Commun 5:3732
42. Valivarthi R et al (2015) Measurement-device-independent quantum key distribution: from idea towards application. J Mod Opt 62(14):1141–1150
43. Djordjevic IB, Varrazza R, Hill M, Yu S (2004) Packet switching performance at 10 Gb/s Across a 4 × 4 optical crosspoint switch matrix. IEEE Photon Technol Lett 16:102–104
44. Varrazza R, Djordjevic IB, Yu S (2004) Active vertical-coupler based optical crosspoint switch matrix for optical packet-switching applications. IEEE/OSA J Lightw Technol 22:2034–2042
45. Tang Y-L et al (2016) Measurement-device independent quantum key distribution over untrustful metropolitan network. Phys Rev X 6(1):011024
46. Valivarthi R et al (2016) Quantum teleportation across a metropolitan fibre network. Nat Photonics 10:676–680
47. Djordjevic IB (2017) Advanced optical and wireless communications systems. Springer International Publishing, Switzerland
48. Pirandola S, Laurenza R, Ottaviani C, Banchi L (2017) Fundamental limits of repeaterless quantum communications. Nat Commun 8:15043
49. Duan L-M, Lukin M, Cirac JI, Zoller P (2001) Long-distance quantum communication with atomic ensembles and linear optics. Nature 414:413–418
50. Qiu J et al (2014) Quantum communications leap out of the lab. Nature 508:441–442
51. Lucamarini M, Yuan ZL, Dynes JF, Shields AJ (2018) Overcoming the rate–distance limit of quantum key distribution without quantum repeaters. Nature 557:400–403
52. Ma X, Zeng P, Zhou H (2018) Phase-matching quantum key distribution. Phys Rev X 8:031043
53. Tamaki K, Lo H-K, Wang W, Lucamarini M (2018) Information theoretic security of quantum key distribution overcoming the repeaterless secret key capacity bound. arXiv:1805.05511 [quant-ph]
54. Lin J, Lütkenhaus N (2018) Simple security analysis of phase-matching measurement-device-independent quantum key distribution. Phys Rev A 98:042332
55. Yin H-L, Fu Y (2019) Measurement-device-independent twin-field quantum key distribution. Sci Rep 9:3045
56. Tamura Y et al (2018) The First 0.14-dB/km loss optical fiber and its impact on submarine transmission. J Lightw Technol 36:44–49
57. Zhuang Q, Zhang Z, Dove J, Wong FNC, Shapiro JH (2016) Floodlight quantum key distribution: a practical route to gigabit-per-second secret-key rates. Phys Rev A 94:012322
58. Zhang Z, Zhuang Q, Wong FNC, Shapiro JH (2017) Floodlight quantum key distribution: demonstrating a framework for high-rate secure communication. Phys Rev A 95:012332
59. Zhang Z, Chen C, Zhuang Q, Wong FNC, Shapiro JH (2018) Experimental quantum key distribution at 1.3 gigabit-per-second secret-key rate over a 10 dB loss channel. Quantum Sci Technol 3:025007
60. Qu Z, Djordjevic IB (2018) Continuous variable-quantum key distribution based on Kramers-Kronig scheme. In: Proceedings of ECOC 2018, Paper Th1G.6, September 23–27, 2018. Rome, Italy
61. Qu Z, Djordjevic IB (2018) High-speed free-space optical continuous variable-quantum key distribution based on Kramers-Kronig Scheme. IEEE Photonics J 10(6):7600807
62. Mecozzi A, Antonelli C, Shtaif M (2016) Kramers-Kronig coherent receiver. Optica 3:1220–1227
63. Haykin S (2001) Communication systems, 4th edn. Wiley, Inc
64. Haykin S (2014) Digital communication systems. Wiley, Inc
65. Kronig RL (1926) On the theory of the dispersion of x-rays. J Opt Soc Am 12:547–557

66. Kramers HA (1927) La diffusion de la lumiere par les atomes. Atti Cong Intern Fis 2:545–557
67. Shen Y, Zou H, Tian L, Chen P, Yuan J (2010) Experimental study on discretely modulated continuous-variable quantum key distribution. Phys Rev A 82:022317

Chapter 10
Covert/Stealth/Low Probability of Detection Communications and QKD

Abstract This chapter is devoted to covert communications, also known as low probability of detection/intercept, as well as stealth communications, and how they can improve secret-key rate for QKD applications. The chapter starts with brief introduction to covert communications, followed by the description of their differences with respect to steganography. One of the key technologies to enable covert communication over wireless channels, the spread spectrum concept, is introduced next. After that, the rigorous treatment of covert communication over an additive white Gaussian noise channel is provided, and the square root law is derived. The importance of hypothesis testing in covert communications is discussed as well. The covert communication over the discrete memoryless channels is discussed after that. The next topic is related to different approaches to overcome the square root law, including the use of friendly jammer that varies the noise power so that the square root law can be overcome, and positive covert rate be achieved. The concept of effective secrecy is introduced next, a recent secrecy measure, whose definition includes both strong secrecy and stealth communication conditions. After that, the covert/stealth communication concept is applied to optical communication systems. We further describe how the covert concept can be applied to information reconciliation step in QKD to simultaneously improve secret-key rate and extend the transmission distance.

10.1 Introduction

Various encryption and physical-layer security (PLS) approaches, discussed in previous chapters, are able to protect the content of the message but are not able to protect the users' privacy from detecting the very existence of the message [1–5]. To solve this problem, the *covert communications' concept* has been introduced [1–5], also known as stealth or low probability of detection/intercept (LPD/LPI) communications, in which the encrypted channel is hidden within the public channel. On such a way, we will be able not only to protect the message but also to protect user's privacy by preventing the detection of transmission attempt. Possible applications include military applications, the social unrest when local or governmental authorities want

I. B. Djordjevic, *Physical-Layer Security and Quantum Key Distribution*,
https://doi.org/10.1007/978-3-030-27565-5_10

to shut down any communication link, in particular, encrypted links, and to prevent unauthorized snooping (this could be both governmental and nongovernmental).

The covert communication concept can find its root in the field of *steganography* [6–8], which is the first covert communication concept introduced by humans. However, the key difference is the fact that steganography is applied on an application layer, while the covert communication on a physical layer, similar to the spread spectrum techniques [9–15]. The key idea behind steganography is to hide the message to be transmitted in some object such as another message, image, video, or file. While the purpose of the cryptography is to protect the contents of a message alone, the purpose of the steganography is to conceal the fact that a secret message is being transmitted as well as to protect its contents. Another key difference between the covert communication and steganography is that in covert communication the secret message to be transmitted is subject to noise, while in steganography the practice of hiding the secret message is noiseless.

The covert communication concept was first introduced to wireless communications [1] and was later extended to optical communications [2]. In optical communications, the same concept comes under different names such as optical encryption [16, 17], optical steganography [18], and stealth communication [19].

10.2 Steganography Basics

The steganography represents a very ancient discipline, described by Herodotus in his book *The Histories* [20], around 440 B.C., in which he provided two examples of steganography. In modern digital steganography, the message is hidden in finite-length, finite-alphabet *covertext* objects [21]. The process of embedding the hidden messages in covertext generates the *stegotext*, which is changing the properties of the covertext. The warden Willie looks for these changes through the process known as a *stegoanalysis*. It is assumed that covertext is not available to Willie; however, it is assumed that Willie has the complete statistical model of covertext available. Now, providing that Alice also possesses the complete statistical model of the stegotext, when secret key in the order of n bits, denoted as $\Theta(n)$, is pre-shared with Bob before the embedding process takes place, as shown in [7], it is possible to hide $\Theta(n)$-bits in an n-symbol covertext before being detected by the warden Willie. On the other hand, also shown in [7], when the complete statistical model of the stegotext is not available to Alice, Alice can safely embed $\Theta\left(\sqrt{n}\log n\right)$ bits by modifying $\Theta\left(\sqrt{n}\right)$ symbols in covertext of length n, providing that Alice and Bob pre-share $\Theta\left(\sqrt{n}\log n\right)$ secret bits. Unfortunately, as $n \to \infty$, the steganography rate tends to zero since $\Theta\left(\sqrt{n}\log n\right)/n \underset{n\to\infty}{\to} 0$. This *square root low* problem can be overcome by the subset selection as shown in [8]. The key idea is to restrict the cover to the subset that grows linearly in n whose distribution is close to the altered data [8]. However, as mentioned in introduction, the covert communication is different from steganography in several key aspects [21], which prevents the direct application of steganographic theory to the covert communications. Two of these aspects are mentioned above already, namely, the stegotext is noiseless and of finite length. Further, during the embedding process Alice replaces the portions of the stegotext. Finally, if we want to use the steganography concepts described in [7] in covert communication, we would need to transmit whole covertext over noisy channel, which is not practical. In practice of covert communication, the secret message to be transmitted is hidden in ambient noise and as such is difficult to be detected by the warden Willie, which prevents the transmission once being detected.

10.3 Spread Spectrum Systems Fundamentals

In wireless communications, the covert communication can be enabled by employing the spread spectrum (SS) techniques [9–12]. Namely, the spread spectrum allows LPD transmission by spreading signal over a huge time-frequency space, such that signal power is below the noise floor, and thus impairing Willie's ability to discriminate between noise and information-bearing signal.

Signals belonging to the spread spectrum class must simultaneously satisfy the following *spreading* and *despreading criterions*. The generic spread spectrum digital communication system is illustrated in Fig. 10.1. The *spreading criterion* indicates that information-bearing signal of bandwidth B_D is spread in frequency domain, on a transmitter side, as shown in Fig. 10.1, with the help of the *spreading signal* so that the resulted signal spectrum of bandwidth B_{SS} is much larger than the bandwidth of the original information-bearing signal, that is, $B_{SS} \gg B_D$. On the other hand, the *despreading criterion* indicates that the recovery of information-bearing signal from noisy spread spectrum received signal is achieved by correlating the received signal with the synchronized replica of the spreading signal, as illustrated in Fig. 10.1.

The channel encoder and decoder, the basic ingredients of a digital communication system, are shown as well. If there is no any interference/jammer present, an exact copy (replica) of the original signal can be recovered on the receiver side, and the spread spectrum system performance is transparent to the spreading–despreading process. In the presence of the narrowband interference signal, introduced during transmission, during the despreading process, the bandwidth of the interference signal will be increased by the receiver side spreading signal. Since the filter that follows

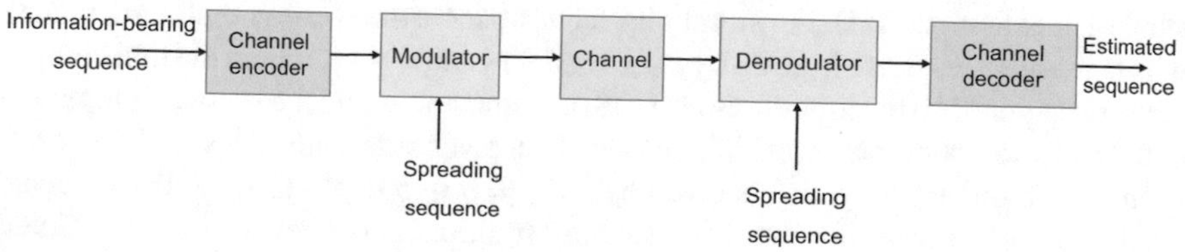

Fig. 10.1 The generic spread spectrum digital communication system. The spreading sequences on transmitter and receiver sides are synchronized

despreading matches the information-bearing signal, the interfering spread signal will be significantly filtered, and the output signal-to-interference ratio (SIR), often called signal-to-noise ratio (SNR), will be improved. This improvement in SIR due to spread spectrum processing is commonly referred to as the *processing gain* (PG), typically determined as $PG \approx B_{\text{SS}}/B_{\text{D}}$.

Depending on how the spread spectrum signals have been generated, the SS systems can be classified into two broad categories [9–12]: (i) *direct sequence spread spectrum (DSSS)* and (ii) *frequency-hopping spread spectrum (FH-SS)*.

In *DSSS* modulation process, the modulated signal $s(t)$ is multiplied by a wideband *spreading signal*, also known as spreading code/sequence, denoted by $s_c(t)$. Typically, the spreading code has a constant value $+1$ or -1 over a time duration T_c, which is much shorter than the symbol duration T_s. The bits in spreading code are commonly referred to as *chips*, while the duration of the bit in spreading code T_c is known as the *chip time*. Finally, the rate $1/T_c$ is called the *chip rate*. The bandwidth of spreading signal, denoted as B_c, is approximately $1/T_c$, so that the processing gain can be estimated as $PG \cong B_{\text{SS}}/B_D \cong (1/T_c)/(1/T_s) = T_s/T_c$. The multiplication of $s(t)$ and $s_c(t)$ in time domain, namely, $s(t)s_c(t)$, can be represented as the convolution in frequency-domain $S(f) * S_c(f)$, where $S(f)$ is the Fourier transform (FT) of $s(t)$, while $S_c(f)$ is the FT of $s_c(t)$. The corresponding bandwidth of the resulting signal is approximately equal to $B_c + B_D$. In the presence of additive white Gaussian noise (AWGN) process $z(t)$, the received SS signal is given by $r(t) = s(t)s_c(t) + z(t)$. After despreading, the resulting signal becomes

$$s_c(t)r(t) = s(t)\underbrace{s_c^2(t)}_{=1} + \underbrace{s_c(t)z(t)}_{z'(t)} = s(t) + z'(t). \tag{10.1}$$

Since $z'(t)$ still has the white Gaussian statistics, the despreading does not have effect on signal transmitted over AWGN. When the jamming (interference) signal is wideband, it can be approximated by an equivalent AWGN process, and the end effect will be reduction of overall SNR. On the other hand, when the jamming signal $j(t)$ is narrowband, in frequency domain after despreading it can be represented as $S_c(f) * J(f)$, where $J(f) = \text{FT}\{j(t)\}$. Therefore, the power of the jamming signal gets distributed over SS bandwidth, and the corresponding jamming signal power after the matched filter is approximately $B_{\text{SS}}/B_D \cong PG$ times lower. This indicates that SS approach is effective in dealing with narrowband interference. We will now show that SS approach is also effective in dealing with multipath fading-induced intersymbol interference (ISI). For simplicity, let us observe the two-ray multipath model, whose impulse response is given by $h(t) = a\delta(t) + b\delta(t - \tau)$, where the first component corresponds to the line-of-sight (LOS) component, while the second component is the reflected component, which arrives at receiver side with delay τ. The corresponding frequency response is given by $H(f) = a + b\exp(-j2\pi f\tau)$. The received signal, in the absence of noise, after two-ray multipath channel can be represented in frequency domain as $[S_c(f) * S(f)]H(f)$. The corresponding time-domain representation is given by $as_c(t)s(t) + bs_c(t - \tau)s(t - \tau)$. After despreading, we can write

$as(t) + bs_c(t)s_c(t - \tau)s(t - \tau)$. Since the reflected component is asynchronous with respect to the spreading signal, it will not be despread, and the power of reflected component after the matched filter will be reduced approximately PG times. This indicates that DSSS systems are effective in dealing with multipath fading-induced ISI.

In *FH-SS*, we occupy a large bandwidth by *randomly hopping* input data-modulated carrier form one frequency to another. Therefore, the spectrum is spread sequentially rather than instantaneously. The term sequentially is used to denote that frequency hops are arranged in pseudo-randomly ordered code sequence $s_c(t)$. The chip time T_c dictates the time elapsed between two neighboring frequency hops. The FH-SS is suitable to combine with M-ary frequency-shift keying (MFSK). When N different carrier frequencies are used in FH-SS systems, the bandwidth occupied is given by NB_D, where B_D is the bandwidth of original information-bearing signal, as before. The FH-SS systems are attractive for defense/military applications. Depending on the rate at which frequency hops occur, we can classify the FH-SS systems into two broad categories:

- The *slow-frequency hopping (SFH)* in which the symbol rate of MFSK $R_s = 1/T_s$ is an integer multiple of the hop rate $R_c = 1/T_c$. In other words, several symbols are transmitted with the same carrier frequency (for each frequency hop).
- The *fast-frequency hopping (FFH)* in which the hop rate R_c is an integer multiple of the MFSK symbol rate R_s. In other words, the carrier frequency gets changed several times during the transmission of the same symbol. Clearly, in the multipath fading environment, FFH scheme provides inherent frequency diversity.

In the presence of AWGN, the FH does not have influence on AWGN performance. Let us consider the narrowband interference at carrier frequency f_n of bandwidth $B = B_D$. When the FH frequency coincides with the interference carrier, the particular symbol transmitted will be affected by the interference. However, this occurs $1/N$ fraction of time. The power of interference is effectively reduced approximately $1/N$ times. However, compared to DSSS systems, where the power of narrowband interference is reduced all the time, the FH-SS systems are affected by the full power of the jammer a fraction of the time. In FFH systems, the interference affects the portion of symbol only, and this problem can be solved by coding, to correct for occasional errors. On the other hand, in SFH systems, many symbols are affected by the interference, so that the coding with interleaving is required to solve this problem.

Regarding the influence of multipath fading-induced ISI, for two-ray channel, on FH-SS systems, whenever $\tau > T_{\text{coh}}$, T_{coh} is the coherence time, the reflected component will exhibit a different carrier frequency than LOS component and will not affect the detection of LOS component. However, when $\tau < T_{\text{coh}}$ the FFH system will exhibit the flat fading, while the SFH system will exhibit either slowly varying flat fading for interference bandwidth $B < 1/\tau$ or slowly varying frequency-selective fading for $B > 1/\tau$.

10.4 Covert Communication Fundamentals

Let us first determine what is the required signal power for the communication process not to be detectable. For this purpose, let us observe the model shown in Fig. 10.2, which corresponds to free-space RF transmission. In this model, the Alice–Bob's (main) channel is modeled as a zero-mean additive white Gaussian noise (AWGN) of variance σ_B^2, that is, $\mathcal{N}(0, \sigma_B^2)$. On the other hand, the Alice–Willie's channel is also modeled as a zero-mean AWGN, but of variance σ_W^2, that is, $\mathcal{N}(0, \sigma_W^2)$. Willie is assumed to be listening to the channel for any communication that occurs between Alice and Bob. If he determines that Alice has transmitted a message to Bob, Willie shuts down the channel and communication is no longer possible.

In this scenario, Willie only needs to use a radiometer to determine the presence of a signal [1], which measures the average detected power by $S = zz^{\mathrm{T}}/n$, where $z = [z_1 \ldots z_n]$ and T denotes the transposition operation. Willie selects the threshold t_{sh} and applies the following decision rule:

$$S = \frac{zz^T}{n} \underset{H_0}{\overset{H_1}{\gtrless}} \sigma_W^2 + t_{\mathrm{sh}}, \tag{10.2}$$

where H_0 is the null hypothesis (only noise observations are present), while H_1 is the alternate hypothesis (assumption that Alice transmits). The probability that Willie fails to detect a message that is transmitted (the *missed detection probability*) is denoted as P_{MD}, while the probability of a *false alarm* by Willie when no message is transmitted is denoted as P_{FA}. The values of these probabilities will depend on the detection threshold, t_{sh}, that Willie chooses. If the threshold t_{sh} is too large, P_{MD} will increase due to the transmitted signal being below it. However, if the threshold t_{sh} is chosen too small, P_{FA} will increase as the noise power is greater than the threshold. It is straightforward to show that the mean and variance of S

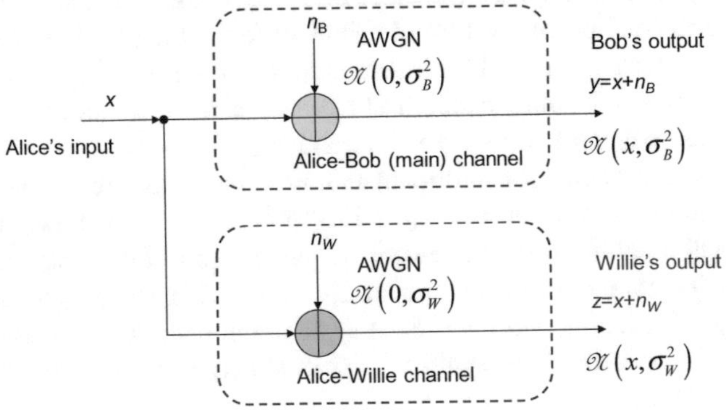

Fig. 10.2 The covert communication scenario assuming that the channel model is additive white noise Gaussian noise (AWGN)

when H_0 is true are given by $E(S) = \sigma_W^2$, $\mathrm{Var}(S) = 2\sigma_W^4/n$. The total power that Willie will observe in n channel observations will be approximately $n\sigma_W^2$, while the observations of total power will be with high probability within the interval $\left(n\sigma_W^2 - \underbrace{cn\sqrt{\sigma_W^4/n}}_{c\sigma_W^2\sqrt{n}},\ n\sigma_W^2 + c\sigma_W^2\sqrt{n}\right)$, where c is a positive constant. Therefore, the total power that Alice can transmit in n channel uses before Willie becomes suspicions is limited to $\mathbb{O}(\sigma_W^2\sqrt{n})$. In other words, the Alice power per single channel use should be $\mathbb{O}(\sigma_W^2\sqrt{n}/n) = \mathbb{O}(\sigma_W^2/\sqrt{n})$. From Shannon's capacity theorem, we can conclude that Alice can maximally transmit

$$n\log\left(1 + \frac{\sigma_W^2/\sqrt{n}}{\sigma_B^2}\right) \overset{\log(1+x)\sim x}{\widehat{\simeq}} \mathbb{O}\left(\frac{\sigma_W^2}{\sigma_B^2}n\frac{1}{\sqrt{n}}\right) = \mathbb{O}\left(\frac{\sigma_W^2}{\sigma_B^2}\sqrt{n}\right), \tag{10.3}$$

covert bits per single channel use, which represents the famous *square root law for covert communications*. For more rigorous derivation, an interested reader is referred to Ref. [1]. The covert rate for square root law, clearly, tends to zero, given that $\lim_{n\to\infty}\mathbb{O}(\sqrt{n})/n = 0$. Compared to the steganographic square root law, there is an additional $\log n$ term, which appears in steganography because the steganographic channel is noiseless. Based on Eq. (10.3), we conclude that Alice and Bob can reliably exchange $\mathbb{O}(\sigma_W^2\sqrt{n}/\sigma_B^2)$ covert bits, by employing the appropriate error correction code, typically known to the Willie, providing that they pre-share $\mathbb{O}(\sqrt{n}\log n)$ bits [1]. This pre-shared secret is related to the positions of covert bits in n channel uses. This subset $\mathbb{S}$ of n channel uses can be generated by flipping n times the biased random coin with probability of heads being $\mathbb{O}(1/\sqrt{n})$. The particular channel use is chosen if the result of coin flipping is the heads. Given that the number of channel uses is n, the cardinality of subset $\mathbb{S}$ will be $|\mathbb{S}| = n\mathbb{O}(1/\sqrt{n}) = \mathbb{O}(\sqrt{n})$. Since we need $\log n$ bits to represent each location, we conclude that we need to pre-share $\mathbb{O}(\sqrt{n}\log n)$ bits. The random one-time pad of size $|\mathbb{S}|$ should also be used to prevent Willie from exploiting the ECC structure to detect the Alice activity [2]. In incoming subsection, we discuss the covert communication concept from the *hypothesis testing* point of view [22–24].

10.4.1 Hypothesis Testing and Covert Communication

The Willie's detection model, from *hypothesis testing* point of view, is illustrated in Fig. 10.3.

Willie classifies all observations on Alice's activities into null hypothesis (H_0) when only noise is present and alternative hypothesis (H_1) when Alice transmits, and her signal is corrupted by noise. Willie's probability of error can be defined as

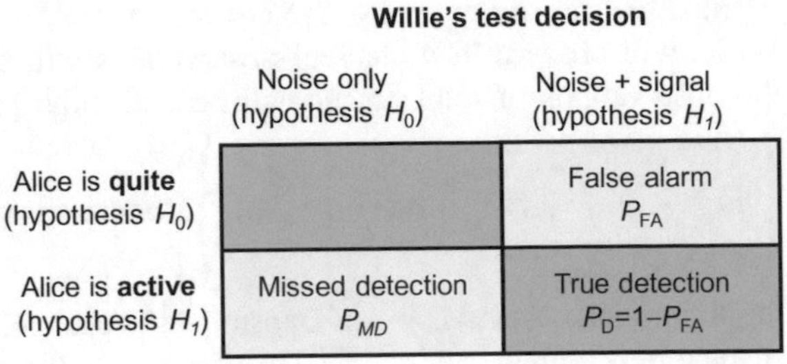
Willie's test decision
Noise only (hypothesis H_0)
Noise + signal (hypothesis H_1)
Alice is **quite** (hypothesis H_0)
Alice is **active** (hypothesis H_1)
False alarm P_{FA}
Missed detection P_{MD}
True detection P_D=1−P_{FA}

Fig. 10.3 The Willie's detection model from hypothesis testing point of view

$$P_e^{(W)} = \frac{1}{2}(P_{FA} + P_{MD}) \leq \frac{1}{2}. \tag{10.4}$$

Alice adjusts the transmit power to ensure that Willie's probability is limited by

$$P_e^{(W)} \geq \frac{1}{2} - \varepsilon, \forall \varepsilon > 0, \tag{10.5}$$

which is known as the *covertness criterion*. Alternatively, the covertness criterion can be expressed as $P_{\text{MD}} + P_{\text{FA}} \geq 1 - \zeta$, for any $\zeta > 0$ [1].

Clearly, the covert communication scenario is similar to *radar detection* problem [22, 23], so that the *Neyman–Pearson criterion* is applicable here, which does not require the knowledge of the prior probabilities for two hypotheses. In this scenario, we fix the false alarm probability to as large as we can tolerate, and then we minimize the missed detection probability; in other words, we can write [22]

$$\min_{D(z)} P(D_0|H_1),\ P(D_1|H_0) \leq P_{FA}, \tag{10.6}$$

where $D(z)$ represents the Willie's decision rule used to process z. Alternatively, we can instead maximize the detection probability $P_{\text{D}} = 1 - P_{\text{MD}} = P(D_1|H_1) = 1 - P(D_0|H_1)$, while maintaining the false alarm probability at prescribed level P_{FA}. So we have to solve the following problem:

$$\max_{D(z)} P(D_1|H_1) = \max \int_{R_1} w_1(z)dz,\ P(D_1|H_0) = \int_{R_1} w_0(z)dz \leq P_{FA}, \tag{10.7}$$

by properly choosing the region R_1 of z-space, in which we decide D_1. We use $w_0(z)$ and $w_1(z)$ to denote the probability density functions of z conditioned on H_0 and H_1, respectively. When the decisions are made based on one observation, we can employ the following decision rule:

$$\Lambda(z) = \frac{w_1(z)}{w_0(z)} \underset{H_0}{\overset{H_1}{\gtrless}} \Lambda_0, \tag{10.8}$$

where $\Lambda(z)$ is the likelihood ratio, while the Λ_0 is the decision threshold to be determined. The false alarm probability will be then

$$P(D_1|H_0) = \int_{\Lambda_0}^{\infty} w_0(\Lambda)d\Lambda \leq P_{FA}, \tag{10.9}$$

where the distribution $w_0(\Lambda)$ is determined from the condition $w_0(z)\mathrm{d}z = \mathrm{w}_0(\Lambda)\mathrm{d}\Lambda$. The threshold is determined from maximum tolerable false alarm probability P_{FA}, by solving Eq. (10.9) with respect to Λ_0, for the equality sign. The detection probability will be then

$$P_D = 1 - P_{MD} = P(D_1|H_1) = \int_{\Lambda_0}^{\infty} w_1(\Lambda)d\Lambda, \tag{10.10}$$

where the distribution $w_1(\Lambda)$ is determined from the condition $w_1(z)\mathrm{d}z = \mathrm{w}_1(\Lambda)\mathrm{d}\Lambda$. The Willie's receiver operating characteristic (ROC) is illustrated in Fig. 10.4. Ideally, Willie would like to ensure that probability of detection P_{D} tends to 1, while false alarm probability tends to zero. However, that is not possible in practice, given that when $P_{\mathrm{FA}} = 0$, the detection probability will be zero as well. If the prior probability of using the channel was 0.5, then $P_{\mathrm{D}} = P_{\mathrm{FA}}$, which corresponds to the line of slope 1 and represents an ideal case for Alice. However, as discussed above in practice the Alice-to-Bob channel is used with certain probability. What Alice can do is to limit Willie's probability of error to $1/2 - \varepsilon$, by properly selecting the distribution $w_1(z)$ to use during transmission of covert symbols, as illustrated in Fig. 10.4. By employing this idea, as shown in [1], Alice can reliably transmit $\Theta(n^{1/2})$ bits in n channel uses over the AWGN Alice-to-Bob channel, while maintaining the probability of error at Willie's detector at $1/2 - \varepsilon$, for any $\varepsilon > 0$. (Of course, this claim is valid under assumption that Alice knows at least lower bound of Willie's variance.)

For the optimum test that minimizes the Willie's probability of error, the following is valid [24]:

Fig. 10.4 The Willie's receiver operating characteristic (ROC)

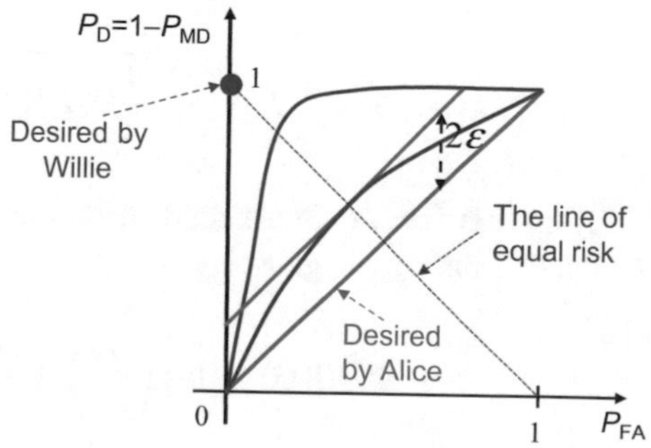

$$P_e^{(W)} = \frac{1}{2}(P_{FA} + P_D) = \frac{1}{2} - \frac{1}{2}\mathcal{V}(P_0, P_1), \tag{10.11}$$

where $\mathcal{V}$ is the *total variational distance* between distributions P_0 and P_1, with corresponding density functions w_0 and w_1, which is defined as

$$\mathcal{V}(P_0, P_1) = \frac{1}{2}\|w_1(z) - w_0(z)\|_1, \tag{10.12}$$

where $\|\cdot\|_1$ represents the $\mathcal{L}_1$-norm. The total variational distance is upper bounded by the Pinsker's inequality [25]:

$$\mathcal{V}(P_0, P_1) \le \sqrt{\frac{1}{2}\mathcal{D}(P_0||P_1)}, \tag{10.13}$$

where $\mathcal{D}(P_0||P_1)$ denotes the *relative entropy*, also known as Kullback–Liebler (KL) divergence, defined as [25]

$$\mathcal{D}(P_0||P_1) = \int_{\mathcal{Z}} w_0(z)\log\left[\frac{w_0(z)}{w_1(z)}\right]dz, \tag{10.14}$$

where $\mathcal{Z}$ is the support of $w_1(z)$. Further, let P^n denote the distribution of a sequence $\{z_i\}_{i=1}^n$, wherein each z_i generated from distribution P is i.i.d, then the relative entropy satisfies the following property [25]:

$$\mathcal{D}\left(P_0^n||P_1^n\right) = n\mathcal{D}\left(P_0||P_1\right). \tag{10.15}$$

When Alice does not transmit, Willie's samples will be from zero-mean Gaussian distribution $\mathcal{N}\left(0, \sigma_W^2\right)$, and we can denote the n samples generated from this distribution as P_0^n. On the other hand, when Alice transmits Willie does not know the exact distribution and can assume that Alice employs a zero-mean Gaussian source with variance $P_x + \sigma_W^2$, where P_x is average power of transmitted signal. We can denote the n samples generated from Gaussian source $\mathcal{N}\left(0, P_x + \sigma_W^2\right)$ as P_1^n. The corresponding variational distance will be upper limited by

$$\mathcal{V}\left(P_0^n, P_1^n\right) \le \sqrt{\frac{1}{2}\mathcal{D}\left(P_0^n||P_1^n\right)} = \sqrt{\frac{n}{2}\mathcal{D}\left(P_0||P_1\right)}. \tag{10.16}$$

Given that for two Gaussian distributions $p(z) = \mathcal{N}\left(\mu_1, \sigma_1^2\right), q(z) = \mathcal{N}\left(\mu_2, \sigma_2^2\right)$ the relative entropy is given by

$$\mathcal{D}(P||Q) = \log\left(\frac{\sigma_2}{\sigma_1}\right) + \frac{\sigma_1^2 + (\mu_1 - \mu_2)^2}{2\sigma_2^2} - \frac{1}{2}, \tag{10.17}$$

the total variational distance (10.16) can be upper bounded by

$$\mathbb{V}\left(P_0^n, P_1^n\right) \le \sqrt{\frac{n}{2}\left(\frac{1}{2}\log\left(\frac{\sigma_W^2 + P_x}{\sigma_W^2}\right) + \frac{1}{2}\frac{\sigma_W^2}{\sigma_W^2 + P_x} - \frac{1}{2}\right)}$$
$$= \frac{1}{2}\sqrt{n}\sqrt{\log\left(1 + \frac{P_x}{\sigma_W^2}\right) - \frac{\frac{P_x}{\sigma_W^2}}{1 + \frac{P_x}{\sigma_W^2}}}. \tag{10.18}$$

Based on previous discussion, we learned that the transmitted power must be $\Theta(n^{1/2})$, and by applying the Taylor expansion for small argument, we obtain

$$\mathbb{V}\left(P_0^n, P_1^n\right) \le \frac{P_x}{2\sigma_W^2}\sqrt{n}. \tag{10.19}$$

Now by limiting the total variation distance to 2ε (see Fig. 10.4), the Willie's probability of error can be made close to 0.5.

10.4.2 Covert Communication Over Discrete Memoryless Channels

The concept of the discrete memoryless channel (DMC) has already been introduced in Chap. 2, and when applied to the covert communication concept it can be represented as shown in Fig. 10.5. The input symbol x_0 represents the situation when Alice does not transmit. Other symbols y_i ($i = 1, 2, \ldots, N_B$) in Alice-to-Bob channel are prohibited by Willie. The Alice-to-Willie channel can also be represented by another DMC channel. In this channel, Willie's nonzero outputs are denoted by z_i ($i = 1, 2, \ldots, N_W$). We can interpret this DMC covert communication channel as obtained from AWGN model provided in Fig. 10.2 by quantization of the corresponding channel outputs. Both Alice-to-Bob and Alice-to-Willie DMCs are characterized by corresponding crossover (transition) $p(y_j/x_i)$ and $p(z_j/x_i)$ probabilities. The simplest version of this DMC covert communication channel model is obtained by setting $N_W = N_B = 2$, with corresponding Alice-to-Bob and Alice-to-Willie channels represented by binary symmetric channels (BSCs) with crossover (transition) probabilities being p_B and p_W, respectively, which is provided in Fig. 10.6. It has been shown in [26] that when Willie's channel is worse than Bob's channel, then no more than $\Theta(n^{1/2})$ covert bits can be reliably transmitted over n channel uses. Moreover, in this scenario, no pre-shared information between Alice and Bob is required.

To generalize the square root law to the DMCs, authors in [3] introduced the channel resolvability. The channel resolvability is related to the minimum entropy of the input required to generate the channel output with the desired distribution. This concept has also been applied to get better information-theoretic secrecy results in [27].

The authors in [28] also show that the square root law applies to wide range of DMCs for LPD applications. Notice that in this scenario authors consider only

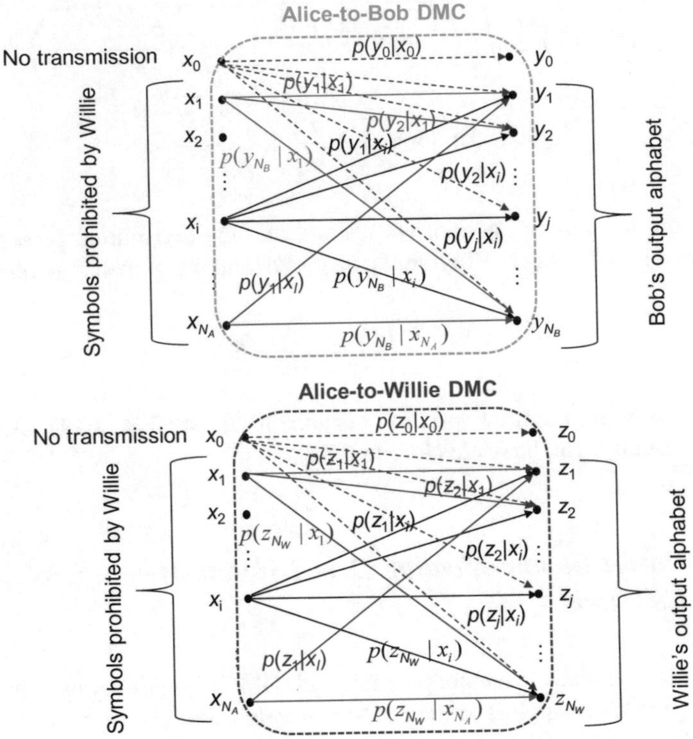

Fig. 10.5 The covert discrete memoryless channel. (Top) Alice-to-Bob DMC, (Bottom) Alice-to-Willie DMC

Fig. 10.6 The covert discrete memoryless channel with Alice-to-Bob and Alice-to-Willie channels being BSCs

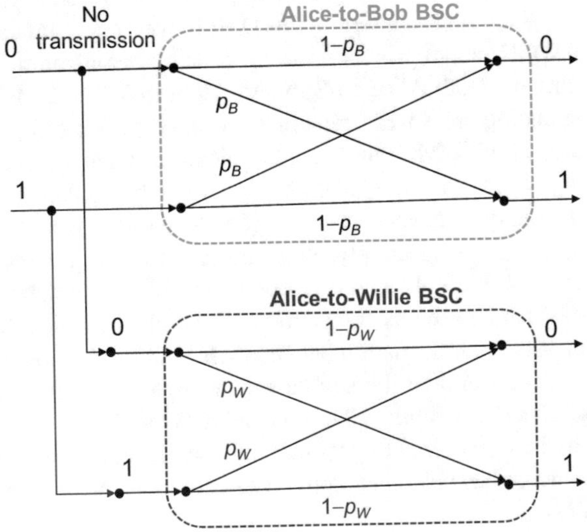

Alice-to-Willie DMC. They consider two cases, when symbol 0 is redundant (the transmitter is off) and when the symbol 0 is not redundant.

In derivation of the square root law, it is assumed that Willie knows when the transmission takes place. However, Alice and Bob can prearrange time interval for transmission during the days as well as the exact location of the covert symbols. On such a way, Willie will need to monitor the channel all day long. It has been shown in [29] that Willie's ignorance of the Alice transmission time can allow her to transmit additional information to Bob. This naturally yields us to the question, is there a way to transmit beyond the square root law, which is the subject of discussion in next section?

10.5 Positive-Rate Covert Communications

Lee and Baxley assumed in [30] that Willie has uncertainty about the channel variance and that he employs the power detector to detect Alice's transmission. In this scenario, when Alice transmits with small power P_{x}, the Willie's probability of error can be limited to $P_{\mathrm{e}}^{(\mathrm{W})} = 1/2 - \varepsilon$, $\varepsilon > 0$. The number of covert symbols for n channel uses can be estimated by $n\log\left(1 + P_x/\sigma_B^2\right) \approx n(P_x/\sigma_B^2) = \mathcal{O}(n)$. However, this scenario imposes the restriction on Willie, and as such might not be of practical importance. More realistic scenario will be to employ a friendly (unfirmed) jammer [4], who varies his noise variance every $k < n$ symbols. Willie cannot employ the empty slots to estimate the friendly jammer's power and thus a power detector will not be able to properly estimate the channel power, and Alice and Bob can transmit $\mathcal{O}(n)$ covert bits in n channel uses. Moreover, the authors in [4] have shown that for $k = n$ the power detector is an optimal detector. The friendly jammer should not vary the noise power too often, because Willie can average out the noise. On the other hand, the friendly jammer should not vary the noise power too slow because Willie can estimate the noise power properly in that case. It has also been shown in [31] that positive-rate covert communication is also possible when Willie is ignorant of the probabilistic structure of the noise channel.

10.6 Effective Secrecy

Wyner introduced the so-called *wiretap channel* [32], now also known as a *degraded wiretap channel model*, in which Eve's channel is degraded version of the Alice–Bob channel (main channel). The wiretap channel gets generalized and refined by Csiszár and Körner [33], and the corresponding model, now known as the *broadcast channel with confidential messages* (BCC), is provided in Fig. 10.7. The broadcast channel is assumed to be discrete and memoryless and characterized by input alphabet X, output alphabets Y and Z (corresponding to Bob and Eve, respectively), and transition PDF $f(yz|x)$. So, the channel itself is modeled by a joint PDF for Bob's and

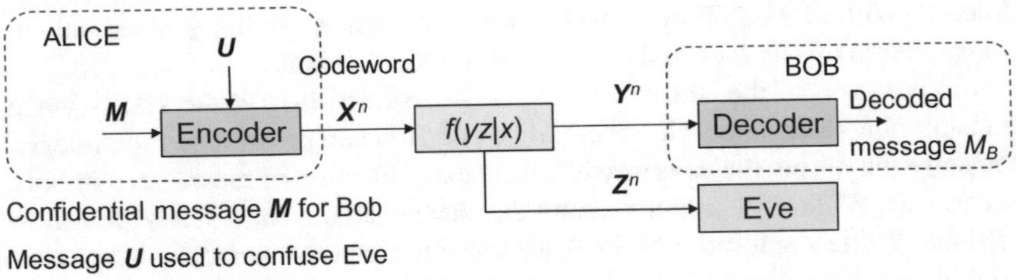

Fig. 10.7 The wiretap channel with confidential message $\boldsymbol{M}$ and random symbol $\boldsymbol{U}$, commonly referred to as the broadcast channel model with confidential messages (BCC)

Eve's observations, $p(\mathbf{y}\mathbf{z}|\mathbf{x})$, given the channel input. In both papers, the *secrecy* was measured as mutual information rate, that is,

$$\frac{1}{n} I\left(\boldsymbol{M}, \boldsymbol{Z}^n\right) \le \varsigma, \tag{10.20}$$

wherein the mutual information $I(\boldsymbol{M}, \boldsymbol{Z}^n)$ was measured between Alice message $\boldsymbol{M}$ and Eve's output string $\boldsymbol{Z}^n = Z_1, \ldots, Z_n$, which can be interpreted as the measure of leakage, while ς can be interpreted as the maximum tolerable leakage rate. As we have discussed in Chap. 4, this definition of secrecy represents the *weak secrecy condition*, namely, the number of bits leaked to Eve is equal to ςn. The *strong secrecy condition*, on the other hand, can be defined as [34–39]

$$I\left(\boldsymbol{M}, \boldsymbol{Z}^n\right) \le \zeta, \tag{10.21}$$

which is defined for any $\zeta > 0$.

It naturally arises the question how the covert/stealth communication and secrecy are related? First of all, the term *stealth* is used here to denote the scenario that the message-carrying signal is properly shaped to achieve the desired channel output distribution, similarly as in [38, 39]. The term covert is used to denote the scenario when the message-carrying signal is hidden in noise. Authors in [38, 39] introduced the new security measure, which they called the *effective secrecy*, and this definition includes both strong secrecy and stealth communication conditions. The effective secrecy, denoted as S_E, is defined in [38, 39] as the *relative entropy* (KL divergency) between the joint distribution of $\boldsymbol{M}\boldsymbol{Z}^n$, $P(\boldsymbol{M}\boldsymbol{Z}^n)$, and product of the distribution of $\boldsymbol{M}$, $P(\boldsymbol{M})$, and the distribution that Eve expects when the source is not transmitting a meaningful message Q_{Z^n}, that is, [38, 39]

$$S_E = \mathcal{D}\left(P_{MZ^n} || P_M Q_{Z^n}\right). \tag{10.22}$$

By using the definition expression for the relative entropy, Eq. (10.14), we can represent previous equations as [38, 39]

$$S_E = \mathfrak{D}\left(P_{MZ^n}||P_M Q_{Z^n}\right) = \sum_{(m,z^n)} P\left(m, z^n\right) \log\left[\frac{\overbrace{P\left(m, z^n\right)}^{P(z^n|m)P(m)}}{P(m)Q(z^n)} \frac{P(z^n)}{P(z^n)}\right]$$

$$= \underbrace{\sum\nolimits_{(m,z^n)} P\left(m, z^n\right) \log\left[\frac{P(z^n|m)}{P(z^n)}\right]}_{I(M,Z^n)} + \underbrace{\sum\nolimits_{(m,z^n)} P\left(m, z^n\right) \log\left[\frac{P(z^n)}{Q(z^n)}\right]}_{\mathfrak{D}(P_{Z^n}||Q_{Z^n})}$$

$$= I\left(M, Z^n\right) + \mathfrak{D}(P_{Z^n}||Q_{Z^n}), \tag{10.23}$$

where the first term is the *measure of strict secrecy*, while the second term is the *measure of stealth*. In equation above, the P_{Z^n} is the distribution that Eve observes, while Q_{Z^n} is the default distribution that Eve observes when Alice is not sending any useful information. Let us assume that Alice's default behavior is to select x^n from the memoryless distribution $Q^n_X(x^n)$. The default output distribution will be [39] $Q(z^n) = \sum_{x^n} P^n_{Z|X}(z^n|x^n)Q^n_X(x^n) = P^n_Z(z^n)$. By ensuring that effective secrecy is small enough, both $I(M, Z^n)$ and $\mathfrak{D}(P_{Z^n}||Q_{Z^n})$ will be small, and Eve will not be able to determine the confidential message M and at the same time will not be able to recognize whether Alice is emitting anything meaningful. We additionally require that the *probability of error* for Alice-to-Bob channel, defined as

$$P_e^{(n)} = \Pr[M_B \neq M], \tag{10.24}$$

is reasonably small. The encoding process on Alice side can be represented as the mapping $f(\cdot)$ that maps M and U to X^n, that is,

$$X^n = f(M, U), \tag{10.25}$$

wherein the confidential message M is uniformly chosen from $\{1, 2, \ldots, 2^{nR}\}$, where R is the code rate, while the auxiliary message U, sent to confuse the Eve, is uniformly chosen from $\left\{1, \ldots, 2^{nR_U}\right\}$.

The authors in [39] consider the scenario when the default distribution is memoryless so that we can write $Q(z^n) = Q^n_Z(z^n)$. We say that the code rate R is *achievable*, when the codeword length n is sufficiently long, while the probability of error and effective secrecy are limited as follows [39]:

$$P_e^{(n)} \leq \varepsilon_1, S_E = \mathfrak{D}\left(P_{MZ^n}||P_M Q_{Z^n}\right) \leq \varepsilon_2; \forall \varepsilon_1 > 0, \varepsilon_2 > 0. \tag{10.26}$$

The supremum of the set of achievable code rates can be referred to as the *effective secrecy capacity* denoted as C_S. If there does not exist Q_X such that $P_Z = Q_Z$ the C_S will be zero; otherwise, it can be determined by [39]

$$C_S = \max_{Q_{VX}:P_Z=Q_Z} [I(V; Y) - I(V; Z)], \tag{10.27}$$

where the maximization is performed over all joint Q_{VX}, wherein V, X, and YZ form the Markov chain V–X–YZ. Clearly, this equation has similarity to secrecy capacity discussed in Chap. 4.

Similar to the covert communications (see Sect. 10.4.1), we can apply the hypothesis testing to the stealth communication. It has been shown in [38, 39] that as long as the inequality,

$$R + R_U > I(X; Z), \tag{10.28}$$

is satisfied the best Eve's strategy to detect Alice's activity would be the guessing. Here we are concerned with the stealth term $\mathcal{D}(P_{Z^n}||Q_Z^n)$, and for every channel output we consider the following two hypotheses:

$$H_0 : Q_Z^n, \quad H_1 : P_Z^n. \tag{10.29}$$

Therefore, when the hypothesis H_1 is accepted Eve decides that Alice transmission is meaningful. We can define two types of errors, the *false alarm* probability $P_{\text{FA}} = \Pr(H_1$ is accepted | H_0 is true) and the *missed detection* probability $P_{\text{MD}} = \Pr(H_0$ is accepted | H_1 is true). We can use the Neyman–Pearson test in which the false alarm probability is kept at the prescribed value, while the P_{MD} is minimized. For every z^n, Eve can apply the following decision rule:

$$\Lambda(z^n) = \frac{P_Z^n(z^n)}{Q_Z^n(z^n)} \underset{H_0}{\overset{H_1}{\gtrless}} \Lambda_0, \tag{10.30}$$

which is illustrated in Fig. 10.8. So, the decision region D_1, corresponding to the case when H_1 is accepted, is defined as the set of points z^n for which the likelihood function is larger than the threshold and can be defined as (similar to [39])

$$D_1^n = \{z^n | \Lambda(z^n) \geq \Lambda_0\}. \tag{10.31}$$

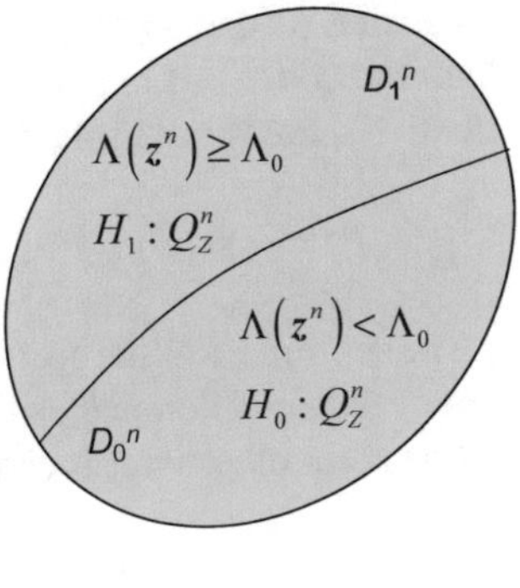

Fig. 10.8 The illustration of selection of the decision regions in hypothesis testing for stealth communications

Fig. 10.9 The illustration of the trade-off between false alarm and missed detection probabilities

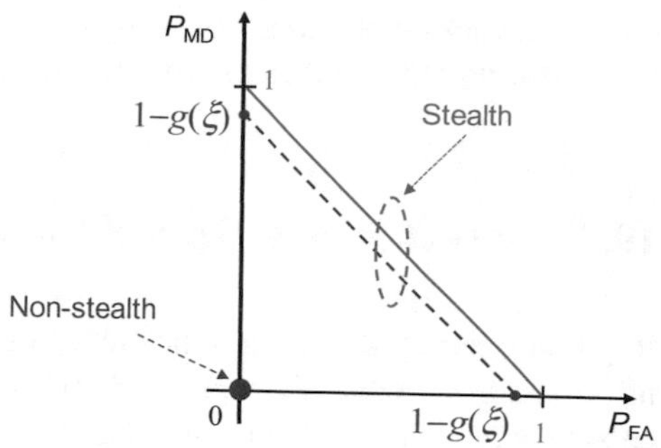

The corresponding false alarm and missed detection probabilities can be then determined by

$$\begin{aligned} P_{FA} &= Q_Z^n(D_1^n) = 1 - Q_Z^n(D_0^n), \\ P_{MD} &= P_Z^n(D_0^n). \end{aligned} \tag{10.32}$$

The authors in [38, 39] have shown that if $\mathfrak{D}(P_{Z^n}||Q_Z^n) \leq \xi,\ \xi > 0$, then the following is valid:

$$1 - g(\xi) \leq P_{FA} + P_{MD} = 1 + g(\xi),\ g(\xi) = \sqrt{2\xi \log 2}. \tag{10.33}$$

The trade-off between the false alarm and missed detection probabilities, when the inequality (10.28) is valid, is illustrated in Fig. 10.9. Namely, as both $n \to \infty$ and $\xi \to 0$, then the stealth measure tends to zero, while Eve's probability of error tends to 0.5; in other words, we can write

$$\begin{aligned} &\mathfrak{D}(P_{Z^n}||Q_Z^n) \to 0, \\ &P_e^{(E)} = \frac{1}{2}(P_{FA} + P_{MD}) \to \frac{1}{2}. \end{aligned} \tag{10.34}$$

If Eve does not want to tolerate the missed detection ($P_{\mathrm{MD}} \to 0$), she needs to pay high price since in that case the false error probability will tend to 1. On the other hand, if eve does not allow the false alarm ($P_{\mathrm{FA}} \to 0$), then her missed detection probability tends to 1. So the best strategy would be to preselect tolerable false alarm probability P_{FA} and determine the threshold from $P_{FA} = Q_Z^n(D_1^n)$. The optimum missed detection probability would be then $P_{\mathrm{MD,opt}} = 1 - P_{\mathrm{FA}}$. Clearly, in stealth communication case, we operate the system near the line $P_{\mathrm{FA}} + P_{\mathrm{MD}} = 1$, which corresponds to Eve's probability of error tending to 0.5.

If the stealth term $\mathfrak{D}(P_{Z^n}||Q_Z^n) > 0$ as $n \to \infty$, Eve detects Alice's action and her probability of error tends to zero, so the operating point for the non-stealth condition would be in the origin in Fig. 10.9.

Effective secrecy enables simultaneous security and stealth. Someone needs to be careful in selection of the scheme, because some schemes might achieve stealth but not secrecy and vice versa. For comparison of covert communication and physical-layer security (PLS) schemes, an interested reader is referred to [40].

10.7 Covert/Stealth Optical Communications

Regarding the optical communications, significant effort has been made to satisfy never-ending demands for higher data rates and to deal with the incoming bandwidth capacity crunch [18]. On the other hand, by taping out the portion of DWDM signal, the huge amount of data can be compromised [41–43]. Therefore, the security of future optical networks is becoming one of the key issues to be addressed sooner rather than later. To solve this problem, we recently proposed several PLS schemes suitable for optical communications [42, 43]. Various optical PLS schemes, as discussed above, are able to protect the content of the message, but are not able to protect the users' privacy from detecting the very existence of the message [1–4, 8].

To solve for this problem, in our recent paper [5], we proposed to use the Slepian sequence-based fiber Bragg gratings (Slepian-FBGs), described initially in [18] in a different context, to enable future positive-rate covert/stealth optical communications. The Slepian sequences [44], which are mutually orthogonal sequences for sequence length N and discrete bandwidth W, have been used here as impulse responses of corresponding FBGs to be employed as encoders for optical encryption and covert/stealth-like communication. A Slepian sequence of the jth order, denoted as $u_n^{(j)}(N, W)$, is defined by the real solution to discrete equation [44]:

$$\sum_{i=0}^{N-1} \frac{\sin 2\pi(n-i)}{\pi(n-i)} u_i^{(j)}(N, W) = \mu_j(N, W) u_n^{(j)}(N, W), \tag{10.35}$$

where i and n represent the particular samples of the Slepian sequence, with $n = 0, \pm1, \pm2, \ldots$ being the sequence sample index and $j = 0, 1, 2, \ldots$ denoting the order of the particular Slepian sequence. The $\mu_j(N, W)$ denote the ordered eigenvalues of the system of Eq. (10.35). The eigenvalues range from $0 < \mu_j \le 1$, corresponding to the concentration of each Slepian sequence within the desired time interval, with a value of 1 being the case where the sequence energy is entirely included inside the desired time interval. With the Slepian-FBGs, we can achieve the positive-rate covert optical communication in which $\mathcal{O}(n)$ covert symbols can be transmitted in n channel uses; therefore, dramatically increasing the covert communication rate.

The encoder scheme for *Slepian FBG-based optical covert communication*, shown in Fig. 10.10(top), is composed of two stages: *encryption stage*, the upper section of the encoder, and the *masking stage*, the lower section of the encoder, serving the role of a friendly jammer in wireless communications. The pulse laser is used in encryption stage, while the ASE noise source, such as EDFA together with variable optical attenuator (VOA), is used in the masking stage. The data sequence is by means of an electro-optical modulator (E/O MOD) such as Mach–Zehnder modulator

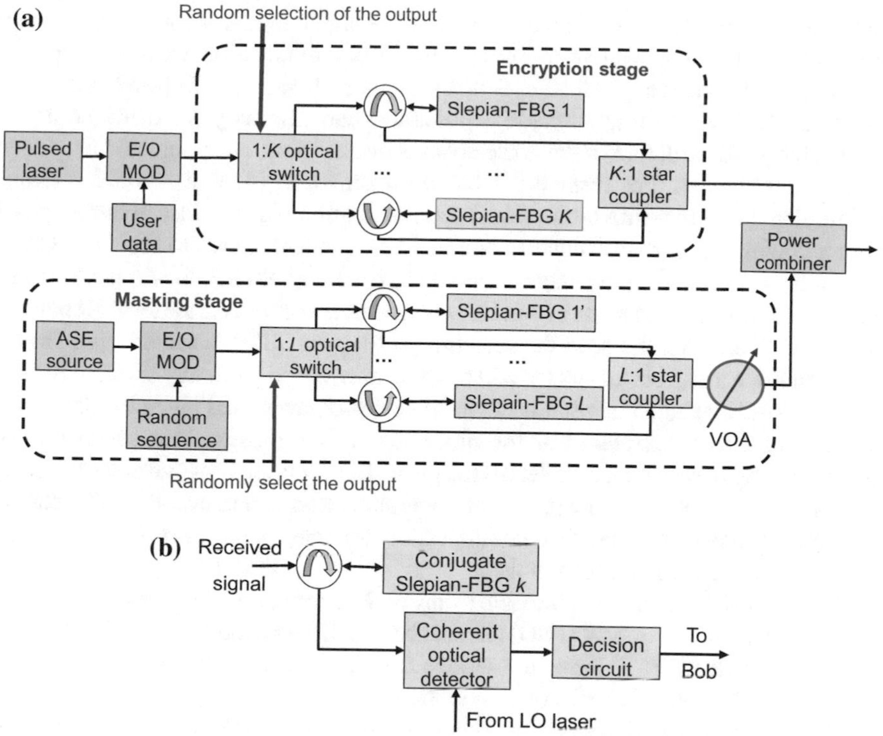

Fig. 10.10 Encoder/decoder configurations for Slepian-FBGs-based optical covert communications: **a** encryption scheme, **b** decryption scheme

(MZM), phase modulator, or optical I/Q modulator converted into optical domain. The 1:K optical switch is used to randomly select one out of K available Slepian-FBGs to be used as an encryption device. On the other hand, an arbitrary sequence, synchronized with data sequence generator, is used in masking stage to ensure that signal sequence and noise sequence are synchronized. The optical input to E/O MOD in masking stage is an additive white Gaussian noise signal generated by the ASE noise source. One of L available Slepian-FBGs is selected at random by control input of the 1:L optical switch. On such a way, the colored Gaussian noise is orthogonal to encrypted data sequence. Both Slepian-FBGs of encryption and masking sections are derived from the same class of Slepian sequences of cardinality $\geq L + K$. On the receiver side, as shown in Fig. 10.10(bottom), the matched conjugate Slepian-FBG has been used to decipher the transmitted sequence. When the transmitter and receiver use complementary Slepian-FBGs, strong autocorrelation peak gets generated at every signaling interval, while zero autocorrelation is obtained after non-matched Slepian-FBGs. Since the masking sequence has been generated by randomly selected Slepian-FBG generated from the same set of Slepian sequences as for the encryption stage of the transmitter, there is no need to use de-masking at the receiver side. Therefore, the masking is used only to hide any data structure to the adversary in both time and spectral domains. At the same time, the signal level is kept below the ASE noise level. To enable positive-rate covert optical communication, we have to operate the scheme shown in Fig. 10.10 as follows. Namely, from [4, 8, 31] we learned that positive-rate covert communication is possible when adversary is either ignorant of the probabilistic structure of the noise channel or does not have complete knowledge of the noise model. For instance, it has been shown in [4] that AWGN variance fluctuations yield to positive-rate covert communication. The key idea of our proposal is in addition to Alice–Bob covert communication model shown in Fig. 10.10(top), to employ also the masking stage shown in Fig. 10.10(bottom), to add the Slepian-FBGs colored Gaussian noise of variable average noise power, whose level is changed by VOA. Given that the seed for selection of Slepian-FBGs is secretly pre-shared between Alice and Bob, adversary will neither know which randomly selected FBG is used in masking stage nor what is the average noise power, and given his ignorance of the probabilistic structure of the noise source, the positive-rate covert optical communication will be possible. Also, given the ignorance of adversary's noise model, the positive-rate covert optical communication is achievable. This scheme can also be used as a stealth communication scheme, if the information-bearing signal is not hidden in ambient noise.

The proposed scheme is applicable to any optical communication system: (i) SMF based, (ii) free-space optical (FSO) channel based, (iii) few-mode fiber (FMF) based, (iv) few-core fiber (FCF) based to mention few. In the rest of this section, we focus on Slepian-FBGs-based FSO covert communications.

The encryption and decryption schemes for covert FSO communications are already provided in Fig. 10.10. The output of the encryption stage is with the help of an expanding telescope, as illustrated in Fig. 10.11, transmitted toward Bob's receiver. After compressing telescope, as shown in Fig. 10.11, the decryption takes place with the help of decryption stage provided in Fig. 10.10(bottom). During FSO propagation, the signal experiences the atmospheric turbulence and Mie scattering effects.

To model turbulence effects, we employ the gamma–gamma distribution of irradiance, which is valid in all turbulence regimes. The Rytov variance is used as a figure of merit to characterize the turbulence strength, since it takes refractive structure parameter C_n^2, FSO link length L, and wavelength λ into account, and it is defined as $\sigma_R^2 = 1.23\,C_n^2(2\pi/\lambda)^{7/6}L^{11/6}$. For $\sigma_R^2 < 1$, the system is affected with weak turbulence, while for $\sigma_R^2 > 1$ with strong turbulence. To demonstrate high potential of proposed Slepian-FBG-based FSO covert communications, we performed Monte Carlo simulations, with results being summarized in Fig. 10.12, in which

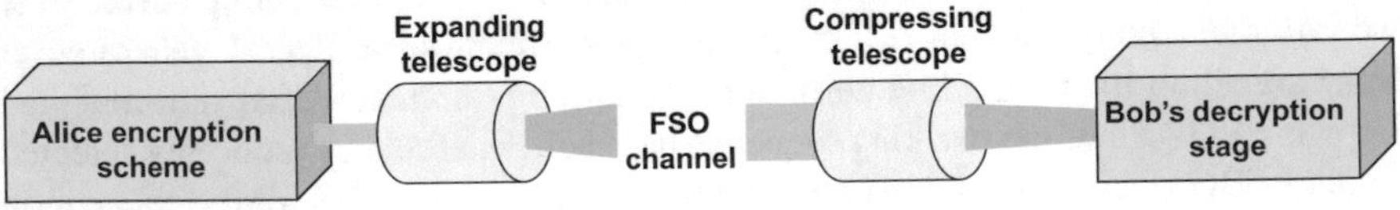

Fig. 10.11 Slepian-FBGs-based covert/stealth scheme suitable for use in FSO links

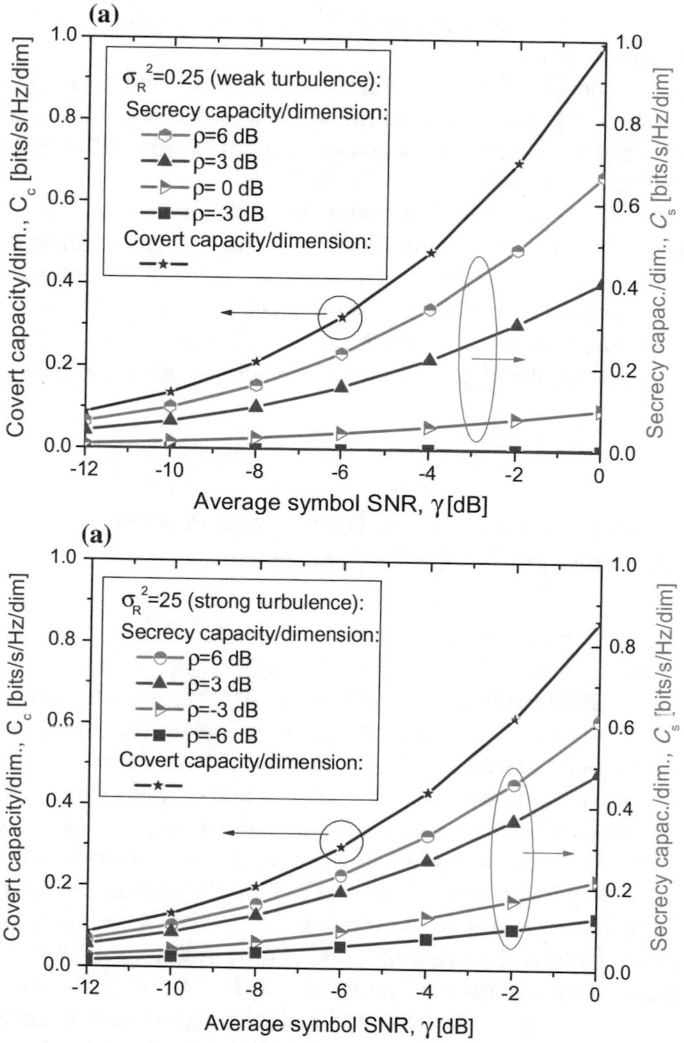

Fig. 10.12 Covert capacity (per dimension) of proposed scheme against secrecy capacities of corresponding PLS scheme over FSO links for weak (top) and strong (bottom) turbulence regimes (modified from [5]; ©IEEE 2018, reprinted with permission)

we compare the covert capacity per single dimension (spatial mode) against secrecy capacity/dimension of the PLS scheme proposed in [42]. For the PLS scheme, the ratio in average SNRs for main (Alice–Bob) and adversary channels, defined as $\rho = \bar{\gamma}_m / \bar{\gamma}_a$, is used as a parameter. Clearly, the Slepian-FBGs-based covert communication scheme provides positive covert capacity for all SNR values in main channel and significantly outperforms the corresponding PLS scheme. While PLS scheme benefits from atmospheric turbulence effects, the covert communication scheme is affected by turbulence effects, but still significantly outperforms the PLS scheme in strong turbulence regime.

The scenario considered here assumes that the warden Willie is passive so that the covert secret-key capacity is equal to the covert capacity of the channel. When warden Willie is active, the covert secret-key capacity is lower. For active warden scenario, an interested reader is referred to [45].

The authors in [2] considered quantum covert communication over an optical channel and concluded that at quantum level the square root law is still applicable. On the other hand, authors in [46] study quantum communication over optical channels, wherein the noise arises from either the environment or is generated by Alice. Authors also study the covert QKD, in which both raw key transmission and information reconciliation are performed covertly, which results in positive rate, but very low SKRs.

10.8 Covert Communication-Based Information Reconciliation for QKD Protocols

Here we describe how the covert communication can be efficiently applied to a QKD protocol for error reconciliation stage only [47–49]. As an illustration, let us modify the conventional BB84 protocol to include the use of covert/stealth classical communications instead of an authenticated public channel and the theoretical upper bound for secure-key rates is presented. The performance of the upper bounds versus the bit error rate is compared for the BB84 protocol using the standard and covert/stealth classical communication channel. A key performance metric of a QKD system is the secret-key rate (SKR). For the polarization entanglement BB84 protocol, the security proof has been presented by Koashi and Preskill [50]. After Alice and Bob have established the unsecured key by performing time sifting and basis reconciliation, the probability of a time slot being included in the unsecured key is given by P_{Key}. By exchanging a small portion of the unsecured key, the bit error rate (BER), δ, is estimated. Then using reverse error reconciliation and an appropriately selected error correction code, the errors between Alice and Bob are corrected. The parity bits generated by the error correction code are transmitted by Bob over a public channel with no errors and Alice corrects her key. Eve is capable of only listening to discussions on the public channel and she is able to gain information about the unsecured key. The amount of information revealed is determined by the error reconciliation inefficiency, $f(\delta) \geq 1$. To finally generate the secure key, privacy amplification is applied employing the universal hash functions. This leads to the generation rate of the secure key (more precisely *secrecy fraction*) given by

$$R_{SKR} = P_{Key}[1 - f(\delta)h_2(\delta) - h_2(\delta)], \tag{10.36}$$

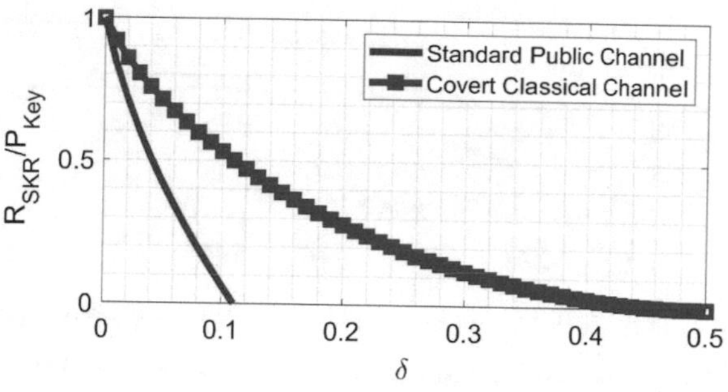

Fig. 10.13 Comparison of the SKR for the BB84 protocol when the parity bits are transmitted over the standard public channel and the covert classical channel

where $h_2(\delta)$ is the binary entropy function. When an error correction code with ideal error reconciliation efficiency ($f(\delta) = 1$) is used, the SKR scales as $R_{\text{SKR}} = P_{\text{Key}}\,[1 - 2h_2(\delta)]$. Using this SKR calculation, the maximum BER where a nonzero SKR can be achieved is 0.11, as illustrated in Fig. 10.13. By modifying the BB84 protocol to perform error reconciliation over classical covert/stealth communication channel, the SKR can be significantly increased. Instead of Bob transmitting the parity bits over a public channel, where Eve is capable of gaining information of the shared key, the parity bits are transmitted over a covert/stealth classical channel. The covert/stealth classical channel prevents Eve from gaining additional information from the error reconciliation phase, and therefore the SKR scales as

$$R_{SKR,\,Cov} = P_{Key}[1 - h_2(\delta)]. \tag{10.37}$$

To see the difference in the performance of the two protocols, the secure generation rate as a function of the BER is shown in Fig. 10.13, while using a public channel and an error correction code with ideal error reconciliation efficiency $R_{\text{SKR}} = 0$ when $\delta \geq 0.11$; now by employing the covert/stealth classical channel instead, the $R_{\text{SKR,Cov}}$ is nonzero for all values of $\delta < 0.5$. Based on Fig. 10.13 we can conclude that SKR for BB84 protocol can be significantly improved when the information reconciliation is performed covertly.

To further study how the application of covert communications during the error reconciliation phase of a QKD protocol affects the protocol, a polarization-entangled QKD system implementing the BB84 protocol is considered over a 30 km maritime channel. Details regarding the beam propagation modeling can be found in [51]. To characterize the channel, atmospheric data as a function of the height above the ocean surface was provided for the turbulence strength and channel attenuation by [52]. The atmospheric turbulence in the channel, which is causing beam wander, intensity fluctuations, and phase distortion to the wavefront, is characterized by the index of refraction structure parameter C_n^2. A single spatial mode QKD system is considered for the 10, 50, and 90% deciles [the 10% decile corresponds to the channel conditions that provided the top 0 to 10% of the power in the bucket for % of the time] of the channel. The apertures are assumed to be 75 cm in diameter and no adaptive optics is used. The normalized secure-key rates (SKR) and bit error rates (BER) are

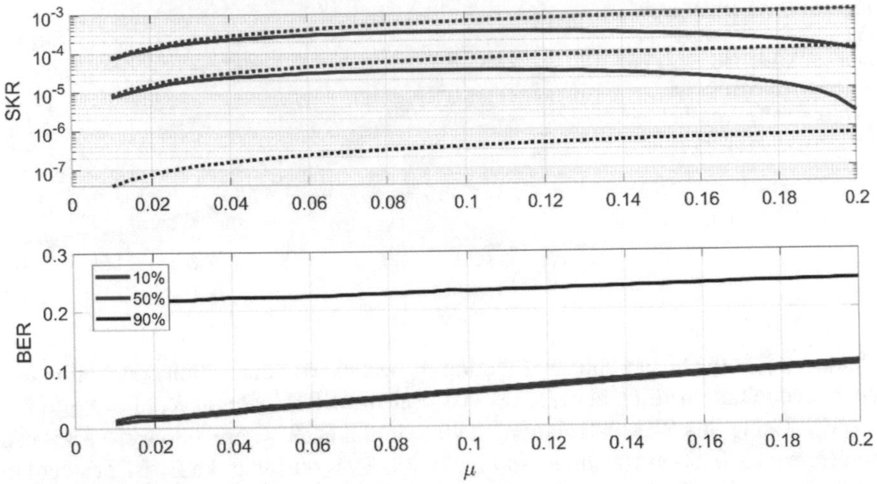

Fig. 10.14 Normalized SKR and BER for a polarization entanglement BB84 protocol, transmitting average μ-entangled photon pairs over a 30 km maritime channel with various strengths. The solid line represents the standard BB84 protocol, while the dashed lines represent when error reconciliation is performed over a covert communication channel

summarized in Fig. 10.14 as a function of the average number of entangled photon pair emitted by the source, μ.

In the 10 and 50% decile, a classical QKD protocol has a positive SKR, as the BER is less than 0.11. There is a slight increase in the BER for the 50% decile due to the increased channel loss. The 90% decile, however, is not able to generate any secure key for the classical QKD protocol, as the BER is greater than 0.2. When error reconciliation is performed over a covert channel, the SKR for all channel conditions can be seen to increase as μ increases. This allows a channel with very high loss or high noise levels to be able to generate a secure key.

10.9 Concluding Remarks

This chapter has been devoted to the covert communications, also known as low probability of detection/intercept, as well as stealth communications, and to describe how they can be used to improve secret-key rate for QKD applications. The chapter starts with brief introduction to covert communications in Sect. 10.1, followed by description of their differences with respect to steganography (Sect. 10.2). One of the key technologies to enable covert communication over wireless channels, the spread spectrum concept, has been introduced next in Sect. 10.3. The rigorous treatment of covert communication over an additive white Gaussian noise channel has been provided in Sect. 10.4, and the square root law is derived. The importance of hypothesis testing in covert communications has been discussed in Sect. 10.4.1. The covert communication over the discrete memoryless channels has been described in Sect. 10.4.2. Section 10.5 has been related to different approaches to overcome the square root law, including the use of friendly jammer that varies the noise power so that the square root law can be overcome, and positive covert rate be achieved. The concept of effective secrecy has been introduced in Sect. 10.6, a recent secrecy measure, whose definition includes both strong secrecy and stealth communication conditions. After that, in Sect. 10.7, the covert/stealth communication concept has been applied to optical communication systems. In Sect. 10.8, we have described how the covert concept can be applied to information reconciliation step in QKD to simultaneously improve secret-key rate and extend the transmission distance.

References

1. Bash BA, Goeckel D, Towsley D (2013) Limits of reliable communication with low probability of detection on AWGN channels. IEEE J Sel Areas Commun 31:1921–1930
2. Bash BA, Gheorghe AH, Patel M, Habif JL, Goeckel D, Towsley D, Guha S (2015) Quantum-secure covert communication on bosonic channels. Nat Commun 6:8626
3. Bloch MR (2016) Covert communication over noisy channels: a resolvability perspective. IEEE Trans Inform Theory 62(5):2334–2354
4. Sobers TV, Bash BA, Guha S, Towsley D, Goeckel D (2017) Covert communication in the presence of an uninformed jammer. IEEE Trans Wireless Commun 16:6193–6206
5. Djordjevic IB (2018) Slepian-FBGs-based optical covert communications. In: Proccedings of 2018 IEEE photonics conference (IPC), 30 Sept.–4 Oct. 2018, Reston, VA, USA
6. Petitcolas FAP, Anderson RJ, Kuhn MG (1999) Information hiding: a survey. Proc IEEE 87(7):1062–1078
7. Fridrich J (2009) Steganography in digital media: principles, algorithms, and applications. Cambridge University Press, Cambridge, New York
8. Craver S, Yu J (2010) Subset selection circumvents the square root law. In: Proceedings of SPIE 7541 media forensics and security II, p 754103
9. Pickholtz R, Schilling D, Milstein L (1982) Theory of spread-spectrum communications—a tutorial. IEEE Trans Commun 30(5):855–884
10. Verdú S (1998) Multiuser detection. Cambridge University Press, Cambridge, New York
11. Djordjevic IB (2017) Advanced optical and wireless communications systems. Springer International Publishing, Switzerland
12. Glisic S, Vucetic B (1997) Spread spectrum CDMA systems for wireless communications. Artech House Inc, Boston, London
13. Goldsmith A (2005) Wireless communications. Cambridge University Press, Cambridge, New York
14. Proakis JG (2001) Digital communications, 4th edn. McGraw-Hill, Boston, USA
15. Haykin S (2014) Digital communication systems. Wiley, Hoboken, NJ, USA
16. Torres P, Valente LCG, Carvalho MCR (2002) Security system for optical communication signals with fiber Bragg gratings. IEEE Trans Microw Theory Tech 50(1):13–16
17. Castro JM, Djordjevic IB, Geraghty DF (2006) Novel super-structured bragg gratings for optical encryption. IEEE/OSA J Lightwave Technol 24:1875–1885
18. Djordjevic IB, Saleh AH, Küppers F (2014) Design of DPSS based fiber bragg gratings and their application in all-optical encryption, OCDMA, optical steganography, and orthogonal-division multiplexing. Opt Express 22(9):10882–10897
19. Wu B, Shastri J, Mittal P, Tait AN, Prucnal PR (2015) Optical signal processing and stealth transmission for privacy. IEEE J Sel Top Signal Process 9:1185–1194

20. Herodotus, The histories, chap. 5—The fifth book entitled Terpsichore, 7—The seventh book entitled Polymnia. London, England: J. M. Dent & Sons, Ltd (1992). ISBN 0-460-87170-6
21. Bash BA, Goeckel D, Towsley D, Guha S (2015) Hiding information in noise: fundamental limits of covert wireless communication. IEEE Commun Mag 53(12):26–31
22. McDonough RN, Whalen AD (1995) Detection of signals in noise, 2nd edn. Academic Press, San Diego, New York, Boston
23. Drajic DB (2003) Introduction to statistical telecommunication theory. Academska Misao, Belgrade (in Serbian)
24. Lehmann E, Romano J (2005) Testing statistical hypotheses, 3rd edn. Springer, New York
25. Cover TM, Thomas JA (1991) Elements of information theory. Wiley, New York
26. Che PH, Bakshi M, Jaggi S (2013) Reliable deniable communication: hiding messages in noise. In: Proceedings of 2013 IEEE international symposium on information theory. Istanbul, Turkey, pp 2945–2949
27. Bloch M, Laneman J (2013) Strong secrecy from channel resolvability. IEEE Trans Info Theory 59(12):8077–8098
28. Wang L, Wornell GW, Zheng L (2015) Limits of low-probability-of-detection communication over a discrete memoryless channel. In: Proceedings of 2015 IEEE international symposium on information theory (ISIT). Hong Kong, pp 2525–2529
29. Bash BA, Goeckel D, Towsley D (2014) LPD communication when the warden does not know when. In: Proceedings of 2014 IEEE international symposium on information theory, Honolulu. Honolulu, HI, USA, pp 606–610
30. Lee S, Baxley RJ (2014) Achieving positive rate with undetectable communication over AWGN and Rayleigh channels. In: Proceedings of 2014 IEEE international conference on communications (ICC). Sydney, NSW, pp 780–785
31. Lee S, Baxley RJ, Weitnauer MA, Walkenhorst B (2015) Achieving undetectable communication. IEEE J Sel Top Signal Process 9(7):1195–1205
32. Wyner AD (1975) The wire-tap channel. Bell Syst Tech J 54(8):1355–1387
33. Csiszár I, Körner J (1978) Broadcast channels with confidential messages. IEEE Trans. Inf. Theory IT 24(3):339–348
34. Csiszár I (1996) Almost independence and secrecy capacity. Prob. Inf. Trans. 32(1):40–47
35. Maurer U, Wolf S (2000) Information-theoretic key agreement: from weak to strong secrecy for free. In: Advances in cryptology—Eurocrypt 2000. Lecture Notes in Computer Science. Springer-Verlag, pp 351–368
36. Bloch M, Barros J (2011) Physical-layer security: from information theory to security engineering. Cambridge University Press, Cambridge
37. Bloch M (2014) Fundamentals of physical layer security. In: Zhou X, Song L, Zhang Y (eds) Physical layer security in wireless communications. CRC Press, Boca Raton, London, New York, pp 1–16
38. Hou J, Kramer G (2014) Effective secrecy: reliability, confusion and stealth. In: 2014 IEEE International Symposium on Information Theory. Honolulu, HI, pp 601–605
39. Hou J, Kramer G, Bloch M (2017) Effective secrecy: reliability, confusion, and stealth. In: Schaefer R, Boche H, Khisti A, Poor H (eds) Information theoretic security and privacy of information systems. Cambridge University Press, Cambridge, pp 3–20
40. Forouzesh M, Azmi P, Mokari N, Wong KK (2018) Covert communications versus physical layer security. https://arxiv.org/abs/1803.06608
41. Sun X, Djordjevic IB (2016) Physical-layer security in orbital angular momentum multiplexing free-space optical communications. IEEE Photonics J 8(1):7901110
42. Djordjevic IB, Zhang S, Wang T (2017) Physical-layer security in optical communications enabled by Bessel modes. In: Proceedings of 2017 IEEE Photonics Conference (IPC 2017). Lake Buena Vista, FL, USA, p ThH1.3
43. Djordjevic IB (2017) OAM-based hybrid free-space optical-terahertz multidimensional coded modulation and physical-layer security. IEEE Photonics J 9(4):7905712
44. Slepian D (1978) Prolate spheroidal wave functions, Fourier analysis, and uncertainty-V: the discrete case. Bell Syst Tech J 57:1371–1430

45. Tahmasbi M, Bloch M (2019) Covert secret key generation with an active warden. https://arxiv.org/abs/1901.02044
46. Arrazola JM, Scarani V (2016) Covert quantum communication. Phys Rev Lett 117:250503
47. Gariano J, Djordjevic IB (2019) Employing covert communications-based information reconciliation and multiple spatial modes to polarization entanglement QKD. Opt Lett 44(1):687
48. Gariano J, Djordjevic IB (2018) Polarization entanglement quantum key distribution with covert classical communications. In: Proceedings of 2018 IEEE photonics conference (IPC 2018)
49. Gariano J, Djordjevic IB (2019) Covert communications-based information reconciliation for quantum key distribution protocols. In: Proceedings of 21th international conference of transparent optical network (ICTON 2019), 9–13 July 2019. Angers, France
50. Koashi M, Preskill J (2003) Secure quantum key distribution with an uncharacterized source. Phys Rev Lett 90:057902
51. Gariano J, Djordjevic IB (2018) Trade study of aperture size, adaptive optics and multiple spatial modes for a polarization entanglement QKD system over a 30 km maritime channel. Appl Opt 57(10):8451–8459
52. Data provided by D. P. SPAWAR Systems Center Pacific

Appendix

This chapter is mostly devoted to the abstract algebra fundamentals [1–17], needed in Chaps. 2, 4, 5, and 7. The chapter is organized as follows. In Sect. A.1, we introduce the concept of groups and provide their basic properties. In Sect. A.2, we introduce the concept of fields, while in Sect. A.3, the concept of vector spaces. Further, Sect. A.4 is devoted to algebra of finite fields. In Sect. A.5, we study the group acting on the set, which is relevant in quantum error correction. Further, the metric spaces are introduced in Sect. A.6. Finally, in Sect. A.7, the concept of Hilbert spaces is introduced relevant in quantum mechanics and quantum information processing.

A.1 Groups

Definition D1 A group is the set G that together with an operation, denoted by "+", satisfies the following axioms:

1. *Closure*: $\forall\, a, b \in \mathrm{G} \Rightarrow a + b \in \mathrm{G}$,
2. *Associative law*: $\forall\, a, b, c \in \mathrm{G} \Rightarrow a + (b + c) = (a + b) + \mathrm{c}$,
3. *Identity element*: $\exists\, e \in \mathrm{G}$ such that $a + e = e + a = a\ \forall\, a \in \mathrm{G}$, and
4. Inverse element: $\exists\, a^{-1} \in \mathrm{G}\ \forall\, a \in \mathrm{G}$, such that $a + a^{-1} = a^{-1} + a = e$.

We call a group ***Abelian*** or a ***commutative*** group, if the operation "+" is also commutative: $\forall\, a, b \in \mathrm{G} \Rightarrow a + b = b + a$.

Theorem T1 The identity element in a group is the unique one, and each element in the group has a unique inverse element.

Proof If the identity element e is not unique one, then there exists another one e': $e' = e' + e = e$, implying that e' and e are identical. In order to show that a group element a has a unique inverse a^{-1}, let us assume that it has another inverse element a_1^{-1}: $a_1^{-1} = a_1^{-1} + \mathrm{e} = a_1^{-1} + (a + a^{-1}) = (a_1^{-1} + a) + a^{-1} = e + a^{-1} = a^{-1}$, thus implying that the inverse element in the group is unique.

I. B. Djordjevic, *Physical-Layer Security and Quantum Key Distribution*,
https://doi.org/10.1007/978-3-030-27565-5

Examples:

- $F_2 = \{0, 1\}$ and "+" operation is, in fact, the modulo-2 addition, defined by $0 + 0 = 0$, $0 + 1 = 1$, $1 + 0 = 1$, $1 + 1 = 0$. The closure property is satisfied, 0 is identity element, 0 and 1 are their own inverses, and operation "+" is associative. The set F_2 with modulo-2 addition forms, therefore, a group.
- The set of integers form a group under the usual addition operation on which 0 is identity element, and for any integer n, $-n$ is its inverse.
- Consider the set of codewords of **binary linear (*N*, *K*) block code**. Any two codewords added per modulo-2 form another codeword (according to the definition of the linear block code). All-zeros codeword is identity element and each codeword is its own inverse. The associative property also holds. Therefore, all group properties are satisfied, and the set of codewords form a group, and this is the reason why this is called the *group code*.

The number of elements in the group is typically called the ***order*** of the group. If the group has a finite number of elements, it is called the ***finite*** group.

Definition D2 Let H be a subset of elements of group G. We call H subgroup of G if the elements of H themselves form the group under the same operation as that defined on elements from G.

In order to verify that the subset S is the subgroup, it is sufficient to check the closure property and that an inverse element exists for every element of subgroup.

Theorem T2 (Lagrange theorem) The order of a finite group is an integer multiple of any of its subgroup order.

Example: Consider the set Y of N-tuples (received words on a receiver side of a communication system), each element of the N-tuple being 0 or 1. It can easily be verified that elements of Y forms a group under modulo-2 addition. Consider a subset C of Y with elements being the codewords of a binary (N, K) block code. Then, C forms a subgroup of Y, and the order of group Y is divisible by the order of subgroup C.

There exist groups, whose all elements can be obtained as power of some element, say a. Such a group G is called the ***cyclic group***, and the corresponding element a is called the *generator* of the group. The cyclic group can be denoted by $G = \langle a \rangle$.

Example: Consider an element α of a finite group G. Let S be the set of elements $S = \{\alpha, \alpha^2, \alpha^3, \ldots, \alpha^i, \ldots\}$. Because G is finite, S must be finite as well, and therefore not all powers of α are distinct. There must be some l, m $(m > l)$ such that $\alpha^m = \alpha^l$ so that $\alpha^m \alpha^{-l} = \alpha^l \alpha^{-l} = 1$. Let k be the smallest such power of α for which $\alpha^k = 1$, meaning that $\alpha, \alpha^2, \ldots, \alpha^k$ are all distinct. We can verify now that set $S = \{\alpha, \alpha^2, \ldots, \alpha^k = 1\}$ is a subgroup of G. S contains the identity element, and for any element α^i, the element α^{k-i} is its inverse. Given that any two elements $\alpha^i, \alpha^j \in S$ the corresponding element obtained as their product, $\alpha^{i+j} \in S$ if $i + j \leq k$. If $i + j > k$ then $\alpha^{i+j} \cdot 1 = \alpha^{i+j} \cdot \alpha^{-k}$, and because $i + j - k \leq k$, the closure property is clearly satisfied. S is, therefore, the subgroup of group G. Given the definition of cyclic group, the subgroup S is also cyclic. The set of codewords of a cyclic (N, K) block code can be obtained as a cyclic subgroup of the group of all N-tuples.

Theorem T3 Let G be a group and $\{H_i | i \in I\}$ be a non-empty collection of subgroups with index set I. The intersection $\cap_{i \in I} H_i$ is a subgroup.

Definition D3 Let G be a group and X be a subset of G. Let $\{H_i | i \in I\}$ be the collection of subgroups of G that contain X. Then the intersection $\cap_{i \in I} H_i$ is called the subgroup of G generated by X and is denoted by $<X>$.

Theorem T4 Let G be a group and X non-empty subset of G with elements $\{x_i | i \in I\}$ (I is the index set). The subgroup of G generated by X, $<X>$, consists of all finite products of the x_i. The x_i elements are known as generators.

Definition D4 Let G be a group and H be a subgroup of G. For any element $a \in G$, the set $aH = \{ah | h \in H\}$ is called the left coset of H in G. Similarly, the set $Ha = \{ha | h \in H\}$ is called the right coset of H in G.

Theorem T5 Let G be a group and H be a subgroup of G. The collection of right cosets of H, $Ha = \{ha | h \in H\}$ forms a partition of G.

Instead of formal proof, we provide the following justification. Let us create the following table. In the first row, we list all elements of subgroup H, beginning with the identity element e. The second column is obtained by selecting an arbitrary element from G, not used in the first row, as the leading element of the second row. We then complete the second row by "multiplying" this element with all elements of the first row from the right. Out of not previously used elements, we arbitrary select an element as the leading element of the third row. We then complete the third row by multiplying this element with all elements of the first row from the right. We continue this procedure until exploiting all elements from G. The resulting table is as follows:

$h_1 = e$	h_2	…	h_{m-1}	h_m
g_2	$h_2 g_2$	…	$h_{m-1} g_2$	$h_m g_2$
g_3	$h_2 g_3$	…	$h_{m-1} g_3$	$h_m g_3$
…	…	…	…	…
g_n	$h_2 g_n$	…	$h_{m-1} g_n$	$h_m g_n$

Each row in this table represents a cosset, and the first element in each row is a cosset leader. The number of cossets of H in G is, in fact, the number of rows, and it is called the index of H in G, typically denoted by $[G:H]$. It follows from table above $|H| = m$, $[G:H] = n$ and $|G| = nm = [G:H]|H|$, which can be used as the proof of Lagrange's theorem. In other words, $[G:H] = |G|/|H|$.

Definition D5 Let G be a group and H be a subgroup of G. H is normal subgroup of G if it is invariant under conjugation, that is, $\forall\ h \in H$ and $g \in G$, the element $ghg^{-1} \in H$.

In other words, H is fixed under conjugation by the elements from G, namely, $gHg^{-1} = H$ for any $g \in G$. Therefore, the left and the right cosets of H in G coincide: $\forall\, g \in G,\, gH = Hg$.

Theorem T6 Let G be a group and H a normal subgroup of G. If $G|H$ denotes the set of cosets of H in G, then the set $G|H$ with coset multiplication forms a group, which is known as the quotient group of G by H.

The coset multiplication of aH and bH is defined as $aH * bH = abH$. It follows from Lagrange's theorem that $|G/H| = [G:H]$. This theorem can straightforwardly be proved by using the table above.

A.2 Fields

Definition D6 A ***field*** is a set of elements F with two operations, addition "+" and multiplication "·", such that

1. F is an Abelian group under addition operation, with 0 being the identity element.
2. The nonzero elements of F form an Abelian group under the multiplication operation, with 1 being the identity element.
3. The *multiplication operation is distributive over the addition operation*:

$$\forall a, b, c \in F \Rightarrow a \cdot (b + c) = a \cdot b + a \cdot c.$$

Examples:

- The set of real numbers, with operation $+$ as ordinary addition and operation $\cdot$ as ordinary multiplication, satisfies above three properties and it is therefore a field.
- The set consisting of two elements $\{0, 1\}$, with modulo-2 multiplication and addition given in the table below, constitutes a field known as **Galois field**, which is denoted by GF(2).

+	0	1
0	0	1
1	1	0

·	0	1
0	0	0
1	0	1

- The set of integers modulo-p with modulo-p addition and multiplication forms a field with p elements, which is denoted by GF(p), providing that p is a prime.
- For any q that is an integer power of prime number p ($q = p^m$, m—an integer), there exists a field with q elements, which is denoted as GF(q). (The arithmetic is not modulo-q arithmetic, except when $m = 1$.) GF(p^m) contains GF(p) as a subfield.

Addition and multiplication in GF(3) are defined as follows:

+	0	1	2
0	0	1	2
1	1	2	0
2	2	0	1

·	0	1	2
0	0	0	0
1	0	1	2
2	0	2	1

Addition and multiplication in GF(2^2):

+	0	1	2	3
0	0	1	2	3
1	1	0	3	2
2	2	3	0	1
3	3	2	1	0

·	0	1	2	3
0	0	0	0	0
1	0	1	2	3
2	0	2	3	1
3	0	3	1	2

Theorem T7 Let Z_p denote the set of integers $\{0, 1, \ldots, p-1\}$, with addition and multiplication defined as ordinary addition and multiplication modulo-p. Then Z_p is a field if and only if p is a prime.

A.3 Vector Spaces

Definition D7 Let V be a set of elements with a binary operation "+" and let F be a field. Further, let an operation "·"· be defined between the elements of V and the elements of F. Then V is said to be a ***vector space*** over F if the following conditions are satisfied $\forall\, a, b \in F$ and $\forall\, \boldsymbol{x}, \boldsymbol{y} \in V$:

1. V is an Abelian group under the addition operation.
2. $\forall\, a \in F$ and $\forall\, \boldsymbol{x} \in V$, then $a \cdot \boldsymbol{x} \in V$.
3. *Distributive law*:

$$a \cdot (\boldsymbol{x} + \boldsymbol{y}) = a \cdot \boldsymbol{x} + a \cdot \boldsymbol{y}$$
$$(a + b) \cdot \boldsymbol{x} = a \cdot \boldsymbol{x} + b \cdot \boldsymbol{x}.$$

4. *Associative law*:

$$(a \cdot b) \cdot \boldsymbol{x} = a \cdot (b \cdot \boldsymbol{x}).$$

If 1 denotes the identity element of F, then $1 \cdot \boldsymbol{x} = \boldsymbol{x}$. Let 0 denote the zero element (identity element under +) in F, and $\mathbf{0}$ the additive element in V, -1 the additive inverse of 1, the multiplicative identity in F. It can easily be shown that the following two properties hold:

$$0 \cdot \boldsymbol{x} = \boldsymbol{0}$$
$$\boldsymbol{x} + (-1) \cdot \boldsymbol{x} = \boldsymbol{0}.$$

Examples:

- Consider the set V, whose elements are n-tuples of the form $\boldsymbol{v} = (v_0, v_1, \ldots, v_{n-1})$, $v_i \in F$. Let us define the addition of any two n-tuples as another n-tuple obtained by componentwise addition and multiplication of n-tuple by an element from F. Then, V forms vector space over F, and it is commonly denoted by F^n. If $F = R$ (Rth field of real number), then R^n is called the Euclidean n-dimensional space.
- The set of n-tuples whose elements are from GF(2), again with componentwise addition and multiplication by an element from GF(2), forms a vector space over GF(2).
- Consider the set V of polynomials whose coefficients are from GF(q). Addition of two polynomials is the usual polynomial addition, addition being performed in GF(q). Let the field F be GF(q). Scalar multiplication of a polynomial by a field element from GF(q) corresponds to the multiplication of each polynomial coefficient by the field element, carried out in GF(q). V is then a vector space over GF(q).

Consider a set V that forms a vector space over a field F. Let $\boldsymbol{v}_1, \boldsymbol{v}_2, \ldots, \boldsymbol{v}_k$ be vectors from V, and $a_1, a_2, \ldots, a_k$ be field elements from F. The ***linear combination*** of the vectors $\boldsymbol{v}_1, \boldsymbol{v}_2, \ldots, \boldsymbol{v}_k$ is defined by

$$a_1\boldsymbol{v}_1 + a_2\boldsymbol{v}_2 + \cdots + a_k\boldsymbol{v}_k.$$

The set of vectors $\{\boldsymbol{v}_1, \boldsymbol{v}_2, \ldots, \boldsymbol{v}_k\}$ is said to be ***linearly independent*** if there does not exist a set of field elements $a_1, a_2, \ldots, a_k$, not all $a_i = 0$, such that

$$a_1\boldsymbol{v}_1 + a_2\boldsymbol{v}_2 + \cdots + a_k\boldsymbol{v}_k = \boldsymbol{0}.$$

Example: The vectors (0 0 1), (0 1 0), and (1 0 0) (from F^3) are linearly independent. However, the vectors (0 0 2), (1 1 0), and (2 2 1) are linearly dependent over GF(3) because they sum to zero vector.

Let V be a vector space and S be subset of the vectors in V. If S in itself a vector space over F under the same vector addition and scalar multiplication operations applicable to V and F, then S is said to ***a subspace*** of V.

Theorem T7 Let $\{\boldsymbol{v}_1, \boldsymbol{v}_2, \ldots, \boldsymbol{v}_k\}$ be a set of vectors from a vector space V over a field F. Then, the set consisting of all linear combinations of $\{\boldsymbol{v}_1, \boldsymbol{v}_2, \ldots, \boldsymbol{v}_k\}$ forms a vector space over F, and is, therefore, a subspace of V.

Example: Consider the vector space V over GF(2) given by the set {(0 0 0), (0 0 1), (0 1 0), (0 1 1), (1 0 0), (1 0 1), (1 1 0), (1 1 1)}. The subset S = {(0 0 0), (1 0 0), (0 1 0), (1 1 0)} is a subspace of V over GF(2).

Example: Consider the vector space V over GF(2) given by the set {(0 0 0), (0 0 1), (0 1 0), (0 1 1), (1 0 0), (1 0 1), (1 1 0), (1 1 1)}. For the subset B = {(0 1 0), (1 0

0)}, the set of all linear combinations is given by $S = \{(0\,0\,0), (1\,0\,0), (0\,1\,0), (1\,1\,0)\}$, and forms a subspace of V over F. The set of vectors B is said to ***span*** S.

Definition D8 A ***basis*** of a vector space V is set of linearly independent vectors that spans the space. The number of vectors in a basis is called the ***dimension*** of the vector space.

In example above, the set $\{(0\,0\,1), (0\,1\,0), (1\,0\,0)\}$ is the basis of vector space V, and has dimension 3.

A.4 Algebra of Finite Fields

A ***ring*** is a set of elements R with two operations, addition "+" and multiplication "·", such that

(i) R is an Abelian group under addition operation, (ii) multiplication operation is associative, and (iii) multiplication is associative over addition.

The quantity a is said to be congruent to b to modulus n, denoted as $a \equiv b \pmod n$, if $a - b$ is divisible by n. If $x \equiv a \pmod n$, then a is called a residue to x to modulus n. A class of residues to modulus n is the class of all integers congruent to a given residue (mod n), and every member of the class is called a representative of the class. There are n classes, represented by (0), (1), (2), …, ($n - 1$), and the representative of these classes are called a complete system of incongruent residues to modulus n. If i and j are two members of a complete system of incongruent residues to modulus n, then addition and multiplication between i and j are defined by

$$\mathrm{i} + j = (i + j)(\mathrm{mod}\; n)\;\mathrm{and}\; i \cdot j = (i \cdot j)(\mathrm{mod}\; n).$$

A complete system of residues (mod n) forms a commutative ring with unity element. Let s be a nonzero element of these residues. Then s possess inverse if and only if n is prime, p. Therefore, if p is a prime, a complete system of residues (mod p) forms a Galois (or finite) field, and is denoted by GF(p).

Let P(x) be any given polynomial in x of degree m with coefficients belonging to GF(p), and let $F(x)$ be any polynomial in x with integral coefficients. Then $F(x)$ may be expressed as

$$F(x) = f(x) + p \cdot q(x) + P(x) \cdot Q(x),\ \mathrm{where}\ f(x) = a_0 + a_1 x + a_2 x^2 + \cdots + a_{m-1} x^{m-1},\ a_i \in \mathrm{GF}(p).$$

This relationship may be written as

$F(x) \equiv f(x) \bmod \{p, P(x)\}$, and we say that $f(x)$ is the residue of $F(x)$ modulus p and $P(x)$. If p and $P(x)$ are kept fixed but $f(x)$ varied, p^m classes may be formed, because each coefficient of $f(x)$ may take p values of GF(p). The classes defined by $f(x)$ form a commutative ring, which will be a field if and only if $P(x)$ is irreducible over GF(p) (not divisible with any other polynomial of degree $m - 1$ or less).

The finite field formed by p^m classes of residues is called a Galois field of order p^m and is denoted by GF(p^m). The function $P(x)$ is said to be minimum polynomial for generating the elements of GF(p^m) (the smallest degree polynomial over GF(p) having a field element $\beta \in$ GF(p^m) as a root). The nonzero elements of GF(p^m) can be represented as polynomials of degree at most $m - 1$ or as powers of a primitive root α such that

$$\alpha^{p^m-1} = 1, \quad \alpha^d \neq 1 \text{ (for } d \text{ divifing } p^m - 1).$$

A primitive element is a field element that generates all nonzero field elements as its successive powers. A primitive polynomial is irreducible polynomial that has a primitive element as its root.

Theorem T8 Two important properties of GF(q), $q = p^m$ are

1. The roots of polynomial $x^{q-1} - 1$ are all nonzero elements of GF(q).
2. Let $P(x)$ be an irreducible polynomial of degree m with coefficients from GF(p) and β be a root from the extended field GF($q = p^m$). Then all the m roots of $P(x)$ are

$$\beta, \beta^p, \beta^{p^2}, \ldots, \beta^{p^{m-1}}.$$

To obtain a minimum polynomial, we divide $x^q - 1$ ($q = p^m$) by the least common multiple of all factors like $x^d - 1$, where d is a divisor of $p^m - 1$. Then we get the **cyclotomic equation**—the equation having for its roots all primitive roots of equation $x^{q-1} - 1 = 0$. The order of this equation is ϕ ($p^m - 1$), where ϕ (k) is the number of positive integers less than k and relatively prime to it. By replacing each coefficient in this equation by least nonzero residue to modulus p, we get the cyclotomic polynomial of order ϕ ($p^m - 1$). Let $P(x)$ be an irreducible factor of this polynomial, then $P(x)$ is a minimum polynomial, which is, in general, not the unique one.

Example: Let us determine the minimum polynomial for generating the elements of GF(2^3). The cyclotomic polynomial is

$$\left(x^7 - 1\right)/(x - 1) = x^6 + x^5 + x^4 + x^3 + x^2 + x + 1 = \left(x^3 + x^2 + 1\right)\left(x^3 + x + 1\right).$$

Hence, $P(x)$ can be either $x^3 + x^2 + 1$ or $x^3 + x + 1$. Let us choose $P(x) = x^3 + x^2 + 1$.

$$\phi(7) = 6, \deg[P(x)] = 3.$$

Three different representations of GF(2^3) are given in table below.

Power of α	Polynomial	3-tuple
0	0	000
α^0	1	001
α^1	α	010
α^2	α^2	100
α^3	α^2+1	101
α^4	$\alpha^2+\alpha+1$	111
α^5	$\alpha+1$	011
α^6	$\alpha^2+\alpha$	110
α^7	1	001

A.5 Group Acting on the Set

We first study the action of a group G on an arbitrary set S, and then consider the action of G on itself. A group G is said to act on a set S if

1. Each element $g \in G$ implements a map $s \rightarrow g(s)$, where s, $g(s) \in S$;
2. The identity element e in G produces the identity map: $e(s) = s$; and
3. The map produced by g_1g_2 is the composition of the maps produced by g_1 and g_2, namely, $g_1g_2(s) = g_1(g_2(s))$.

The set of elements from S that are images of s under the action of G is called the **orbit** of $s \in S$, and often denoted as orb(s). The orbit is formally defined by orb(s) $= \{g(s)|g \in G\}$. Another important concept is that of stabilizer. **Stabilizer** of $s \in S$, denoted as S_s, is the set of elements from G that fixes s: $S_s = \{g \in G|g(s) = s\}$. It can be shown that (i) each orbit defines an equivalence class on S so that the collection of orbits partitions S and (ii) the stabilizer S_s is a subgroup of G.

The Pauli group G_N on N-qubits (consisting of all Pauli matrices together with factors $\pm 1, \pm$j) and the set S with the subspace spanned by the basis codewords. Each basis codeword $|i\rangle$ defines its own stabilizer S_i, and the code stabilizer is obtained as intersection of all S_i.

If $S = G$, in quantum mechanical applications, the action is usually conjugation action, namely, $g \rightarrow xgx^{-1}$. Corresponding orbit of g is called the *conjugacy class* of g, and corresponding stabilizer is called the **centralizer** of g in G, denoted as $C_G(g)$. Centralizer of g in G is formally defined by $C_G(g) = \{x \in G|xg = gx\} = \{x \in G|xgx^{-1} = g\}$. Therefore, the centralizer of g in G, $C_G(g)$, contains all elements in G that commute with g.

If $S \subseteq G$, the centralizer of S in G, denoted as $\in G$, is defined by $C_G(S) = \{x \in G|xS = Sx\}$.

The center of G, denoted as $Z(G)$, is the set of all elements from G that commutes with all elements from G, that is, $Z(G) = \{x \in G|xg = gx, \forall\, g \in G\}$.

The normalizer of S in G, denoted as $N_G(S)$, is defined by $N_G(S) = \{x \in G | xSx^{-1} = S\}$. Let H be a subgroup of G. The **self-normalizing** subgroup of G is one that satisfies $N_G(H) = H$.

A.6 Metric Spaces

Definition D9 A linear space X, over a field K, is a **metric** space if there exists a nonnegative function $u \to \|u\|$, which is called the **norm** such that ($\forall\, u, v \in X, c \in K$):

1. $\|u\| = 0 \Leftrightarrow u = 0$ (0 is the fixed point),
2. $\|cu\| = |c| \cdot \|u\|$ (homogeneity property), and
3. $\|u + v\| \leq \|u\| + \|v\|$ (triangle inequality).

Definition D10 The metric is defined as $d(u, v) = \|u - v\|$.

The metric has the following important properties:

1. $d(u, v) = 0 \Leftrightarrow u = v$ (reflectivity),
2. $d(u, v) = d\,(v, u)$ (symmetry), and
3. $d\,(u, v) + d\,(v, w) \geq d\,(u, w)$ (triangle inequality).

Examples: $K = R$ (the field of real numbers), $X = \mathrm{R}^n$ (Euclidean space). The various norms can be defined as follows:

$$\|\boldsymbol{x}\|_p = \left(\sum_{k=1}^{n} |x_k|^p\right)^{1/p} \quad 1 \leq p \leq \infty \text{(p-norm)}$$

$$\|\boldsymbol{x}\|_\infty = \max_k |x_k| \quad \text{(absolute value norm)}$$

$$\|\boldsymbol{x}\|_2 = \left(\sum_{k=1}^{n} |x_k|^2\right)^{1/2} \quad \text{(Euclidean norm)}$$

Example: For $X = C[a, b]$ (complex signals on interval $[a, b]$), the norm can be defined by either way as shown below:

$$\|u\| = \max_{a \leq t \leq b} |u(t)|$$

$$\|u\| = \int_a^b |u(t)| dt.$$

Example: For $X = L^r[a, b]$, the norm can be defined by

$$\|u\| = \left(\int_a^b |u(t)|^r dt \right)^{1/r}.$$

Definition D11 Let $\{u_n\}_{n \in N}$ be a sequence of points in the metric space X and let $u \in X$ be such that

$\lim_{n \to \infty} \|u_n - u\| = 0$, we say that u_n converges to u.

Definition D12 A series $\{u_n\}_{n \in N}$ satisfying

$$\lim_{n,m \to \infty} \|u_n - u_m\| = 0$$

is called the Cauchy series.

Definition D13 Metric space is complete if every Cauchy series converges.

Definition D14 Complete metric space is called Banach space.

A.7 Hilbert Spaces

Definition D15 A vector space X over the complex field C is called unitary (or space with a dot (scalar) product) if there exists a function $(u, v) \to C$ satisfying

1. $(u, u) \geq 0$,
2. $(u, u) = 0$ iff $u = 0$,
3. $(u + v, w) = (u, w) + (v, w)$,
4. $(cu, v) = c(u, v)$, and
5. $(u, v) = (v, u)^*$.

$\forall\, u, v, w \in X$ and $c \in C$. The function (u, v) is a dot (scalar) product.

This function has the following properties:

$$(u, cv) = c^*(u, v)$$
$$(u, v_1 + v_2) = (u, v_1) + (u, v_2)$$
$$|(u, v)|^2 \leq (u, u)(v, v).$$

Definition D16 Define a norm:

$$\|u\| = \sqrt{(u, u)}.$$

Unitary space with this norm is called a pre-Hilbert space. If it is complete then it is called a **Hilbert space**.

Example: R^n is a Hilbert space with dot product introduced by

$$(\boldsymbol{y},\boldsymbol{x}) = \sum_{k=1}^{n} x_k y_k = \left(\boldsymbol{x}^{\mathrm{T}}, \boldsymbol{y}^{\mathrm{T}}\right).$$

Example: C^n is a Hilbert space with dot product introduced by

$$(\boldsymbol{y},\boldsymbol{x}) = \sum_{k=1}^{n} x_k y_k^* = \left(\boldsymbol{x}^{\dagger}, \boldsymbol{y}^{\dagger}\right)^*.$$

In this space, the following important inequality, known as Cauchy–Schwarz–Bunyakovsky inequality, is valid:

$$\left|\sum_{k=1}^{n} x_k y_k^*\right|^2 \le \left(\sum_{k=1}^{n} |x_k|^2\right)\left(\sum_{k=1}^{n} |y_k|^2\right).$$

Example: $L^2(a, b)$ is a Hilbert space with dot product introduced by

$$(v, u) = \int_a^b u(t) v^*(t) dt.$$

Orthogonal Systems in Hilbert Space

Definition D17 A set of vectors $\{u_k\}_{k \in I}$ in the Hilbert space forms an orthogonal system if

$$(u_n, u_k) = \delta_{n,k} \|u_k\|^2 \ \forall n, k \in J$$
$$\delta_{n,k} = \begin{cases} 1, \ n = k \\ 0, \ n \neq k \end{cases} \quad \|u_k\| = \sqrt{(u_k, u_k)}.$$

The set J can be finite, countable, or uncountable. If $\|u_k\| = 1$, then the system is orthonormal.

Gram–Schmidt Orthogonalization Procedure

Here we adopt the Dirac notation [1] and use the $|u\rangle$ to denote the column vector, in quantum mechanics known as the ket, and $\langle u|$ to denote the corresponding dual vector, known as bra, which is Hermitian conjugate of $|u\rangle$. The dot (scalar) product of two vectors $|u\rangle$ and $|v\rangle$ is denoted by $\langle v|u\rangle$. The Gram–Schmidt procedure assigns an orthogonal system of vectors $\{|u_0\rangle, |u_1\rangle, \ldots\}$ starting from the countable many linearly independent vectors $\{|v_0\rangle, |v_1\rangle, \ldots\}$. For the first basis vector, we select $|u_0\rangle = |v_0\rangle$. We then express the next basis vector as $|v_1\rangle$ minus projection along basis $|u_0\rangle$:

$$|u_1\rangle = |v_1\rangle - \lambda_{10}|u_0\rangle.$$

The projection λ_{10} can be obtained by multiplying the previous equation with $\langle u_0|$ and by determination of dot product as follows:

$$\underbrace{\langle u_0|u_1\rangle}_{=0} = \langle u_0|v_1\rangle - \lambda_{10}\langle u_0|u_0\rangle.$$

By solving this equation per λ_{10}, we obtain

$$\lambda_{10} = \langle u_0|v_1\rangle / \langle u_0|u_0\rangle.$$

Suppose that we have already determined the basis vectors $|u_0\rangle$, ..., $|u_{k-1}\rangle$. We can express the next basis vector $|u_k\rangle$ in terms of vector $|v_k\rangle$ and projections along already determined basis vectors as follows:

$$|u_k\rangle = |v_k\rangle - \lambda_{k0}|u_0\rangle - \lambda_{k1}|u_1\rangle - \cdots - \lambda_{k,k-1}|u_{k-1}\rangle = |v_k\rangle - \sum_{j=0}^{k-1} \lambda_{kj}|u_j\rangle.$$

By multiplying the previous equation with $\langle u_i|$ and by evaluating the dot product, we obtain

$$\underbrace{\langle u_i|u_k\rangle}_{=0} = \langle u_i|v_k\rangle - \sum_{j=0}^{k-1} \lambda_{kj}\langle u_i|u_j\rangle; \ i = 0, 1, \ldots, k-1.$$

By invoking the principle of orthogonality, the summation in equation above is nonzero only for $j = i$ so that the projection along ith basis function can be obtained by

$$\lambda_{k,i} = \frac{\langle u_i|v_k\rangle}{\langle u_i|u_i\rangle}.$$

Finally, kth basis vector can be obtained by

$$|u_k\rangle = |v_k\rangle - \sum_{j=0}^{k-1} \frac{\langle u_j|v_k\rangle}{\langle u_j|u_j\rangle}|u_j\rangle; \ k = 1, 2, \ldots$$

References

1. Djordjevic IB (2012) Quantum information processing and quantum error correction: an engineering approach. Elsevier/Academic Press, Amsterdam-Boston
2. Djordjevic I, Ryan W, Vasic B (2010) Coding for optical channels. Springer Science + Business Media, New York
3. Ryan WE, Lin S (2009) Channel codes: classical and modern. Cambridge University Press, Cambridge, New York
4. Lin S, Costello DJ (2004) Error control coding: fundamentals and applications, 2nd edn. Pearson Prentice Hall, Upper Saddle River, NJ
5. Pinter CC (2010) A book of abstract algebra. Dover Publications, New York (reprint)
6. Anderson JB, Mohan S (1991) Source and channel coding: an algorithmic approach. Kluwer Academic Publishers
7. Grillet PA (2007) Abstract algebra. Springer
8. Chambert-Loir A (2005) A field guide to algebra. Springer
9. Lang S (1993) Algebra. Addison-Wesley Publishing Company, Reading
10. Gaitan F (2008) Quantum error correction and fault tolerant quantum computing. CRC Press, Boca Raton, FL
11. Raghavarao D (1988) Constructions and combinatorial problems in design of experiments. Dover Publications, Inc., New York (reprint)
12. Drajic DB, Ivanis PN (2009) Introduction to information theory and coding. Akademska Misao, Belgrade, Serbia (in Serbian)
13. Cvijetic M, Djordjevic IB (2013) Advanced optical communications and networks. Artech House, Boston, London
14. Djordjevic IB (2017) Advanced optical and wireless communications systems. Springer International Publishing, Switzerland
15. Wicker SB (1995) Error control systems for digital communication and storage. Prentice Hall, Englewood Cliffs, NJ
16. Vucetic B, Yuan J (2000) Turbo codes-principles and applications. Kluwer Academic Publishers, Boston
17. Vasic B (2005) Digital communications I. Lecture Notes. University of Arizona

I. B. Djordjevic, *Physical-Layer Security and Quantum Key Distribution*,
https://doi.org/10.1007/978-3-030-27565-5

Index

I. B. Djordjevic, *Physical-Layer Security and Quantum Key Distribution*,
https://doi.org/10.1007/978-3-030-27565-5

D

Q

T

Y

Z

Printed in the United States
By Bookmasters